Super-Intense Laser–Atom Physics

NATO ASI Series

Advanced Science Institutes Series

A series presenting the results of activities sponsored by the NATO Science Committee, which aims at the dissemination of advanced scientific and technological knowledge, with a view to strengthening links between scientific communities.

The series is published by an international board of publishers in conjunction with the NATO Scientific Affairs Division

A   **Life Sciences**   Plenum Publishing Corporation
B   **Physics**   New York and London

C   **Mathematical and Physical Sciences**   Kluwer Academic Publishers
D   **Behavioral and Social Sciences**   Dordrecht, Boston, and London
E   **Applied Sciences**

F   **Computer and Systems Sciences**   Springer-Verlag
G   **Ecological Sciences**   Berlin, Heidelberg, New York, London,
H   **Cell Biology**   Paris, Tokyo, Hong Kong, and Barcelona
I   **Global Environmental Change**

Series B: Physics

Super-Intense Laser–Atom Physics

Edited by

Bernard Piraux

Université Catholique de Louvain
Louvain-la-Neuve, Belgium

Anne L'Huillier

Service des Photons, Atomes et Molécules
Centre d'Etudes de Saclay
Gif-sur-Yvette, France

and

Kazimierz Rzążewski

Polish Academy of Sciences
Centrum Fizyki Teoretycznej
Warsaw, Poland

Plenum Press
New York and London
Published in cooperation with NATO Scientific Affairs Division

Proceedings of a NATO Advanced Research Workshop on
SILAP (Super-Intense Laser–Atom Physics),
held January 8–14, 1993,
in Han-sur-Lesse, Belgium

NATO-PCO-DATA BASE

The electronic index to the NATO ASI Series provides full bibliographical references (with keywords and/or abstracts) to more than 30,000 contributions from international scientists published in all sections of the NATO ASI Series. Access to the NATO-PCO-DATA BASE is possible in two ways:

—via online FILE 128 (NATO-PCO-DATA BASE) hosted by ESRIN, Via Galileo Galilei, I-00044 Frascati, Italy

—via CD-ROM "NATO Science and Technology Disk" with user-friendly retrieval software in English, French, and German (©WTV GmbH and DATAWARE Technologies, Inc. 1989). The CD-ROM also contains the AGARD Aerospace Database.

The CD-ROM can be ordered through any member of the Board of Publishers or through NATO-PCO, Overijse, Belgium.

Library of Congress Cataloging in Publication Data

Super-intense laser–atom physics / edited by Bernard Piraux, Anne L'Huillier, and Kazimierz Rzazewski.
    p.   cm.—(NATO ASI series. Series B, Physics; vol. 316)
    "Published in cooperation with NATO Scientific Affairs Division."
    "Proceedings of a NATO Advanced Research Workshop on SILAP (Super-Intense Laser–Atom Physics), held January 8–14, 1993, in Han-sur-Lesse, Belgium—T.p. verso.
    Includes bibliographical references and index.

    1. Laser manipulation—Congresses. 2. Laser pulses, Ultrashort—Congresses. 3. Atoms—Congresses. I. Piraux, Bernard. II. L'Huillier, Anne. III. Rzazewski, Kazimierz. IV. North Atlantic Treaty Organization. Scientific Affairs Division. V. NATO Advanced Research Workshop on SILAP (1993: Han-sur-Lesse, Belgium) VI. Series: NATO ASI series. Series B, Physics; v. 314.
QC689.5.L35S86 1993   93-26357
539.7—dc20   CIP

ISBN 978-1-4615-7965-6   ISBN 978-1-4615-7963-2 (eBook)
DOI 10.1007/978-1-4615-7963-2

©1993 Plenum Press, New York
Softcover reprint of the hardcover 1st edition 1993
A Division of Plenum Publishing Corporation
233 Spring Street, New York, N.Y. 10013

PREFACE

The rapid development of powerful pulsed lasers is at the origin of a considerable interest in studying the response of an atom, a molecule (or a solid) to a strong electromagnetic field. It is now possible to produce at the laboratory scale, ultra-short pulses with a duration of 100 femtoseconds (10^{-13} second) and a power of the order of 1 terawatt (10^{12} Watt). Under these conditions, very high peak intensities may be obtained and electric fields exceeding typical electron binding fields in atoms are generated.

The interaction of an atom or a molecule with such electromagnetic fields has a highly non-linear character which leads to unexpected phenomena. Amongst them,
- above-threshold ionization (ATI) *i.e.* the absorption of additional photons in excess of the minimal number necessary to overcome the ionization potential and its molecular counterpart, above-threshold dissociation (ATD);
- generation of very high harmonics of the driving field;
- stabilization of one-electron systems in strong fields.

These processes were the main topics of two international meetings which were held in 1989 and 1991 in the United States under the common name SILAP (Super-Intense Laser-Atom Physics).

This volume contains the lectures and oral communications presented at the third SILAP meeting; it has been organized as a NATO Advanced Research Workshop (NATO ARW 920638) and was held at the "Domaine des Masures", Han-sur-Lesse (Belgium) from the 8th to the 14th of January 1993. The objective of the meeting was to bring together experimentalists and theoreticians in order to evaluate current understanding of the above processes and to discuss new developments from both the theoretical and the experimental side. Recently, considerable efforts have been made to tackle the challenging problem of the interaction of a short and intense laser pulse with a system having two active electrons. The present volume reflects this trend and is divided into the following parts:

1. generation of high harmonics

2. dynamics of multiphoton ionization of atoms

3. dynamics of photoionization and photodissociation of molecules

4. interaction of a two-electron system with a strong field

Very high harmonics of order larger than 125 have been detected recently in several laboratories as the result of the interaction of light atoms (ions) with strong low frequency laser pulses. These processes are the most non-linear ones to date and their observation opens the route to the development of (high brightness) coherent sources in the vuv and x-ray ranges. A full understanding of harmonic generation requires the knowledge of both the single atom response to a strong radiation field and the propagation properties of the harmonics field in dense partly ionized gazeous media. From the experimental point of view, the major effort has focused on three important issues: the optimization of the harmonic conversion efficiency (which is essentially controlled by the phase matching effect), polarization effects and the measurements of the angular distribution of very order harmonics. Theoretical studies, on the other hand, concentrated essentially on the mechanism of harmonic emission by a single atom. First attempts however, have been made at taking into account the macroscopic (coherent) response of the medium by means of one-dimensional models.

The process of ionization of a single atom or an ion exposed to a strong electromagnetic field is one of the most fundamental problems. Considerable progress has been made in understanding various aspects like the stabilization of one-electron systems in strong fields, the role of intermediate resonances during the interaction of the atom with the pulse and tunneling. As mentioned already, the most important theoretical achievement concerns the interaction of a two-electron system with a strong field. The first results about the role of electron-electron correlation in the ionization mechanism have been presented during the conference. Besides this, many other questions stay unsolved like for instance the exact relationship between the electron and photon emission process and the problem of the simultaneous emission of two electrons.

The experience developed in super intense laser atom physics is at the origin of similar investigations in molecules. Above-threshold ionization as well as a new, though very similar process, above-threshold dissociation, have been found to occur in molecules. Coherences play also a crucial role; they lead to trapping and creation of new molecular states and may induce stabilization.

This workshop was generously supported by the NATO Scientific Affairs Division. It has also benefitted from the financial support of the "Fonds National de la Recherche Scientifique", the "Université Catholique de Louvain" and the "Ministère de l'Education, de la Recherche et de la Formation" of the French Community of Belgium. The authors want to thank Alfred Maquet for his overview of the conference in which he shares his personal views about this rapidly expanding field which is now becoming a new discipline. The authors are grateful to all whose efforts made this workshop a great success. Particular thanks go to Leticia Villalobos de Piraux, Marie-Cecile Passot and to F. Herman and his staff of the "Domaine des Masures".

B. Piraux
A. L'Huillier
K. Rzążewski

Louvain-la-Neuve, April 1993

CONTENTS

PART III : DYNAMICS OF PHOTOIONIZATION AND PHOTODISSOCIATION OF MOLECULES

PART IV : INTERACTION OF A TWO-ELECTRON SYSTEM WITH A STRONG FIELD

SILAP III: AN OVERVIEW

Alfred Maquet

Laboratoire de Chimie Physique-Matière et Rayonnement
Université Pierre et Marie Curie (PARIS VI)
11, Rue Pierre et Marie Curie,
F 75 231 Paris, CEDEX 05, France

Beyond the fact that it has permitted the gathering of senior and junior scientists active in the field of Super-Intense Laser-Atom Physics, the NATO Advanced Research Workshop held in Han-sur-Lesse, has displayed several new interesting features in the development of this very active branch of atomic and molecular physics. I present here an account of my personal views on the recent development of the field, as it appears in the light of the lectures delivered during the workshop.

The field is mature. By this, I mean that the theory is now at a stage to be able to reproduce details of experimental photoelectron spectra, namely the so-called Above-Threshold Ionization (ATI) spectra, obtained through multiphoton ionization in short laser pulses. This was a challenge which has attracted much attention since the first observation of ATI spectra more than ten years ago. More important is the fact that this has been achieved in two distinct atomic systems, namely hydrogen and cesium, through the use of entirely different approaches; see below.

The field is rapidly expanding. New processes have been demonstrated, such as high-order harmonic generation in rare gases (up to 135th harmonic of Nd:YAG laser photons in Ne !) or predicted (atomic stabilization against ionization in ultra-intense fields). They are now the subject of very active (if not heated!) discussions. Note that, by new processes, I mean here those which were almost not discussed in the literature five years ago. In fact, the possible observation of these processes was either completely unexpected, as for high-order harmonic generation, or counter-intuitive, as for stabilization. Another new feature is that the possible observation of the latter in molecules is also discussed.

New theoretical tools are being created. The advent of laser sources delivering ultra-short, "femtosecond", pulses has evidenced the need for the development of new methods to solve the time-dependent Schrödinger equation. Within this context, not only single-active electron systems have been considered but the first attempts to tackle the formidable task to

Super-Intense Laser-Atom Physics, Edited by
B. Piraux *et al.*, Plenum Press, New York, 1993

solve the time-dependent problem for fully correlated two-electron systems have been reported.

I turn now to a brief description of the main topics addressed during the course of the workshop.

HIGH-ORDER HARMONIC GENERATION (HHG)

Most experimental groups (Livermore, London, Lund, Rochester, Saclay,....), active in the field of intense laser-atom physics, have reported their latest results on HHG in rare gases and in metals (Budapest). One of the main reason why this process attracts so much interest lies in its potential applications, as it represents a possible route towards the implementation of new coherent VUV or soft-X ray pulsed sources. The prospects are that the characteristics of these sources will compare very favourably with those of existing ones. Among the topics discussed are the angular distribution and phase matching problems which are of crucial importance in determining the practical efficiency of HHG. The contribution of the harmonics originating from the ions created inside the laser field has been also discussed. New interesting features, including the generation of intense pulses of circularly polarized light, have been demonstrated via HHG in the presence of two different laser fields (two-colour HHG). This opens, in particular, the possibility to develop X-ray magnetic dichroism studies.

The experimental observations on HHG have motivated a large number of theoretical studies. Progress reports by most groups active in the field have been presented. One of the main conclusion of the lively discussions held during the workshop was that HHG is not a very sensitive test of the validity of theoretical simulations. The reason is that power spectra associated with almost every strongly driven anharmonic system, either quantum or classical, exhibit a conspicuous plateau followed by a marked decline, much similar to the ones observed in experiments.

A challenge was however to account for the upper limit of the plateau which provides the maximum harmonic frequency $\omega_{max} = N_{max}\omega_L$, where ω_L is the laser frequency. Convincing arguments, supported by classical simulations, have been put forward, implying that the limit of the plateau is set by the maximum energy an atomic electron can acquire in the combined atom+laser potential. This would set a limit on the maximum energy the electron can release by emission of a photon when returning close to the nucleus. This simple model suggests that $N_{max}\omega_L = E_{ion} + nU_p$, where E_{ion} is the atomic ionization energy and $U_p = F_0^2/(4\omega_L^2)$ is the quiver energy of a free electron in the laser field with amplitude F_0. Numerical simulations indicate that $2 < n < 3.2$, in rough agreement with experiments.

Amongst the topics also addressed in the lectures were:
-the possibility of observing harmonic generation in the stabilization regime, i.e. at ultra-high laser intensities and for relatively high frequencies;
-the role of phase matching and the propagation of harmonic light in the macroscopic medium;
-the possibility to observe even harmonics as a signature of a laser-induced symmetry breaking in the target:

-the dominant role of the bound states in reproducing the main features of harmonic spectra;
-non-perturbative calculations of higher-order susceptibilities;
-the possible occurrence of Raman-like lines in harmonic spectra.

ATOMIC (MOLECULAR) STABILIZATION AGAINST IONIZATION (DISSOCIATION) IN ULTRA-INTENSE FIELDS

Theory suggests that stabilization might occur in two distinct regimes:

i) adiabatic stabilization in the high frequency regime. It is characterized by the presence of a minimum in the variation of atomic states lifetimes as a function of the laser field strength;

ii) "dynamic" or " wave-packet" stabilization, which could occur at lower frequencies, in the presence of a laser pulse with a steep turn-on. As a result of the steep turn-on, Rydberg states can be populated, which can be very stable against ionization (population trapping).

No experimental evidence of "adiabatic" stabilization has been reported here. However, population trapping in Rydberg states and also the creation of atomic "anti wave-packets", have been discussed.

On the theoretical side, two important advances have been reported: on the one hand, a new, fully non-perturbative Floquet calculation, globally confirms the early predictions by Gavrila *et al.* on adiabatic stabilization in hydrogen. On the other hand, a new Coulomb-sturmian technique has been implemented to solve the time-dependent Schrödinger equation, which allows one the discussion of dynamic stabilization.

It has been also suggested that Rydberg states, strongly coupled to the continuum by the laser field, could experience "interference stabilization" as a result of coupled Raman-like processes, leading to an overall stabilization or "coherent trapping" of the atomic population in excited states.

Another interesting feature of the workshop has been to present evidence, thanks to an excellent session dedicated to molecules in strong (not super-intense!) laser fields, that most of the processes discussed in atoms have their counterpart in molecules. For instance, to ATI in atoms corresponds Above-Threshold Dissociation (ATD) in molecules and to the coherent trapping of atomic population in excited states would correspond the so-called "bond hardening" in molecules, which results in the population of laser-induced bound states.

STRONG FIELD EXPERIMENTS

Amongst the new experimental achievements reported are:

-the first multiphoton ionization experiment with a half-cycle width (!!) pulse;

-the evidence for population trapping in Rydberg wave packets;

-a detailed study of resonant multiphoton ionization. In particular, the problem of the dependence of the ATI spectra on the spatial and temporal characteristics of the pulse has been addressed and the "onion vs. potato" controversy has been partially resolved, which, according to the experts present, was already a remarkable achievement!

-new very precise ATI spectra in hydrogen (the theoretician's favorite atom) and in cesium. In both cases, theoreticians have been able to incorporate into their calculations most of the parameters characterizing the pulse shape (duration, spatial dependence, peak intensity,...), in order to account for the observed spectra.

-new experiments on multiphoton photodetachment in negative ions. Among the new features reported are the evidence for Above Threshold Detachment in Cl⁻ and the confirmation of an *a priori* counter-intuitive prediction made by M. Crance several years ago, of a decrease of the saturation intensity when the number of absorbed photons increases. In another experiment, performed in Cs⁻, Feshbach resonances have been observed in the Cs^+ ion signal.

The latter observation of multiphoton double detachment of electrons from a negative ion, raises the question of the importance of correlations in multielectron systems.

TWO-ELECTRON SYSTEMS

The problem of describing electronic correlations in complex systems, within the context of intense field physics, has become a very active area of investigation. This workshop is probably the first one in which this topic has been so widely covered. Indeed most calculations, reported up to now, were performed within a single-active electron picture, even for multielectron systems. The limitations of such an approach have clearly been evidenced by comparing frozen-core and configuration-interaction calculations in two-active electron systems such as Mg. However, the integration of the time-dependent Schrödinger equation for a fully correlated two-electron system remains a formidable task, which requires huge amounts of computing time and storage, even for the simpler systems like He or H⁻.

The first attempts in this direction have been reported, either using massively parallel computers such as the Connection Machine, or introducing simplified 1-D models. Although such studies can help to discuss issues such as the relative importances of sequential vs. simultaneous double ionization, they are still in their early stages and do not yet provide a complete description of two-electron processes.

More conventional calculations have also been reported, including two-photon ionization of H⁻ and the single-photon, double ionization, of He. The occurrence of Wigner cusps in the photoelectron spectra has also been predicted and the use of fully correlated wave functions in H⁻ has led to the uncovering of the existence of laser-induced bound states, as a result of the effect of the electron correlation.

Another interesting development has been to realize that classical approaches can be of some help, not only to discuss multiphoton ionization, HHG and stabilization in single-electron atoms, but also in two-electron systems. The key step is to introduce "soft Coulomb" or short range repulsive potentials in order to regularize the Coulomb interaction which, otherwise leads to highly unstable (chaotic) motion of classical systems of charged particles.

METHODOLOGICAL ADVANCES

As a theorist, I have been most impressed by the methodological advances achieved in the past few years since, say, last ICOMP in summer 1991.

-The Fourier-Floquet method, coupled to a Sturmian basis expansion of the Floquet components of the wave functions, is now completely reliable and allows very precise calculations and explanation of details of ATI spectra in hydrogen.

-Sturmian basis set expansions have been used in computation schemes to solve the time-dependent Schrödinger equation, not only in hydrogen but also in two-electron systems.

-Promising preliminary results obtained from an important effort to combine the Floquet method with a R-matrix approach, in order to deal with complex atoms interacting with strong laser fields, have been presented.

-The so-called spectral method has been implemented for the first time in the context of a time-dependent problem, in order to extract the "Floquet content" of the wave function of an atomic system, in the presence of a field with a time-varying enveloppe.

-Two-electron, fully correlated, wave functions, have been constructed using massively parallel computers...

Many other efforts, conceptual as well as computational, aiming at a better understanding of the still unsolved problem of finding a general description of the response of microscopic (atomic or molecular) systems in a (possibly ultra-) intense laser field, have been reported. They have demonstrated the vitality of this active branch of atomic and molecular physics.

Before letting the reader discovering by himself all these new results, I wish, on behalf of all participants, to thank the organizers and specially Bernard and Lætitia, for their warm hospitality, which made everyone enjoyed an excellent Workshop.

PART I

GENERATION OF HIGH HARMONICS

RECENT RESULTS IN HIGH-ORDER HARMONIC GENERATION

Ph. Balcou, P. Salières and Anne L'Huillier

Service des Photons, Atomes et Molécules
Centre d'Etudes de Saclay, 91191 Gif-sur-Yvette, France

INTRODUCTION

High order harmonic generation by atoms in intense laser fields has drawn considerable attention in the last few years. Experiments[1-5] have demonstrated the possibility to reach wavelengths shorter than 8 nm with reasonable conversion efficiencies, thus opening the way to the development of high brightness coherent sources in the vuv and soft x-ray ranges. The understanding of these processes involves both the single atom response to a strong electromagnetic field (strongly non-perturbative) and in particular its emission spectrum[6-8], and the generation and propagation of the harmonic fields in dense, partly ionized, gaseous media[9]. A quite fascinating aspect of this field is this interplay between atomic physics in a strong field regime and nonlinear optics in a rather unusual situation. Since the early results[10], progress has been made in three directions. Shorter wavelengths and much higher nonlinear orders have been generated by using light atoms (or ions) and strong (low frequency) radiation fields[1-5,11]. Detailed studies have been undertaken for the moderately high harmonics and have led to a better understanding of some aspects of these processes, in particular of phase matching effects[12,13]. Finally, some first attempts at characterizing the harmonic emission have been made[14-16].

The present report intends to be a brief presentation of the recent experimental results obtained at Saclay. We articulate our discussion around the three questions: *How far* (in energy) can harmonics be generated ? *How many* photons can be produced ? *How well* is the radiation emitted, or in other words, what is the spatial and temporal coherence of the harmonic radiation ?

HOW FAR ?

Our recent results obtained in the rare gases with a 1 ps 1053 nm Nd-Glass laser[5] are presented in Fig.1. We compare the results obtained in xenon, argon, neon and helium at an intensity at best focus equal to 1.5×10^{15} W/cm^2. In xenon, the intensity of the harmonics is constant from the 5th harmonic to the 23rd (see Ref. 17) and then falls off abruptly over four orders of magnitude. In argon, there seems to be two successive plateaus, one up to the 27th harmonic, the other from the 39th to the 49th harmonic. In neon and helium, we see very long and quite similar plateaus,

Super-Intense Laser-Atom Physics, Edited by
B. Piraux *et al.*, Plenum Press, New York, 1993

whose limit is experimental, due to our monochromator's lack of resolution. The highest resolved order is the 135th, i.e. a photon energy of 160 eV.

The absolute scale gives an estimation of the number of photons generated at each laser shot. The pressure was adjusted from 10 Torr in xenon to 70 Torr in helium in order to optimize the signal over noise ratio and to increase the harmonic emission in Ne and He. The number of photons obtained is found to vary approximately as the square of the atomic density over the range investigated. A more spectacular way of increasing the conversion efficiency is to use a weaker focusing geometry or simply here to defocus the laser beam in the medium (see below). By moving the focus relative to the gas jet by $z = \pm 3.5$ mm in Xe and $z = \pm 2$ mm in Ar, the number of photons is increased by a factor of 50 in Xe and 10 in Ar. Thus we get quite high numbers of photons, almost 10^{10} photons per pulse for the 17th harmonic in xenon.

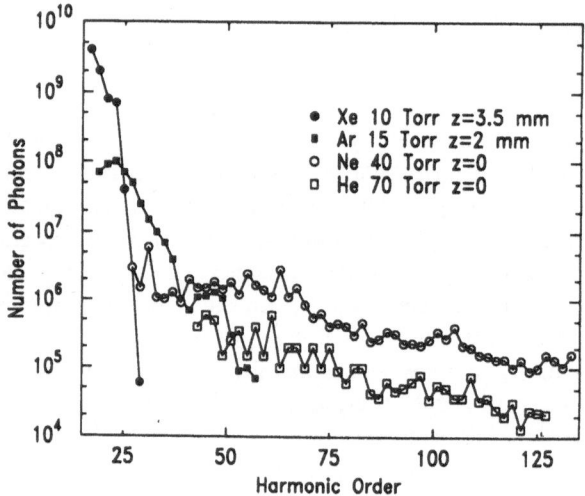

Figure 1. Number of photons obtained in xenon, argon, neon and helium as a function of the harmonic order using a 1-ps 1053-nm 10^{15} W/cm^2 laser pulse

The results shown in Fig.1 agree well with the predictions of Krause, Schafer and Kulander[6] for the extension of the plateau. By performing time-dependent calculations of the response of a single atom for a number of atomic systems in various conditions of laser intensity and frequency, these authors predict that the width of the plateau scales as $I_p + 3U_p$, where I_p is the ionization energy and $U_p = e^2 E^2 / 4 m \omega^2$ the ponderomotive energy (e, m are the electron's charge and mass, E is the field strength and ω the laser frequency). The same scaling can also be found by using simpler models such as the δ- potential employed by Becker and coworkers[18,4], or even the "simpleman" theory which provides a simple classical picture to ionization (and also to harmonic generation) in intense low-frequency laser fields[19].

In our experiments, we find that the maximum energy of the plateau in Xe and Ar at 1053 nm are 27 and 62 eV, to be compared with $W_p = 25$ and 60 eV for the saturation intensities respectively of 7×10^{13} and 1.5×10^{14} W/cm^2 in our experimental conditions[5]. In neon and helium, the rule gives $W_p = 170$ and 230 eV ($I_s = 5$ and 7×10^{14} W/cm^2). These values are higher than the 160 eV corresponding to the 135th harmonic, the highest detectable harmonic in our experiment. We have also recently performed experiments with the Ti:Sapphire laser system of the Lund Institute of Technology[20], the results of which are in good agreement with the above interpretation.

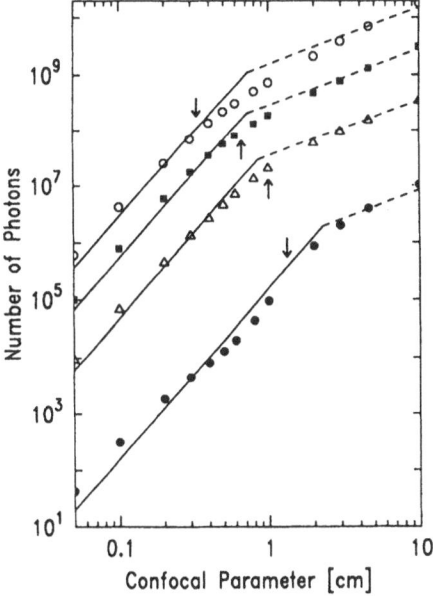

Figure 2. Number of photons calculated in xenon at the 7th, 13th 19th and 25th harmonic frequencies as a function of the laser confocal parameter.

HOW MUCH ?

The second important question, especially for the application of the high harmonics is how many photons can be produced and how can this number be optimized? The obvious answer to the latter question is by exploiting the coherence of the process i.e., by increasing the number of emitting dipoles, provided they remain in phase. The strength of the harmonic field that exits the nonlinear medium is proportional to the length over which it can be built up coherently. This coherence length L_{coh} can be due to dispersion in the medium : $L_{coh}^{disp} = \pi/\Delta k$, where Δk is the phase mismatch, proportional to the atomic density, and/or to focusing of the incident radiation : $L_{coh}^{foc} = b \tan[\pi/(q-1)]/2$, where q denotes the harmonic order and b the laser confocal parameter. The easiest way to enhance the number of photons is to increase the atomic density. Indeed, the number of photons varies as the square of the atomic density, as long as dispersion effects due to the atoms or to the electrons are not important, i.e. $L_{coh}^{disp} > L$, L denoting the medium's length. Moreover, in the regime $L_{coh}^{disp} > L > L_{coh}^{foc}$, one can also optimize the conversion efficiency by using a loosely focused geometry.

In Fig. 2, we show calculated numbers of photons[17] obtained at the 7th, 13th, 19th and 25th harmonic frequencies from the top to the bottom as a function of the confocal parameter b. The calculations have been performed in xenon for a 1.2 ps incident laser pulse and using parameters close to the experimental ones. The propagation equations for the harmonic fields in the medium are solved in the slowly-varying envelope and paraxial approximations, using as a source the single-atom dipole moment obtained from time-dependent calculations (see Ref. 9 for more details). The atomic density imposed by the pulsed gas jet is taken to be a truncated Lorentzian distribution with a 0.08 cm width at half maximum and a 0.16 cm total width. The peak atomic density is taken to be $5.3 \ 10^{17}$ atoms/cm^3 (15 Torr). The laser intensity is 4×10^{13} W/cm^2, below the saturation intensity estimated to 6×10^{13} W/cm^2 for a 1.2 ps pulse. The incident laser beam is assumed to be Gaussian and not modified by propagation in the nonlinear medium.

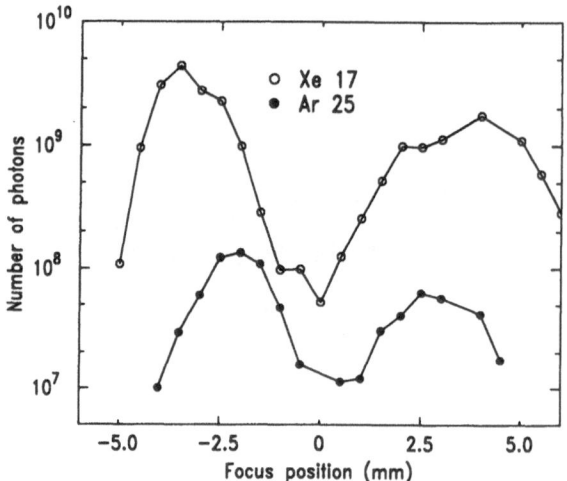

Figure 3. Number of photons obtained in xenon and argon at the 17th and 25th harmonic as a function of the position of the focus in the gas jet.

These results indicate that the number of photons varies first as the cubic power of the laser confocal parameter, and then linearly. The interpretation of this simple dependence is straigthforward : when $L > L_{coh}^{foc}$, the harmonic field $E_q \propto L_{coh}^{foc} \propto b$ so that the intensity of the harmonic field $|E_q|^2 \propto b^2$ and the spatially integrated intensity I_q (or number of photons N_q) varies as b^3. On the other hand, when $L < L_{coh}^{foc}$, E_q is independent of b and the integrated intensity becomes proportional to b.

As b/L increases, one switches from the b^3 regime (coherence length smaller than the length of the medium) to a linear regime (weak focusing ; the harmonic field remains in phase with the driving polarization term).

Another way to increase the coherence length is to defocus the laser beam relative to the atomic jet. The coherence length increases away from the focus (one gets closer to a plane wave approximation) until $L_{coh}(z) \approx \pi b(1 + 4z^2/b^2)/2(q - 1)$ becomes larger than $L/2$, z denoting the coordinate on the propagation axis. To observe a significant effect, the intensity at best focus must be much higher than the saturation intensity. In Fig. 3, the numbers of photons for the 17th harmonic in xenon and the 25th harmonic in argon are plotted as a function of the position of the focus (in mm) in the jet. The maximum emission is obtained at $z = \pm\, 3.5$ mm in Xe and at

Table 1

- Photon energy	20 eV
- Pulse width[*]	0.5 ps
- Spectral width[*]	0.5 Å
- $\Delta E / E$[*]	10^{-3}
- Photons / pulse	$4\ 10^9$ phot
- Photon Flux[*]	10^{22} phot/s
- Conversion efficiency	$5\ 10^{-6}$
- Instantaneous power[*]	30 kW
- Energy	15 nJ
- Repetition rate	1 shot / 10 s
- Brightness[*]	10^{21} phot/sÅ(mrad)^2

$z = \pm\, 2$ mm in Ar (i.e. $I \approx 3I_s$ in both cases). We estimate our best focus intensity to be equal to 1.5×10^{15} W/cm^2, i.e. 10, and 20 times the saturation intensity in Ar and Xe respectively. The number of photons increases by a factor of 10 in Ar and of 50 in Xe as the laser is defocused. Thus we get quite high numbers of photons, particularly in Xe.

We indicate in Table 1 the characteristics of the emitted radiation at 20 eV. It shows that this XUV source has fairly interesting properties. From our numerical calculations in xenon, in the experimental conditions, we find that the pulse width of the 17th harmonic is 0.5 ps, its spectral width about 0.5 Å. The photon flux (N_q/τ_q, τ_q being the qth harmonic pulse width) is very high, about 10^{22} photons/s. Although these data (apart from the total number of photons) are mostly numerical (those are indicated by the star in the table) and not yet measured, they should provide valuable estimates.

HOW WELL ?

It is important to characterize thoroughly the harmonic radiation not only to understand the basic nonlinear physics but also to facilitate the use of the radiation in a number of possible applications. Measurements of temporal profiles of the radiated harmonics have been performed at Imperial College[16], using a 1053-nm 50-ps Nd-Glass laser. Another essential study for the understanding of the physics of high order harmonic generation processes is that of the angular distribution of the harmonics (i.e. their spatial profiles)[14,15].

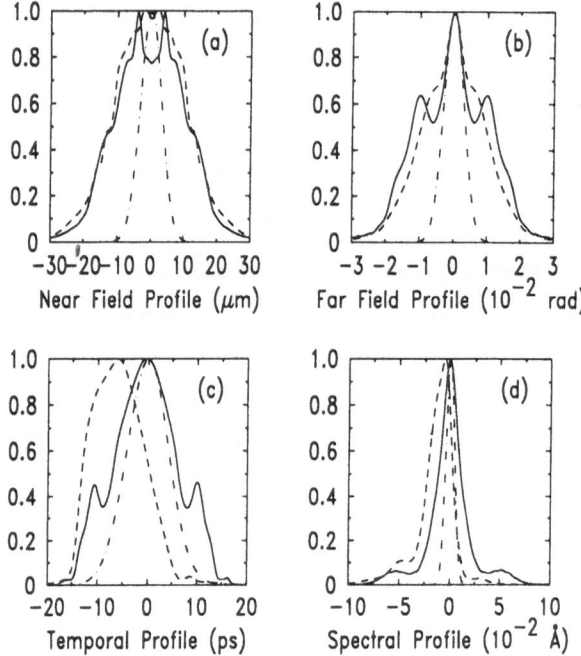

Figure 4. Calculated near-field, far-field, temporal and spectral profiles of the 13th harmonic of a 36-ps 1064-nm Nd-YAG laser, generated in a 15-Torr xenon gas jet.

We first present results of numerical calculations[9] for the spatial, temporal and spectral profiles of the 13th harmonic at three intensities 5×10^{12} (perturbative limit), 2×10^{13} and 4×10^{13} W.cm^{-2}. All the curves have been normalized to have the same maximum value. The near field profile (at the end edge of the medium, i.e. at 0.8 mm from the focus) is shown in Fig. 4(a). Fig. 4(b) displays the far field profile calculated at 30 cm from the focus and indicated in radians. Both spatial profiles have been time-integrated over the pulse width. Finally, Figs. 4(c) and (d) show respectively the temporal profile and spectral lineshape of the 13th harmonic outside the medium. The harmonics are defocused compared to that expected in the weak field (perturbative) limit. They exhibit some structures (rings). Note that using a weaker focusing geometry leads to smoother (more coherent) spatial profiles. In the same way, the temporal profiles are not always regular, presenting structures which

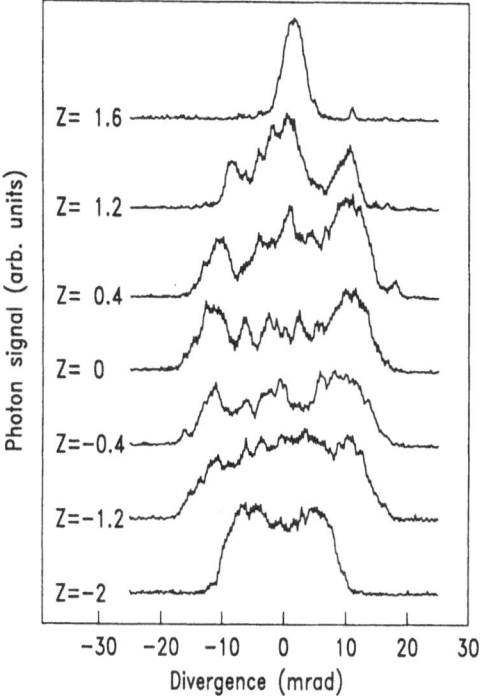

Figure 5. Experimental far field profiles of the 13th harmonic at different focus positions, using a 36-ps 1064 nm Nd-YAG laser in a 15 Torr- gas jet of xenon. The laser intensity is estimated to be 7×10^{13} W/cm^2, i.e. about twice the saturation intensity.

are due in part to resonance effects in the single atom response. Ionization leads to an asymmetry in the profiles as seen in particular in the result at 4×10^{13} W.cm^{-2}. The depletion and therefore the asymmetry is more significant for higher intensities. Finally, the spectral lineshapes exhibit a significant shift to higher frequencies at intensities high enough to begin to ionize the medium. This blue shift is induced by the fundamental blue shift and is approximately equal to q times that of the fundamental. At high intensities, the 13th harmonic lineshape broadens and shifts compared to the perturbative limit. However, it remains quite narrow, with a $\Delta\lambda/\lambda$ less than 10^{-4}.

We now present measurements[15] of far field profiles of low-order harmonics, performed with a 36-ps 1064 nm Nd-YAG laser, focused here by a 300 mm lens in a

15-Torr gas jet of xenon atoms. We use microchannel plates placed at 30 cm from the output slit of the spectrometer, coupled to a phosphorus screen which is read by a CCD camera. The 2-dimensional image recorded by the camera displays the spectrum in the horizontal direction and the spatial profile in the far field in the vertical direction (parallel to the grating's grooves). Fig.5 shows measurements of the 13th harmonic profile for several positions Z of the focus in the gas jet (with the same convention as before : the laser propagates from negative to positive Z) at an intensity estimated to 7×10^{13} W/cm^2, i.e. about twice the saturation intensity. The divergence of the fundamental is estimated to be 30 mrad at $1/e^2$. The profile varies from a nice Gaussian shape ($Z = 1.6$ mm) to a broader and more accidented

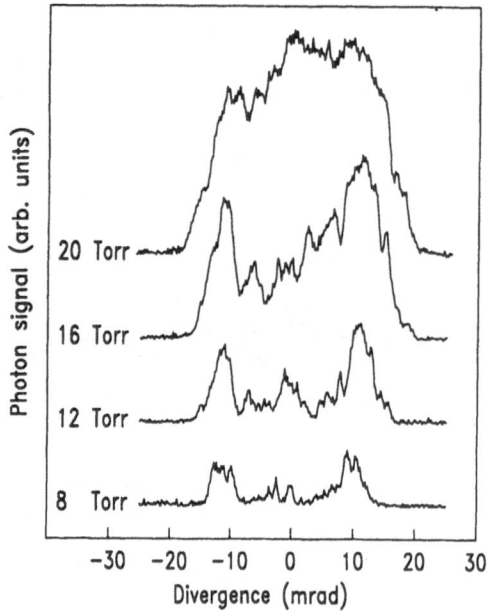

Figure 6. Experimental far field profiles of the 13th harmonic at different pressures at the best focus position. Same conditions as in Fig.5.

distribution when the focus position coincides with the center of the jet, and finally to a kind of "flat top" shape. The middle curves exhibit oscillations, 3 for $Z = 2$ and 5 or 6 closer to $Z = 0$. The other harmonics investigated (from the 11th to the 19th) present similar profiles. The distribution is broadest when the laser is focused into a high density region of the nonlinear medium and for a long propagation length after the best focus region (i.e. for negative Z). This is probably due to the influence of ionization on phase matching. Both the influence of free electrons and the change of the interaction volume depleted by ionization lead to formation of rings in the near field and far field profiles of the generated radiation.

In Fig.6, we show the variation of the 13th harmonic profile with pressure at $Z = 0$ at the same intensity as in Fig.5. The width of the spatial distribution does not change much with pressure, indicating that it is determined by a volume effect, rather than by the dispersion. From 8 to 15 Torr, the distribution presents a large outer ring and a smaller central peak. However, from 16 Torr, the profile becomes more uniform, and the oscillations are completely blurred at 20 Torr. It happens when the dependence of the signal with the pressure saturates, indicating the influence of dispersion due to free electrons.

High order optical harmonic generation might soon become a useful radiation source in the XUV, soft X-ray ranges. Its optimization and characterization are in rapid progress. Theoretical calculations predict that energies much higher than the 160-170 eV measured so far could be produced, e.g. using ions as nonlinear media. Experiments are planned to extend further the limits of these phenomena.

REFERENCES

1. N. Sarukura, K. Hata, T. Adachi, R. Nodomi, M. Watanabe and S. Watanabe, Phys. Rev. A **43** 1669 (1991).
2. K. Miyazaki and H. Sakai, J. Phys. B, **25** L83 (1992).
3. J. K. Crane, M. D. Perry, S. Herman and R. W. Falcone, Opt. Lett. **17**, 1256 (1992).
4. J. J. Macklin, J. D. Kmetec and C. L. Gordon III, Phys. Rev. Lett., 1 february 1993.
5. A. L'Huillier and Ph. Balcou, Phys. Rev. Lett., 1 february 1993.
6. J. L. Krause, K. J. Schafer and K. C. Kulander, Phys. Rev. Lett. **68** 3535 (1992).
7. H. Xu, X. Tang and P. Lambropoulos, Phys. Rev. A, **46**, R2225 (1992).
8. V. C. Reed and K. Burnett, Phys. Rev. A **46**, 424 (1992).
9. A. L'Huillier, Ph. Balcou, S. Candel, K. J. Schafer and K. C. Kulander, Phys. Rev. A, **46**, 2778 (1992).
10. A. L'Huillier, L. A. Lompré, G. Mainfray and C. Manus, in "Atoms in Intense Laser Fields", Ed. M. Gavrila (Advances in Atomic, Molecular and Optical Physics, Supplement 1, Academic Press, New York, 1992) p 139; see also other articles in the same book.
11. Y. Akiyama, K. Midorikawa, Y. Matsunawa, Y. Nagata, M. Obara, H. Tashiro and K. Toyoda, Phys. Rev. Lett. **69**, 2176 (1992).
12. A. L'Huillier, Ph. Balcou and L. A. Lompré, Phys. Rev. Lett. **68**, 166 (1992).
13. Ph. Balcou and A. L'Huillier, Phys. Rev. A, february 1993.
14. D. D. Meyerhofer, S. Augst, C. I. Moore and J. Peatross, SPIE Vol. 1551, "Ultrashort Wavelength Lasers (1991) ; see also this volume.
15. A. L'Huillier and Ph. Balcou, Laser Physics, in press.
16. M. E. Faldon, M. H. R. Hutchinson, J. P. Marangos, J. E. Muffet, R. A. Smith, J. W. G. Tisch and C. G. Wahlström, J. Opt. Soc. Am B, **9**, 2094 (1992).
17. Ph. Balcou, A. S. L. Gomes, C. Cornaggia, L. A. Lompré and A. L'Huillier, J. Phys. B **25**, 4467 (1992).
18. W. Becker, S. Long and J. K. McIver, Phys. Rev. A **41**, 4112 (1990)
19. H. B. van Linden van den Heuvell and H. G. Muller, Cambridge Studies in Modern Optics **8**, Multiphoton Processes, eds. S. J. Smith and P. L. Knight (Cambridge University Press, 1988); T. F. Gallagher, Phys. Rev. Lett. **61**, 2304 (1989); P. B. Corkum, N. H. Burnett and F. Brunel, Phys. Rev. Lett. **62**, 1259 (1989); for the application to harmonic generation, see the contribution of K. C. Kulander et al., this volume.
20. C.-G. Wahlström, J. Larsson, A. Persson, S. Svanberg, Ph. Balcou, P. Salières and A. L'Huillier, postdeadline paper and to be published.

ANGULAR DISTRIBUTIONS OF HIGH-ORDER HARMONICS

D. D. Meyerhofer[1] and J. Peatross[2]

Laboratory for Laser Energetics
[1]Department of Mechanical Engineering
[2]Department of Physics and Astronomy
University of Rochester
250 East River Road
Rochester, NY 14623-1299

INTRODUCTION

High-order harmonics have been observed during laser-atom interactions at a number of different laser wavelengths with short-pulse, high-intensity laser systems. Harmonics with frequencies in excess of 100 times the laser frequency have been reported.[1-3]

An important issue in the generation of high-order harmonics during laser-atom interactions has been the role of phase matching in the atomic medium. Phase-matching effects have been extensively studied by L'Huillier and co-workers.[1,4-7] They have made many detailed comparisons of their calculations with their experiments and find excellent agreement. Their studies have included the effect of the intensity, pressure, and focusing characteristics (both confocal parameter and jet position relative to the focus) on the spatially and temporally integrated harmonic output. Faldon et al.[8] have been able to explain the temporal history of the harmonic emission by considering the effects of ionization and phase matching.

We believe that all of the high-order harmonic generation experiments to date have used gas target pressures in excess of 10 Torr. These high pressures have been used to enhance the harmonic signal, which increases quadratically with the pressure. The phase-mismatch factors (due to both neutral and ionized media) increase with the pressure as does the effect of refraction of the incident laser when the medium becomes ionized. Very few measurements of the angular distributions of the high-order harmonics have been reported.[1,3] The far-field pattern can provide additional insight into the laser propagation, phase matching, and the atomic response.

In this article, we present observations of the angular (far-field) distribution of high-order harmonic generation with a low density gas target with gas pressures of order 0.5 T. With the f/70 focal system and 1-mm-thick target, the phase-mismatch factors from neutral gas, free electrons, and geometric effects are all less than π. Thus in these experiments, the high-order harmonics are generated under conditions where the effects of phase mismatch on the harmonic generation should be minimized. The goals of this work are to separate the atomic response from the effects of phase matching. Our current experimental configuration allows the observation of the far-field distribution of the 11th to 21st harmonics of the incident laser radiation. The article is organized as follows: the laser system, the gas target, and the detector are described first. The onset of refraction, which constrains the experimental conditions, is also described. The angular distributions of the high-order harmonics are then presented. The measurements are compared with the predictions of lowest order perturbation theory. The conclusions are then presented.

EXPERIMENTAL CONDITIONS

The high-order harmonics are observed by focusing a 1-μm, 1.5-ps laser into a moderate density gas target (p ~ 0.5 T, thickness ~1 mm) and then detecting the harmonics with a transmission grating coupled to an MCP-phosphor image intensifier. A schematic of the experiment is shown in Fig. 1. The various components of the experimental setup are discussed in this section, as is the effect of refraction on the operating conditions.

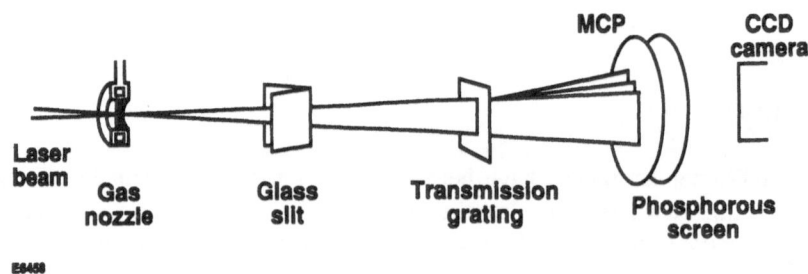

Figure 1. Schematic of the experimental setup, including the gas target and the angularly resolved spectrometer.

The 1-μm, 1.5-ps laser pulses used to generate the harmonics are generated with an Nd:glass, chirped-pulse-amplification laser system,[9,10] which has been described in detail elsewhere.[11] We have recently upgraded the energy capability and focusing quality of the system.

For these experiments, a long Rayleigh range is desired to minimize the geometric phase-matching effects on the high-order harmonics. To increase the f-number of our focusing system a smaller than usual, 1.1-cm radius, beam is produced by the laser system. Under these conditions, the laser energy is limited to less than 300 mJ to avoid B-integral effects and to operate significantly below the compression grating damage threshold. This beam is focused with a 153-cm lens to produce an f-number of 70. While the use of the long focal-length lens sacrifices peak field intensity, the system is still able to achieve 3×10^{15} W/cm^2 with relatively large focal volumes and low field gradients. We are currently limited to much lower energies due to plasma formation in the spectrometer.

The focal spot diameter is 1.2 times diffraction limited. The focal characteristics of the laser were measured by direct imaging, with four-times magnification, onto a CCD camera. The focal spot diameter as a function of the axial distance along the focus is shown in Fig. 2, as is the calculated diameter for a diffraction limited f/70 beam.

The spectrometer used in these experiments consists of a slit with approximate dimensions of 500 μm × 2.5 cm followed by a bare gold wire transmission grating. The slit is aligned parallel to the grating wires. The grating consists of 0.5-μm-diam gold wires, which are separated by 1 μm (center to center). Following the grating is a microchannel plate (MCP) image intensified phosphor screen (Galileo model 8081). The microchannel

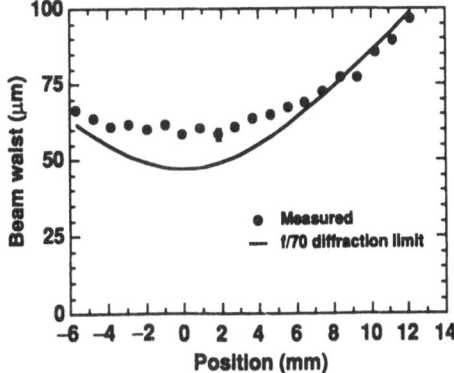

Figure 2. Plot of the measured laser beam waist as a function of axial position along the focus, showing that the beam is 1.2 times diffraction limited with f/70 optics.

plate intensifier is not UV enhanced so it cannot detect harmonics lower than the ninth. The light incident on the slit and grating is dispersed by the grating and produces spectral lines on the phosphor screen. The different harmonic orders are separated in the direction perpendicular to the grating lines and their angular distributions are observed parallel to the slit. Thus we simultaneously obtain information about the far-field pattern as well as the spectral content of the harmonics. At this time, the spectrometer is not absolutely calibrated so that only relative levels of harmonic emission are presented. In all cases, the same relative units are used. The relatively large slit allows a larger photon flux to reach the detector at the cost of reduced spectral resolution. In the present configuration, the harmonic lines begin to overlap with the 23d harmonic.

The gas target is designed to create well-characterized, narrow gas distributions at low densities (1–2 Torr or less with backing pressures up to 5–10 Torr).[12] This low-density regime is desirable to reduce the phase-matching effects during the harmonic generation[5] and the effects of refraction of the focused laser beam. Also, the low backing pressure has the advantage of reducing the possibility of dimer formation in gases such as Xe.[13]

The gas target is a small cylindrical hole through which the focused laser passes and inside of which gas enters from the middle. Because the laser beam goes through the target, the alignment of the device is comparatively simple. The gas target operates on the principles of molecular flow rather than of fluid flow as in the gas jet. Since the flow rate is relatively low, the target can be operated in a continuous rather than in a pulsed mode. The density of the gas within the hole remains relatively high while the gas outside the hole disperses quickly (inverse square of the distance from the hole edge). The target operation is limited to low densities just as the gas jet is limited to high densities. If the gas in the target hole is at too high a pressure, plumes may develop out its ends that would lie on top of the incoming and outgoing laser beam. The jet and the target are thus complementary in the sense that they operate in opposite ranges of pressure.

The gas target consists of two identically machined cylindrical aluminum pieces that are glued together with a thin layer of vacuum epoxy. Aluminum is chosen because of its ease of machining. Figure 3(a) shows a cut-away portion of the two pieces (upper and lower), which are already attached around the outer rim. Figure 3(b) shows the inside of a single piece so that the cylindrical symmetry is observable. Gas flows from the outer ring-

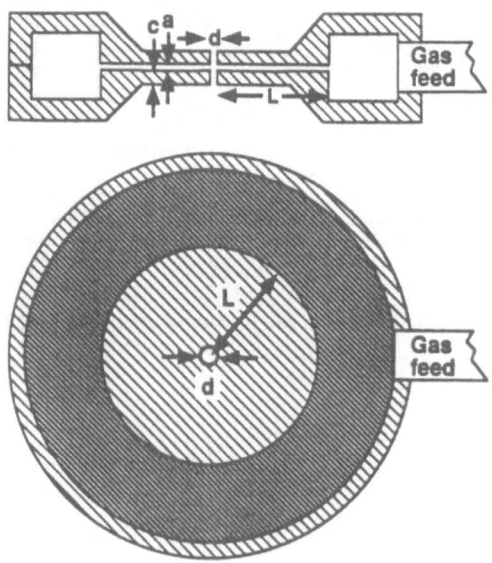

Figure 3. Schematic of the machined gas target used in the experiments.

shaped pocket into the gap between the thinly spaced plates and toward the hole at the center. The gas then escapes out both ends of the hole. When gluing the pieces together, the drill bit that made the holes is inserted through both pieces to ensure alignment. As shown in Fig. 3, a is the separation of the plates, c is the thickness of the plates, d is the hole diameter, and L is the length from the hole center to the inside edge of the outer gas pocket. Typical values for these are $a = 0.2$ mm, $c = 0.4$ mm, $d = 0.5$ mm, and $L = 4$ mm.

The gas distribution in the target is characterized experimentally by observing the recombination emission.[13] It is perhaps more difficult to characterize than the jet since the gas densities are much lower, and the off-axis line of sight to the interaction region is obstructed by the target itself. It is possible to calculate the gas distribution in the target using a Monte-Carlo computer simulation of free molecular flow. We have been able to experimentally verify the predicted density profile for the region just inside the target opening and outward. The Monte-Carlo simulation also predicts the gas flow rate from the target. There is good agreement between the predicted and the measured gas flow rates for various target designs. An example of the measured density profile is shown in Fig. 4. The measurement is significantly broadened by the detector resolution of 0.5 mm. The detector is unable to see the target center because it is aligned 45° off axis. The full width at half maximum of the gas target is of the order of 1 mm with low peak pressure and with a $1/R^2$ fall-off outside the target region.

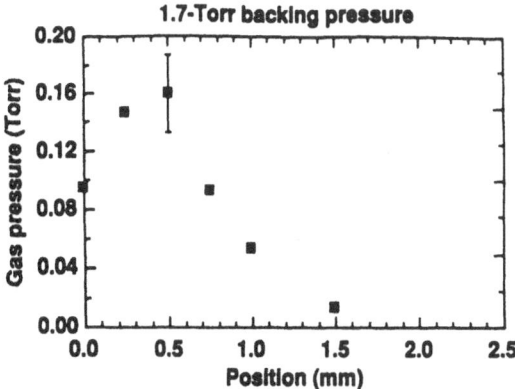

Figure 4. Measured gas pressure in the target as a function of axial distance from the center of the target. The axial resolution is approximately 0.5 mm.

One important goal of our experiments is to operate in a regime where the laser propagation through the target is unaffected by the medium. This makes the interpretation of the experimental results easier. A laser propagating through an ionized medium may undergo refraction or self-focusing. The laser was focused through the gas target and imaged onto a CCD camera with various xenon gas pressures in the target and with various peak laser intensities. The results are shown in Fig. 5. Figures 5(a)–5(c) show the change in effective focal spot with increasing laser intensity at 0.5-Torr pressure. At the highest intensity the xenon fully ionized within the focal volume.[14] At this low pressure, the focal spot is identical to that observed in the vacuum. When the pressure is increased to 2 Torr as in Figs. 5(d)–5(f), an apparent decrease in the focal volume is observed. This is due to refraction, which decreases the f-number of the laser as it leaves the gas target. Preliminary calculations of the refraction under these conditions are consistent with the observed results.

HARMONIC FAR-FIELD DISTRIBUTIONS

The experimental conditions described above have been chosen to minimize the phase-mismatch factors.[5] The combination of f/70 focusing and a thin gas target (1 mm) means that the target thickness is much less than the 12-mm confocal parameter. In addition, the low gas pressures ~0.5 T, mean that the phase mismatch from neutral Xe and free electrons is minimal. Table 1 shows the phase-mismatch factors for various harmonic orders are less than or of the order of 1 radian.

Table 1. Phase-mismatch factors ΔkL (radians).

Harmonic order	9	13	17	21
neutral Xe	0.08	-0.05	-0.08	-0.08
free electrons	0.5	0.7	0.9	1.1

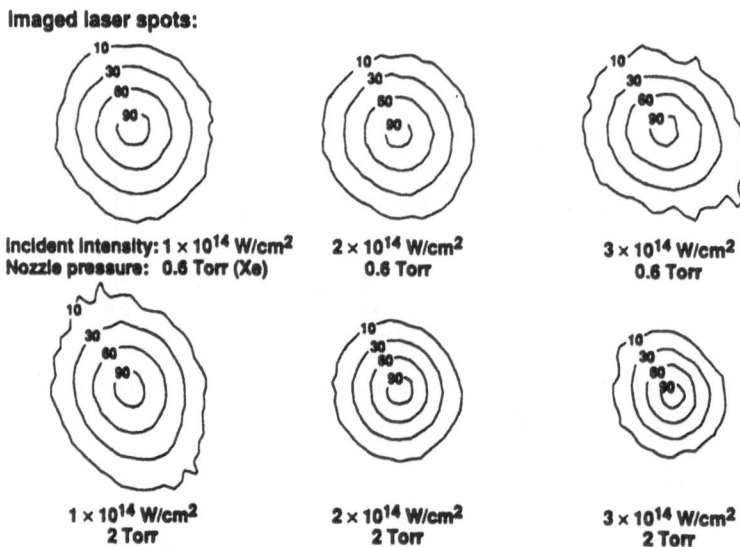

Imaged laser spots:

Incident Intensity: 1×10^{14} W/cm^2 2×10^{14} W/cm^2 3×10^{14} W/cm^2
Nozzle pressure: 0.6 Torr (Xe)   0.6 Torr   0.6 Torr

1×10^{14} W/cm^2 2×10^{14} W/cm^2 3×10^{14} W/cm^2
2 Torr   2 Torr   2 Torr

Figure 5. Images of the laser focus as a function of laser intensity and gas target pressure. Figures 5(a)–5(c) show increasing laser intensity at 0.5 Torr of xenon pressure, and Figs. 5(d)–5(f) show the same for 2.0 Torr of xenon.

Figure 6 shows the far-field (angular) distribution of the 11th to 21st harmonic of the laser at an intensity of 9×10^{13} W/cm^2 and a gas target peak pressure of 0.3 Torr xenon. This result is typical of the harmonic profiles under these conditions. Most of the high-order harmonics have a very narrow angular structure. For xenon under these conditions, large-scale shoulder structures appear on the 13th harmonic. The shoulders are observed on the 15th harmonic in krypton and the 17th in argon.

The relative energy in the different harmonics is determined by integrating the harmonic emission radially, taking the circularly symmetric nature of the emission into account,

$$E_q = \int_0^\infty 2\pi I(r)r\,dr.$$

The emission is symmeterized around the peak signal level. Figure 7 shows the relative energy in the 11th to 23d harmonics in xenon at a gas pressure of 0.5 T for three different intensities. The development of a plateau in the harmonic emission with increasing intensity is evident.

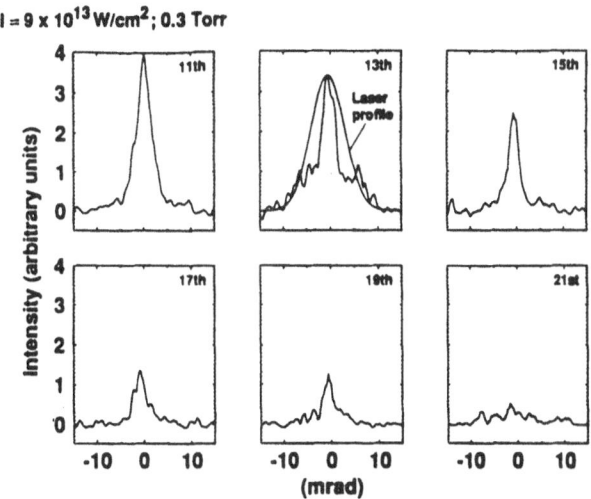

Figure 6. Typical far-field distribution of the 11th to 21st harmonic observed for 0.3 Torr of xenon at 9 × 10^{13} W/cm^2.

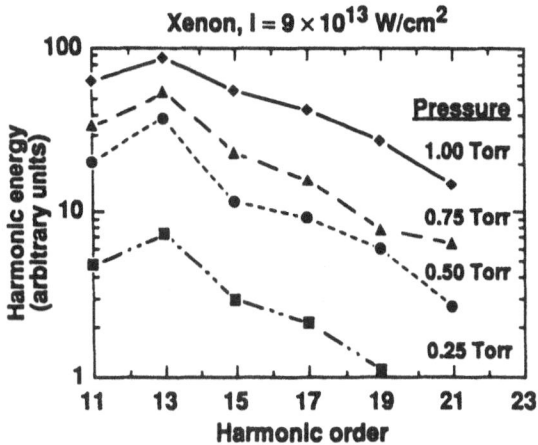

Figure 7. Relative harmonic energy in the 11th to 21st harmonic in xenon at 0.5 Torr and three different intensities.

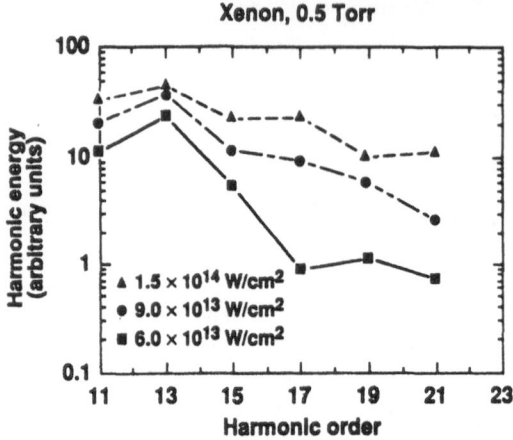

Figure 8. Integrated energy for the 11th through 21st harmonic for xenon at 9×10^{13} W/cm^2 for four different gas pressures.

Under perfect phase-matching conditions, the integrated harmonic intensity should scale as N^2, where N is the number of atoms in the focal volume for a fixed intensity Figure 8 shows the integrated energy for the 11th to 21st harmonic in xenon at 9×10^{13} W/cm^2 at four different gas pressures. The overall shape of the harmonic energy spectrum is maintained and the energy of each of the harmonics scales quadratically with pressure (N^2) as expected.

One of the features, which is common to all of the low pressure far-field distributions, is the development of significant shoulders, which appear most pronounced on one particular harmonic first. This is evident in the 13th harmonic in Fig. 6. The far-field emission for Xe, Kr, and Ar is shown in Fig. 9. For Kr and Ar, the shoulders appear on the 15th and 17th harmonics respectively. In this case, the intensity is approximately the barrier suppression ionization threshold (BSI),[15] and the gas target density has been scaled so that the level of emission is similar in each case. At the BSI threshold, the ionization probability has been observed to be approximately 1% for 1.5-ps, 1-μm laser pulses.[14,15] I is important to note the similarities among the far-field patterns of the different harmonics Each of the harmonics shows the same shapes. In fact, the angular distributions are almost identical. The primary difference is the harmonic at which the shoulders appear. These results are summarized in Table 2. The pronounced structure appears on the harmonic which is one order ($2\omega_L$) higher than the first with greater energy than the field-free atomic ionization potential.

Table 2. Comparison of harmonic structure in different target gases.

Target gas	BSI intensity	Experimental intensity	First harmonic showing structure
Xenon	8.6×10^{13} W/cm^2	9×10^{13} W/cm^2	13
Krypton	1.5×10^{14} W/cm^2	1.2×10^{14} W/cm^2	15
Argon	2.5×10^{14} W/cm^2	2.1×10^{14} W/cm^2	17

COMPARISONS WITH LOWEST ORDER PERTURBATION THEORY

With the exception of the harmonic, which shows the shoulder-like structure, the far-field patterns are quite narrow and seem to get narrower with increasing harmonic order. We can compare the far-field widths with the predictions of lowest order perturbation theory (LOPT). The harmonic intensity is assumed to vary as I^q. The angular width of the harmonics is then expected to vary as $\delta\theta_0 / \sqrt{q}$, where $\delta\theta_0$ is the angular distribution of the laser. The observed harmonic emission for Kr at 0.5 Torr and 10^{14} W/cm^2 is compared with the LOPT predictions in Fig. 10. Good agreement with the far-field patterns of all of the harmonics is observed. It should be noted though, that the relative intensity of each harmonic does not scale with intensity as LOPT would predict.

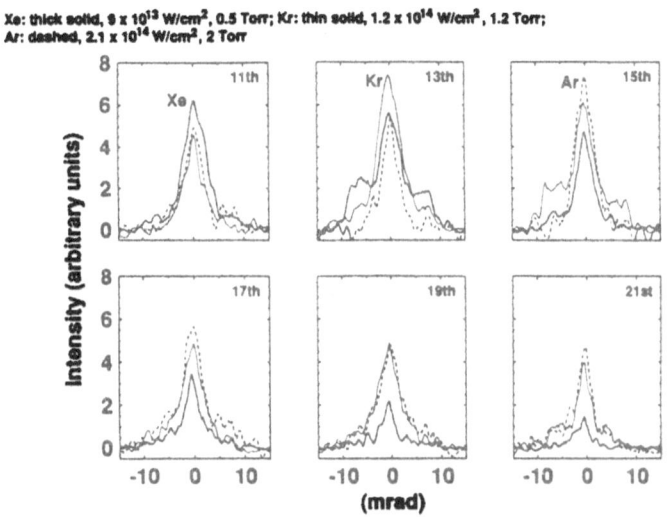

Figure 9. Comparison of the far-field emission for Xe, Kr, and Ar, with the intensity at the BSI threshold.

CONCLUSION

We have made the first observations of the far-field pattern of high-order harmonics under conditions where the geometric and propagation phase-mismatch factors should be negligible. Under these conditions, we find that the far-field pattern (though not the intensity scaling) is consistent with lowest order perturbation theory, except for one harmonic (the 13th in xenon, the 15th in Kr, and the 17th in Ar). The harmonic emission scales quadratically with the pressure and, even under these low pressure conditions, a plateau develops, as has been observed in many other experiments. We are currently studying the details of the far-field profile.

Kr: I = 1 x 10^{14} W/cm^2; 0.5 Torr

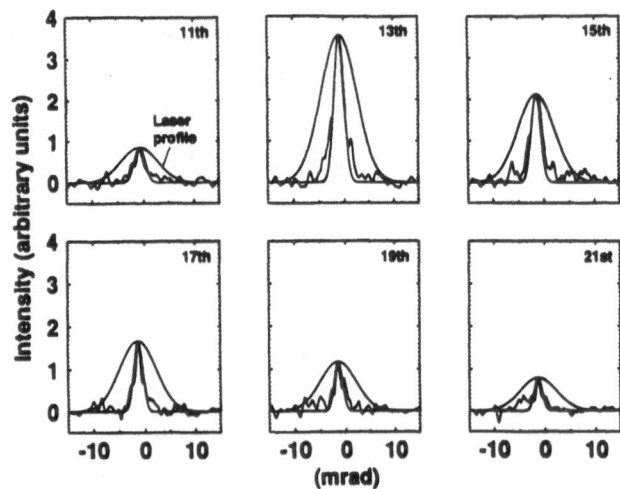

Figure 10. Comparison of the far-field emission for Kr at 0.5 Torr and 1×10^{14} W/cm^2 with the predictions of lowest order perturbation theory.

ACKNOWLEDGEMENTS

This work is supported by the National Science Foundation under contract PHY-9200542. Additional support was provided by the U.S. Department of Energy Office of Inertial Confinement Fusion under Cooperative Agreement No. DE-FC03-92SF19460 and the University of Rochester. The support of DOE does not constitute an endorsement by DOE of the views expressed in this article.

REFERENCES

1. A. L'Huillier, and Ph. Balcou, High-order harmonic generation in rare gases with a 1-ps 1053-nm laser, *Phys. Rev. Lett.* 70:774–777 (1993).
2. J.J. Macklin, J.D. Kmetec, and C.L. Gordon III, High-order harmonic generation using intense femtosecond pulses, *Phys. Rev. Lett.* 70:766–769 (1993).
3. R.A. Smith, J.W.G. Tisch, M. Ciarrocca, S. Augst, and M.H.R. Hutchinson, Recent ultra-high harmonic generation experiments with picosecond laser pulses, *in*: "SILAP," B. Piraux, ed., Belgium, NATO (1993), this volume.
4. A. L'Huillier, K.J. Schafer, and K.C. Kulander, High-order harmonic generation in xenon at 1064 nm: The role of phase matching, *Phys. Rev. Lett.* 66:2200–2203 (1991).
5. A. L'Huillier, K.J. Schafer, and K.C. Kulander, Theoretical aspects of intense field harmonic generation, *J. Phys. B: At. Mol. Opt. Phys.* 24:3315–3341 (1991).
6. A. L'Huillier, P. Balcou, and L.A. Lompré, Coherence and resonance effects in high-order harmonic generation, *Phys. Rev. Lett.* 68:166–169 (1992).
7. A. L'Huillier, P. Balcou, S. Candel, K.J. Schafer, and K.C. Kulander, Calculations of high-order harmonic-generation processes in xenon at 1064 nm, *Phys. Rev. A* 46:2778–2790 (1992).

8. M.E. Faldon, M.H.R. Hutchinson, J.P. Marangos, J.E. Muffett, R.A. Smith, J.W.G. Tisch, and C.G. Wahlstrom, Studies of time-resolved harmonic generation in intense laser fields in xenon, *J. Opt. Soc. Am. B* 9:2094–2099 (1992).

9. D. Strickland, and G. Mourou, Compression of amplified chirped optical pulses, *Opt. Commun.* 56:219–221 (1985).

10. P. Maine, D. Strickland, P. Bado, M. Pessot, and G. Mourou, Generation of ultrahigh peak power pulses by chirped pulse amplification, *IEEE J. Quantum Electron.* 24:398–403 (1988).

11. Y.-H. Chuang, D.D. Meyerhofer, S. Augst, H. Chen, J. Peatross, and S. Uchida, Suppression of the pedestal in a chirped-pulse-amplification laser, *J. Opt. Soc. Am. B* 8:1226–1235 (1991).

12. J. Peatross and D.D. Meyerhofer, Novel gas target for use in laser harmonic generation, submitted to *J. Appl. Phys.* (1993).

13. L.A. Lompré, M. Ferray, A. L'Huillier, X.F. Li, and G. Mainfray, Optical determination of the characteristics of a pulsed-gas jet, *J. Appl. Phys.* 63:1791–1793 (1988).

14. S. Augst, D.D. Meyerhofer, D. Strickland, and S.L. Chin, Laser ionization of noble gases by coulomb-barrier suppression, *J. Opt. Soc. Am. B* 8:858–867 (1991).

15. S. Augst, D. Strickland, D.D. Meyerhofer, S.L. Chin, and J.H. Eberly, Tunneling ionization of noble gases in a high-intensity laser field, *Phys. Rev. Lett.* 63:2212–2215 (1989).

ANGULARLY RESOLVED ULTRA HIGH HARMONIC GENERATION EXPERIMENTS WITH PICOSECOND LASER PULSES

R. A. Smith, J.W.G. Tisch, M. Ciarrocca, S. Augst
and M. H. R. Hutchinson

The Blackett Laboratory
Imperial College of Science, Technology and Medicine
Prince Consort Road
London SW7 2BZ
UK

INTRODUCTION

Ultra-high harmonic generation is a field currently attracting considerable theoretical and experimental interest. The generation of harmonics of orders well in excess of 100 from noble gas targets has recently been demonstrated by a number of groups[1,2,3] using ultra-high intensity pump lasers. Typically only the odd harmonics are observed (due to parity considerations) and the harmonic spectrum is characterised by an initial decline in intensity with increasing harmonic order, followed by a plateau region which may include many harmonics of similar brightness, and finally a sharp cut-off, beyond which no further harmonic emission is seen.

This area of work has been driven both by a desire to understand and use the harmonic generation process itself as a means of studying basic atomic physics at high intensities (10^{13} - 10^{18} Wcm^{-2}) and by the potential use of the harmonics themselves as probes of other physical processes. Clearly a laboratory scale source of bright, coherent, short-pulse xuv radiation would be a powerful tool in its own right for investigating fast phenomena such as picosecond laser-plasma interactions. Harmonics of wavelengths below 100 Å have been seen experimentally, and provide a possible route to a coherent soft x-ray source which is technically less demanding and less energy intensive than current plasma based x-ray laser schemes.

The high-harmonic generation process is currently not well understood, involving as it does both the non-perturbative response of an individual atom to an intense laser field and the propagation and phase matching of the harmonics and pump laser in a rapidly ionising medium.

At intensities above 10^{13} - 10^{14} Wcm^{-2}, with moderate gas jet densities (a few Torr) the problem may also involve aspects of laser-plasma interactions. The free electrons generated by field or multi-photon ionisation couple to the laser field in a complex range of processes including self-focusing and parametric instabilities such as Raman scattering. These are driven by the ponderomotive potential of the pump laser and can result in strong modification of the laser intensity and frequency as well as the free electron distribution.

There are a number of tools which can be applied to the understanding of this area of physics. Clearly computer models are important in solving the non-perturbative single atom problem for high laser fields and the phase matching and propagation of the harmonics. Both the single atom response[4,5] and macroscopic effects such as phase matching [6,7] in gas jets have been investigated using detailed simulations. However a realistic time-dependent 3D atomic model coupled to a 3D propagation code will be both difficult to implement and expensive in terms of computer time. There are a number of experimental techniques which can also be applied to the problem. Simply varying the laser intensity, focusing geometry, or target gas can provide useful insights into the harmonic generation process, and these have been described elsewhere[8]. More sophisticated techniques such as time-resolution of the harmonics[9] may also be used to study the interplay of ionisation, neutral depletion and phase matching.

In general, lower Z noble gases are less efficient at generating a given harmonic, but the plateau region extends to higher order. The more weakly bound an outer electron is, the more easily polarisable by the laser field it is. Such electrons radiate more strongly at a given order but are also more easily ionised. The contribution of ions to the harmonic generation process is still unclear but it seems likely that for the highest orders currently being studied the neutrals dominate. Increasing the laser intensity extends the plateau region to higher order, and Kulander and others[10,11] have developed a semi-empirical scaling law describing this effect.

$$E_{max} \sim Ip + n\, Up \qquad (1)$$

where E_{max} is the energy in eV of the highest harmonic, Ip the ionisation energy, Up the ponderomotive shift due to the laser and n a factor in the range 3-3.5. The physical basis for this scaling law is described in more detail elsewhere in this volume[11]

In the experiment described below, our aim was to investigate the angular distribution of the high harmonics (the far field pattern) when moving from the plateau to the cut-off region of the spectrum.

ANGULAR RESOLUTION EXPERIMENT

Substantial laser intensities (> 10^{12} Wcm^{-2}) are required for high harmonic generation from gas targets. To generate intensities up to 5×10^{14} Wcm^{-2} we employed an Nd:Glass Chirped Pulse Amplification (CPA) system capable of delivering up to 1 J of 1.053 μm light in a 1.2 ps pulse, focusable to ~ 3 times the diffraction limit. This system is similar to a number of other high power CPA lasers described in the literature[12,13], and a simplified diagram is shown in figure 1.

An actively mode locked Nd:YLF oscillator produces 60 ps, 5 nJ transform limited pulses which are injected into 1.3 km of single mode, non-polarisation preserving fibre.

Self phase modulation generates additional bandwidth (which we require to support picosecond pulses) while group velocity dispersion chirps and stretches the pulse to 150 ps. The 45 Å bandwidth pulse from the fibre is stretched further to 600 ps in a 4 pass grating / telescope delay line, and a single pulse from the 35 MHz pulse train injected into a regenerative amplifier. The regenerative amplifier is a linear cavity two glass system, optimised to produce minimal spectral modulation and narrowing of the pulse. This amplifier typically provides a total gain of 10^6 in $\sim$ 30 round trips, giving a 1 mJ output pulse of 22 Å bandwidth. After transiting a pulse slicer Pockels cell placed after the regenerative amplifier, the contrast ratio of switched out to leakage pulses is better than 5×10^8.

The pulse from the regenerative amplifier is then amplified further in a series of Nd:Glass rods. One double pass 9 mm, one single pass 9 mm and one single pass 16 mm amplifier increase the energy up to a maximum of 2 J in a 600 ps stretched pulse. The pulse is then recompressed to 1.2 ps using a 4 pass grating dispersive delay line, to give a maximum compressed energy of 1 J in a 35 mm diameter beam. The laser system is extensively diagnosed to allow both ease of operation and accurate characterisation of the pulse length, energy, bandwidth and spatial profile on a single shot basis.

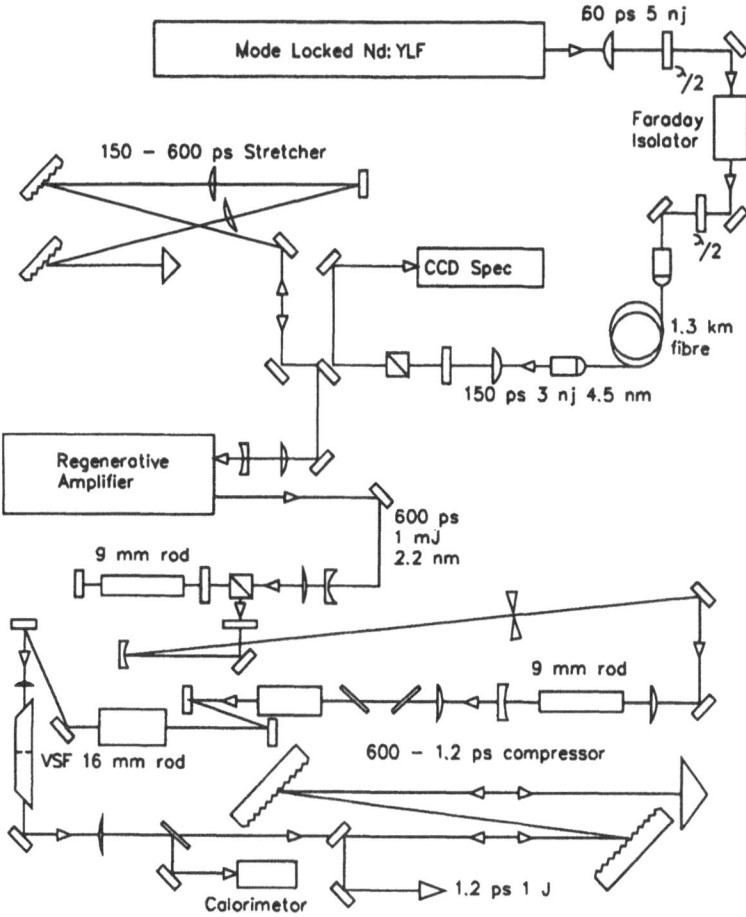

Figure 1. System layout for the CPA laser used for harmonic generation experiments.

The experimental arrangement for angularly resolved harmonic studies is shown in figure 2. Laser pulses were focused by an f = 1.7 m plano-convex lens (f/50) through a wedged AR coated window fronting a vacuum extension tube into our target chamber.

A relatively long focal length lens was chosen to increase the harmonic signal which scales as b^3 where b is the confocal parameter[14]. This has the additional advantage of minimising the effects of geometric phase mis-match. For these conditions the focal spot radius (at the e^2 intensity point) is 140 μm with a measured confocal parameter b ~ 3.9 cm. Imaging of the focal spot with a CCD camera has shown the distribution to be very close to gaussian.

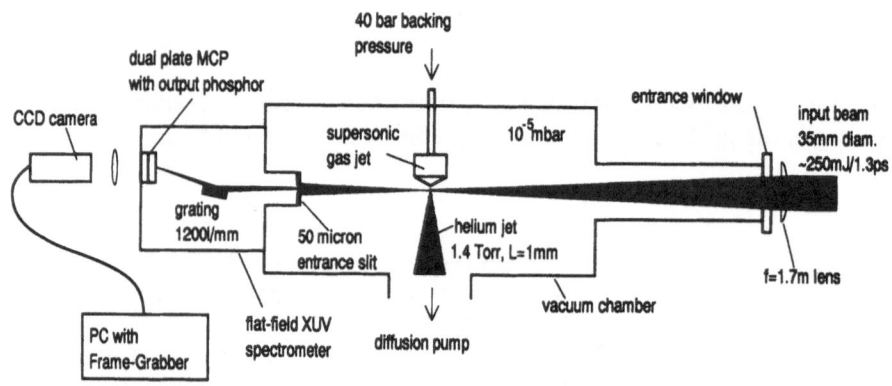

Figure 2. Target chamber, focusing optics and detector system for the angularly resolved experiments.

The medium in which harmonics were generated was a supersonic gas jet produced by a solenoid valve backed to 40 atmospheres, firing into vacuum. The supersonic flow (specified as ~ Mach 5) ensures that the interaction medium is shock free, thus eliminating any strong density perturbations. Recent experiments on laser-gas interactions and ionisation channel formation[15] using this device show that the flow is well collimated below the nozzle, with a top hat density distribution of width L ~ 1 mm and a number density (neutral He atoms) N ~ $5x10^{16}$ cm^{-3} corresponding to a pressure of 1.4 Torr. The harmonics produced in the gas jet propagate colinearly with the laser beam, and were observed with a grazing incidence (87^0) flat-field xuv spectrometer[16] incorporating an absolutely calibrated[17] Hitachi 1800 l/mm grating[18]. The collinear propagation of the pump light and harmonics places an upper limit on the total energy we were able to use due to breakdown on the spectrometer entrance slit. The 50 μm (spectral width) 15 mm (spatial width) entrance slit of the spectrometer was 60 cm away from the focus, and under these conditions energies were limited to below 400 mJ to avoid breakdown of the slit.

The spectrometer grating used[18] covers a wavelength range from ~ 350 to 35 Å, and in the focal plane of the instrument the reciprocal dispersion varies from 4.3 to 7.7 Å/mm. A 50 μm entrance slit gives a resolution of 0.5 Å, and this demands a 300 μm spatial resolution from the detector to be able to resolve neighboring harmonics around the order q = 115.

The detection system was a double channel plate device (20 μm channels) with the front surface sensitised for the xuv with a thin coating of CsI and a spatial resolution of ≈ 50 μm. The 40 mm circular aperture of this detector limited the spectral range on any given shot to around 120 Å. The channel plate phosphor was read out with an intensified CCD camera, giving a measured gain for the total system of 10^8. A PC and frame-grab card were used to acquire and store the CCD image for subsequent analysis.

The geometry of this instrument which is shown in figure 3 provides spectral resolution in one dimension and spatial or angular resolution in the other. In the spatial direction, the grating behaves simply as a flat mirror and any angular information contained in light from a point source is projected undistorted onto the detector. This effect is demonstrated in the split filter calibration shots detailed in figures 4 and 5.

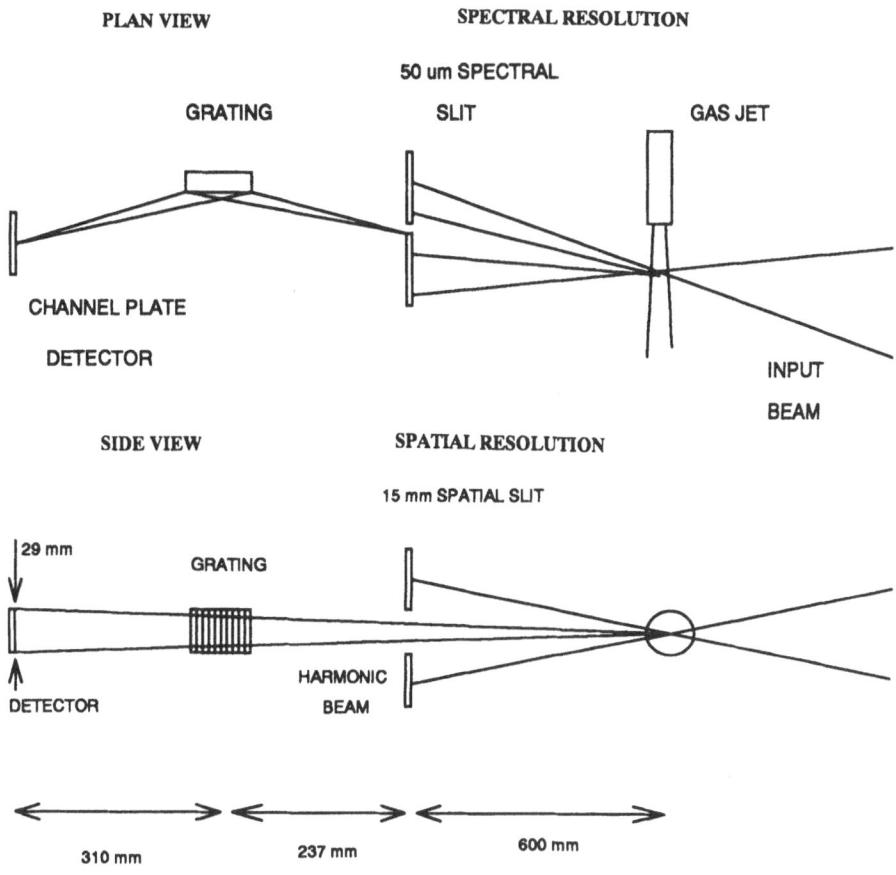

Figure 3. Geometry of the flat field xuv spectrometer.

The limited long wavelength coverage of the spectrometer did not permit us to observe the lowest harmonics, from which a spectral calibration could be derived. A number of techniques were employed to both calibrate absolutely the wavelength scale of our instrument, and to characterise its spatial response in the direction perpendicular to the dispersion of the grating. Moreover, an absolute calibration enabled the wavelengths of the harmonics to be measured directly, so that any possible shifts due to effects such as chirping of the pump by free electrons could be observed.

To obtain absolute wavelength fiducials, we used the L shell vuv emission from a solid copper target irradiated at an intensity of 2×10^{14} Wcm^{-2}. A split filter containing a null section (no filtering), a 1500 Å aluminium filter, and 1000 Å formvar (CH) filter was placed over the spectrometer slit, and the cut-off wavelengths of the aluminium L edge at 173 Å and carbon K edge at 44 Å observed. The geometry of this arrangement is shown in figure 4. Figure 5 shows a typical calibration shot in which the spectral cut off due to the aluminium L edge is clearly visible. It is important to note that the boundaries between the different filters are clear and straight, showing that the instrument does not distort spatial/angular information. The circular outline of the image is due to the cut off at the edge of our (circular) channel plate detector.

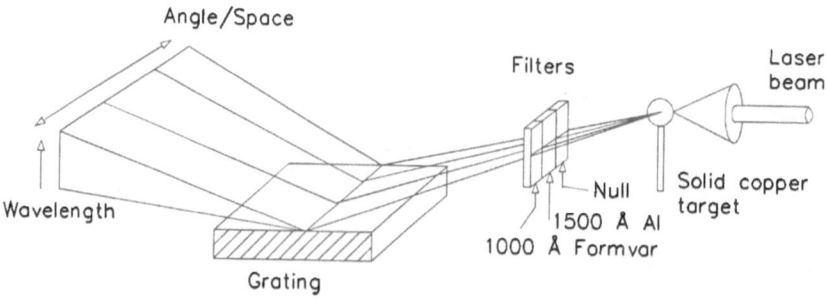

Figure 4. Geometry for multiple filter wavelength calibration shots.

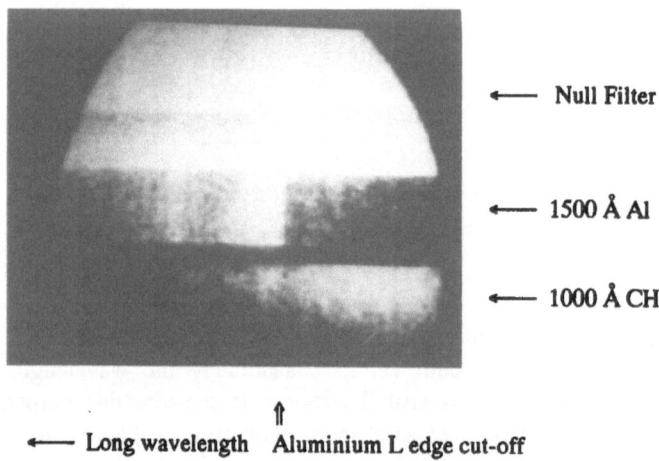

Figure 5. A typical copper continuum / split filter calibration shot showing the aluminium l edge.

Having obtained two absolute wavelength fiducials using this method, we then shot solid carbon, boron and LiF targets, identifying hydrogenic, helium-like and lithium-like lines in multiple order, and using them to obtain an accurate dispersion curve for the instrument. The grating we used was mechanically ruled and blazed, and as such is very efficient in multiple orders[17]. This allows us to use, for example, the very bright carbon Heα line (2-1 transition in helium-like carbon) as a calibration point at 40.27 Å in first order, 80.54 Å in second order and so on. (The dispersion curve measured using this technique is show later in figure 8, along with various harmonic orders).

RESULTS FROM He GAS JET TARGETS

With both helium and neon targets we were able to see clear harmonic spectra on a single shot basis using the conditions and diagnostics previously outlined. The results we now present are all from helium, since in the case of neon we found there to be some contribution from recombination lines on higher intensity shots. It was simple to determine if a given spectral feature was a harmonic, by placing a quarter wave plate in the input beam to change the pump laser polarisation from linear p to circular. No harmonics are seen with circularly polarised light, a result which follows from simple angular momentum conservation. Figure 6 shows a typical single shot harmonic spectrum from helium, with 215 mJ of energy on target, at a focused intensity of 4.4×10^{14} Wcm^{-2}. Due to the limited aperture of our detector only harmonic orders from the 69th up to the 123rd are visible. The "lower harmonics" which are well within the plateau are seen to fully fill the aperture of the detector, while the "higher harmonics" closer to the cut off at 123rd order show a very marked decrease in angular extent. This is a result which we see for practically all shots, although there is some considerable shot to shot variation in the fine structure within the overall angular envelope. This is to be expected for such a non-linear process where the details of the laser / gas jet interaction are likely to differ quite strongly even for very similar laser shots[15].

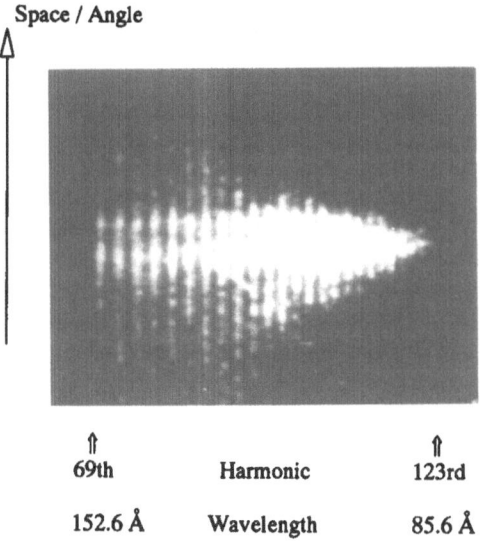

Space / Angle

⇑		⇑
69th	Harmonic	123rd
152.6 Å	Wavelength	85.6 Å

Figure 6. Single shot harmonic spectrum from a helium gas jet, irradiated at an intensity of 4.4×10^{14} Wcm^{-2}.

Figure 7 shows a harmonic spectrum obtained from averaging 18 line outs taken through the centre of spectra recorded at similar laser intensities. The peaks correspond to harmonic intensity on a linear scale, and while no detailed comparisons of relative intensity can be made, we know from solid target data that the response of our detection system is relatively wavelength independent over this spectral range. The fall off which begins at around 117th order is a real effect and not a detection artifact. For our experimental conditions ($I = 3 \times 10^{14}$ Wcm^{-2}, $\lambda = 1.053$ μm) $U_p \approx 31$ eV, and so from Kulander's scaling law[10] we would expect $E_{max} \approx 118$ eV for He ($E_i = 24.6$ eV) and $E_{max} \approx 148$ eV for He$^+$ ($E_i = 54.4$ eV) corresponding to q = 113 and q = 139 respectively. The former is closer to our observed cut off, which suggests that He$^+$ plays a minor role, if any, in harmonic production under our experimental conditions. However care must be taken when making such interpretations due to the uncertainty in the intensity at the laser focus, which can be strongly modified by propagation effects, and when free electrons are present[15,19]. We find that the experimental wavelengths of the harmonics fit to their expected positions to within the spectral resolution of our instrument (see figure 8 below).

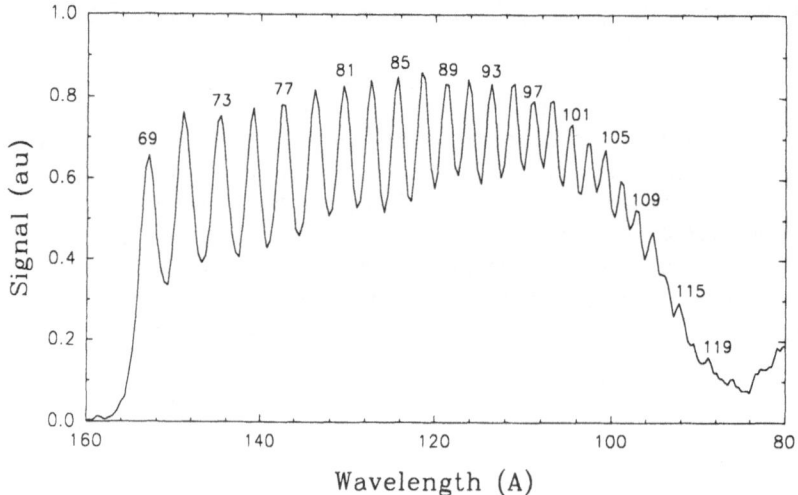

Figure 7. 18 shot average harmonic spectrum , with the long wavelength (lower harmonics) range limited by the detector aperture.

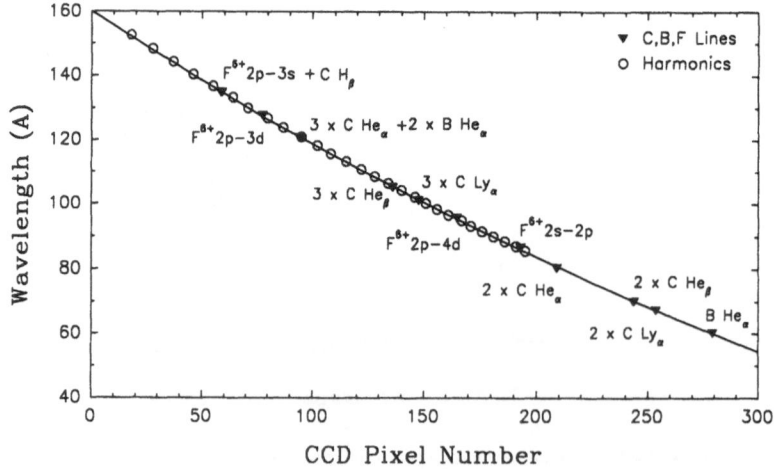

Figure 8. Comparison of experimental harmonic wavelengths with an absolute dispersion curve.

Figure 9 compares the measured single shot angular distributions of four harmonic orders (q = 71, 91, 101 and 111) from the spectrum shown in figure 6. The peak heights have been normalised to make comparison easier. The solid curve is a gaussian best fit to the data points, while the dashed curve is the distribution expected from lowest order perturbation theory (LOPT). The 71st harmonic shows considerable fine structure, particularly in the wings, and the gaussian is a very poor fit to the experimental profile. For the higher orders the angular distribution becomes progressively smoother, and for the harmonics just at the cut off, is a very good fit to a gaussian. Interestingly, the LOPT curve also seems to be in good agreement with the experimental data near the cut off. In each case however, the LOPT peak fits well to the central peak in the measured distributions. This is a somewhat surprising result, since our experimental conditions are clearly far outside the regime where perturbation theory is valid.

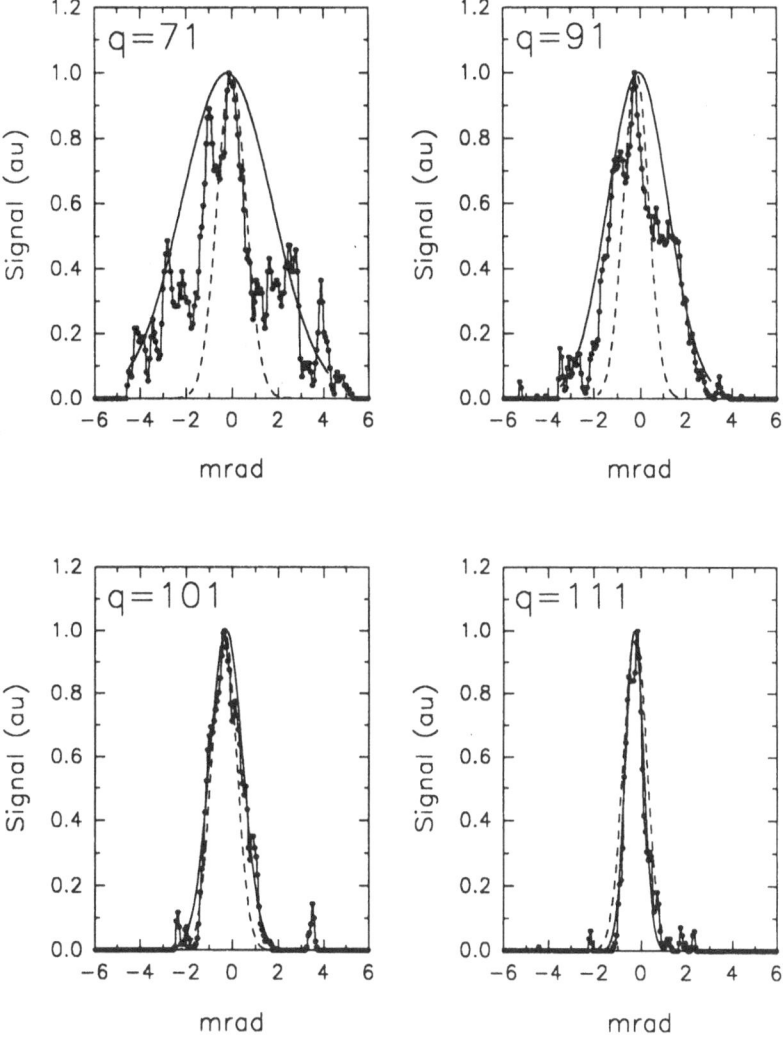

Figure 9. Experimental angular distributions for the 71st, 91st, 101st and 111th harmonics compared to an ideal gaussian (solid line) and LOPT (dashed line)

Figure 10 shows the trends described above for the 79th to 115th order harmonics on the same shot. We plot the harmonic order q against the measured harmonic half-widths W(q) normalised to the pump laser half-width W(1) at the detector (solid circles) taken from gaussian fits to the data where such a fit was appropriate. The lower curve shows the theoretical $W(1)/\sqrt{q}$ half angles given by LOPT. The data points clearly lie on a smooth curve which decreases much more rapidly than $1/\sqrt{q}$, and the two curves intersect close to q = 115, which was the end of the plateau for this shot.

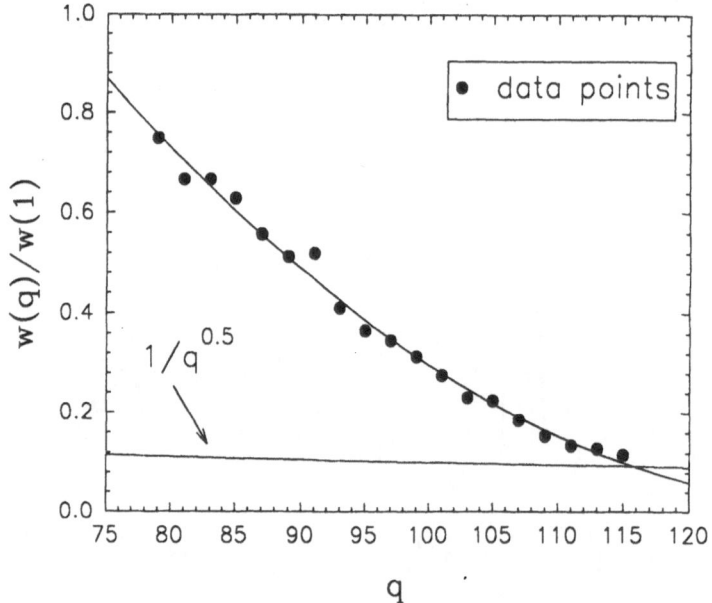

Figure 10. Harmonic order q versus the harmonic width W(q) normalised to the width W(1) of the pump laser at the detector

In conclusion, we have measured the angular distribution of very high harmonics for the first time. We find that harmonics in the plateau region have an angular distribution much broader than that expected from LOPT, and show a rich fine-scale structure. This is a not unexpected result; clearly this regime is highly non-perturbative and ionisation and macroscopic effect such as the influence of free electrons on phase matching will strongly modify the perturbative profiles. In addition we might expect there to be some saturation effects in the atomic response. However, we find a surprising result towards the plateau cut off. Here the angular distribution of the high harmonics loses its fine structure, becomes narrower, and finally becomes very similar to that expected from simple LOPT at the cut off. Clearly this is a highly non-perturbative regime and we currently do not have a good physical picture for this effect. We are in the process of carrying out detailed computer simulations of this experiment in an effort to reproduce these results, and understand the principles behind them.

ACKNOWLEDGEMENTS

We would like to thank Dr. O. Willi for the use of the gas jet, and for his advice on identification of xuv spectra. We would also like to thank the Rutherford Appleton Laboratory for the use of the flat field spectrometer employed on this experiment. This work was supported by the UK Science and Engineering Research Council.

REFERENCES

1. A. L'Huillier and Ph Balcou. To be published in Phys. Rev. Lett. Feb. (1993)
2. J. J. Macklin, J. D. Kmetec, C. L. Gordon III and S. E. Harris. To be published in Phys. Rev. Lett. Feb. (1993)
3. Ph. Balcou and A. L'Huillier. This issue SILAP proceedings. (1993)
4. J. L. Krause, K. J. Schafer and K. C. Kulander. Phys. Rev. A **43**, 4998 (1992)
5. V. C. Reed and K. Burnett. Phys. Rev. A. 46, 424 (1992)
6. A. L'Huillier, X. F. Li and L. A. Lompré. J. Opt. Soc. Am. B Vol. 7, No. 4 (1990)
7. S. C. Rae and K. Burnett. Submitted to J. Phys. B.
8. X. F. Li, A. L'Huillier, M. Ferray, L. A. Lompré and G. Mainfray. Phys. Rev. A Vol. 39, No. 11 (1989)
9. M. E. Falden, M. H. R. Hutchinson, J. P. Marengos, J. E. Muffett, R. A. Smith, J. W. Tisch and C. G. Wahlstrom. J. O. S. A. B Vol. 9 No. 9 (1992)
10. J.L. Krause, K. J. Schafer and K. C. Kulander. Phys. Rev. Lett. Vol **24**, No. 24 3535 (1992)
11. K. C. Kulander et-al. This issue SILAP proceedings. (1993)
12. D. Strickland, and G. Mourou. Opt. Comm. **56**, 219 (1985); P. Maine, D. Strickland, P. Bado, M. Pessot and G. Mourou. IEEE J. Quantum Electron. **24**, 398 (1988)
13. F. G. Patterson and M. D. Perry. J. Opt. Soc. Am. B/Vol. 8 No. 11, (1991)
14. L. A. Lompré, A. L'Huillier, M. Ferray, P. Monot, G. Mainfray and C. Manus. J. Opt. Soc. Am. B Vol. 7, No. 5 (1990)
15. M. Dunne, T. Afshar-Rad, J. Edwards, A. J. MacKinnon, S.M. Viana and O. Willi. "Experimental observation of the creation and expansion of an optical-field-induced ionisation channel in a gas jet target" Sumitted to Phys. Rev. Lett. January 1993.
16. G. Kiehn, O. Willi, A. Damerell and M. Key. Appl. Opt. Vol. **26**, No. 8 pp 425-426 (1981)
17. G. Kiehn, t. Garvey, R. A. Smith, O. Willi, A. Damerell and J. West. Invited Paper, SPIE Vol 831 "X-rays from laser produced plasmas" San Diego (1987)
18. T. Harada, T. Kita. Appl. Opt. Vol. **19**, No.23 pp 3987-3993 (1980)
19. S. C. Rae, Accepted for publication, Opt. Comm.

HIGH-ORDER SUM AND DIFFERENCE-FREQUENCY GENERATION IN HELIUM

John K. Crane and Michael D. Perry

Laser Program, P. O. Box 808, L-493
Lawrence Livermore National Laboratory
Livermore, California 94550

INTRODUCTION

High-order harmonic generation provides a new method for generating coherent, XUV radiation.[1,2] These harmonics are characterized by a rapid, perturbative drop at low orders, followed by a broad plateau extending to photon energies of 150 eV in the lighter, rare gas atoms.[3,4] An experimentally observed limit coincides with the theoretical limit for harmonic generation in neutral atoms given by the expression $E_c(eV)=IP(0)+3U_p(I)$ [5], where E_c is the energy cutoff of the harmonic plateau, $IP(0)$ is the field-free ionization potential and U_p is the electron quiver energy at the maximum intensity, I, seen by the atom. As part of a broad effort to develop this technique into a general purpose, XUV source, extensive work to understand the phase-matching between the harmonic and driving fields, and the resulting effect on the conversion efficiency, angular distribution and spectral brightness has been undertaken at several laboratories.[6-8] Nevertheless, certain aspects of the harmonically generated radiation such as the polarization, relative strength of a given harmonic, and the plateau extent, are defined by the single atom-field interaction. Specifically, the single-atom harmonic spectrum is determined primarily by the interaction of a driven, quasi-free electron with the atomic potential. Using two, independent fields one can affect the electron motion by controlling the relative strength, polarization, and phase of the fields and alter the harmonic spectrum.

In this paper we discuss initial, two-color experiments where we drive the atom with two fields of different, but related frequencies: 1053 nm (1ω) and 526 nm (2ω). In addition to the high-order, odd harmonics, we observe sets of three additional peaks that we attribute to sum and difference-frequency generation between the two fields. By controlling the relative polarization between the two fields we can control the relative strength of the harmonic and mixing components, as well as the polarization of the output XUV photon.

Previous work on the interaction between intense, two-color radiation and atoms has concentrated on the influence of the relative phase of the fields on the ionization rate and electron spectrum. Muller et al.[9] investigated above-threshold ionization (ATI) in krypton produced from two-color illumination by intense (10^{13} W/cm^2) 1064 and 532 nm laser pulses. They observed some additional structure in the ATI spectrum when the two fields were present with their relative polarization parallel. Several groups have examined[10,11] multiphoton ionization using two-color excitation and have shown that the relative phase between the two fields affects the ionization rate. Schafer and Kulander[10] also described calculated harmonic spectra from two-color illumination that contain both odd and even orders. In addition, much attention has been focused recently on manipulating chemical reactions and molecular dynamics by controlling the phase of the coherent field driving the interaction[12].

DESCRIPTION OF THE EXPERIMENT

Our laser is a neodymium:glass system[13] that can deliver up to 8 Joules of 1053 nm radiation in a 800 femtosecond pulse. The laser is sent into a vacuum chamber, as shown in figure 1, and focused immediately below the output of a high-pressure (up to 1500 psi) pulsed valve equipped with a supersonic (Mach 8) nozzle that can produce densities in excess of 10^{19} atoms/cm^3.[14] Light created in the interaction region is collected by a grazing incidence, x-ray spectrometer and imaged onto the input photocathode of an x-ray streak camera. The spectrometer-streak camera combination produces time-resolved, single-shot spectra covering a 20 nm range, allowing us to separate the high-order harmonics from the spontaneous emission produced in the plasma[15]

We frequency double the 1053 nm laser light (1ω) with a thin (4 mm) KD*P crystal that is cut for type I phase matching. Although the laser can produce up to 4 Joules of frequency-doubled light, we operated with 1.5 Joules of 1053 nm light input to produce 300 mJ output at 526 nm (2ω). The laser light is focused into the vacuum chamber with a single element, fused silica, aspheric lens that has a 200 cm focal length at 1053 nm. Chromatic dispersion in the lens causes the 526 nm light to focus 5 cm before the 1053 nm light. In these experiments we placed the gas jet at the focus of the 2ω beam. We operated with 200-300 mJ of 2ω light focused to a 50 μm waist (diameter at $1/e^2$ value in intensity) and a 600 femtosecond pulse width (FWHM), measured with a single-shot autocorrelator. The peak intensity based on these parameters is 5×10^{16} W/cm^2. This value is confirmed by observing fluorescence from Ar VIII transitions which occur only above 3×10^{16} W/cm^2.[*] The beam

---

[*]We calculated the appearance irradiance for Ar VIII lines using the tunneling ionization formalism described by M. V. Ammosov, N. B. Delone, and V. P. Krainov, Sov. Phys. *JETP* . 64:1191 (1986).

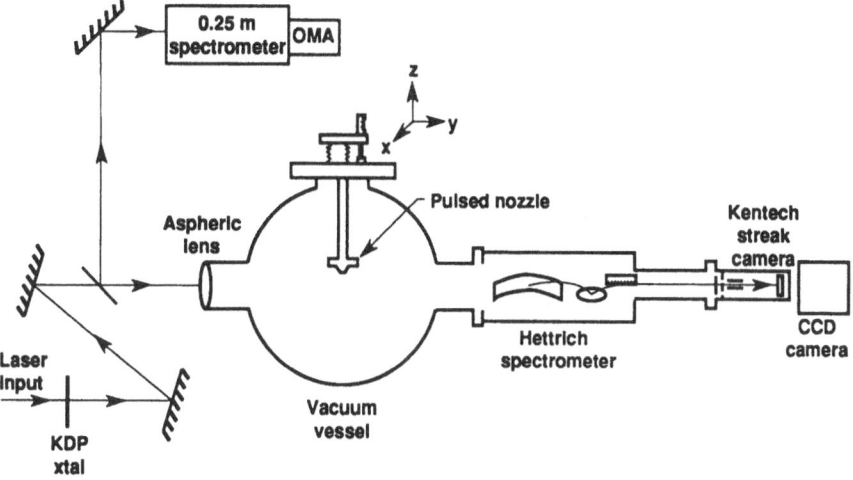

Figure 1. Experimental apparatus for producing and detecting short-pulse XUV radiation.

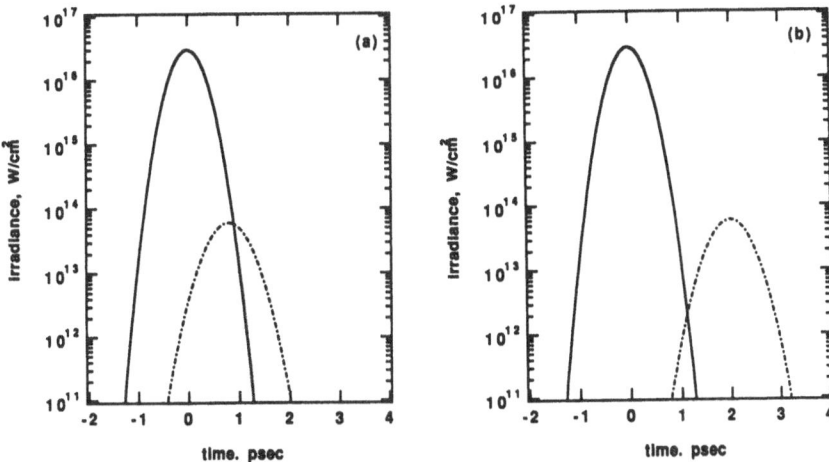

Figure 2. a) Temporal relationship between 1ω and 2ω pulses where the group delay is produced by the lens and KD*P crystal. b) Additional delay produced by 15 mm thick fused silica optical flat.

diameter of the 1ω light at the gas nozzle is 1.0 mm with a corresponding peak intensity at the jet of 6×10^{13} W/cm^2 in an 800 femtosecond pulse.

A temporal walkoff between the 1ω and 2ω pulses occurs because of the difference in group velocity between the two pulses propagating in the fused silica lens and the KD*P doubling crystal. We use this temporal walkoff as a means of varying the overlap between the two colors. We calculate a walkoff parameter, $d_{12}=(v_g(1\omega))^{-1}-(v_g(2\omega))^{-1}=0.79$ psec/cm, in fused silica. The combination of the focusing lens and doubling crystal separate the pulses by approximately 0.8 psec. To further separate the pulses we add a 15 mm optical flat, so that the combined effect of the two pieces of fused silica plus KD*P crystal produces a 2.0 psec time separation between the peak of the two pulses. Figures 2a and 2b shows the

45

temporal relationship between the two pulses for the two different cases. Since the intensity of the 2ω pulse is many times our measured saturation intensity for ionization of neutral helium (6.5×10^{14} W/cm^2)[16], the harmonics are produced during the rising edge of the 2ω pulse. Although the peak intensity of the 2ω field is 3×10^{16} W/cm^2, essentially all harmonic production from neutral helium will occur below 1×10^{15} W/cm^2. This intensity occurs approximately 0.8 psec before the peak of the 2ω pulse, nearly coincident with the peak of the 1ω pulse with only the group delay of the lens and the KD*P present. When the extra 15 mm fused silica optic is added the total delay is 2.0 psec and the 1ω peak coincides with a 2ω intensity less than 1×10^{12} W/cm^2, well below the intensity for harmonic production in helium.

We use a type I doubling crystal to convert the 1053 nm light to 526 nm. The frequency-doubled field is orthogonally polarized with respect to the fundamental (1ω), however we can rotate the polarization of the 1ω field after doubling using a half-wave plate with little effect on the 2ω polarization. We use the half-wave plate to continuously vary the angle between the polarization of the two colors in our experiments.

RESULTS

Sum and Difference Frequency Generation

We measured the harmonic spectrum at various laser irradiances for the following four cases: (a) 2ω only; (b) 1ω only; (c) $1\omega+2\omega$ without the 15 mm optical flat; and (d) 1ω and 2ω with the optical flat. By measuring the spectrum with and without the optical flat, we could insure that everything was held constant, except the temporal overlap between the pulses. In figure 3 we display the spectra for three of the four cases. Figure 3a shows the harmonic spectrum produced when only 2ω is focused into the chamber (1ω is blocked by a thin absorbing filter placed after the doubling crystal). In this figure, we show the region covering 34 to 69 eV. The 15-27 odd harmonics are observed in first order along with weaker second order peaks at the 29th and 31st harmonics at this spectrometer setting (the spectrum extends up to the 35th harmonic of the 526 nm light).This result is identical to our previously reported results.[16] In the second case, we send only 1ω light into the chamber (detune or remove the doubling crystal) and observe no harmonics. The absence of odd harmonics of 1ω light is not surprising at this intensity (6×10^{13} W/cm^2), which is an order of magnitude below the intensity where we normally observe harmonics in this spectral range (29th-55th of 1ω)[16]. We do not show this null result. Figure 3b shows the third case where both colors are sent into the chamber, separated by 800 femtoseconds. Again we observe the odd harmonics of 2ω, however, between pairs of 2ω harmonics we observe three additional peaks. This pattern repeats itself throughout the range of observed 2ω odd harmonics, extending to the plateau cutoff. In the final case, shown in figure 3c, we place the additional 15 mm optical flat in the beam, separating the two colors in time by 2.0 psec. The result is identical to figure 3a- odd harmonics of 2ω only. We repeated this sequence of four experiments throughout the

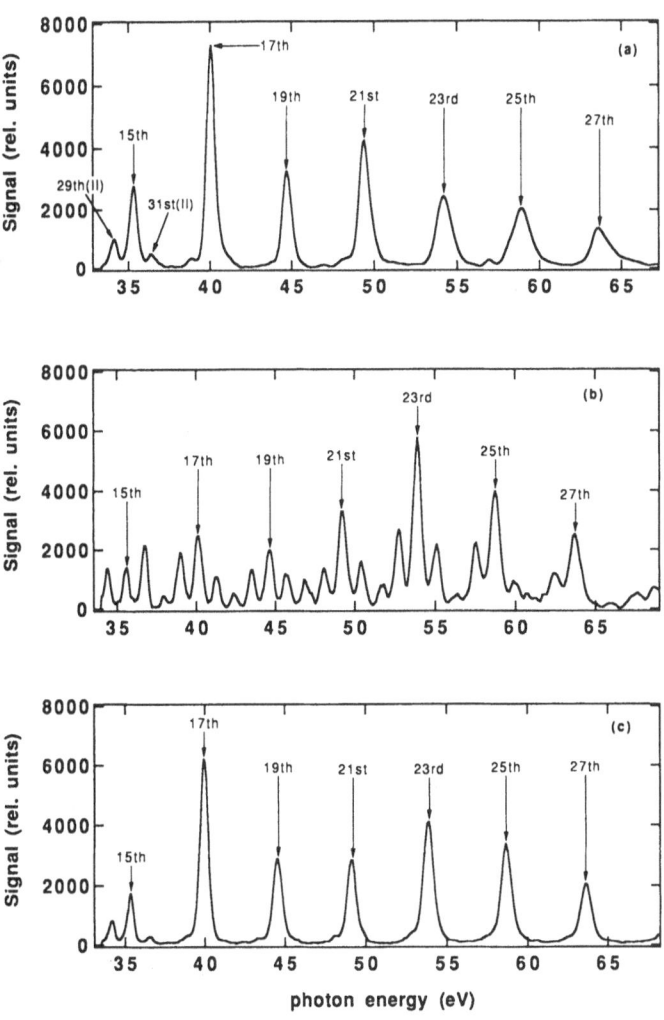

Figure 3. a) Harmonic spectrum with 2ω field only shows the odd harmonics 15-27. b) Spectrum from two-color excitation, where the 1ω and 2ω fields are orthogonally polarized and separated in time by 800 fs as shown in figure 2a. c) Spectrum with two-colors separated in time by 2.0 ps as shown in figure 2b.

spectral range of our spectrometer, spanning the region from the 11th harmonic of 2ω (26 eV) to beyond the cutoff for 2ω harmonics (35th order at 83 eV), observing the same result as that shown in figure 3.

Of the three secondary peaks in between the usual odd harmonics of the 526 nm field, we concentrate first on those centered between the odd harmonics. These peaks appear at even multiples of the 526 nm field. Upon initial inspection, we would label these as the 18th, 20th, 22nd, etc. harmonics of the 2ω field. These *even* harmonics can only be produced by breaking the symmetry of the light-atom Hamiltonian, e.g. with a second field. Alternatively, a sum-frequency mechanism can be used to explain the peaks that appear at even multiples of the 526 nm field. In this case, an odd number of 2ω photons combine with two 1ω photons to produce a single photon with an energy equal to an even multiple of a 2ω photon. This mechanism is fully dipole allowed and the overall harmonic order is still odd. *It is only the energy of the photon which makes it appear as an even harmonic.* Clearly, harmonics produced by this sum frequency mechanism are only produced when both the 1ω and 2ω fields are present.

The secondary peaks on either side of the principal odd harmonics ($q\omega$) of the 2ω field have an energy equal to odd multiples of the 1ω field. However, these are not 1ω harmonics in the usual sense as they only appear when both fields are present. In this case, the emission results from either sum or difference-frequency generation with an *even* number ($q+1$) of 2ω photons. The peak just below the qth odd harmonic is formed by the absorption of $q-1$ photons of the strong 2ω field plus one additional 1ω photon. The peak just above the qth odd harmonic is formed by the absorption of $q+1$ photons minus a single 1ω photon.

Figure 4 is an expansion of the portion of figure 3b that spans the 17th through 21st harmonics of 526 nm with the various peaks labeled as described in the previous text. Higher order processes that are degenerate with the three lowest order schemes are also possible and likely as the intensity of the 1ω field is increased. For example we indicate that the ladder $[q+1](2\omega)-(1\omega)$, which contains 21 photons (20-2ω photons minus one 1ω photon), is responsible for the peak at 45.9 eV. This scheme is degenerate with $[q-1](2\omega)+3(1\omega)$; both involve 21 photons for the excitation and are dipole allowed. We do not believe these higher order diagrams in the 1ω field are contributing significantly at the 1ω irradiance used here. This is based on the fact that the peak which definitely contains two 1ω photons is significantly lower than the peaks which contain only a single 1ω photon.

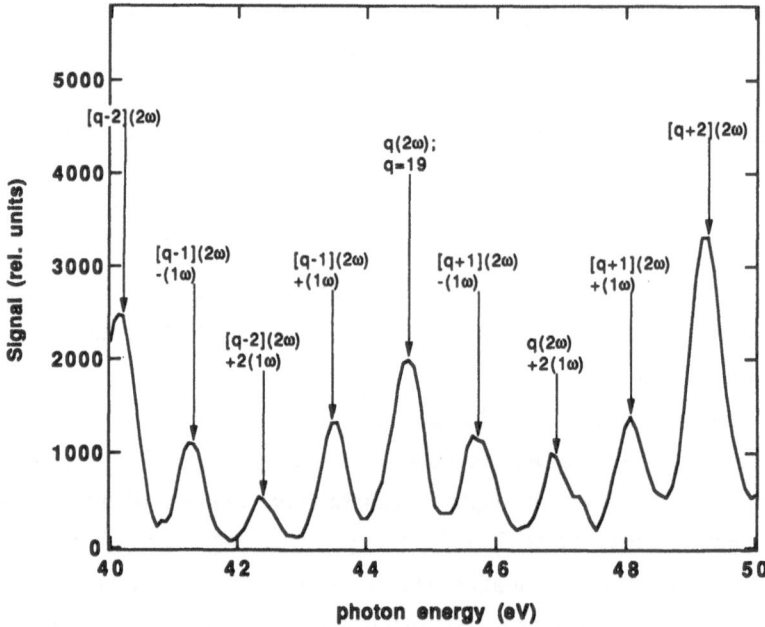

Figure 4. Enlargement of a portion of the spectrum from figure 3b showing the sum and difference-frequency peaks labelled according to the description given in the text.

Polarizations effects

Figure 5a shows a harmonic spectra for the case where the two driving fields are orthogonal to each other. There appears to be a consistent relationship among the three sets of peaks. The odd harmonics of 2ω are the strongest -an order of magnitude larger than the even harmonic peaks. The two sum and difference-frequency peaks that lie to either side of the even harmonic peaks are 2-5 times weaker than the odd harmonic peaks, but consistently stronger than the even harmonic peaks. In figure 5b we show a similar spectrum for the case that the 1ω polarization was rotated after doubling to align it with the 2ω polarization, i.e. the 1ω and 2ω polarizations are parallel. Now all three sets of peaks are comparable in relative strength. We have also rotated the polarization of the 1ω field to intermediate angles between these two extremes and observe that the relative intensities of the sets of peaks can be controlled by this method. Becker et. al.[17] recently showed that by using circularly polarized light for the two-colors certain sets of peaks can be suppressed completely.

Although it may be possible to determine the polarization of the output XUV photons, it is not a trivial measurement to make using standard optical techniques. However by using

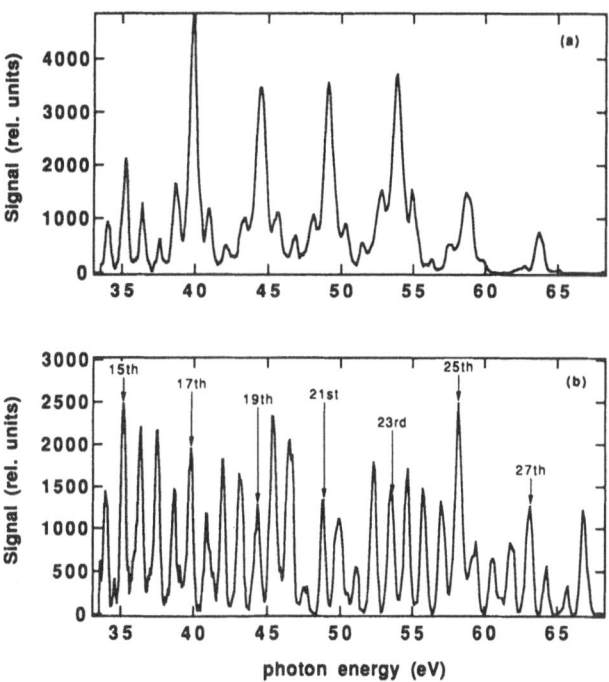

Figure 5. a) Spectrum of harmonic and sum and difference-frequency peaks where the polarizations of the 1ω and 2ω driving fields are perpendicular. b) Spectrum generated by the two-colors polarized parallel with respect to each other.

some simple diagrams we can show the relationship between the polarization of the output XUV photon and the relative polarization of the 1ω and 2ω fields. For the case where both input fields are linearly polarized along the same axis, there will be no change in the orientation between the ground and final states ($\Delta m=0$, where m are the magnetic sublevels), and the polarization of the output, XUV photon will be linear and parallel to the input polarizations. We can explain the sum and difference-frequency peaks for orthogonally polarized fields by a series of dipole allowed transitions. For the even harmonics we add two 1ω photons to an odd number of 2ω photons as described in the previous section. The orthogonal field (1ω) can be expressed as a linear sum of right and left-hand circularly polarized light using the helicity basis set. In this case a two-photon transition can produce a net change in angular momentum, $\Delta m=0$, where the first photon causes a $\Delta m=\pm1$ transition and the second photon, $\Delta m=+1$, as shown in figure 6a. This sum-frequency mechanism is fully dipole allowed with two fields of orthogonal linear polarization. The two peaks on either side of the harmonic frequencies (either even or odd) are produced by adding (sum-frequency generation) or subtracting (difference-frequency generation) a single, 1ω photon to an even number of 2ω photons. Since the 1ω field is polarized orthogonally to the 2ω field, the upper state will be a coherent superposition of $\Delta m=\pm1$ states as shown in figure 6b. This last example suggests that by adding (or subtracting) a circularly polarized, 1ω photon, the final state will be either $\Delta m =+1$ or -1 with respect to the ground state, and the output XUV photon will be circularly polarized. Therefore, by controlling the polarization of the two input fields it may be possible to produce XUV photons of any desired polarization, as well as control the relative strength of the sum and difference-frequency components.

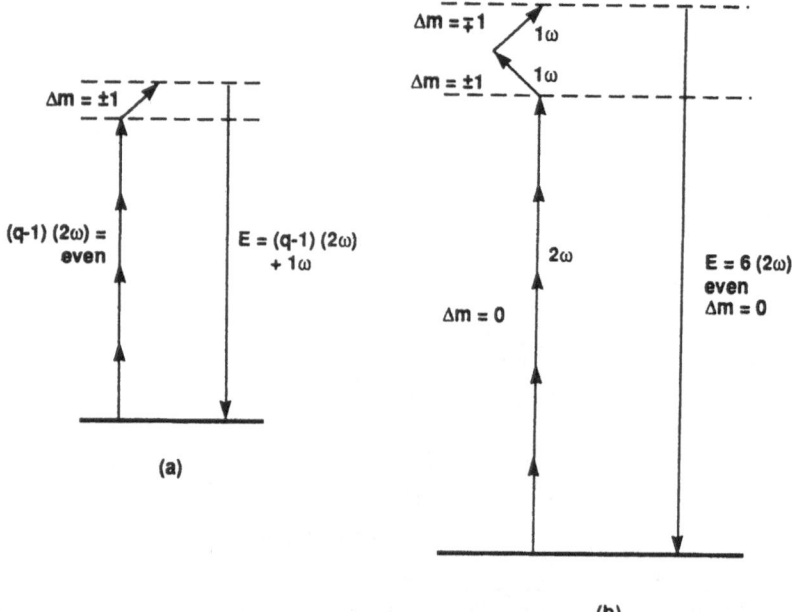

Figure 6. a) Excitation ladder for two, orthogonally polarized fields: odd number of 2ω photons plus 2- 1ω photons. b) Excitation ladder for orthogonally polarized fields: even number of 2ω photons minus a single, 1ω photon.

Role of relative intensity, phase, and polarization on plateau extension

In the tunnelling limit where $\gamma^2 = IP/2U_p \ll 1$, the harmonic spectra is primarily the result of the field-driven, quasi-free electron interacting with the ion core potential;[18] consequently, high-order harmonic generation, as well as ATI, can be described classically.[18,19] In this approach, an ensemble of electrons is launched at phases of the laser field determined by the probablilty for ionization. The trajectory of the electron then determines the probability for harmonic emission by scattering from the core potential. These models are successful at predicting the energy spectrum and cutoff of high-order harmonic generation and ATI when a single field drives the electron. To see how the electron dynamics is affected by two independent fields we derived expressions for the velocity, displacement, and instantaneous kinetic energy starting from the Lorentz force equation. We consider the two simplest cases of parallel and perpendicular polarization for the two fields and write the equations for the instantaneous kinetic energy, T(t). For 1ω perpendicular to 2ω,

$$T(t) = 2U_p(\omega)\{\sin\omega t - \sin\omega t_0\}^2 + 2U_p(2\omega)\{\sin(2\omega t + \phi) - \sin(2\omega t_0 + \phi)\}^2, \tag{1}$$

and for 1ω parallel to 2ω,

$$T(t) = 2U_p(\omega)\{\sin\omega t - \sin\omega t_0\}^2 + 2U_p(2\omega)\{\sin(2\omega t + \phi) - \sin(2\omega t_0 + \phi)\}^2 +$$
$$2U_p(\omega)[I(2\omega)/I(1\omega)]^{1/2}\{\sin\omega t - \sin\omega t_0\}\{\sin(2\omega t + \phi) - \sin(2\omega t_0 + \phi)\}. \tag{2}$$

In these expressions U_p is the time-averaged quiver energy evaluated at the laser intensity, I, ϕ is the relative phase between the two fields, and t_0 is initial time at which the electron is born. Note the additional interference term when the fields are parallel polarized. Consider an example where the 1ω and 2ω fields are of equal intensity, corresponding to the saturation field strength in helium for a 1 ps pulse ($I(1\omega) = I(2\omega) = 1.6 \times 10^{14}$ W/cm^2 for $1\omega \parallel 2\omega$) and ($I(1\omega) = I(2\omega) = 3.2 \times 10^{14}$ W/cm^2 for $1\omega \perp 2\omega$). For an electron born at the peak intensity, i.e. $t_0 = 0$, the quiver energy ranges from 66 to 196 eV for orthogonal polarization, depending on the phase angle. For parallel polarization the maximum quiver energy range is 48 to 58 eV. In both cases, the harmonic cutoff is pushed beyond that which would be obtained from either field independently at these intensities. However, when compared to the cutoff of the single, long wavelength (1ω) field at the saturation strength (6.4×10^{14} W/cm^2), the extension of the cutoff associated with mixed fields is not significant.

CONCLUSIONS

We report the observation of high-order, sum and difference-frequency generation in helium. We produce these high-order frequencies as well as the usual odd harmonics by mixing two-colors, 526 nm and 1053 nm, in the high density output of a pulsed supersonic nozzle. What appear as even harmonics can be explained by sum-frequency generation between the two fields. The relative strength and polarization of these different, XUV mixing products is dependent on the relative polarization between the two fields yielding a form of coherent control. As shown in the two-color work on multiphoton ionization[9-11], further control may be possible via the relative phase and strength of the two fields.

We gratefully acknowledge useful discussions with H. T. Powell, J. A. Paisner, S. Dixit, K. C. Kulander, K. Schafer, and T. Ditmire. We would have been unable to perform these experiments without the technical assistance of S. Herman, H. Nguyen, and B. Adams.

REFERENCES

1. A. McPherson, G. Gibson, H. Jara, U. Johann, T. S. Luk, I. McIntyre, K. Boyer and C. K. Rhodes, Studies of multiphoton production of vacuum-ultraviolet radiation in the rare gases, *J. Opt. Soc. Am. B* . 4:595 (1987).
2. M. Ferray, A. L'Huillier, X. F. Li, L. A. Lompre, G. Mainfray, and C. Manus, Multiple-harmonic conversion of 1064 nm radiation in rare gases, *J. Phys. B*. 21:L31 (1988).
3. J. J. Macklin, J. D. Kmetec, C. L. Gordon, High-order harmonic generation in neon, "Proceedings from the Conference on Quantum Electronics and Laser Science", Anaheim, CA (1992).
4. A. L'Huillier and P. Balcou, High-order harmonic generation in rare gases with a 1 ps 1053 nm laser, *Phys. Rev. Lett.* (to be published).
5. J. L. Krause, K. J. Schafer, and K. C. Kulander, High-order harmonic generation from atoms and ions in the high intensity regime, *Phys. Rev. Lett.* 68:3535 (1992).
6. A. L'Huillier, L. A. Lompre, G. Mainfray, and C. Manus, High-order harmonic generation in rare gases, *in:* "Atoms in Intense Fields," M. Gavrila, ed. Academic, Orlando (1992).
7. D. D. Meyerhofer and J. Peatross, Angular distributions of high order harmonics, *in:* "Proceedings on SILAP III," B. Piraux, ed. Plenum, New York (1993).
8. R. A. Smith, J. W. G. Tisch, M. S. N. Ciarrocca, S. Augst, and M. H. R. Hutchinson, Space-resolved ultra-high harmonic generation with picosecond pulses, *in:* "Proceedings on SILAP III," B. Piraux, ed. Plenum, New York (1993).
9. H. G. Muller, P. H. Bucksbaum, D. W. Schumacher and A. Zavriyev, Above-threshold ionisation iwth a two-colour laser field, *J. Phys. B* . 23:2761 (1990).
10. K. J. Schafer and K. C. Kulander, Phase-dependent effects in multiphoton ionization induced by a laser field and its second harmonic, *Phys. Rev. A* . 45:8026 (1992).
11. R. M. Potvliege and P. H. G. Smith, Two-colour multiphoton ionization of hydrogen by an intense laser field and one of its harmonics, *J. Phys. B*. 25:2501 (1992).
12.. M. Shapiro, J. W. Hepburn, and P. Brumer, Simplified laser control of unimolecular reactions: simultaneous $(\omega 1, \omega 3)$ excitation, *Chem. Phys. Lett.* 149:51 (1988).
13. F. G. Patterson, M. D. Perry, and J. T. Hunt, Design and performance of a multiterawatt, subpicosecond neodymium:glass laser, *J. Opt. Soc. Am. B* . 8:2384 (1991).
14. M. D. Perry, C. Darrow, C. Coverdale, and J. K. Crane, Measurement of the local electron density by means of stimulated Raman scattering in a laser-produced gas jet plasma, *Opt. Lett.* 17:523 (1992).
15. J. K. Crane, M. D. Perry, S. Herman, and R. W. Falcone, High-field harmonic generation in helium, *Opt. Lett.* 17:1256 (1992).
16. J. K. Crane, M. D. Perry, D. Strickland, S. M. Herman, and R. W. Falcone, Coherent and incoherent XUV emission in helium and neon, laser-driven plasmas, *IEEE Trans. Plasma Science* (to be published Feb. 1993).
17. W. Becker, S. Long, and McIver, Two-color higher harmonic production in a zero-range potential, *in:* "Proceedings on SILAP III," B. Piraux, ed. Plenum, New York (1993).
18. P. B. Corkum, et. al., "Short Wavelength V: Physics with Intense Laser Pulses", OSA, Washington D.C. (1993).
19. K. C. Kulander, Dynamics of short-pulse excitation, ionization and harmonic generation, *in:* "Proceedings on SILAP III," B. Piraux, ed. Plenum, New York (1993).

CONTRIBUTION OF BOUND-BOUND TRANSITIONS TO HIGH-ORDER HARMONIC GENERATION

Luis Plaja,[1] Luis Roso[2]

[1]Departament de Matemàtica Aplicada, Escola d'Estudis Empresarials,
Universitat Pompeu Fabra, 08001 Barcelona, Spain.
[2]Departamento de Física Aplicada,
Universidad de Salamanca, 37008 Salamanca, Spain.

INTRODUCTION

We study three different numerical models describing high-order harmonic generation in atomic hydrogen induced by a linearly polarized light of intensity below the atomic unit. These three models are progressively more simple. The comparison between the computed results —that essentially coincide— shows that bound-bound transitions account for most of the important features in harmonic generation.

High-order harmonic generation has been observed accompanying multiphoton ionization by intense laser fields. A lot of work has been done to relate harmonic spectra and free electron energy spectra, showing in certain cases some relationships between them. It is not clear, however, which are the fundamental mechanisms leading to high-order harmonic generation in situations where fast ionization occurs. In this contribution we try to propose that—for hydrogen or other atoms with Rydberg ladders— the essential states to generate harmonics are the bound-states. Since the most intense harmonics are generated by bound-bound transitions, at high laser intensities ionization is an inseparable, but in some sense competing, process. Our approach to show the role of the different states is to compute harmonic generation in three numerical models that are progressively more simple. Comparison between these three results will allow us to show that harmonic generation is mostly due to bound-bound couplings.

THEORY

We numerically study the photoionization of atomic hydrogen by a linearly polarized laser field, in the electric dipole approximation. To identify the roles of the different states, we present three theoretical models that are progressively more simple. These models are schematically presented in Fig. 1. We start with a realistic model, expanding the atomic wavefunction in its spherical harmonics components, and taking advantage of the cylindrical symmetry in the linearly polarized case. We also study a simplified

Super-Intense Laser-Atom Physics, Edited by
B. Piraux *et al.*, Plenum Press, New York, 1993

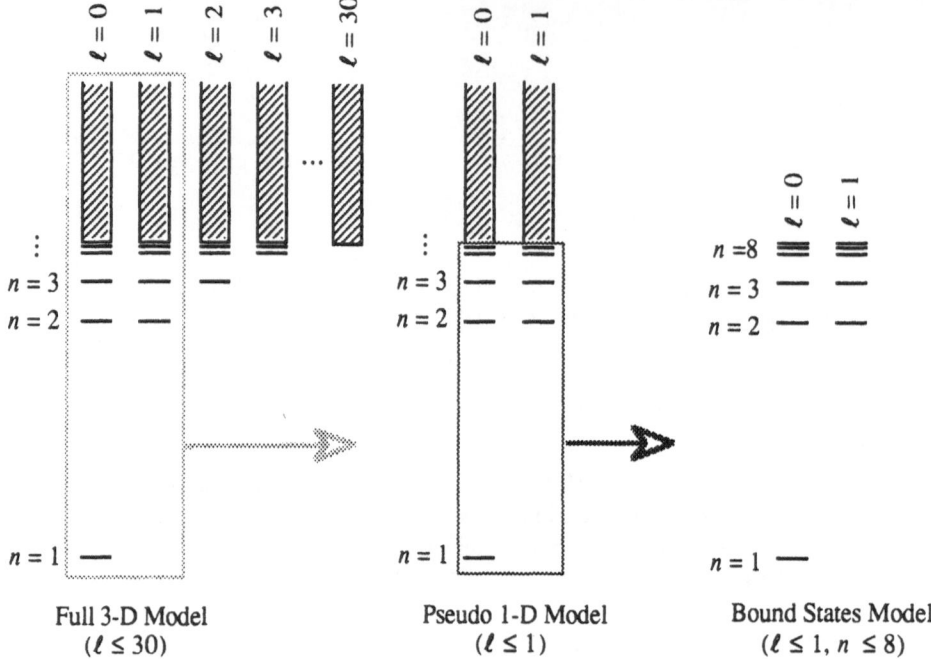

Figure 1. *Scheme of the models of hydrogen used throughout this paper.*

model with only two angular momentum components. And, finally, we present an extremely simple model with only a few bound states.

Three-Dimensional Model

As our first model, we consider a quite realistic description of atomic hydrogen. We describe the laser field as a monochromatic plane-wave, and the laser-atom interaction in the electric-dipole approximation. For simplicity we assume that the laser light is linearly polarized (calling z the polarization axis), and therefore there is a cylindrical symmetry that reduces the dynamics to only two effective dimensions[1].

The time-dependent Schrödinger equation describing the interaction of a hydrogen atom with a linearly polarized laser field, in the electric dipole approximation is,

$$i\frac{d}{dt}\psi(\vec{r},t) = \left[-\frac{1}{2}\nabla^2 - \frac{1}{r} + \vec{r}\vec{E}_0\sin(\omega_L t)\right]\psi(\vec{r},t) \tag{1}$$

where ω_L is the laser frequency and $\vec{E}_0$ is the electric field envelope. In this research we will consider only square pulses of 25 field-periods long preceded by a 5 period turn on with a sinus-squared envelope. We also consider a laser linearly polarized along the z-axis, $\vec{r}\vec{E}_0 = zE_0 = rE_0\cos(\theta)$, being θ the azimuthal angle, and we take advantage of the expansion in angular momentum basis, as standard in this kind of problems[2,3],

$$\psi(\vec{r},t) = \sum_{\ell=0}^{\infty}\frac{1}{r}\chi_\ell(r,t)Y_\ell^0(\theta) \tag{2}$$

We consider that the hydrogen atom is initially in the 1s state, therefore only spherical harmonics with $m = 0$ are necessary, due to the linear polarization along the z-axis. Substituting the expansion of the wavefunction into the Schrödinger equation, and projecting on particular spherical harmonics, one gets a system of coupled partial diferential equations, one for each value of ℓ,

$$i\frac{\partial}{\partial t}\chi_\ell(r,t) = \left[-\frac{1}{2}\frac{\partial^2}{\partial r^2} - \frac{1}{r} + \frac{\ell(\ell+1)}{2r^2}\right]\chi_\ell(r,t) +$$
$$rE_0\sin(\omega_L t)\left[C_\ell^+\chi_{\ell+1}(r,t) + C_\ell^-\chi_{\ell-1}(r,t)\right] \tag{3}$$

for $\ell > 0$, and for $\ell = 0$

$$i\frac{\partial}{\partial t}\chi_0(r,t) = \left[-\frac{1}{2}\frac{\partial^2}{\partial r^2} - \frac{1}{r}\right]\chi_0(r,t) + rE_0\sin(\omega_L t)C_0^+\chi_1(r,t) \tag{4}$$

the $C_\ell^\pm$ coefficients account for the coupling between consecutive spherical harmonics components, and are closely related with the Clebsch-Gordan coefficients,

$$C_{\ell-1}^+ = C_\ell^- = \left[\frac{\ell^2}{(2\ell+1)(2\ell-1)}\right]^{1/2} \tag{5}$$

To obtain numerical solutions it is necessary to truncate the spherical harmonics expansion at a maximum value of the angular momentum, ℓ_{max}. Therefore one has to deal with a system of ℓ_{max} partial differential equations (in time t, and radius r).

Harmonic generation spectra is computed in the length gauge. From the expectation value of the dipole moment, $d(t) = \langle\psi(\vec{r},t)|z|\psi(\vec{r},t)\rangle$ we compute the Fourier transform, $d(\omega)$, and the spectrum of the coherently scattered light is associated with $|d(\omega)|^2$. We will refer to it as the harmonic intensity spectrum.

As an artificial tool to separate the contributions of different angular momentum components we define the partial contribution of the $(\ell-1) \leftrightarrow \ell$ transition to the dipole moment:

$$d_{\ell-1,\ell}(t) = C_\ell^- \int \chi_\ell^* r\chi_{\ell-1}dr + c.c. \tag{6}$$

It is easy to show that $d(t) = d_{0,1}(t) + d_{1,2}(t) + \cdots$.

Pseudo-One-Dimensional Model

To understand the contribution of the different angular momentum components to high-order harmonic generation we consider now other model. Instead of introducing a large enough number of angular momentum components, we restrict ourselves to only two, i.e. to $\ell_{max} = 1$. When only s and p states are considered, and other states with larger angular momentum are neglected, Schrödinger equation is approximated to

$$i\frac{\partial}{\partial t}\chi_s(r,t) = \left[-\frac{1}{2}\frac{\partial^2}{\partial r^2} - \frac{1}{r}\right]\chi_s(r,t) + rE_0\sin(\omega_L t)C_0^+\chi_p(r,t) \tag{7}$$

$$i\frac{\partial}{\partial t}\chi_p(r,t) = \left[-\frac{1}{2}\frac{\partial^2}{\partial r^2} - \frac{1}{r} + \frac{1}{r^2}\right]\chi_p(r,t) + rE_0\sin(\omega_L t)C_1^-\chi_s(r,t) \tag{8}$$

where the notation χ_s, χ_p has been introduced instead of χ_0 and χ_1 respectively.

We have named this model as pseudo-one-dimensional. This name was given because it has exactly the same complexity as the one-dimensional models[1]. Now electron

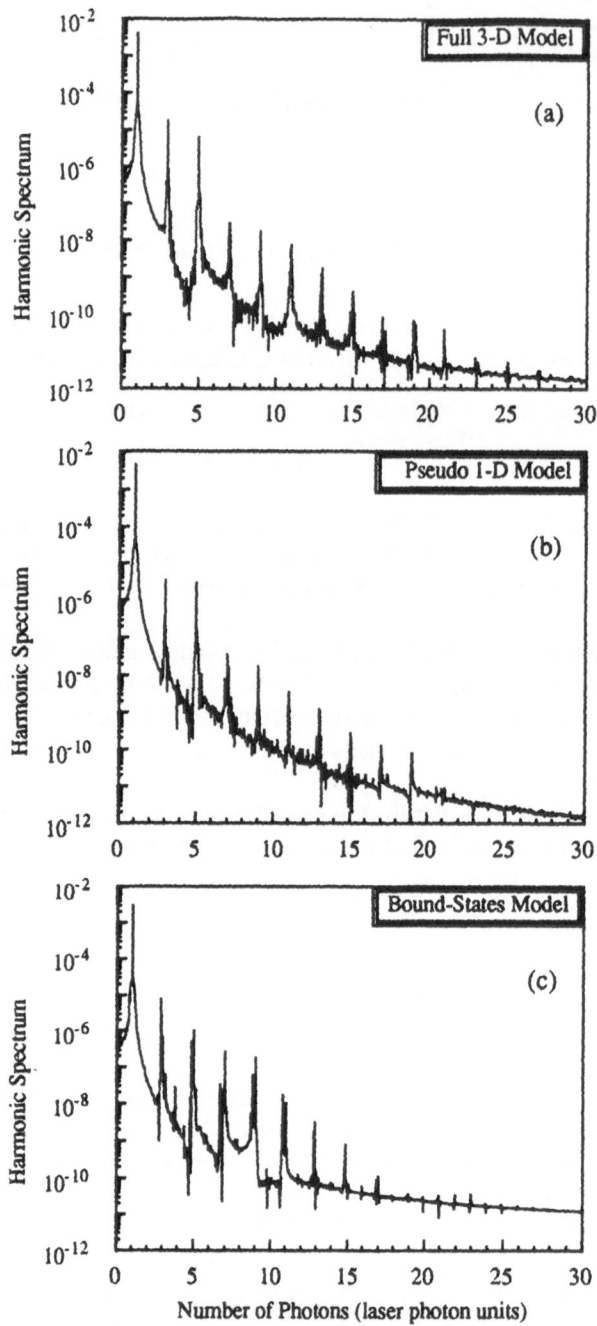

Figure 2. *Harmonic spectra for $E_0 = 0.03au.$ and $\omega_L = 0.1au.$ calculated with (a) the full three-dimensional model ($\ell_{max} = 30$), (b) the pseudo-one-dimensional model ($\ell_{max} = 1$), and (c) the bound-states model ($\ell = 0, 1; n \leq 8$).*

population is forced to be in the s or p states. Therefore excitation is slightly changed because electrons cannot jump to other higher angular momentum states.

Few-Bound-Levels Model

The two previous models imply the numerical solution of partial differential equations. Now we introduce a new model, more simplified than these two, that only implies the solution of a system of a few ordinary differential equations. We will consider only a few bound states, assuming that all electron population is forced to be in these states. Ionization is therefore neglected.

We call b_{ns} and b_{np} the amplitudes of the bound states $|ns\rangle$ and $|np\rangle$, respectively, n being the principal quantum number. All free states are neglected. In this case, the time-dependent Schrödinger equation becomes,

$$i\frac{d}{dt}b_{ns}(t) = \omega_n b_{ns}(t) + E_0 \sin(\omega_L t) \sum_{n'=2}^{N} d_{ns}^{n'p} b_{n'p}(t) \tag{9}$$

$$i\frac{d}{dt}b_{np}(t) = \omega_n b_{np}(t) + E_0 \sin(\omega_L t) \sum_{n'=1}^{N} d_{np}^{n's} b_{n's}(t) \tag{10}$$

where $d_{ns}^{n'p} = \langle ns|z|n'p\rangle$ is the dipole matrix element between the $|ns\rangle$ and $|n'p\rangle$ bound states. We cut both Rydberg series at a given number N. Therefore we are considering a system on $2N - 1$ equations.

RESULTS

We compare harmonic generation spectra computed for these three different models. We propose that the comparison between the results may give a clue of which processes are fundamental for harmonic generation. Figure 2 shows this comparison for a square pulse of 25 cycles long, preceded by a 5 cycle sinus squared turn-on. The laser frequency is $\omega_L = 0.1$ au, i.e. five photons are necessary to reach the ionization threshold. The amplitude is $E_0 = 0.03$ a.u. In all cases the population is initially in the ground state. It is surprising that the three models give harmonic generation patterns so similar. Its is remarkably surprising the coincidence between the three computed harmonic intensity spectra shown in Fig. 2. Figure 2a has been calculated with the three-dimensional model, using 31 angular momentum components ($\ell_{max} = 30$). Figure 2b has been calculated with only two angular momentum components (with what we called the pseudo-one-dimensional model). Finally Fig. 2c has been calculated within our bound states model, using only 15 states (namely $|1s\rangle, \cdots, |8s\rangle$ and $|2p\rangle, \cdots, |8p\rangle$). We want to understand the reasons for these coincidences step by step. First we will analyze the role of the maximum angular momentum component, ℓ_{max}, and second we will study the contribution of the bound states, neglecting the ionization dynamics.

Role of the Maximum Angular Momentum Component

Comparison between Fig. 2a ($\ell_{max} = 30$) and Fig. 2b ($\ell_{max} = 1$) indicates clearly that states with $\ell = 2, 3, \ldots$ are not essential for harmonic generation. To understand the partial contribution to the dipole moment shown in Fig. 2a of each $(\ell - 1) \leftrightarrow \ell$ transition we can use the decomposition (6). Figure 3 shows the spectra computed from a three-dimensional calculation. While 31 angular momentum components have been used ·to calculate the time-evolution, the dipole moment is calculated only from

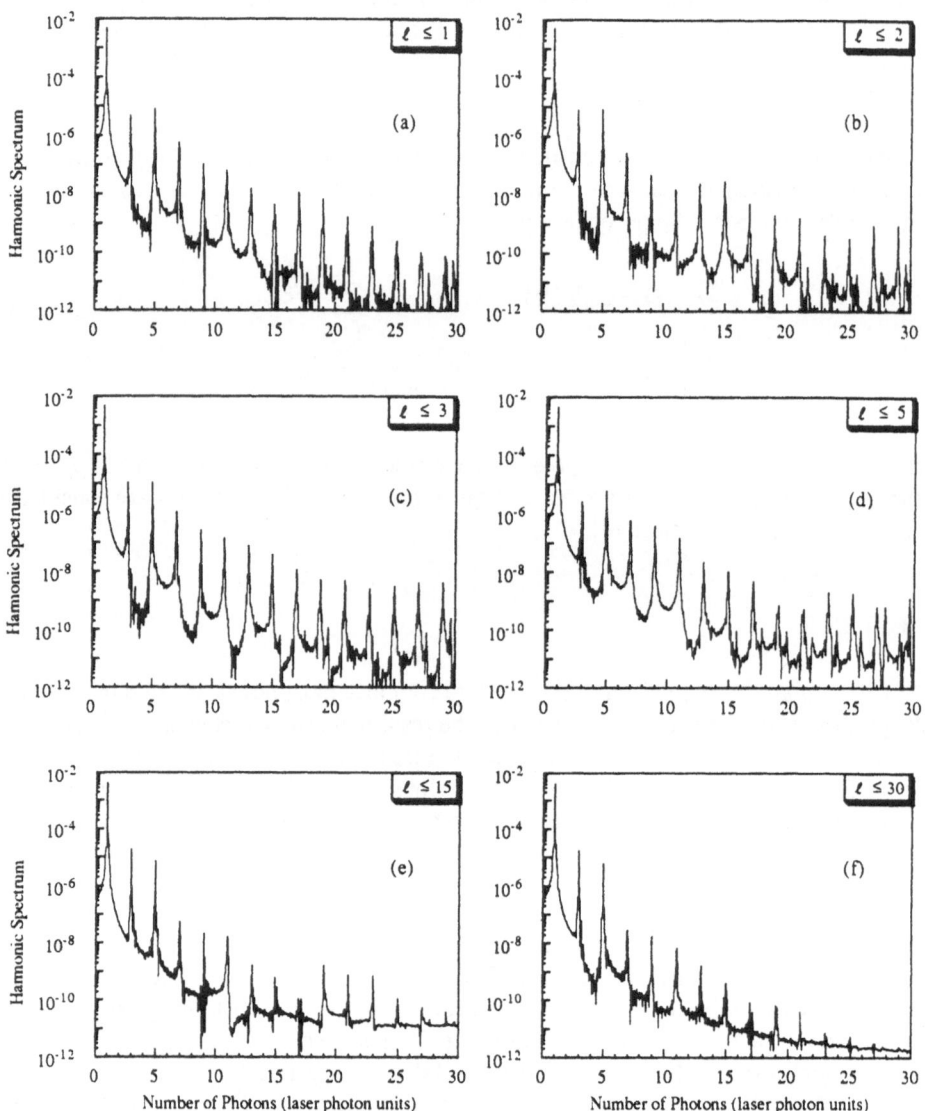

Figure 3. *Contribution of the different angular momentum states to the harmonic generation calculated with the full three-dimensional model. From (a) to (f) an increasing number of angular momentum states are taken into account to calculate the electron dipole momentum. The calculation is done with the same laser parameters as in Fig. 2.*

the contribution of some states. The first spectrum, Fig. 3a, shows the contribution of the $\ell = 0 \leftrightarrow \ell = 1$ transitions, ie. the spectrum obtained from $d_{0,1}(t)$. Comparing with Fig. 2b, it is surprising that calculating correctly the wavefunction, with many angular momentum components, and the considering only a part of the dipole moment —the part corresponding to the $s \leftrightarrow p$ transitions— gives a totally incorrect result, while a truncated calculation of the wavefunction as the pseudo-one-dimensional case ($\ell_{max} = 1$) is able to give a correct harmonic intensity spectrum. Fig. 3a shows more intense harmonic peaks that Fig. 2b. Probably the reason for this are the fast oscillations induced in $\chi_1(r,t)$ due to the coupling with $\chi_2(r,t)$. Because of the not very strong dependence with ℓ of the $C_\ell^{\pm}$ coefficients, these fast oscillations are associated to other fast oscillations in $\chi_2(r,t)$, and so on. Therefore in the three-dimensional case when all the contributions are considered, interference between them eliminates the higher peaks. Thus Fig. 3b shows the harmonic spectra obtained from $d_{1,0}(t) + d_{1,2}(t)$, Fig. 3c from $d_{1,0}(t) + d_{1,2}(t) + d_{2,3}(t)$, Fig. 3d from $d_{1,0}(t) + \cdots + d_{4,5}(t)$, and so on. In Fig. 3e, that corresponds to the harmonic spectra obtained from $d_{1,0}(t) + \cdots + d_{14,15}(t)$, the destructive interference of the highest harmonic peaks begins to be apparent, and it is total in Fig. 3f (the same as Fig. 2a).

Role of the Bound States

Comparison between Fig. 2b and Fig. 2c shows that the bound states are essential to understand harmonic generation spectra. The spectrum shown in Fig. 2c has been calculated with only a few bound states ($|1s\rangle, \ldots, |8s\rangle$, and $|2p\rangle, \ldots, |8p\rangle$), and all other states have not been considered. Therefore ionization has been neglected completely, and still harmonics with frequencies larger than the ionization threshold appear. We find surprising the coincidence in some higher harmonics. For example the peak at $15 \, \omega_L$ is well described, and for this peak its frequency is equal to three times the ionization threshold energy. This means that for generation of a coherently scattered photon is not necessary to pass over an intermediate resonant state in the continuum. Therefore we want to suggest the idea of harmonic generation spectra and free electron population spectra correspond to different physical effects. Figure 4 justifies the choice of $n \leq 8$ as cutoff: Fig. 4a shows the harmonic intensity spectrum obtained with a three level model (states $|1s\rangle$, $|2s\rangle$ and $|2p\rangle$), it is different from the real spectra. Figure 4b corresponds to five bound states (states $|1s\rangle$, $|2s\rangle$, $|3s\rangle$, $|2p\rangle$ and $|3p\rangle$), and so on. At this laser intensity it is enough to cut the number of bound states at $n = 8$, ($|8s\rangle$ and $|8p\rangle$).

CONCLUSIONS

We have presented a comparison between three numerical calculations describing high-order harmonic generation in atomic hydrogen. The three results give essentially the same result regardless the fact that they are very different. The conclusion is that the essential features of harmonic generation are contained in the most simple model of these three, and other couplings included in the more realistic models are not fundamental to study harmonic generation.

Since the simplest of the three models considered deals only with a finite number of bound states, it is reasonable to conclude the high-order harmonic generation is mostly due to bound-bound couplings, and ionization plays no role. We might speculate a little bit about the physical origin of this result. The processes that lead to harmonic generation are not the same that allow electrons to ionize. If we Fourier transform the

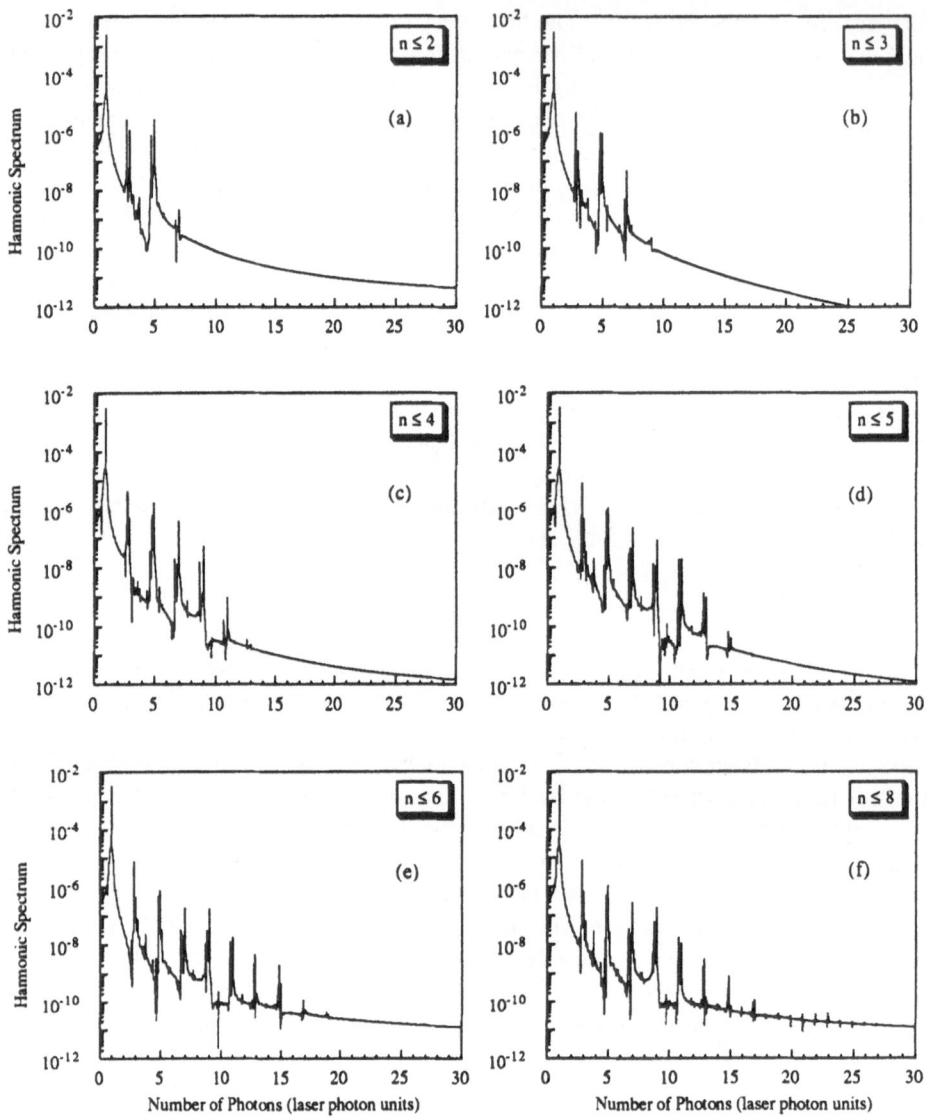

Figure 4. *Spectra calculated with the bound-states model. From (a) to (f) an increasing number of bound states are considered when integrating the Schrödinger equation. The calculation is done with the same laser parameters as Fig. 2.*

time-dependent wavefunction, fast oscillations may appear, indicating that electrons are promoted to high energy states (i.e. ionized). However, harmonic generation is not related directly to a projection of the wavefunction but to the expected value of the dipole operator. When calculating the expectation value, some of this oscillations may destructively interfere. An idea in this line has been recently pointed out [4], showing that situations with symmetrical couplings may lead to cancellations of the harmonic components. Pushing our conclusions, we can propose that in general *symmetry kills harmonic generation.* Unless abrupt changes occur in the density of free electron states, harmonic generation is not expected from ionizing electrons. In the case of hydrogen, the transition from bound to free states is smooth due to the presence of a Rydberg ladder, and therefore free states almost play no role. In the case of negative ions, for example, where an abrupt ionization threshold appears, without closely lying bound states, some harmonic generation is possible due to the ionizing electrons with energies close to the threshold. In this case the interference exists but is not complete. Therefore we propose the the best systems for intense high-order harmonic generation are those with more than one bound state, and this is even valid to generate frequencies much larger than the ionization threshold energy.

ACKNOWLEDGMENTS

The authors acknowledge the rest of participants of this SILAP III Workshop for their questions, comments and fruitful discussions on this and other related topics. We also acknowledge CESCA (Centre de Supercomputació de Catalunya) for the computer facilities, and partial support from the Spanish DGICYT (grant PB-90-0725) and from NATO (grant CRG-900352).

REFERENCES

1. L. Roso-Franco, A. Sanpera, M.Ll.Pons and L.Plaja, Photoionization of the hydrogen atom: Three-dimensional results and pseudo-one-dimensional model, *Phys. Rev. A* **44**:4652 (1991).
2. P.L. De Vries, Calculation of harmonic generation during the multiphoton ionization of the hydrogen atom, *J. Opt. Soc. Am. B* **7**:517 (1990).
3. K.J. LaGattuta, Laser effects in photoionization: Numerical solution of coupled equations for a three-dimensional coulomb potential, *J. Opt. Soc. Am. B* **7**:639 (1990).
4. M.Yu.Ivanov and K.Rzążewski, Are free-free transitions a good basis for nonlinear optics?, *J. Mod. Opt.* in press (1993).

SYMMETRY BREAKING AND THE CONTROL OF HARMONICS WITH STRONG SHORT LASER PULSES

M.Yu. Ivanov and P.B. Corkum

National Research Council of Canada
Montreal Rd, M-23A, Ottawa, Ontario, K1A 0R6

Since the beginning of nonlinear optics it has been thought that a system with inversion symmetry cannot radiate even-order harmonics of the incident laser light. The purpose of this paper is to show that this is not always true. In fact, emission of even harmonics from a system with inversion symmetry will be typical for the interaction with intense ultrashort laser pulses.

Another property of a system with inversion symmetry is the absence of a constant dipole moment. Inducement of a constant dipole moment in such a system was always considered as a peculiarity of the interaction with a constant electric field. We will show, however, that short intense laser pulses also can induce constant dipole moment in a symmetric system, breaking the inversion symetry on a time-scale of the laser pulse. As a result, the system begins to radiate even harmonics of laser light.

In retrospect, the fact that intense ultrashort laser pulses can break the symmetry of a quantum system looks natural. When the peak pulse intensity is high and the pulse rise time is short (on a scale of a field period), the interaction with quantum system is essentially aperiodic: the interaction strength changes significantly over one field period. It is then, when the symmetry of the quantum system can be broken and new effects like emission of even harmonics can arise. It is also clear that under these conditions symmetry breaking and those phenomena that occur as a result of symmetry breaking will be highly sensitive to the parameters determining the front of the ultrashort laser pulse: rise time, peak intensity, and phase.

Consider a quantum system with a symmetric Hamiltonian $\hat{H}_0(x, \hat{p})$:

$$\hat{H}_0(x, \hat{p}) = \hat{H}_0(-x, -\hat{p}) \tag{1}$$

The eigenstates φ_n of this Hamiltonian are parity eigenfunctions: $\varphi_n(-x) = \pm\varphi_n(x)$ (plus for even and minus for odd states). According to the Floquet theorem, when this system interacts with a monochromatic electric field $\mathcal{E}(t) = \mathcal{E}_0 exp(-i\omega_L t) + c.c.$,

Super-Intense Laser-Atom Physics, Edited by
B. Piraux *et al.*, Plenum Press, New York, 1993

the time-dependent Shroedinger equation with a periodic Hamiltonian $\hat{H} = \hat{H}_0 - d\mathcal{E}(t)$ has solutions in a form [1-2]:

$$\Psi^{(n)}(x, t) = e^{-iE^{(n)}t}\phi^{(n)}(x, t) \quad , \quad \phi^{(n)}(x, \, t + 2\pi/\omega_L) = \phi^{(n)}(x, t) \tag{2}$$

The states $\phi^{(n)}(x, t)$ are the Floquet states and $E^{(n)}$ are the quasienergies.

The new time-dependent Hamiltonian $\hat{H}$ has no inversion symmetry, and the Floquet states $\phi^{(n)}(x, t)$ are not parity eigenfunctions. Thus, laser field breaks the standard symmetry of the system. However, the new Hamiltonian has the following symmetry:

$$\hat{H}(x, \hat{p}, t) = \hat{H}(-x, -\hat{p}, t + \pi/\omega_L) \tag{3}$$

As a consequence, the Floquet wavefunctions $\phi^{(n)}(x, t)$ are the eigenfunctions of the generalized parity operator, which changes (x, p, t) to $(-x, -p, t + \pi/\omega_L)$, and can be classified into the states of even and odd generalized parity [3,4]:

$$\phi^{(n)}(-x, \, t + \pi/\omega_L) = \pm\phi^{(n)}(x, t) \quad , \tag{4}$$

and symmetry breaking due to laser field is incomplete. This property has a straightforward consequence, which we now describe.

Suppose the quantum system was initially in the field-free state φ_{n_0}. If the electric field is turned on adiabatically at $t < 0$ and then at $t \geq 0$ its amplitude is kept constant, at $t > 0$ we will find the system in the Floquet state $\phi^{(n_0)}$ which evolved from φ_{n_0} [1,2,5]. The polarization of the quantum system (or of a rarefied medium of quantum systems) is given by the dipole moment

$$d(t) = <\phi^{(n_0)}|\hat{d}|\phi^{(n_0)}> \tag{5}$$

The function $\phi^{(n_0)}(x, t)$ is time-periodic and also obeys Eq.(4). One can easily show that owing to the symmetry Eq.(4) the Fourier decomposition of $d(t)$ Eq.(5) contains neither even harmonics, nor constant dipole component d_0:

$$d(t) = \sum_m d_{2m+1} exp(-i(2m + 1)\omega_L t) \tag{6}$$

One might say that this result closes the question of even harmonics emission from a symmetric system. But it is not so for the following two reasons.

First, the Fourier decomposition of the dipole moment Eq.(5) gives the spectrum of coherent emission only [6-9]. There is also an incoherent part of emission due to spontaneous transitions between $\phi^{(n_0)}$ and other Floquet states $\phi^{(n)}$. Its spectrum is determined by the Fourier transform of the corresponding off-diagonal transition moment $d_{nn_0}(t) = <\phi^{(n)}|\hat{d}|\phi^{(n_0)}>$ [6-9].

Second, and more important, is that the result Eq.(6), which gives only odd harmonics in the spectrum of coherent emission, is based on the assumption that the turn-on of the laser field is adiabatically slow. Consequently, for $t > 0$ the system is in a single Floquet state $\phi^{(n_0)}$. In contrast, interaction with ultrashort intense laser pulses will always put the quantum system in a superposition of Floquet states, and even the coherent part of emission will contain the Fourier spectrum of the off-diagonal transition moments $d_{mn}(t) = <\phi^{(m)}(t)|\hat{d}|\phi^{(n)}(t)>$.

Using the symmetry relation Eq.(4), one can show that if $\phi^{(n)}$ and $\phi^{(m)}$ have different generalized parity, the Fourier decomposition of $d_{mn}(t)$ is

$$d_{mn}(t) = \sum_k d_{mn}^{(k)} exp(-i(2k\omega_L + \omega_{mn}(\mathcal{E}_0))t) \qquad (7)$$

where $\omega_{mn}(\mathcal{E}_0)$ is the field-dependent transition frequency between the Floquet states with quasienergies $E^{(n)}$ and $E^{(m)}$.

Eq.(7) gives a key to the problem of even harmonics emission from a symmetric quantum system in an intense laser field. As we will show below, strong laser field can induce the degeneracy of the Floquet states *of different generalized parity*, leading to $\omega_{mn}(\mathcal{E}_0) = 0$ for certain m and n. If at least two Floquet states $\phi^{(1)}$ and $\phi^{(2)}$ are degenerate in the laser field, and if at least one of them is populated, the noncoherent part of emission will contain even harmonics of laser light. Moreover, if both $\phi^{(1)}$ and $\phi^{(2)}$ are populated, the term $d_{12}(t) = <\phi^{(1)}|\hat{d}|\phi^{(2)}>$ will appear in the expression for the field-induced dipole moment of the system $d(t) = <\Psi|\hat{d}|\Psi>$. As a result, the spectrum of coherent emission will contain even harmonics and the dipole moment will also have a constant component (see Eq.(7)) $d_0 = d_{12}^{(0)} + c.c.$

Even more important is that, in the field of an ultrashort laser pulse, even harmonics emission and inducement of a constant component d_0 of the dipole moment $d(t)$ will occur without exact field-induced degeneracy of the Floquet states. Indeed, if the pulse is so short that $\omega_{12}(\mathcal{E}_0)\tau << 1$, the Raman shift $\omega_{12}(\mathcal{E}_0)$ of the emission lines will not be resolved. In other words, when the Fourier-limited bandwidth τ^{-1} of the pulse exceeds the distance $\omega_{12}(\mathcal{E}_0)$ between almost degenerate Floquet states, the symmetry of the quantum system will be broken on a time-scale of the ultrashort pulse. Other broadening mechanisms, e.g. ionization, can also make the emission at the frequencies $2k\omega_L + \omega_{12}(\mathcal{E}_0)$ and $2k\omega_L + \omega_{21}(\mathcal{E}_0) = 2k\omega_L - \omega_{12}(\mathcal{E}_0)$ indistinguishable from a single line at the frequency $2k\omega_L$.

Let us now specify the conditions under which these phenomena can occur. Obviously, we have to go beyond the limits of the perturbation theory. Moreover, we also have to go beyond the limits where rotating wave approximation (RWA) and essential states models (ESM) [10] are valid. Indeed, within the limits of the RWA and the ESM the wave function of the quantum system can be written in a form:

$$\Psi(x, t) = e^{-iE_g t}\left[a_g(t)\varphi_g(x) + \sum_{n_1} a_{n_1}(t)\varphi_{n_1}(x)e^{-i\omega_L t} + \sum_{n_2} a_{n_2}(t)\varphi_{n_2}(x)e^{-i2\omega_L t} + ...\right]$$
$$(8)$$

where the index g denotes the ground (initial) state, φ_m represents the field-free states and the amplitudes $a_m(t)$ are slow (compared to $exp(i\omega_L t)$) functions of time. The sets of states denoted by indexes $n_1, n_2, ...$ are separated by ω_L and are excited by one-photon, two-photon, ... transitions. Hence, the states φ_{n_1} have parity opposite to that of φ_g, the states φ_{n_2} have the same parity as φ_g, and so on. Calculating $d(t)$ with the wavefunction Eq.(8) we immediately see that as far as $a_m(t)$ are slow functions of time, $d(t)$ contains only odd harmonics. Only beyond the RWA can the amplitudes $a_m(t)$ have significant fast oscillating components and emission of even harmonics becomes possible.

To illustrate the general conclusions made above we will consider the example of a two-level system in a very strong laser field. A dressed two-level system has exactly two Floquet states – just the minimum amount we need to observe the effect of symmetry breaking and the resulting phenomena. Still, a two-level system is simple enough and allows us to get insight into the physical essence of the phenomena. According to the previous paragraph, we need a strong field $d_{21}\mathcal{E}_0 >> \omega_{21}^{(0)}$, where d_{21} is the dipole matrix element, $d_{21} = <\varphi_2|\hat{d}|\varphi_1>$, and $\omega_{21}^{(0)}$ is the field-free transition frequency. As we will discuss later, it is possible to realize the two-level system under such conditions in practice. We assume that the laser frequency is not too small compared to the transition frequency, that is, $\omega_L \sim \omega_{21}^{(0)}$ or $\omega_L >> \omega_{21}^{(0)}$. The field is turned on either smoothly or abruptly, but the intensity is constant at $t > 0$: $\mathcal{E}(t > 0) = \mathcal{E}_0 cos\omega_L t$.

An analytical solution for the intense-field regime is obtained in Refs.[11,12]. It is valid if either $\omega_{21}^{(0)}/\sqrt{d_{21}\mathcal{E}_0\omega_L} << 1$ or $\omega_{21}^{(0)} << \omega_L$. According to this solution, the transition frequency between the Floquet states in the limit of strong or high-frequency field is (see also Ref.[13]):

$$\omega_{21}(\mathcal{E}_0) = \omega_{21}^{(0)} J_0(2d_{21}\mathcal{E}_0/\omega_L) \tag{9}$$

The wave functions of Floquet states are:

$$|\Psi_1(x,t)> \approx (cosF|1> - isinF|2>)exp(i\Theta/2)$$

$$|\Psi_2(x,t)> \approx (cosF|2> - isinF|1>)exp(-i\Theta/2) \tag{10}$$

where $F(t) \equiv \int^t d_{21}\mathcal{E}(t')dt' = d_{21}A(t)$, $A(t)$ the vector-potential of the laser field, and

$$\Theta(t) = \omega_{21}^{(0)} \int_0^t cos2F(t')dt' \tag{11}$$

For abrupt turn-on $F(t) = d_{21}(\mathcal{E}_0/\omega_L)sin\omega_L t$. If the turn-on is smooth and includes many field periods, the vector-potential is $A(t) \approx A_0 f(t)sin\omega_L t$, and $F(t)$ is approximately $F(t) \approx d_{21}(\mathcal{E}_0/\omega_L)f(t)sin\omega t$.

According to Eq.(9), there are points of exact degeneracy of the Floquet states $\phi^{(1)}$ and $\phi^{(2)}$, given by the zeros of the Bessel function J_0. When the Floquet states are degenerate, the two-level system radiates even harmonics and the dipole moment $d(t)$ can have a constant component. More important is that the transition frequency $\omega_{21}(E)$ decreases with increasing the field strength: $\omega_{21}(E) \propto \omega_{21}^{(0)}\sqrt{d_{21}\mathcal{E}_0/\omega_L}^{-1}$. Hence, for sufficiently short and strong laser pulse with $\omega_L > \omega_{21}^{(0)}$ the inequality $\omega_{21}(\mathcal{E}_0)\tau << 1$ can be fulfilled. As a result, the system will radiate even harmonics not only at certain field values corresponding to $\omega_{21}(\mathcal{E}_0) = 0$, but in the field of any sufficiently strong and short laser pulse.

Using the results of our analytical solution (Eqs.(9-11) and Refs.[11,12]), we can calculate the dipole moment $d(t)$ and the spectrum of coherent emission for arbitrary initial conditions. According to Eqs.(10-11), $d(t)$ has a low-frequency component

$$d_0(t) \approx d_{21}(\alpha_1\alpha_2^* exp(i\omega_{21}(\mathcal{E}_0)t) + c.c.) \tag{12}$$

where $\alpha_{1,2}$ are the probability amplitudes to find the system in the Floquet states $\phi^{(1,2)}$ at $t > 0$, i.e. when the pulse is turned on. According to Eqs.(10-11), at $t = 0$ the Floquet states coincide with bare states, and therefore the amplitudes $\alpha_{1,2}$ are also $a_{1,2}(t = 0)$.

At $\omega_{21}(\mathcal{E}_0) = 0$ the dipole moment has a constant component $d_0 \approx d_{21}(\alpha_1 \alpha_2^* +$ c.c.). Now we see what direction is favoured in the initially symmetric system. For a pulse that is abruptly turned on it is the direction towards which the dipole moment points at the turn-on moment. In other words, the strong field *freezes* the dipole moment of the system. If at $t = 0$ one of the states was empty, the constant component is absent: $d_0 \approx d(0) = 0$ if $a_i(0) = 0$. However, if the pulse turn-on is fast but not abrupt, the empty state will be populated at the pulse front, creating nonzero $d(t = 0)$ and, hence, nonzero constant component of the dipole moment at $t > 0$. The dipole moment will be frozen in the direction it has at $t = 0$. Obviously, the Floquet states populations created at the steep front of the pulse will depend on the details of the pulse rise. Thus, d_0 should depend on the parameters characterizing the pulse rise: the peak field strength $\mathcal{E}_0$, the rise time τ_r and, in the limit of short τ_r, the phase ϕ of the field $\mathcal{E}(t) = \mathcal{E}_0 f(t) \cos(\omega t + \phi)$. Fig.1 shows the results for $d(t)$ and Gaussian pulse rise.

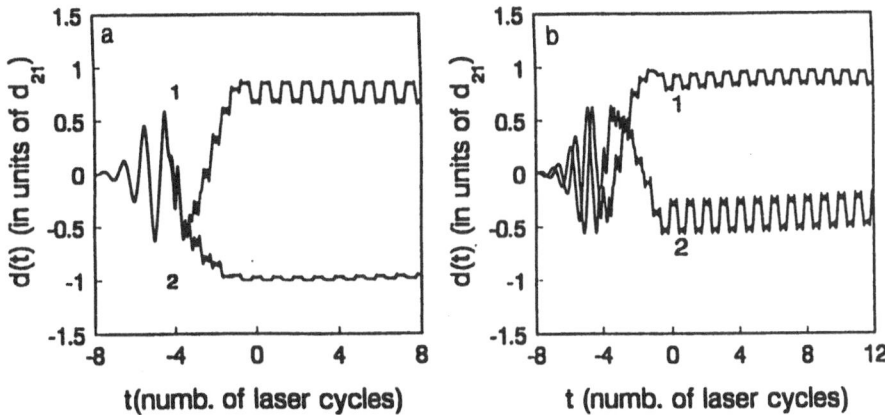

Fig.1. Dipole moment induced in a two-level system by a linearly polarized intense laser pulse $\mathcal{E}_0 f(t) \cos(\omega_L t + \phi)$, where $f(t) = exp(-t^2/\tau^2)$ for $t < 0$, $f(t) = 1$ for $t > 0$. The transition frequency is $\omega_{21}^{(0)} = 0.5\omega_L$. The initial conditions are $a_1 = 1$, $a_2 = 0$.
Fig.1a: $\tau = 3 T_{\omega_L}$ (T_{ω_L} the laser period), $\phi = 0$. Curve 1: $V = d_{21}E = 10.6\omega_L$. Curve 2: $V = 12.195\omega_L$.
Fig.1b: $\tau = 3 T_{\omega_L}$, $V = d_{21}E = 10.6\omega_L$. Curve 1: $\phi = \pi/4$. Curve 2: $\phi = 3\pi/4$.

We can control both the absolute value and the direction of $d(t)$ by controlling $V = d_{21}\mathcal{E}_0$, τ_r, or ϕ. In particular, one can change the direction of d_0 by changing V (Fig.1a). Fig.1b shows how the constant component depends on the phase ϕ. In particular, d_0 changes its sign when ϕ passes $\pi/2$. The constant component disappears if the pulse rise is long, and thus we recover the traditional picture of long laser pulses (Fig.1c).

As was mentioned above, the dipole moment $d(t)$ has a purely constant component only when $\omega_{21}(\mathcal{E}_0) = 0$. However, from Eq.(12) it is clear that if $\omega_{21}(\mathcal{E}_0)\tau << 1$, the dipole moment cannot change its direction during the pulse duration τ and the symmetry of the system will be broken on a time scale of the short pulse for any sufficiently high peak intensity.

Apart from the low-frequency component, there are two groups of the Fourier components of $d(t)$ and, therefore, two groups of lines in the coherent emission spectrum. Both have been identified in computer simulations [14,15]. Our analytical results show how their amplitude and phase depend on laser pulse parameters and initial conditions, and how they can be controlled. The first group of lines are the odd harmonics of the incident light, $\Omega_N = N\omega_L = (2K+1)\omega_L$.

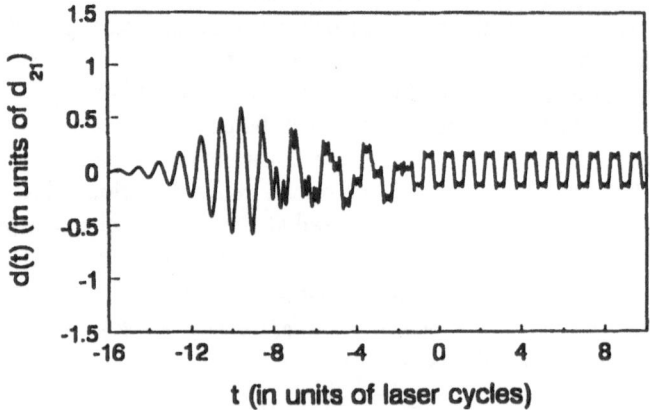

Fig.1c. Same as in Fig.1a,1b, but for $\tau = 6T_{\omega_L}$, $\phi = 0$, $V = 10.6\omega_L$.

The corresponding Fourier components of $d(t)$ are [11,12]:

$$d_{2K+1} = d_{21}\left(|\alpha_1|^2 - |\alpha_2|^2\right)\frac{\omega_{21}^{(0)}}{(2K+1)\omega_L}J_{2K+1}(2d_{21}\mathcal{E}_0/\omega_L)e^{i(2K+1)\omega_L t} \qquad (13)$$

According to Eq.(14), the two-level system in a strong linearly polarized laser field turns out to be a very good source of harmonics, the number of harmonics generated is $N_{max} \approx 2d_{21}\mathcal{E}_0/\omega_L \gg 1$, see Fig.2. The harmonic spectrum exhibits the typical behavior: long plateau and a sharp cut off at $N > N_{max}$. All these features where observed in numerical calculations [6].

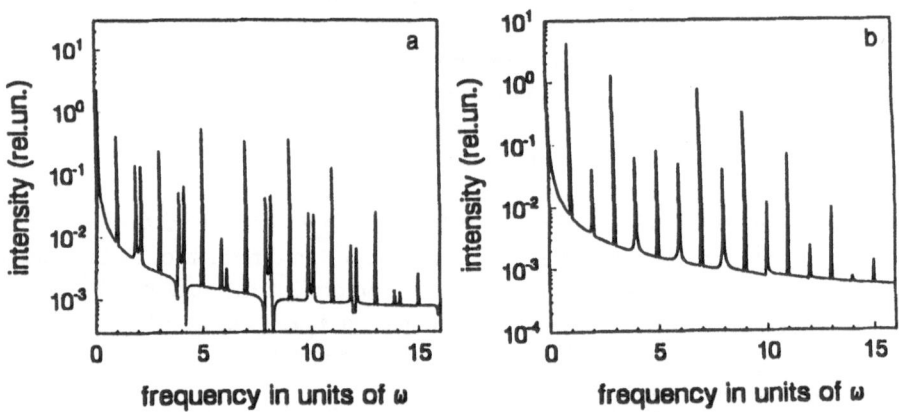

Fig.2. Emission spectrum of a two-level system in intense laser field: $\omega_{21}^{(0)} = 0.5\omega_L$, $\tau = 3T_{\omega_L}$. Initial conditions are $a_1 = 1$, $a_2 = 0$.
Fig.2a: $V = 5\omega_L$, $\phi = \pi/4$
Fig.2b: $V = 4.325\omega_L$, $\phi = 0$

The second group of transitions are the hyper-Raman type processes, i.e. transitions between the two different quasienergy states. They lead to the emission at frequencies $\Omega_N^\pm = N\omega_L \pm \omega_{21}(\mathcal{E}_0) = 2K\omega_L \pm \omega_{21}(\mathcal{E}_0)$. The corresponding Fourier components of $d(t)$ are:

$$d_{2K} = d_{21} \left(\alpha_1 \alpha_2^* e^{i\omega_{21}(\mathcal{E}_0)t} - \alpha_2 \alpha_1^* e^{-i\omega_{21}(\mathcal{E}_0)t} \right) \frac{\omega_{21}^{(0)}}{2K\omega_L} J_{2K}(2d_{21}\mathcal{E}_0/\omega_L) e^{i2K\omega_L t} \quad (14)$$

There are several features of the emission spectrum which are clearly seen from Eqs.(13,14).

First, in a strong field the Stokes and anti-Stokes lines $2K\omega_L - \omega_{21}(\mathcal{E}_0)$ and $2K\omega_L + \omega_{21}(\mathcal{E}_0)$ are very close to the even harmonic frequency $2K\omega_L$, Fig.2a. The shift $\omega_{21}(\mathcal{E}_0)$ decreases with increasing $\mathcal{E}_0$ and becomes much less than $\omega_{21}^{(0)}$ and, hence, much less than ω_L. When the quasienergies are degenerate ($\omega_{21}(\mathcal{E}_0) = 0$), there is a single line with frequency exactly equal to $2K\omega_L$, Fig.2b. Thus, the laser field breaks the symmetry of the system.

Even more important is the fact that, in a sufficiently strong and short laser pulse, even-order harmonics will be generated at almost any peak intensity of the pulse. Indeed, if the pulse duration τ is short and $\omega_{21}(\mathcal{E}_0)\tau << 1$, the double-line structure of $\Omega_{2K} = 2K\omega_L \pm \omega_{21}(\mathcal{E}_0)$ will not be resolved. Thus, short intense pulses always break symmetry.

Second, Eqs.(13,14) show the possibility of controlling the emission spectrum by varying the parameters of the laser pulse (peak field strength $\mathcal{E}_0$, pulse rise time τ_r, phase ϕ) and initial conditions. Indeed, if initially only one state of the system is populated and the pulse rise is long, only one quasienergy state (which evolves from the initially populated bare state) will be populated and only odd harmonics will be present in the spectrum of $d(t)$, Fig.3a. On the other hand, if the initial conditions and (or) the shape of the pulse front provide equal populations of the Floquet states at $t = 0$, odd harmonics will be suppressed. If at the same time $\omega_{21}(\mathcal{E}_0)\tau << 1$ and $Im[\alpha_1 \alpha_2^* \neq 0]$, the system will efficiently emit even harmonics instead of odd ones, Fig.3b.

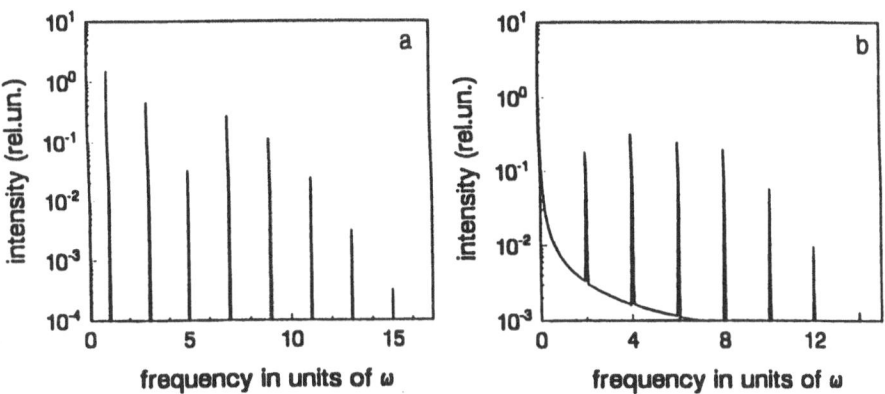

Fig.3. Emission spectrum of a two-level system in intense laser field: $\omega_{21}^{(0)} = 0.25\omega_L$, $V = 4.325\omega_L$, $\tau = 5T_{\omega_L}$.
Fig.3a: initial conditions $a_1 = 1$, $a_2 = 0$.
Fig.3b: initial conditions $a_1 = 1/\sqrt{2}$, $a_2 = i/\sqrt{2}$.

The study of a two-level system in a strong-field regime $d_{21}\mathcal{E}_0 >> \omega_{21}^{(0)}$ is not only of theoretical, but also of practical importance and the regime itself is experimentally accessible. For example in a double-well potential the lowest two levels are usually well separated in energy from other levels. Therefore, in the absence of field-induced resonances with higher states the problem is reduced to the dynamics of a laser driven two-level system. Simple estimate show that for a double quantum well like AlGaAs, where the dipole matrix element of transition between the two lowest states is $d_{21} \sim 10^2 Bohr$ and the transition frequency is $\omega_{21}^{(0)} \approx 1.7 \ 10^{-2} eV$, the strong field regime $d_{21}\mathcal{E}_0 >> \omega_{21}^{(0)}$ is realized at intensities as low as $I \sim 10^7 \ W/cm^2$.

For a double-well quantum system inducement of a constant dipole moment means that the charge distribution is localized in either left or right well. In contrast, in the absence of laser field the dipole moment would change its sign after the time $\tau_0 = \pi/\omega_{21}^{(0)}$ (see Eq.(12)), reflecting the tunneling of the charge to the opposite well. Inducement of constant dipole moment means that intense laser field suppresses this process [13-17].

Double-well Kramers-Henneberger (KH) potential always appears in the problem of atomic stabilization in superintense laser field [18]. When the separation of the wells $\alpha = \mathcal{E}_0/\omega_L^2$ is large, $\alpha >> 1$, in the zeroth approximation each well can be considered separately and supports its own ground state. As usually, the two lowest states of the complete Hamiltonian will be their symmetric and the antisymmetric combinations. Hence, in the limit $\alpha >> 1$ we will have two nearly degenerate Floquet states of oppposite parity, broadened by ionization and ultrashort interaction time. As a result, in the limit $\alpha >> 1$ interaction of atom with intense ultrashort laser pulses will result in the emission of even-order harmonics of the incident light, as was numerically observed in [19] (see also [20]).

In conclusion, we have shown that sufficiently intense laser field is able to induce the degeneracy of the Floquet states of different generalized parity, breaking the symmetry of the dressed quantum system. As a result, the quantum system begins to radiate even harmonics and acquires a constant dipole moment. In the field of intense ultrashort laser pulses exact degeneracy of Floquet states is not required: the transition frequency between nearly degenerate states is not resolved owing to ionization and large Fourier-limited bandwidth $1/\tau$ of the laser pulse. Being in a single Floquet state, the quantum system cannot radiate even harmonics *coherently*. Hence this effect is highly sensitive to the initial conditions and the parameters of the laser pulse (peak intensity, rise time, phase), as they determine the populations of the Floquet states. Controlling these parameters, one has complete control of the field-induced polarization and the spectrum of coherent emission.

Acknowledgements

We acknowledge helpful discussions with P.Dietrich and H.Metiu.

REFERENCES

1. N.B.Delone, V.P.Krainov, *Atoms in Strong Light Fields*, Berlin, Springer, 1985
2. N.L.Manakov, V.D.Ovsiannikov, L.P.Rapoport, Phys.Rep., v.141, 319, (1986).
3. H.P.Bruer, K.Dietz, M.Holthaus, Z.Phys.D, v.8, 349, (1988)
4. A.Peres, Phys.Rev.Lett, **67**, 158, (1991)
5. M.V.Kuzmin, V.N.Sazonov, Sov.Phys.JETP,v.52, 889,(1981)
6. B.Sundaram. P.Milonni, Phys.Rev.A, v.41, 6571, (1990)
7. J.H.Eberly, M.V.Fedorov, Phys.Rev.A, v.45, 4706, (1992)
8. K.Rzazewski, J.H.Eberly, Journ.Mod.Opt., v.39, 795, (1992)
9. K.LaGattuta, Journ.Mod.Opt., v.39, 1181, (1992)

10. J.H.Eberly, J.Javanainen, K.Rzazewski, Phys.Rep.,v.204, 331, (1991)

11. M.Yu.Ivanov, P.B.Corkum, Phys.Rev.A, v.48, 000, (1993)

12. M.Yu.Ivanov, P.B.Corkum, P.Dietrich, Laser Physics, v.3, 375, (1993)

13. Jose M. Gomez Llorente, J.PLata, Phys.Rev.A, **45**, R6958, (1992)

14. R.Bavli, H.Metiu, Phys.Rev.Lett, v.69, 1986, (1992)

15. R.Bavli, H.Metiu, Phys.Rev.A, to be published

16. F.Grossmann, T.Dittrich, P.Jung, P.Hanggi, Phys.Rev.Lett, **67**, 516, (1991)

17. Y.Dakhnovskii, H.Metiu, submitted to Phys.Rev.A.

18. M.Gavrila, J.Z.Kaminski, Phys. Rev. Lett, v.52, 613, (1984); M.Pont, N.R. Walet, M.Gavrila, C.W.McCurdy, Phys. Rev. Lett, v.61, 939, (1988); K.Kulander, K.Schafer, J.Krause, Phys. Rev. Lett, v.66, 2601, (1991);

19. K.Kulander, K.Schafer, J.Krause, Phys. Rev. Lett, v.66, 2601, (1991);

20. M.Mittleman, Phys.Rev.A, v.46, 4209, (1992)

PROPAGATION OF INTENSE ULTRASHORT
PULSES AND HARMONICS IN GASES

Keith Burnett and Stuart C. Rae

Clarendon Laboratory, Department of Physics
University of Oxford, Parks Road
Oxford OX1 3PU, United Kingdom

INTRODUCTION

Developments in the last few years have led to focal intensities achievable by ultrashort-pulse lasers exceeding 10^{18} W/cm^2. Although relatively few experiments have been performed under such extreme conditions, there is a substantial amount of experimental data on the interaction of matter, particularly gases, with slightly less intense pulses, in the range 10^{13}–10^{16} W/cm^2. As a result of this concentrated experimental effort, there is a growing need for theories which can describe the overall interaction of an ultrashort intense laser pulse with a target material, rather than only isolated aspects. This is particularly the case for harmonic generation at high intensities,[1-5] where the problems associated with atomic dynamics are coupled with those of plasma formation and propagation.

Inevitably, a unified approach is only possible using time-consuming numerical integration techniques, but the ready availibility of high-performance workstations is bringing such power within more general reach. Combined models have already been pursued to some extent, for example, in the Livermore/Saclay calculations of harmonic generation,[6-8] which use the separately-calculated atomic response of a three-dimensional atom in a two-dimensional slowly-varying-envelope propagation code. While this method works well at intensities of 10^{13}–10^{14} W/cm^2, there are difficulties in going above this regime, where the atoms suffer a significant degree of ionization. In this paper we will attempt to present a self-consistent picture of the interaction of an intense ultrashort laser pulse with a target material. We will concentrate on the case of gaseous rather than solid targets, because we wish specifically to address the high harmonic generation problem.

There are four important aspects of the interaction problem which ultimately must be dealt with in combination. These are: the high-harmonic generation process itself, spectral shifts and broadening, spatial modifications to propagation, and the creation of highly non-equilibrium plasmas. As this last point is more relevant to proposed schemes for novel recombination x-ray lasers, which lie somewhat outside the scope of this volume, we will not be discussing this particular topic further here, but instead refer the reader to the references.[9-13] In the following sections we will discuss the remaining three topics and the connections between them.

SPATIAL MODIFICATIONS

Focusing an intense ultrashort pulse into a gas generates free electrons, either by multiphoton absorption, tunneling or over-the-barrier ionization, and these electrons, through their effect on the refractive index, can influence the propagation of the pulse. Due to the intensity variation across the beam profile, any intensity dependence in the refractive index will have a lensing effect, either convergent or divergent depending on the sign of the refractive index gradient. Self-focusing can be caused by a nonlinearity in the neutral gas,[14] expulsion of electrons from regions of high intensity by the ponderomotive force,[15] or a relativistic increase in the electron mass.[16,17] Self-defocusing is caused by an enhanced ionization rate, and hence a reduction in refractive index, near the axis.[18-20] Although many theoretical calculations have predicted the formation of stable waveguide-like structures,[21-26] the experimental evidence for such structures is limited,[27] and instabilities and the formation of multiple foci are often observed.[28,29] In certain cases, plasma-induced defocusing appears to dominate, and the maximum plasma density is clamped to a value much less than critical.[30]

Here we will decribe a two-dimensional (2D) time-dependent model we have developed to describe the propagation of an intense laser pulse through a gas, which includes diffraction, tunneling ionization and self-defocusing.[20] In order to closely match experimental conditions, we assume a 1 μm, 1 ps Gaussian envelope laser pulse focused into a cell filled with hydrogen at a pressure of up to 1 atm.

We assume a cylindrically symmetrical, linearly polarized laser pulse propagating in the z-direction through an ionizing gas. Starting from the 1D wave equation,

$$\nabla^2 E - \frac{1}{c^2}\frac{\partial^2 E}{\partial t^2} = \mu_0 \frac{\partial J}{\partial t} \ , \tag{1}$$

where E is the transverse electric field and J is the plasma current, we introduce a complex slowly-varying envelope function,

$$E = u(r, z, t) \ \exp\{i(kz - \omega_0 t)\} \ , \tag{2}$$

and make the paraxial approximation, in which the second-order derivatives of u with respect to z and t are neglected. Eq. (1) thus reduces to

$$\frac{\partial u}{\partial t} = \frac{ic}{2k}\left(\frac{\partial^2 u}{\partial r^2} + \frac{1}{r}\frac{\partial u}{\partial r}\right) - c\frac{\partial u}{\partial z} - \frac{ikc}{2}\frac{N_e}{N_{cr}}u \ . \tag{3}$$

Here N_e is the free electron density, N_{cr} is the critical density, ω_0 is the laser frequency and $k = \omega_0/c$ is the vacuum wavenumber. We solve Eq. (3) numerically on a 2D spatial grid. The time-dependent electron density is obtained at each point by simultaneously solving the ionization rate equation, using a simple tunneling expression.[31,32]

It is easy to derive a simple estimate of when defocusing becomes significant. If we assume a homogeneous plasma, with a refractive index given by $n = (1 - N_e/N_{cr})^{1/2}$, then the phase change over a length L is approximately $\Delta\varphi = (\pi L/\lambda)(N_e/N_{cr})$. A phase change of $\pi/2$ corresponds to a doubling of the diffraction-limited beam divergence, and the corresponding length, the 'defocusing length', is given by $L_D = (\lambda/2)(N_{cr}/N_e)$. When the defocusing length is less than half the confocal parameter, $L_D \lesssim z_c/2$, we expect defocusing to be dominant. Thus, we expect the maximum plasma density to be roughly limited to

$$(N_e/N_{cr})_{\text{max}} \lesssim \lambda/z_c . \tag{4}$$

Figure 1 shows the maximum electron density as a function of gas pressure for $f/10$ focusing of a 10^{15} W/cm^2 pulse in atomic hydrogen. At low pressures the gas is fully ionized, but at pressures approaching 1 atm the electron density is clearly saturating. We have also performed calculations for different f-numbers, and a tighter focus allows a higher density to be achieved, as expected. The limiting density given by Eq. (4) in this case is 4×10^{18} cm^{-3}, which is about a significant factor less than that found from the full numerical calculation.

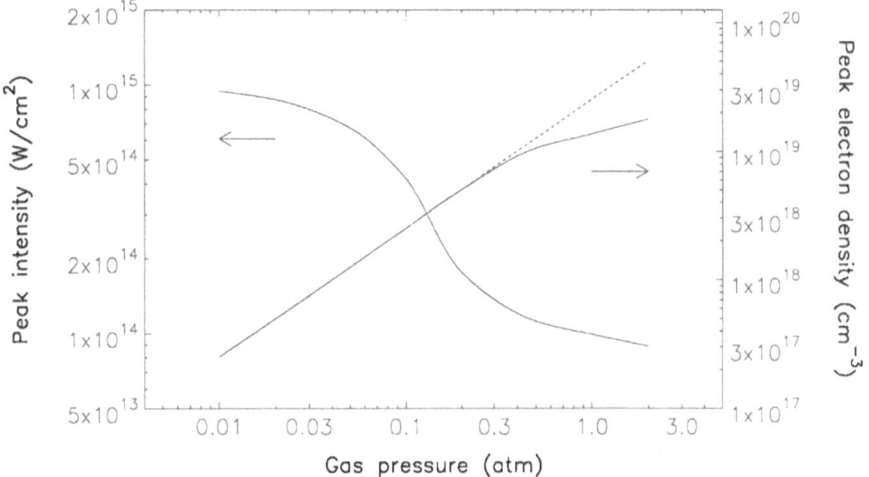

Figure 1. Peak electron density and maximum focused intensity as functions of atomic hydrogen pressure for $f/10$ focusing. Laser pulse is 1 μm, 1 ps, Gaussian-envelope with peak vacuum intensity 10^{15} W/cm^2.

In addition to limiting the peak electron density, Fig. 1 also shows how defocusing clamps the maximum intensity to a value lower than the vacuum focus intensity. At high pressures and for long confocal parameters, the intensity is reduced to the threshold for significant ionization in the gas. Clearly, for different species this limiting intensity would vary, depending on the ionization potential. What is perhaps more surprising is that even relatively low pressures can have a significant effect on the peak intensity, particularly in a long focal length geometry, such as that used in harmonic generation experiments.

SPECTRAL BLUESHIFTING

The spectral broadening and frequency shifting of intense laser pulses propagating through nonlinear materials has been studied for many years.[33] Various mechanisms can cause self-phase-modulation and frequency shifting, including, if the laser pulse is sufficiently intense to ionize the atoms, plasma formation. In this case the spectral broadening is highly asymmetric, as recombination occurs on a relatively slow timescale, and only a blueshifted wing is generated. This effect was predicted by Bloembergen[34] in the early 1970s, and early experiments were performed by Yablonovitch,[35] using nanosecond CO_2 laser pulses. In recent years, similar experiments have been performed by Downer and co-workers[36] at the University of Texas at Austin, using 100 fs dye laser pulses, by LeBlanc and Sauerbrey[37] at Rice University using 400 fs KrF laser pulses, and by Hutchinson and co-workers[38] at Imperial College, London, using a 1 ps, 1 μm Nd:YLF/Nd:glass laser.

In the simple case of a homogeneous plasma and cw laser, for an electron density much less than critical, the wavelength shift of a monochromatic beam due to plasma formation is given by[39]

$$\Delta\lambda = -\frac{e^2 N_i L \lambda_0^3}{8\pi^2 \epsilon_0 m_e c^3} \frac{dZ}{dt} \ .$$
(5)

Here N_i is the ion density, L is the interaction length and Z is the degree of ionization. While this equation is useful in establishing scaling rules, the complete spectrum can only be obtained by solving Maxwell's equations for the propagation of the pulse, in parallel with coupled ionization rate equations to determine dZ/dt at each point.

In 1D, we can numerically solve the wave equation, Eq. (1), for the propagation of the laser pulse through the medium, and the plasma current can simultaneously be obtained from a set of coupled ionization rate equations and the equation of motion for the free electrons in the laser field. Other mechanisms, such as collisional ionization, can also be included. This was the approach adopted by the present authors and others in interpreting experiments by Downer et al.,[39,40] and LeBlanc and Sauerbrey.[37]

A propagation model in 1D is necessarily limited because it cannot include effects due to the focusing geometry. The self-phase-modulation which results from plasma formation affects both the spatial and spectral aspects of the pulse, and the coupling between these effects is of particular interest. In 2D, using the slowly-varying-envelope model outlined in the previous section, we can investigate this coupling directly. In this case, the spectrum can be written as a Fourier transform,[41]

$$F(\omega) = \tfrac{1}{2\pi} \int_{-\infty}^{\infty} u(t)\, e^{i(\omega-\omega_0)t}\, dt \ .$$
(6)

Here $u(t)$ is the envelope function defined in Eq. (2), which contains information about both the amplitude of the pulse envelope, and the phase shift due to propagation and the presence of free electrons. Given knowledge of $u(r,t)$ for some z well beyond the focal region, it is possible to construct a two-dimensional map of the blueshifted pulse, showing frequency against radial distance.[42] Such a map, an example of which is shown in Fig. 2, reveals a wealth of detail, much of which is lost in a typical experiment where the entire beam is collected and analysed.

In the case of high harmonic generation in a strongly-ionizing gas, each of the harmonics will show some degree of spatial and spectral modification, and careful measurement of these effects may be able to indicate the plasma conditions under which the harmonics were generated. Blueshifting of high-order harmonics has already been reported by a group at Stanford.[43]

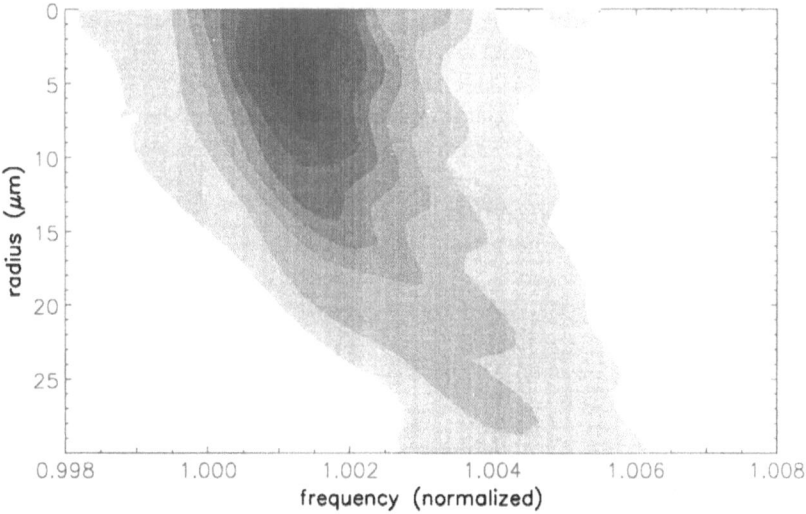

Figure 2. Blueshifting of a 1 μm, 1 ps laser pulse focused at $f/10$ to a vacuum intensity of 10^{15} W/cm^2 in 1 atm of atomic hydrogen. The spectrum is calculated for a point 300 μm beyond the vacuum focus, where the degree of ionization is negligible.

HIGH HARMONIC GENERATION

The task of calculating harmonic spectra divides naturally into two parts: the single-atom response, and phase matching effects during propagation. At the lower end of the intensity range (10^{13}–10^{14} W/cm^2), L'Huillier et al.[6-8] have carried out detailed simulations of harmonic spectra by combining atomic calculations for a 3D atom with a 2D slowly-varying-envelope propagation code. Here we take a different approach, and address the self consistent problem, in which each atom in the plasma responds differently to the combined fundamental and harmonic fields as they propagate through the interaction region. This goes outside the slowly-varying-envelope approximation, and allows us to investigate both very short pulses, and also situations in which there is strong ionization.

Performing self-consistent calculations is extremely demanding computationally, and we are thus limited to propagation in 1D and, at best, a 1D model for the atomic dynamics. Nevertheless, such calculations are extremely useful for studying some of the unresolved issues in high harmonic generation, in particular, the effect of the large phase mismatch which results from significant free electron formation, the role of ions in the generation of very high harmonics, and the width and structure of the harmonic peaks.

Tunneling Harmonics

Very recently Akiyama et al. have reported harmonic generation from rare-gas-like ions using an ultrashort-pulse KrF laser.[44] Here we describe calculations of harmonic generation in ions using a simple model which allows the direct investigation of harmonic propagation through an ionizing medium.[45] We do not use a full multi-level description for the ions, but instead assume that the tunneling is the dominant ionization mechanism. Although this neglects all the dynamical behaviour of the partially-bound electron, tunneling ionization rates generally agree well with experiment.[46–48]

A tunneling model for harmonic generation was initially proposed by Brunel.[49] In this limit the ionization rate can be calculated using a dc tunneling formula,[31,32] which is a highly nonlinear function of the electric field. Thus, the ionization is strongly peaked near the peaks of the laser cycles. This non-uniform ionization rate can generate a large series of odd harmonics, although there is no plateau: the relative intensity decreases monotonically with the harmonic order.

For a single atom, the harmonic spectrum can be obtained through a Fourier transform of the dipole acceleration.[50] In the tunneling approximation, the acceleration is given by

$$\ddot{d}(t) = p_i(t)\,\mathcal{E}(t)\,\sin\omega_0 t \ , \tag{7}$$

where $\mathcal{E}(t)$ is the electric field envelope of the laser pulse and $p_i(t)$ is the ionization probability.

In order to investigate phase matching effects, it is necessary to incorporate the single-atom tunneling current into a full time-dependent propagation model. We will use a 1D propagation model very similar to that used previously to describe plasma-induced spectral blueshifting.[39] Unfortunately, such a model does not allow us to simultaneously investigate effects due to plasma-induced defocusing.

By neglecting the geometric effects, we are effectively assuming that the phase matching is dominated by dispersion due to plasma formation. This is a reasonable approximation if the intensity-dependence of the harmonics is fairly small, as is the case with the harmonics generated by tunneling ionization.

Computational restrictions mean that in order to simulate realistic conditions we must use a relatively short propagation distance (typically 10–20 wavelengths) and a high density to achieve a similar density-length product, and hence the same degree of dispersion, as the corresponding experiments. This scaling is valid as long as the density in the model remains significantly below critical density.

The presence of free electrons in the plasma results in a difference in the refractive indices for the fundamental, n, and the harmonic field, n_q. Assuming that $\omega_p \ll \omega_q$ (plasma density much less than critical), the phase mismatch introduced by this dispersion can be written

$$\delta\phi = q\omega_0 \frac{L}{c}(n_q - n) \simeq \frac{L\omega_p^2}{2\omega_0 c}\frac{q^2 - 1}{q} \ , \tag{8}$$

where L is the length of plasma and ω_p is the plasma frequency. For $q \gg 1$, the phase mismatch depends linearly on the harmonic order. For an infinitely narrow fundamental line, the phase mismatch leads to a reduction in the efficiency of harmonic generation by a factor which is given by the sinc^2 function. For a line with finite width, the phase matching factor must be integrated across the lineshape. In the limit $\delta\phi \gg 2\pi$, the phase matching factor varies as the inverse square of the density and thus cancels the coherent gain, leading to complete saturation of the harmonic efficiency.

By calculating the harmonic intensities as functions of density, this phase matching behaviour can be clearly shown. Figure 3 shows the results for several different harmonic orders in a 10-wavelength Li$^+$ plasma, for an 80-cycle Gaussian envelope pulse of wavelength 250 nm and peak intensity 5×10^{16} W/cm^2. The straight lines on the graph have a slope of 2, which fits the data perfectly at low densities. As the phase mismatch increases, the conversion efficiency falls below this ideal line, and the characteristic sinc^2 oscillations are observed.

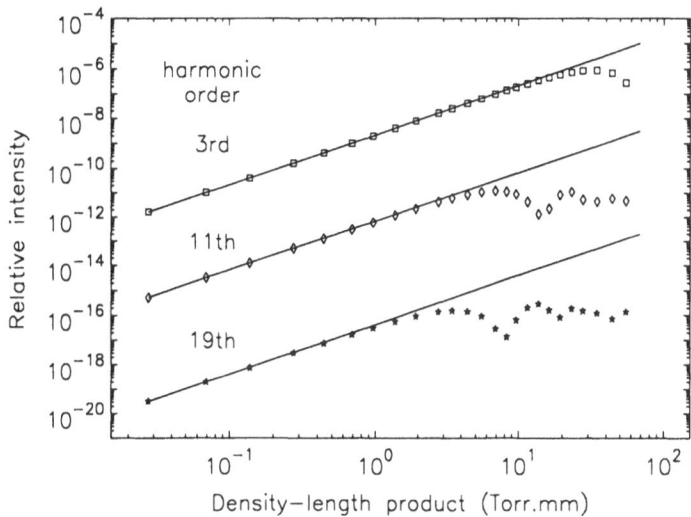

Figure 3. Relative efficiency as a function of density-length product for the 3rd, 11th and 19th harmonics, generated by tunneling ionization in a Li^+ plasma. See text for details.

In this example, even for a fully-ionized plasma, the effect of phase matching on the lower-order harmonics is fairly small, given typical experimental pressures of $\lesssim 15$ Torr. However, for higher harmonic orders or longer fundamental wavelengths, the problem can be far more serious. The presence of plasma dispersion means that such harmonics will not propagate efficiently in an ionized medium. Thus, the crucial question for high harmonic generation using an intense laser pulse is: when are the harmonics produced relative to the threshold for ionization? The investigation of this point requires a more detailed model for the atomic dynamics.

Extremely High-Order Harmonic Generation

The simple model presented in the previous section, whilst strictly accurate in the high-intensity, low-frequency limit, fails to include any of the structure of the atom, and consequently does not reproduce the almost ubiquitous feature of harmonic generation: a plateau of middle-order harmonics followed by a steep drop in conversion efficiency. The next step in improving the description of the atomic dynamics is to incorporate the solutions to the Schrödinger equation for a 1D model atom.

Such a model has been extensively studied in the past,[51-53] although the harmonics generated by the 1D model atom have never been investigated at intensities where there is significant or complete ionization. A 3D code is currently not capable of running in this regime, for two reasons: it is restricted to a very limited number of cycles, so realistic pulse shapes cannot be described; and the calculation must be performed in a frame of reference where the wavefunction oscillates in the laser field, leading to problems with reflections at the edge of the computational grid, particularly at high intensities.[54,55] In contrast, the drastically reduced number of grid points required by the 1D model means that pulses of hundreds of cycles can be studied, and by performing the integration in the Kramers frame (moving with the oscillation), the wavefunction which hits the edge of the grid is freely-diffusing, and thus is easier to absorb without causing spurious reflections.

In our approach, we numerically solve the Schrödinger equation in the form

$$i\frac{\partial \psi}{\partial t}(x,t) = -\frac{1}{2}\frac{\partial^2 \psi}{\partial x^2}(x,t) + V(x + \alpha(t))\psi(x,t) \,, \qquad (9)$$

where the potential is given by

$$V(x) = \frac{-1}{\sqrt{1 + x^2}} \,, \qquad (10)$$

and $\alpha(t)$ is the displacement caused by the laser field. The potential is an approximation to a true Coulomb potential: it has the correct asymptotic dependence and hence a good Rydberg series, but avoids the singularity at the origin. There are a large number of bound states, and the ground state has binding energy of 18.2 eV. By suitably scaling the photon energy and peak electric field of the laser, the model potential can be scaled to match any desired 'real' atom (though of course the detailed level structure, which determines the resonance behaviour, will not directly correspond).

We initially solve the time-independent Schrödinger equation to obtain the ground state wavefunction, then evolve this under the influence of the laser field. In the Kramers frame, the electron quiver motion is transferred to the potential, which sweeps back and forth through the wavefunction. The spectrum is obtained from the Fourier transform of the dipole acceleration.

Figure 4 shows a calculated harmonic spectrum in the strongly ionizing regime, for a 1.053 μm, 1 ps fwhm laser pulse with a sine-squared electric field envelope, and the parameters scaled for a neon atom. The peak intensity of the pulse is 10^{15} W/cm^2, and the atom actually ionizes during the rising edge, as shown by the inset figure.

The details of such spectra will be more fully explained in a forthcoming paper,[56] but it appears that the long tail of extremely high-order harmonics are actually gen-

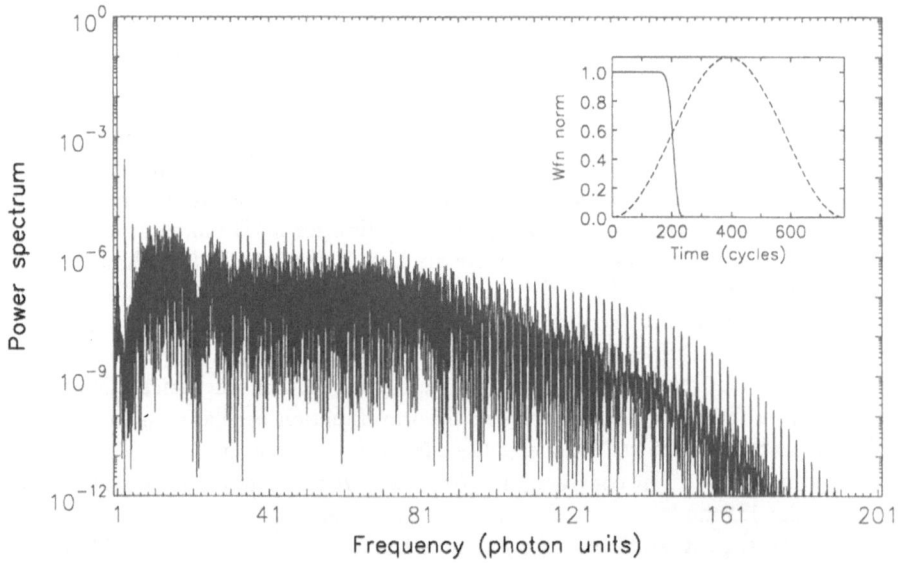

Figure 4. Single-atom harmonic spectrum for a model neon atom, and a 1.053 μm, 1 ps fwhm laser pulse with a sine-squared electric field envelope, at a peak intensity of 10^{15} W/cm^2.

erated very close to the point of ionization. In the previous section we showed that a simple tunneling atom could be incorporated into a 1D propagation code, and that in this case ionization had a serious effect on the propagation of the high harmonics. We are now able to include a much improved atomic model into the same propagation code. Although this a 1D+1D model could not be expected to quantitatively reproduce experimental results, it would be ideal for investigating the conditions required to optimize the production of high-order harmonic radiation.

CONCLUSIONS

We are working towards the development of a fully self-consistent model for the interaction of an intense ultrashort laser pulse with a gaseous target. The model needs to include a realistic description of the atomic dynamics, together with ideally a multi-dimensional, non-slowly-varying-envelope description of propagation. In this paper we have outlined the basic ingredients of such a model, and discussed the progress which we have achieved so far towards a 1D+1D implementation. A comprehensive interaction model of this sort would be an invaluable tool for investigating many aspects of atomic behaviour in strong laser fields.

ACKNOWLEDGEMENTS

This work is part of a program supported by the Science and Engineering Research Council.

REFERENCES

1. A. L'Huillier, L. A. Lompré, G. Mainfray, and C. Manus, *in:* "Atoms in Intense Laser Fields," M. Gavrila, ed., Academic, Boston (1992).
2. P. Balcou, C. Cornaggia, A. S. L. Gomes. L. A. Lompré, and A. L'Huillier, *J. Phys. B* 25:4467 (1992).
3. J. K. Crane, M. D. Perry, S. Herman, and R. W. Falcone, *Opt. Lett.* 17:1256 (1992).
4. K. Miyazaki and H. Sakai, *J. Phys. B* 25:L83 (1992).
5. N. Sarukura, K. Hata, T. Adachi, R. Nodomi, M. Watanabe, and S. Watanabe, *Phys. Rev. A* 43:1669 (1991).
6. A. L'Huillier, K. J. Schafer, and K. C. Kulander, *Phys. Rev. Lett.* 66:2200 (1991).
7. A. L'Huillier, K. J. Schafer, and K. C. Kulander, *J. Phys. B* 24:3315 (1991).
8. A. L'Huillier, P. Balcou, S. Candel, K. J. Schafer, and K. C. Kulander, *Phys. Rev. A* 46:2778 (1992).
9. N. H. Burnett and P. B. Corkum, *J. Opt. Soc. Am. B* 6:1195 (1989).
10. N. H. Burnett and G. D. Enright, *IEEE J. Quantum Electron.* 26:1797 (1990).
11. P. Amendt, D. C. Eder and S. C. Wilks, *Phys. Rev. Lett.* 66:2589 (1991); D. C. Eder, P. Amendt and S. C. Wilks, *Phys. Rev. A* 45:6761 (1992).
12. B. M. Penetrante and J. N. Bardsley, *Phys. Rev. A* 43:3100 (1991).
13. S. C. Rae and K. Burnett, *Phys. Rev. A* 46:2077 (1992).
14. P. L. Kelley, *Phys. Rev. Lett.* 15:1005 (1965).
15. H. Hora, *Phys. Fluids* 12:182 (1969).
16. C. E. Max, J. Arons, and A. B. Langdon, *Phys. Rev. Lett.* 33:209 (1974).
17. W. B. Mori, C. Joshi, J. M. Dawson, D. W. Forslund and J. M. Kindel, *Phys. Rev. Lett.* 60:1298 (1988).
18. R. Rankin, C. E. Capjack, N. H. Burnett and P. B. Corkum, *Opt. Lett.* 16:835 (1991).
19. W. P. Leemans, C. E. Clayton, W. B. Mori, K. A. Marsh, A. Dyson and C. Joshi, *Phys. Rev. Lett.* 68:321 (1992); W. P. Leemans, C. E. Clayton, W. B. Mori, K. A. Marsh, P. K. Kaw, A. Dyson, C. Joshi and J. M. Wallace, *Phys. Rev. A* 46:1091 (1992).

20. S. C. Rae, accepted by *Opt. Commun.*
21. G.-Z. Sun, E. Ott, Y. C. Lee and P. Guzdar, *Phys. Fluids* 30:526 (1987).
22. J. C. Solem, T. S. Luk, K. Boyer and C. K. Rhodes, *IEEE J. Quantum Electron.* 25:2423 (1989).
23. T. Kurki-Suonio, P. J. Morrison and T. Tajima, *Phys. Rev. A* 40:3230 (1989).
24. A. B. Borisov, A. V. Borovskiy, V. V. Korobkin, A. M. Prokhorov, C. K. Rhodes and
    O. B. Shiryaev, *Phys. Rev. Lett.* 65:1753 (1990); A. B. Borisov, A. V. Borovskiy,
    O. B. Shiryaev, V. V. Korobkin, A. M. Prokhorov, J. C. Solem, T. S. Luk, K. Boyer and
    C. K. Rhodes, *Phys. Rev. A* 45:5830 (1992).
25. P. Sprangle, E. Esarey and A. Ting, *Phys. Rev. A* 41:4463 (1990).
26. P. Sprangle, E. Esarey, J. Krall, and G. Joyce, *Phys. Rev. Lett.* 69:2200 (1992).
27. Y. M. Li, J. N. Broughton, R. Fedosejevs, and T. Tomie, *Opt. Commun.* 93:366 (1992).
28. A. B. Borisov, A. V. Borovskiy, V. V. Korobkin, A. M. Prokhorov, O. B. Shiryaev, X. M. Shi,
    T. S. Luk, A. McPherson, J. C. Solem, K. Boyer and C. K. Rhodes, *Phys. Rev. Lett.*
    68:2309 (1992);
29. X. Liu and D. Umstadter, to be published.
30. T. Auguste, P. Monot, L.-A. Lompré, G. Mainfray and C. Manus, *Optics Comm.* 89:145 (1992);
    P. Monot, T. Auguste, L.-A. Lompré, G. Mainfray and C. Manus, *J. Opt. Soc. Am. B* 9:1579
    (1992).
31. A. M. Perelomov, V. S. Popov, and M. V. Terentev, *Sov. Phys. JETP* 23:924 (1966).
32. M. V. Ammosov, N. B. Delone, and V. P. Krainov, *Sov. Phys. JETP* 64:1191 (1987).
33. R. R. Alfano, ed. "The Supercontinuum Laser Source," Springer Verlag, Berlin (1989).
34. N. Bloembergen, *Opt. Commun.* 8:285 (1973).
35. E. Yablonovitch, *Phys. Rev. Lett.* 31:877 (1973); *Phys. Rev. Lett.* 32:1101 (1974); *Phys. Rev. A*
    10:1888 (1974).
36. W. M. Wood, G. Focht, and M. C. Downer, *Opt. Lett.* 13:984 (1988); M. C. Downer,
    W. M. Wood, and J. I. Trisnadi, *Phys. Rev. Lett.* 65:2832 (1990); W. M. Wood,
    C. W. Siders, and M. C. Downer, *Phys. Rev. Lett.* 67:3523 (1991).
37. S. P. LeBlanc, R. Sauerbrey, S. C. Rae, and K. Burnett, submitted to *J. Opt. Soc. Am. B.*
38. M. H. R. Hutchinson, private communication.
39. S. C. Rae and K. Burnett, *Phys. Rev. A* 46:1084 (1992).
40. B. M. Penetrante, J. N. Bardsley, W. M. Wood, C. W. Siders, and M. C. Downer,
    J. Opt. Soc. Am. B 9:2032 (1992).
41. R. H. Stolen and C. Lin, *Phys. Rev. A* 17:1448 (1978).
42. S. C. Rae and K. Burnett, to be published.
43. J. J. Macklin, J. D. Kmetec, C. L. Gordon, and S. E. Harris, to be published.
44. Y. Akiyama, K. Midorikawa, Y. Matsunawa, Y. Nagata, M. Obara, H. Tashiro, and K. Toyoda,
    Phys. Rev. Lett. 69:2176 (1992).
45. S. C. Rae and K. Burnett, submitted to *J. Phys. B.*
46. G. Gibson, T. S. Luk, and C. K. Rhodes, *Phys. Rev. A* 41:5049 (1990).
47. S. Augst, D. D. Meyerhofer, D. Strickland, and S. L. Chin, *J. Opt. Soc. Am. B* 8:858 (1991).
48. T. Auguste, P. Monot, L. A. Lompré, G. Mainfray, and C. Manus, *J. Phys. B* 25:4181 (1992).
49. F. Brunel, *J. Opt. Soc. Am. B* 7:521 (1990).
50. K. Burnett, V. C. Reed, J. Cooper, and P. L. Knight, *Phys. Rev. A* 45:3347 (1992).
51. J. H. Eberly, Q. Su, and J. Javanainen, *Phys. Rev. Lett.* 62:881 (1989); *J. Opt. Soc. Am. B*
    6:1289 (1989).
52. J. H. Eberly, Q. Su, J. Javanainen, K. C. Kulander, B. W. Shore, and L. Roso-Franco,
    J. Mod. Opt. 36:829 (1989).
53. V. C. Reed and K. Burnett, *Phys. Rev. A* 42:3152 (1990); *Phys. Rev. A* 43:6217 (1991);
    Phys. Rev. A 46:424 (1992).
54. J. L. Krause, K. J. Schafer, and K. C. Kulander, *Phys. Rev. A* 45:4998 (1992).
55. J. L. Krause, K. J. Schafer and K. C. Kulander, *Phys. Rev. Lett.* 68:3535 (1992).
56. S. C. Rae and K. Burnett, to be published.

MULTIPLE HARMONIC RADIATION FROM METAL SURFACES INDUCED BY STRONG LASER PULSES

Gy. Farkas

Hungarian Academy of Sciences
Research Institute for Solid State Physics
Department of Laser Physics
H-1525 Budapest, Hungary

INTRODUCTION

The demonstration of the second harmonic radiation from metal (solid) surfaces played a very important role in the first verification of the fundamental laws of nonlinear optics[1] already in the early years of laser physics. Although since that time both fundamental and practical physics of the surface harmonic generation have been developed, it remained restricted to the second (rarely to the third) order case[2] only. Correspondingly the theoretical efforts in its interpretation were concentrated and limited to the use of the standard second order perturbation calculation[3] methods until recent times.

The last decade's researches have revealed[4] that the general phenomena of the nonlinear polarization of the single atoms are governed by the fundamental laws of the intense field QED, therefore neither the low nor the high order perturbative approaches are capable for their description[5]. For their correct understanding in the single atom case the use of nonperturbative theories are required leading to the experimentally verified well-known extremely wide spectrum consisting of a very high number of odd harmonics with a wide plateau region terminated with a very sharp cutoff.

Considering the general validity of these fundamental QED laws, the obvious question arises: in what form manifest themselves the general laws of the multiple harmonic generation phenomenon in the case of a solid surface.

When treating this latter problem, however, we have to take into account the differences between the case of single atoms and the use of an ensemble of atoms of a solid surface, respectively.

1. PECULIARITIES OF THE SURFACE OF A SOLID CONSISTING OF AN ENSEMBLE OF ATOMS.

The expected harmonic orders in the case of a nonlinear system can be determined from the usual expansion of the polarization P with respect to the incoming E field

$$P = \chi^{(1)} : E + \chi^{(2)} : EE + ...,$$

and from the values of the $\chi^{(n)}$ susceptibilities.

From general symmetry considerations several evident consequences immediately follow.

Super-Intense Laser-Atom Physics, Edited by
B. Piraux *et al.*, Plenum Press, New York, 1993

For the case of single atoms $\chi^{(n)}$-s are scalars and the pure dipole approximation can be applied which results in the appearance of odd harmonics only. For solids $\chi^{(n)}$-s are in general high rank tensors; therefore using multipole expansion the even orders also may give contributions, therefore the appearance of both odd and even harmonics are expected, as we can see easily below, e. g., in the case of second order. In this case $P_2 = \chi^{(2)}:EE$.

For atoms, being $\chi^{(2)}$ a scalar, the symmetry does not remain, if the sign of P_2 and E is changed simultaneously: the second order harmonic generation is forbidden; the same considerations are true therefore for each higher even order that all vanish. A solid, however, may have both isotropic or anisotropic structure, respectively. Therefore we have to consider separately these two cases. In the first case for a system with lack of inversion symmetry $\chi^{(2)}$ has a vectorial character and

$$P_2 = \chi^{(2)} \text{ (vectorial)}:EE \neq 0,$$

therefore second harmonic may appear. In the second case for a system with inversion symmetry when $\chi^{(2)}$ is a scalar

$$P_2 = \chi^{(2)} \text{ (scalar)}:EE = 0 \quad,$$

consequently no second harmonic is expected, as is the case of atoms.

However, experiments performed for surfaces always do show 2ω light. The explanation is connected with the contribution of the higher order multipole (the electric quadrupole and the magnetic dipole) terms, as we can easily illustrate it in the following qualitative argumentation.

Consider the Sommerfeld-type "free" electrons of a metal of n electron density in a semi-infinite $V(z)$ single step potential. We illuminate them with an

$$E_1 \propto H_1 \propto e^{i\omega t}$$

incoming e. m. field. The force acting on the electrons is

$$\vec{F} = e\vec{E}_1 - \frac{e}{c}\left(\vec{v} \times \vec{H}_1\right) .$$

The induced charge density ρ, the velocity $\vec{v}$ and the current

$$\vec{j} = \rho\vec{v} = \partial\vec{P}/\partial t$$

all may be expanded into Fourier-series as

$$\sim [\quad] e^{-i\omega t} + [\quad] e^{-i2\omega t} + [\quad] e^{-i3\omega t} + \dots$$

Retaining only the "2ω" terms, we have for $\vec{P}_2$

$$\vec{P}_2 = ine^3/\left(4m^2c\omega^3\right)\left(\vec{E}_1 \times \vec{H}_1\right) + e/\left(8\pi m\omega^2\right)\vec{E}_1\nabla\vec{E}_1 \quad .$$

It can be seen that symmetry conditions are satisfied, considering that both terms on the right side represent 3 polar vectors, being E_1 and (symbolically) ∇ polar vectors and H_1 correspond to 2 polar vectors.

We recognize the magnetic dipole contribution in the first term and the electric quadrupole effect in the second term. Both of them can radiate at the surface only, being the first a longitudinal dipole, while the second surface term has nonzero very high gradient value at the surface only:

$$\nabla\vec{E}_1 = 4\pi\rho = (1-\varepsilon)E_{1z}\delta(z) \neq 0,$$

where the metal (of dielectric constant ε) is assumed to occupy the half space $z \geq 0$. These considerations are valid for <u>any</u> surface electron "layer": ion core electrons also may contribute to the harmonic generation, when the incoming light incidence is a grazing one, i.e., E_1 is perpendicular to the surface (p-polarization)[6].

2. PREDICTIONS OF THE EARLY SEMICLASSICAL THEORIES

These qualitative considerations, outlined in the previous chapter, are formulated by simple, but correct second order perturbation theories which calculate the laser induced surface harmonics, coherently integrating the second order radiation contribution of all electrons participating in the interaction process. The general result of these calculations[3] predicts that the second order intensity $I_{2\omega}$ induced by the incoming E_1 laser field strength depends on the angle of incidence Θ and on the polarization angle Φ in the following form (at least in the cases of grazing incidence[3,8]):

$$I_{2\omega} \propto \left(E_\perp\right)^4 \sin^4\Phi\cos^4\Theta \ .$$

This relation has been proven to be quite general and valid in practice for different surfaces of metallic materials both with and without inversion symmetry when p-polarization and grazing incidence was used in the course of a series of earlier and recent investigations[7,8].

We conclude therefore that for the determination of the multiharmonic fields for solid surface it is necessary to determine first the

$$\vec{j} = \partial\vec{P}/\partial t$$

source term, but not only in dipole, but in multipole approximation. In the classical phenomenology this source term is

$$\vec{j} = \frac{\partial\vec{P}}{\partial t} + \frac{\partial\vec{P}_{NL}}{\partial t} + c\left(\vec{\nabla}\times\vec{M}\right) - \frac{\partial}{\partial t}\left(\vec{\nabla}\times\vec{Q}\right) + ...$$

where $\vec{P}$ is the linear, $\vec{P}_{NL}$ is the nonlinear polarization, $\nabla\vec{Q}$ is the electric quadrupole, $\left(\vec{\nabla}\times\vec{M}\right)$ is the magnetic dipole term.

Similarly, in the quantum mechanical description we have to include besides the usual $\hat{P} = -e\hat{x}$ electric dipole operator, the $\hat{Q} = -(1/2)\,\hat{x}\hat{x}$ electric quadrupole and the $\hat{M} = -(e/mc)(\hat{x}\times\hat{p})$ magnetic dipole operator in the Hamiltonian in the usual form[1]:

$$\hat{\mathcal{H}} = \frac{\hat{p}^2}{2m} + v\left(\hat{\vec{r}}\right) - \hat{\vec{P}}\cdot\hat{\vec{E}} - \hat{\vec{M}}\cdot\hat{\vec{H}} - \hat{Q}\!:\!\nabla\hat{\vec{E}} + ...$$

Then with the use of the Maxwell's equations

$$\nabla\times\vec{E} = -\frac{1}{c}\frac{\partial\vec{H}}{\partial t}$$

$$\nabla\times\vec{H} = \frac{1}{c}\frac{\partial\vec{E}}{\partial t} + \frac{4\pi}{c}\frac{\partial}{\partial t}\left(\vec{P} + \vec{P}_{NL} - \nabla\vec{Q} + ...\right) + 4\pi\left(\nabla\times\vec{M}\right) + ...$$

we have to calculate the

$$\nabla\times\nabla\times\vec{E} + \frac{1}{c^2}\frac{\partial^2\left(\varepsilon\vec{E}\right)}{\partial t^2} = -\frac{4\pi}{c^2}\frac{\partial^2\vec{P}^{NL}}{\partial t^2}$$

propagation equation which is essentially the expression of the Huygens-principle for the harmonic light propagation[1].

3. RECENT QUANTUM THEORIES FOR ARBITRARY ORDER HARMONIC GENERATION FROM METAL

As we mentioned, the second (third) order semiclassical perturbation calculation describes quite satisfactorily the experimental result of second (third) order harmonic generation from surfaces in the low (MW/cm^2) laser intensity ranges.

Our general question is, however, to investigate the multiple harmonic generation from metal surfaces, similarly to the case of a single atom, at those higher laser intensities, at which the perturbation approach becomes questionable and the appearance of all harmonic orders may be expected.

The nonperturbative generalization of the problem in question up to arbitrary high intensity and harmonic order has been published only recently[9]. This work[9] tackles the problem analytically as a single -particle-quantum-mechanical process by using a semiinfinite rectangular single step asymmetric model potential (Sommerfeld-model) and therefore the dipole approximation is allowed in this case. Consequently it can not be applied directly to a realistic experiment, considering that a single particle treatment can not furnish macroscopic quantities as, e.g., the correct (specular) harmonic emission angle, which has to be determined by the collective coherent contributions of all electrons. Nevertheless, it does predict the appearance of both even and odd harmonics as the consequence of the asymmetric potential used. This calculation determines the radiation field as the field of dipole oscillation of the single electron investigated, the oscillation is perpendicular to the surface. This resulting field corresponds to the well known dipole radiation distribution, i.e., most of its energy is radiated roughly parallel to the metal surface, symmetrically around the dipole axis. In this work the multiharmonic yield is presented even for the case, when the perturbation parameter x >1. For lower orders it reobtains the former semiclassical results: the dependences of angle incidence Θ, polarization Φ, etc. The formula obtained for the n-th harmonic yield (i.e. the number of emitted harmonic photons divided by the incident intensity) for E incoming amplitude is

$$Y_n \propto \frac{n}{24\pi^3}\left(\frac{e^2}{\hbar c}\right)^4 \left(\frac{eE}{2mc^2\omega^2}\right)^{2n-2} \sin^{2n}\Theta \sin^{2n}\Phi \quad ; \qquad n=1,2,3...$$

It predicts all harmonics. The authors stress that it is valid only for the case when the perturbation parameter x <1. To obtain the whole harmonic beams reflected coherently and specularly from the surface, coherent summation of this single particle yield over all participating electrons and the use of the propagation equations presented above is necessary.

A partly familiar one dimensional problem is solved, however already exactly in work[10], where the harmonics spectrum is determined from the electron scattering on a rectangular (i.e. symmetric) potential well. Therefore the obtained nonperturbative computational results furnishes the total integrated power spectrum of both even and odd harmonic light.

The recent work[11] determined the harmonic spectrum from the crystalline silicon bands by nonperturbative computational methods. By nature only odd harmonics were obtained up to n=9 order. The possibility of the use of 10^{11} W/cm^2 intensities in femtosecond laser pulses for metals is discussed and suggested.

Therefore we concluded from these predictions that it was reasonable to perform experiments to find the eventual analogy between the multiple harmonic generation in gas atoms and metal surfaces, respectively.

4. EXPERIMENTAL RESULTS

In the experiment to be described high order harmonics were observed by us (detailed description can be found in Ref. 12.) from a gold surface; gold was chosen as an appropriate metal to realize the single-step rectangular potential case.

The experimental arrangement is shown in Fig. 1. Linearly polarized, single, bandwidth limited pulses were selected from the leading part of an actively mode-locked pulse train of a Nd:YAG or Nd:phosphate glass laser (L) (pulse duration: ~35 psec, intensity: ~ 5 GW/cm^2 in the cross section of the beam, repetition rate: 10 Hz, wavelength: 1.064 µm). The light beam was directed through the slit (D) and a polarization rotator (R) onto the appropriately polished and treated[13] surface of a 2 mm thick polycrystalline gold sample (Au) kept under ~10^{-7} torr in a vacuum vessel (V). A grazing incidence of 70° was used - at which the laser intensity was roughly $5\cos(70°){\sim}2$ GW/cm^2 on the surface - to prevent the heating and plasma effects and to favour the predicted[3,6] surface excitations. (The sample was prepared in the same traditional way, as in all our previous multiphoton electron emission experiments[13-19].) The reflected light passed through a quartz prism (P) which deviated and separated in direction the different harmonics. The first one of them, i.e., the high intensity fundamental laser light was blocked by the stop (S). Then by turning the prism (P) in the appropriate direction, the further harmonic light components entered the slit (D) of a monochromator (M) (capable to measure down to 200 nm) one by one and were detected by a photomultiplier (PM). The spectrum was taken by the wavelength scan of the monochromator M simultaneously rotating the prism (P) into the appropriate direction. The most crucial disturbing effects which may occur in these type of experiments[2] had been controlled and taken into account. For example, the possible hazard lights which might be induced in elements other than the gold surface (diaphragms, windows, prism, etc.) were filtered and completely eliminated by inserting different density color filters (F1, F2, F3) in appropriate combinations between the different elements of the set up. A further similar control was made realizing the configuration, in which the light beam passed the whole set up without touching the gold surface. The signal of the photomultiplier was analyzed by a data processing system (DA) containing a boxcar averager and an acquisition and detecting set up.

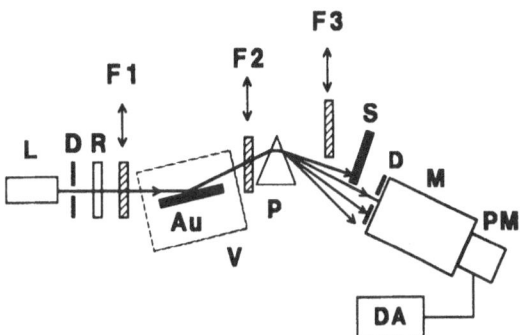

Figure 1. Experimental arrangement. L: passive mode-locked Nd:YAG or Nd:phosphate glass laser; D: diaphragms; R: polarization rotator, λ/2 plate; F1,2,3: spectral filters; Au: polished gold target; V: vacuum vessel; P: quartz prism; S: beam stop for blocking the fundamental; M: monochromator; PM: photomultiplier; DA: data acquisition system

In the following we summarize the experimental results. Fig. 2 shows, that the 2nd, 3rd, 4th and 5th harmonics, which were emitted in the specularly reflected laser beam in solid angles roughly equal to the 0.7 mrad laser divergency, appeared at the correct expected harmonic wavelengths, i.e., at 533 nm, 355 nm, 266 nm and 213 nm within the ~1 nm instrumental resolution of the monochromator. Special care was taken during the whole experiment to avoid any plasma formation on the surface by keeping the laser intensity value lower than 10 GW/cm^2 in the beam cross section, i.e., I<3 GW/cm^2 on the surface.

As can be seen in Fig. 2, the 2nd and the 3rd harmonic lines appeared in a clear form, without any additional lines or considerable background in this part (700-300 nm) of the spectrum. Besides the harmonic lines, with less reproducibility, sharp additional weaker lines appeared also in time coincidence with the laser pulse, however, only in this latter shorter wavelength spectral range (A and B in Fig. 2). They correspond to transitions of the neutral Au atom, as it was identified from table data[20]. In contrast to the similar appearance of additional lines in noble gas experiments which are performed always in inherent plasma background[4,5], here any plasma process can be excluded as a possible origin.

Our experimental set up being roughly the same as used in our multiphoton electron emission experiments[17-19], the absence of plasma and any surface damage might be routinely checked (as we do it always[13-18]) by observing the multiphoton photocurrent, which is sensitive in high order for any disturbing effects. In any case, the appearance of the weak A, B, lines is irrelevant from the point of view of our original objective.

The observation of higher than fifth order was not possible with the monochromator available in this experiment.

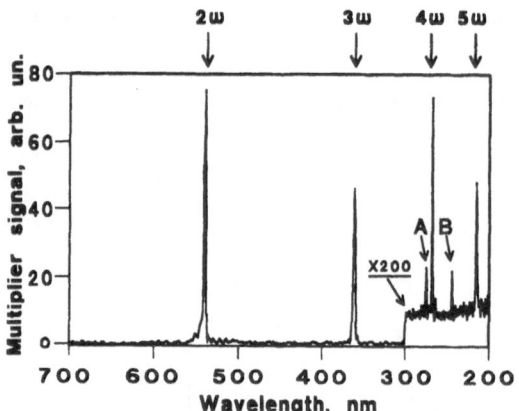

Figure 2. Spectrum of the light emitted from the gold surface illuminated by picosecond Nd:YAG laser pulses at $I \sim 5$ GW/cm^2 intensity. Shown are the 2nd, 3rd, 4th and 5th harmonics and the satellite lines A, B. (Direct readings without the overall spectral response calibration). Note the enhanced sensitivity of the detecting system in the 300-200 nm wavelength region.

The maximum harmonic signal was obtained at the p-polarization of the incoming laser light. The produced harmonics also were always p-polarized. Measuring the yields of each harmonic as a function of the laser polarization we obtained an unambiguous proof that only the laser electric field component $E_\perp$ (ω), which is perpendicular to the gold surface causes the effect. This fact proves that the effect is predominantly of surface origin, i.e., it comes from the laser induced nonlinear current of surface electrons[2,3,10]. We measured the laser light polarization dependence curves of all detected harmonics by changing the angle Φ between the laser field strength and the plane of incidence. Figure 3 shows a typical polarization dependence curve for the third harmonic (n=3) and expected (cos^2 Φ)^3 curve with the fitted n_{exp}=3.2±0.2 value. The measured curves for n=2, 3, 4, 5, correspond to the expected cos$^{2n}\Phi$ dependencies. (This polarization measurement furnished a basic control in that geometrical configuration, in which the laser beam did not touch the gold surface, in this case no effect was found.)

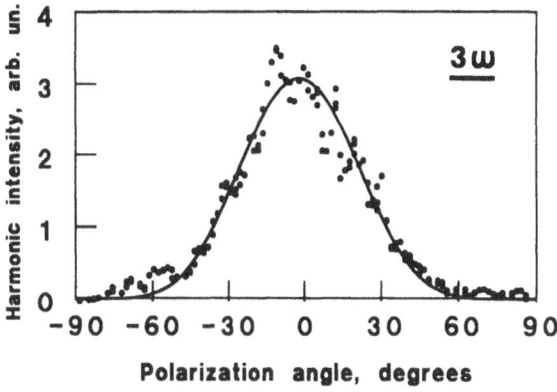

Figure 3. Harmonic light intensity dependence on the polarization angle, relative to the plane of incidence ($\Phi = -90°$ and $90°$ correspond to the s-polarization; $\Phi = 0°$ corresponds to the p-polarization) for the 3ω case.

From the polarization dependence we estimated that the harmonic intensity $I(n\omega)$ depended on the incoming laser intensity $I(\omega) \propto |E_{\perp}(\omega)|^2$ in the form of n-th order power law, i.e. as $I(n\omega) \propto I^n(\omega)$ or $|E(n\omega)|^2 \propto |E_{\perp}(\omega)|^{2n}$, where $E(n\omega)$ is the field of the n-th harmonic (see Table 1.). It was found that the polarization of $E(n\omega)$ was lying also in the plane of incidence. The experimental data led to typical harmonic production efficiency of 10^{-10} for n=2 at I~5 GW/cm^2.

Table 1. Experimental exponents (n_{exp}) of the power dependences of the harmonic light found experimentally by us.

Harmonic order, n	2	3	4	5
n_{exp}	1.8 ± 0.2	3.2 ± 0.2	4.1 ± 0.2	4.7 ± 0.5

The experimental results obtained can not be compared directly to the existing theories. The first class of them, i.e., the early semiclassical theories[2,3] are restricted to the determination of the characteristics of the low (2nd and rarely the 3rd) order harmonic production only, but they use a very realistic, collective electron model. They account for the basic features observed by us: the polarization dependence, the laser intensity dependence, the specular reflection direction of the harmonic light. Their ~ 10^{-10} efficiency value predicted for n=2 agrees well with our measured experimental value of ~ 10^{-10}. The observed higher - n - order polarization and power law dependences strongly suggest that the generalization of these low order theories up to higher n orders is possible, similarly to the case of the atoms[4,5].

As for the non-perturbative theories[10,11], at sufficiently low laser intensities they are equivalent with the second order perturbative calculations. As for the very high order perturbative approaches[9], for n>2, these might be considered as the above mentioned generalizations of the second order perturbative process (apart from the summation of the collective contributions of each individual electron). Although they are not capable by nature to account directly for the correct emission angle, our experiment verified their prediction for the possibility of production of all harmonics above the second up to the fifth. However, they predict a stronger decrease of the efficiencies with the harmonic order n, than found experimentally by us (Fig. 4.). This fact seems to be surprising for the first sight, being the perturbative parameter, x, defined in[9], $x=3\times10^{-3}$ in our experiment, i.e., we were in the perturbative range. It is well known, however, that theories - elaborated especially for multiphoton electron emission from macroscopic metal surfaces using a modified

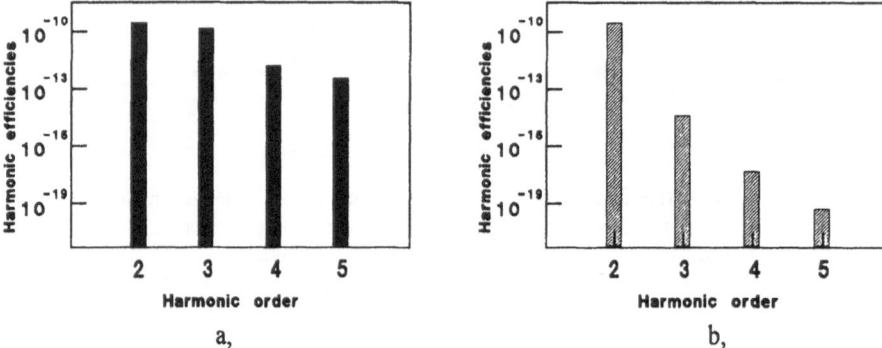

Figure 4. Multiharmonic efficiencies: a- experimental values found by us;
b- theoretical values predicted by Ref. 9.

perturbation parameter and based on the above mentioned correct coherent summation[21] - predict the breakdown of the perturbation approach at considerably lower ($\sim$10 GW/cm^2) laser intensities. This was confirmed also experimentally[17,18]. Therefore, a similar decrease of the upper limit of the validity region of the perturbation calculation can not be excluded also in the case of the laser induced high order harmonic generation processes at metal surfaces.

In conclusion, we demonstrated for the first time the laser induced multiharmonic generation from a metal surface up to 5th order. The laser intensity dependences of the harmonics showed approximately a power law character; the form of the polarization dependence followed the $\cos^{2n}(\Phi)$ law around the p-polarization direction. The comparison of the theoretical considerations and our experimental results suggests that our observed multiharmonic generation arises mostly from the Sommerfeld conduction electrons at the surface of the metal. The rough estimation shows that the measured efficiency (10^{-10} for n=2 at 5 GW/cm^2) coincides quite well with predictions of the theories based on collective many electron models[3]. Although the measured efficiencies decrease much weaker with the harmonic order n, than predicted by the single particle perturbative theory[9], other kind of realistic theories for metals[21] and their experimental verifications[17,18] furnish explanations, which may support our observations.

As for the question of the introduction, we can state that these results and further theoretical and experimental considerations suggest that - in a certain extent similarly to the case of single atoms - considerably higher order harmonic components may be induced in metal surfaces in the future with intense femtosecond laser pulses.

The multiharmonic generation from a metal surface, apart from its fundamental importance, as a possible future application, may be the base of a simple and very effective vacuum ultraviolet source due to these experimentally found high efficiency values. The harmonic production efficiencies found here at already 5 GW/cm^2 can be achieved in noble gases[4,5] only at about four orders of magnitude higher inducing laser intensities ($\sim$10^{13} W/cm^2).

This work was partly supported by the OTKA Foundation of the Hungarian Academy of Sciences (No 2936) and partly by the Large Installation Plan of the European Community (DG XII). The experiments were performed in a cooperation program with the participation of Prof. C. Fotakis, Dr. Cs. Tóth, Dr. S. D. Moustaizis and Mr. N. A. Papadogiannis and the author. The theoretical advices of Prof. T. Nagy and Dr. S. Varró and the technical help of Mr. A. Kőházi-Kis are greatly acknowledged.

References

1. N. Bloembergen, Non-Linear Optics, *in* "Quantum Optics and Electronics" C. DeWitt, A. Blandin, C. Cohen-Tannoudji ed., Gordon and Breach, New-York London Paris, (1965).
2. W.K. Burns, N. Bloembergen, Third-harmonic generation in absorbing media of cubic or isotropic symmetry, *Phys. Rev.* B4:3437 (1971).
3. S.S. Jha, Second-order processes and harmonic fields in solids, *Phys. Rev.* 145:500 (1966); S.S. Jha, Theory of optical harmonic generation at a metal surface, *Phys. Rev.* 140A:2020 (1965).
4. A. McPherson, G. Gibson, H. Jara, U. Johann, T.S. Luk, I. McIntyre, K. Boyer, C.K. Rhodes, Studies of multiphoton production of vacuum-ultraviolet radiation in the rare gases, *J.Opt.Soc.Am.* B4:595 (1987); J. Bokor, P.H. Bucksbaum, R.R. Freeman, Generation of 35.5.-nm coherent radiation, *Opt. Lett.* 8:217 (1983) M. Ferray, A. L'Huillier, X.F. Li, L.A. Lompré, G. Mainfray, C. Manus, Multiple-harmonic conversion of 1064 nm radiation in rare gases, *J.Phys.* B21:L31 (1988); X.F. Li, A. L'Huillier, M. Ferray, L.A. Lompré, G. Mainfray, Multiple harmonic generation in rare gases at high laser intensity, *Phys. Rev.* A39:5751 (1989).
5. A. L'Huillier, K.J. Schafer, K.C. Kulander, Theoretical aspects of intense field harmonic generation, *J.Phys.* B24:3315 (1991); J. H. Eberly, Q. Su. and J. Javanainen, High-order harmonic production in multiphoton ionization, *J. Opt. Soc. Am.* B6:1289 (1989).
6. N. Bloembergen, R. K. Chang, Second-harmonic generation of light from surface layers of media with inversion symmetry, *in* "Physics of Quantum Electronics," P. L. Kelley, B. Lax, P. E. Tannenwald ed., McGraw-Hill Book Company, New-York, (1966).
7. R. Murphy, M. Yeganeh, K. J. Song and E. W. Plummer, Second-harmonic generation from the surface of a simple metal, Al, *Phys. Rev. Lett.* 63:318 (1989).
8. J. Rudnick and E. Stern, Second harmonic radiation from metal surfaces, *Phys. Rev.* B4:4274 (1971).
9. A. Mishra, J.I. Gersten, Theory of multiharmonic generation and multiphoton electron emission at a metal surface, *Phys. Rev.* B43:1883 (1991); A. Mishra, J.I. Gersten, Multiharmonic generation in the high frequency limit at a metal surface, *Phys. Rev.* B45:8665 (1992).
10. R.A. Sacks, A. Szôke, Generation of harmonic radiation during electron scattering from a picewise-constant potential in an intense electromagnetic field, *J. Opt. Soc. Am.* B8:1987 (1991).
11. L. Plaja, L. Roso-Franco, High-order harmonic generation in a crystalline solid, *Phys. Rev.* B45:8334 (1992).
12. Gy. Farkas, Cs. Tóth, S. D. Moustaizis, N. A. Papadogiannis and C. Fotakis, Observation of multiple-harmonic radiation induced from a gold surface by picosecond neodymium-doped yttrium aluminium garnet laser pulses, *Phys. Rev.* A46:R3605 (1992).
13. Gy. Farkas, S.L. Chin, P. Galarneau and F. Yergeau, A new type of intense CO_2 laser induced electron emission from a gold surface, *Optics Commun.* 48:275 (1983)
14. Gy. Farkas, Multiphoton phenomena in photoelectron emission processes metals and high laser intensities, *in* "Multiphoton Processes", J.H. Eberly and P. Lambropoulos ed., Wiley, New York (1978).
15. Gy. Farkas, The problem of laser-induced tunnel ionization, *in* "Photons and Continuum States of Atoms and Molecules", N.K. Rahman, G. Guidotti, and M. Allegrini ed., Springer, Berlin, (1987).
16. Gy. Farkas and Cs. Tóth, Energy spectrum of photoelectrons produced by picosecond laser-induced multiphoton photoeffect, *Phys. Rev.* A41:4123 (1990).
17. L.A. Lompré, J. Thebault, and Gy. Farkas, Intensity and polarization effects of a single 30-psec laser pulse on five-photon surface photoeffect of gold, *Appl. Phys. Lett.* 27:110 (1975).
18. Cs. Tóth, Gy. Farkas, K.L. Vodopyanov, Laser-induced electron emission from an Au surface irradiated by single picosecond pulses at $\lambda=2.94$ µm. The intermediate region between multiphoton and tunneling effects, *Appl. Phys.* B53:221 (1991).

19. D. Charalambidis, E. Hontzopoulos, C. Fotakis, Gy. Farkas and Cs. Tóth, High current, small divergence electron beams produced by laser-induced surface photoelectric effect, *J. Appl. Phys.* 65:2843 (1989).
20. C.E. Moore, *in* "Atomic Energy Levels", Vol. III C. C. Keiths ed., U.S. National Bureau of Standards Circular No. 467. U. S. GPO, Washington (1971).
21. A. P. Silin, *Fiz. Tverd. Tela.* (Leningrad) 12:3553 (1970) [Multiphoton surface photoeffect in metals, *Sov. Fiz. Solid. State.* 12:2886 (1970-71)]; A. M. Brodskii, Specific features of many-photon photoemission from metal and semiconductors in strong fields, *Zh. Eksp. Theor. Fiz.* 60:1452 (1971) [*Sov. Phys. JETP* 33:286 (1971)].

PART II

DYNAMICS OF MULTIPHOTON IONIZATION OF ATOMS

DYNAMICS OF SHORT-PULSE EXCITATION, IONIZATION AND HARMONIC CONVERSION

K. C. Kulander, K. J. Schafer and J. L. Krause[*]

Physics Department
Lawrence Livermore National Laboratory
Livermore, CA 94551

INTRODUCTION

In recent years there have been very significant advances in short pulse, high intensity laser technology. Lasers with pulse lengths of 0.1-1 ps and wavelengths from 0.2-1 μm can be focused to produce intensities from 10^{12} to above 10^{18} W/cm^2. One major use of these systems has been for studies of the response of atoms and molecules to such intense, well characterized electromagnetic fields. Because these pulses are very short, neutral atoms can survive to experience intensities where theoretical treatments based on the traditional perturbation expansion of the wave function in terms of the field-free states will fail completely to describe the dynamics of the system. An explicit, non-perturbative time-dependent calculation is one approach which can represent these strong field effects.

Evidence for non-perturbative ionization can be found, for example, in the photoelectron energy spectra which contain numerous peaks separated by the photon energy. This spectrum demonstrates a high probability for the absorption of *many* more photons than the minimum required for ionization. This phenomenon, called above threshold ionization (ATI), has been observed in several experiments[1] and can be obtained theoretically from an energy analysis of the final state wave function obtained from a time-dependent calculation.[2] A second remarkable observation in high intensity experiments is the efficient production of *very* high-order harmonic radiation during the laser pulse. In the direction of propagation of the incident beam, strong coherent emission at odd multiples of the driving frequency is produced. The even harmonics are forbidden by symmetry. Harmonics up to the 109th order of a 140 fs Ti-sapphire (806 nm) laser[3a] and 133rd harmonic of a 1053 nm 1 ps Nd-YLF laser[3b] have been obtained in neon. Again, the time-dependent wave function contains the information that explains this highly nonlinear, non-perturbative behavior. The Fourier transform of the dipole induced in the electronic charge distribution gives the spectrum of emitted radiation. For many atomic systems, the spectra calculated using this technique agree quantitatively with the observed harmonic emission.[4]

In this paper we discuss recent calculations for xenon in pulsed laser fields which illustrate the multiphoton processes described above. In these calculations we have employed a single-active-electron (SAE) model which explicitly follows the separate time evolution of each of the valence electrons in the frozen, mean-field of the remaining, unexcited electrons, the nucleus and the laser field. This model works well for the rare gas atoms, at least partially because the neglected double or higher excitations involve states well above the ionization threshold. In the next section we give a brief description of the SAE model, emphasizing in particular the development of ion-core specific effective potentials. We present ionization, photoelectron and photoemission rates for xenon at 1.06 μm and compare them to recent experimental results in the third section. We show that harmonics and ATI although related are not identical. The electron energy distributions do not show the cutoff observed in harmonics. After describing the quantum mechanical results and comparing them to recent experiments, we will present a simple two-step, quasi-classical model which explains these differences and which gives a more complete understanding of the dynamics of non-perturbative excitation.

METHOD AND MODEL

A many electron atom in an intense, pulsed laser field is a formidable computational problem. This is because the Hamiltonian,

$$H(t) = H_o + V_I(t),\tag{1}$$

is explicitly time dependent. H_0 is the (non-relativistic) atomic Hamiltonian and $V_I(t)$ is the interaction between the electron and the field. In many of the short-pulse experiments carried out today, V_I is comparable in strength to the Coulombic interactions within the atom. Therefore the method of solution must be able to treat these different interactions on an equal footing. Because the full multielectron problem is beyond present computational capabilities, we have developed a model based on the Hartree-Fock method in which the time-dependent wave function in approximated by a product of single particle orbitals. This approach can provide considerable insight into the excitation dynamics as well as quantitative predictions for the observed multiphoton processes.

The laser is assumed to be linearly polarized along the z-axis and the field is strong enough to be treated semiclassically. Then

$$V_I(t) = -e \sum_i z_i \mathcal{E}_o f(t) \sin(\omega t).\tag{2}$$

Here ε_0 is the magnitude of the field and ω is the frequency. The pulse envelope $f(t)$ is typically chosen to be either a sine-squared or trapezoidal pulse. If we are interested in calculating quantities relevant to a particular intensity, $I_0 = c\varepsilon_0^2/8\pi$, we choose a pulse envelope that rises over several optical cycles to its maximum value (1.0) and then is held constant for 20-30 cycles. The pulse rise must involve at least a few cycles or the calculated results can be contaminated by unphysical transients.

The calculations reported here have been performed using the SAE approximation. The details of these calculations have been presented elsewhere[5] and will be repeated here only to the extent necessary to describe a recent improvement we have made in the effective potentials. In this model we hold the orbitals of all but one of the electrons in the atom fixed and allow the active electron to respond to the laser in the mean field of the remaining electrons and the nucleus.

The effective potentials for the active electron are generated according to the prescription given previously.[6] The potentials are constructed from valence orbitals of the atom obtained from Hartree-Slater calculations for the ground and singly-excited states of the atom. The exchange-correlation parameter is adjusted so that the orbital energy of interest agrees with the spectroscopic value. The valence orbital, which has a particular value of the orbital angular momentum, ℓ, is then used to construct an ℓ-dependent effective potential whose lowest eigenvalue is the ionization energy of that state. A separate calculation is carried out for each ℓ-value to generate the complete potential. We use the $\ell = 2$ potential for $\ell \geq 2$. This results in quite accurate excitation energies, much better than normally obtained in either Hartree-Fock or Hartree-Slater calculations. This procedure will provide a model of xenon which accounts only for the manifold of singly excited states based on the lowest ionic core, $^2P_{3/2}$. For all rare gases, a second manifold of states converges to the next spin-orbit component of the ion, the $^2P_{1/2}$ state. Therefore, we repeat the above procedure to obtain effective potentials for this second set of excited states. We further assume that multiphoton excitations between these two manifolds are very weakly coupled on the time scale of the multiphoton processes of interest so they can be treated separately. This assumption is reasonable because once one of the electrons is excited outside a particular core configuration, transitions into states with the other core requires two electrons to change state. Thus it is within the spirit of the SAE approximation to neglect such excitation pathways. However, in calculating the induced dipole responsible for the harmonic emission we add coherently the contributions from the separate calculations for the two manifolds.

We have constructed such core-specific effective potentials for all the rare gases. In xenon the splitting between the ion core limits is quite large, approximately 1.3 eV. Therefore the contributions from its second manifold are significant only in photoemission for the range of intensities and wavelengths considered here. However, for very short laser pulses, the atom can experience higher laser intensities and the contributions from these states will become relatively more important. We have found that for all the lighter rare gases, including krypton where the ionic cores are separated by 0.66 eV, the additional manifold of excited states is very important, and for some wavelengths and intensities, can be dominant.[7,8]

Another important effect due to the spin-orbit coupling comes into play whether the upper ionic core is specifically involved or not. This is because the excitation dynamics is very sensitive not only to the ionization potential or binding energy of the active electron but also to m, the projection of the orbital angular momentum along the polarization axis. Since no spin-orbit terms are explicitly included in our Hamiltonian, we solve the Schrödinger equation in LS-coupling. This means that for a linearly polarized laser field, m is a good quantum number. Therefore, we need to assign weights to the valence orbitals according to their weights in the spin-orbit coupled states. In earlier calculations on xenon, the valence shell p-electrons with $m = 0$ were found to provide the dominant contribution to the ionization[9] and photoemission[10] rates. There are two $m = 0$ valence electrons in the p-shell, 4/3 associated with the $^2P_{3/2}$ core and 2/3 with $^2P_{1/2}$. Contributions to the ionization rates and harmonic emission strengths from the other four p-electrons in the valence shell of krypton which have $m = \pm 1$ are found to be unimportant for the intensities considered here. Thus we solve the time-dependent Schrödinger equation,

$$ i\,\hbar \frac{\partial}{\partial t} \psi_j(t) = H_j(t)\psi_j(t) \tag{3} $$

for the 4/3 electrons with $m = 0$ that leave the ion in the $^2P_{3/2}$ state and the 2/3 electrons with $m = 0$ that leave the ion in the $^2P_{1/2}$ state. In Eq. (3), $\psi_j(t)$ is the active-electron orbital which at $t = 0$ is the valence pseudo-orbital with total angular momentum j, and H_j contains the appropriate effective potentials. Ionization and photoemission rates are determined during the last part of the pulse which rises to its maximum intensity over five optical cycles and then has a constant intensity for the next 20-30 cycles. After the ramp, the transient excitations decay by ionization over the next few cycles. We obtain an ionization rate by monitoring the norm of the wave function. Our finite difference integration of Eq. (3) is carried out in a finite box with absorbing boundaries. As flux reaches the edges of the grid it is removed. The rate at which this occurs is defined to be the ionization rate.

The emission strengths at the harmonic frequencies are proportional to the square of the Fourier transform of the total induced dipole, which we typically calculate over the last five cycles of the pulse described above:

$$ d_q = \frac{1}{T_f - T_i} \sum_j \int_{T_i}^{T_f} dt\; e^{-iq\omega t} \langle \psi_j(t)|z|\psi_j(t) \rangle. \tag{4} $$

In fact we have found it better to use the acceleration form of the dipole, which is numerically more tractable, for determining the time-dependent dipole.[11,12] For energy and angular distributions of the ejected electrons, we choose a pulse which also turns off over a few additional cycles (a trapezoidal pulse) and then perform an analysis of the final wave function using an energy window function.[2]

RESULTS

In this section we illustrate the methods we have defined above by considering the response of a xenon atom to an intense 1.06 μm laser field. We present electron and photon emission rates for laser intensities well into the non-perturbative regime, above 10^{13} W/cm^2. In particular, we will show that although the harmonics and ATI are related they are not identical. We then present a simple two-step, quasi-classical model which addresses these differences.

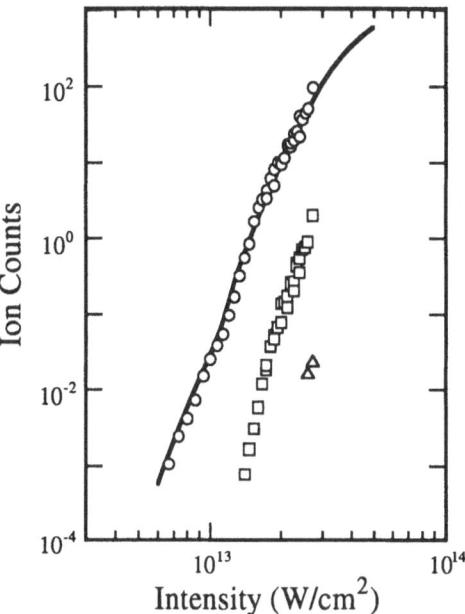

Fig. 1 Calculated (line) and measured (symbols) xenon ion yields as functions of intensity: Xe^{+1} (circles), Xe^{+2} (squares) and Xe^{+3} (triangles).

Ionization and Photoelectron Emission from Xenon

We have studied the excitation of xenon by 1.06 μm radiation for intensities between 5×10^{12} and 5×10^{13} W/cm^2. In Fig. 1 we show the measured[13] ion yields for the first three ion charge states as functions of intensity for a 50 ps pulse. The calculated yield for Xe^{+1} is given by the solid line. Here some small shift of the reported experimental intensity (much less than the maximum expected uncertainty of 50 per cent) has been made to obtain the agreement shown. Having calibrated our results against ion yield curves, we can then compare our electron or photon emission strengths in a consistent manner. The calculated ion yield curve has been averaged over the temporal and spatial distributions in the focal volume. In Fig. 2 we show the ATI spectra for intensities below saturation both calculated (single-atom) and measured. Because of uncertainties in detector efficiency the vertical scale for the experimental data has been shifted so the the two spectra agreed for the 24th order peak at the intermediate intensity.

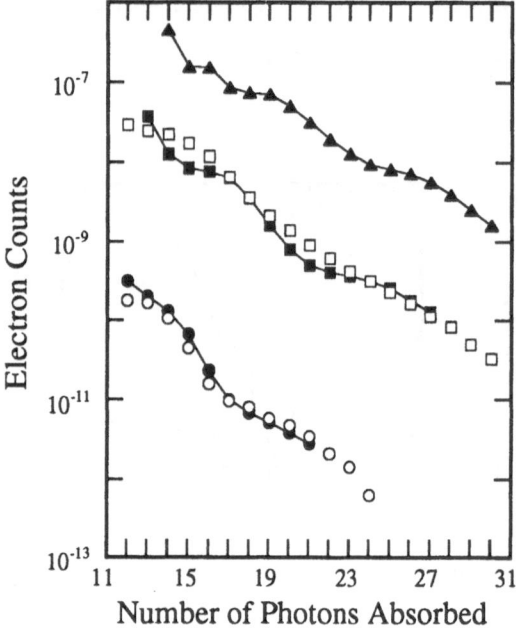

Fig. 2 Photoelectron kinetic energy (ATI) spectrum for Xe at 1.05 μm at 1 (circles), 1.5 (squares) and 2 (triangles) $\times 10^{13}$ W/cm^2: Calculated (filled symbols), Measured (open symbols).

Finally, we complete our presentation of the predicted and observed emission rates for xenon by showing the intensity dependence of the harmonic yields for a 36 ps YAG laser. In Fig 3 we show an absolute comparison[14] between the calculated and measured number of harmonic photons as functions of order for four different incident intensities. One can see that as the intensity increases the spectrum develops an increasingly broad plateau of harmonics which have approximately equal strength and then an abrupt cutoff. The calculated spectra include the macroscopic effects of propagation through the excited, and ionized, medium within the focal volume. Although these propagation effects can affect the overall strength of the emission, the spectral shape observed is very similar to the single-atom

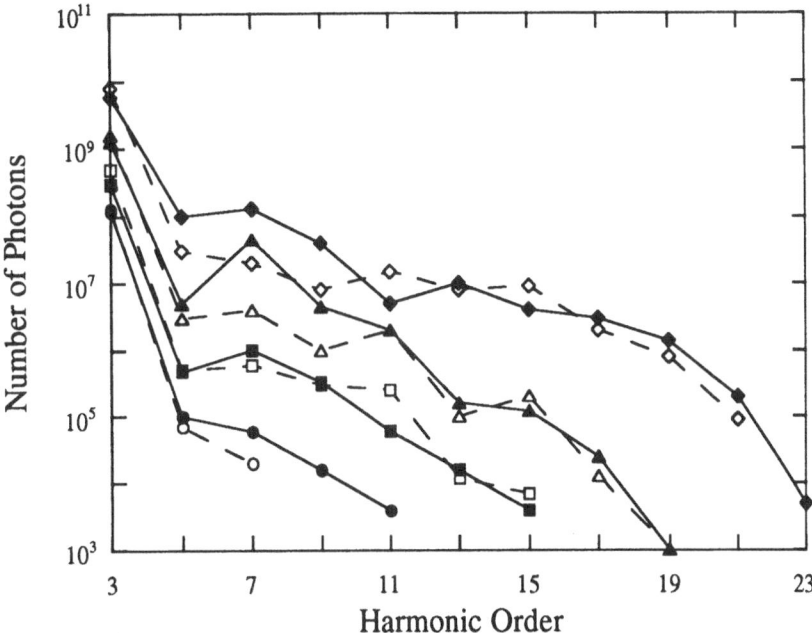

Fig. 3 Harmonic photon yields for Xe at 1.06 μm at 1 (circles), 1.3 (squares), 2 (triangles) and 4 (diamonds) $\times 10^{13}$ W/cm^2: Calculated (filled symbols), Measured (open symbols).

emission strengths. In the high intensity regime phase matching of the harmonics within the plateau is almost order-independent.[4,15]

Two-step quasiclassical model for high field ionization

In the high intensity regime we find that a new contribution to harmonic emission becomes possible. At lower intensities the familiar nonlinear susceptibilities which exist because of the anharmonicity of the atomic potential result from relatively small displacements of the electronic charge density. The intensity dependences of these components of the overall induced polarization are provided by perturbation theory. Thus the q-th order susceptibility scales as the q-th power of the driving field. We show in Fig. 4a a schematic "perturbative" harmonic spectrum for two intensities which differ by a factor of 2. At higher intensities a new source of high energy photoemission becomes probable. In a strong, low frequency field electrons can escape from the vicinity of the ion core by tunnelling through the barrier formed by the Coulomb attraction of the core and the instantaneous electric field due to the laser. This barrier is narrowest along the axis of polarization, so that electrons will be released near this axis, approximately at the point on the outer edge of the barrier where the binding energy is equal to the initial state's ionization potential. We show a cut of the combined potentials for a particular electric field strength and the position, z_{init}, of the newly released electron in Fig. 5.

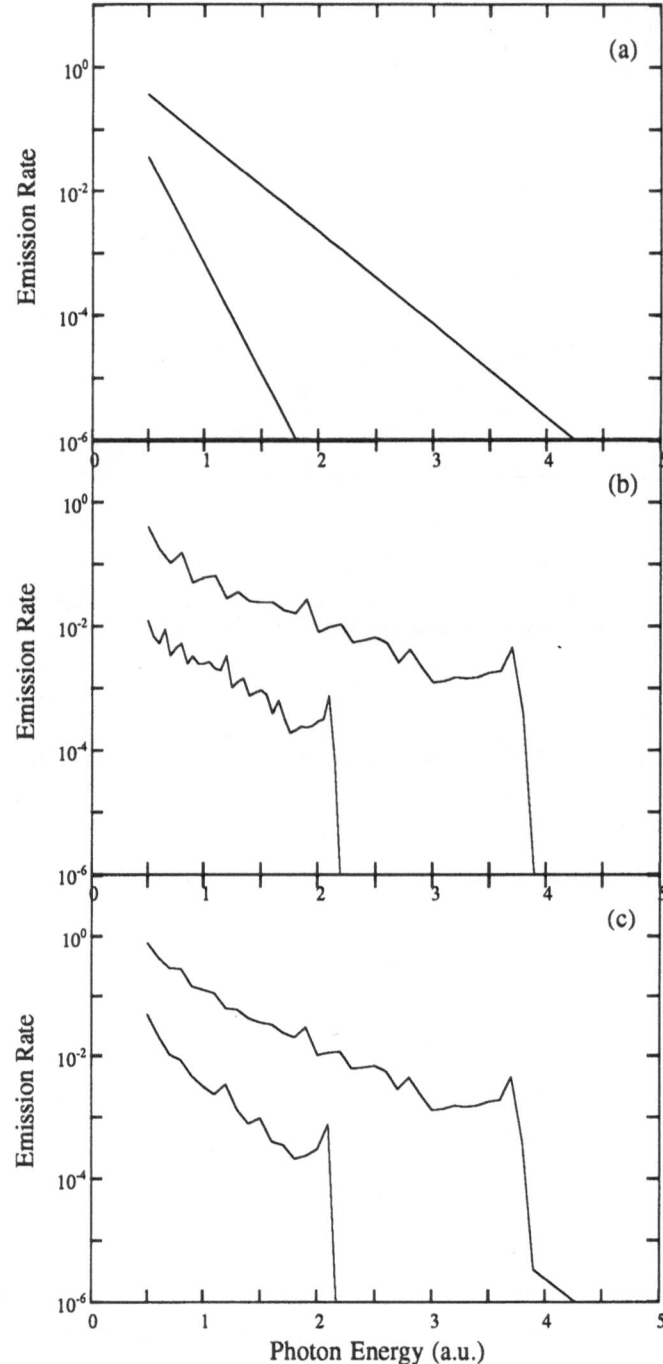

Fig. 4 (a) Schematic perturbative harmonic emission strengths. (b) Quasi-classical emission strengths from trajectory calculations. (c) Sum of two contributions. In all three cases intensity for the upper curve is twice that for the lower.

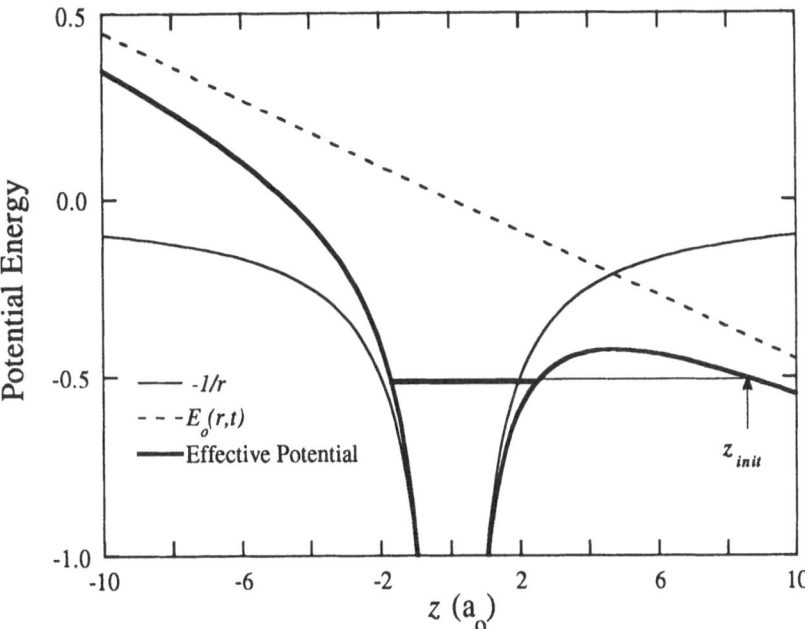

Fig. 5 Instantaneous effective potential for hydrogen in a field polarized along the z-axis. The ground state wave function (indicated by bold line within well) can ionize by tunnelling through the suppressed barrier. Electrons are released at z_{init}.

Tunnelling is most probable when the field is at its maximum ($\omega t = (n+1/2)\pi$ in Eq. (2)) which occurs twice during each optical cycle. Because the amplitude of the field varies slowly with the phase, there will be a finite interval around these maxima during which electrons will be freed. Using a simple tunnelling formula[16] we find the interval increases as the intensity increases. The relevance of this process to photon emission is that most of the electrons which escape through the barrier will be driven by the laser field back across the ion core at which time they may emit a photon. While the electron is near the nucleus it can make a transition back to the ground state, emitting a high energy photon. Obviously the energy of the resulting photon will correspond to the energy the electron has when it returns to vicinity of the core. This return energy depends strongly on the particular phase during the optical cycle at which the electron escaped through the barrier.

After tunnelling the density of states available to the electron is large enough that its subsequent evolution should be accurately characterized by classical mechanics. Therefore in the high intensity regime we can employ a two-step quasi-classical model to represent the ionization dynamics. The first step is the tunnelling process which frees some well defined fraction of the total electron wave function at each instant during the cycle. The second step is the classical motion of the electrons born outside the barrier. In Fig. 6 we show a phase space plot of a representative trajectory which was released with a phase which results in multiple returns to the nucleus. The initial conditions of the trajectory were defined by zero velocity and the displacement along the direction of polarization corresponding to the outer point on the barrier produced by the field at that phase of the cycle. The potential used in the calculation is hydrogenic (soft-Coulomb[17]). The wavelength (1.82 μm) and intensity (1×10^{14} W/cm^2) were chosen to be well into the tunnelling regime. The amplitude of quiver motion of a free electron in this field is 84 a_0. The figure shows that the electron mostly moves as a free electron, occasionally experiencing abrupt collisions with the nucleus. Only when the electron is close to the nucleus can it release its energy by emitting a photon. We keep track of the energies at the return points and identify these as possible harmonic photons. Symmetry and the conservation of energy will require the photon frequency be an odd integral multiple of the driving frequency. The emission is dominated by transitions back to the ground state.

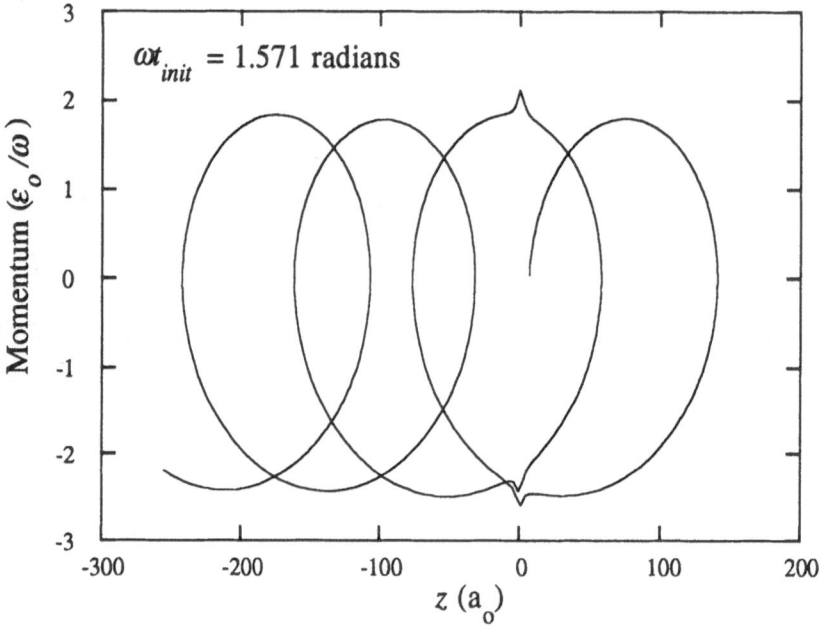

Fig. 6 Trajectory of an electron which tunnelled to freedom just before the maximum in the oscillating electric field.

We repeat the above procedure for each phase of the optical cycle and record the return energies of the electrons. We show in Fig. 7 the distribution of return energies obtained for 180 trajectories equally spaced in time over the first half of the cycle. The results are identical to that obtained during the second half by symmetry. The trajectories are followed for four cycles of the field. We plot the return energy in excess of the field-free binding energy in units of U_p, the ponderomotive (or quiver) energy of a free electron in the oscillating field. U_p is given by $I/4\omega^2$ in atomic units. We have also plotted the tunnelling rate for hydrogen in this field as a function of the phase. A small fraction of the emitted electrons (about 25 percent in this case) initiated in the interval just after the phase $\pi/4$ drift away from the nucleus without a collision. Phases earlier than $\pi/4$ produce trajectories which do not recross the nucleus, but the tunnelling barrier is too broad for any significant release to occur. The rest of the trajectories return at least once. We see that for initial conditions near the maximum of the field the electron can undergo multiple collisions with the nucleus. This is because these initial conditions give the minimum in the initial drift velocity of the released electron. It is the drift velocity which is measured in ATI experiments, either directly in short pulse experiments or indirectly in long pulses due to the additional acceleration of the electron by the gradient of the focussed laser field. In the absence of collisions the maximum possible electron energy in long pulse experiments is $3U_p$. Electrons born with high drift energy generally will have at most one collision the nucleus. Because of this rescattering the final energy of the detected electron can be much larger than $3U_p$.

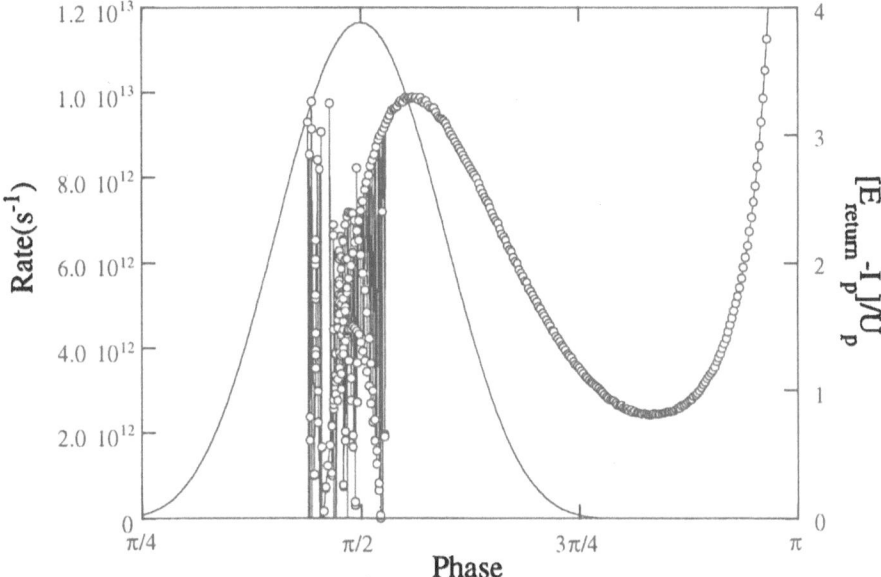

Fig. 7 Hydrogen tunnelling rate (solid line-left axis) and energies of returning trajectories at the nucleus (symbols-right axis) as a functions of the phase of the oscillating field ($I = 10^{14}$ W/cm^2).

In Fig. 8 we show a histogram of the return energies weighted by the tunnelling rate appropriate to the initial conditions for each trajectory. The result is a broad, flat distribution out to approximately $3.2U_p$ followed by an abrupt cutoff. One can see from Fig. 7 that return energies much above the cutoff are possible for phases near π, but these again are the result of initial displacements which are very large and therefore due to such thick barriers that the tunnelling rate is negligible. For probable initial conditions, the return energies do not exceed $I_p + 3.2U_p$. *We believe this energy distribution provides the source for the plateau in the harmonic spectrum.* As each trajectory returns to the nucleus, it will have a probability for emitting a photon which is proportional to the oscillator strength of transition back to the ground state at the return energy. The optical field is varying so slowly during the high energy collision that one can reasonably approximate the transition strength by its field-free value. For hydrogen at these energies, the oscillator strength[18] is decreasing by approximately ω^{-3}. Weighting the probabilities in Fig. 8 by this factor we obtain this second contribution to the high-order harmonic spectrum which is shown in Fig. 4b. In the lowest portion of this figure we have combined, again schematically, the two components of the spectrum. We have assumed the harmonics in the plateau scale approximately as the fifth power of the incident intensity and those from the short-range "perturbative" effects of the anharmonic, atomic potential according to their individual power laws. Although the real atomic spectrum is the result of a coherent sum of the contributions from these different sources, the figure does illustrate the relative strengths and distributions from the two processes which are involved.

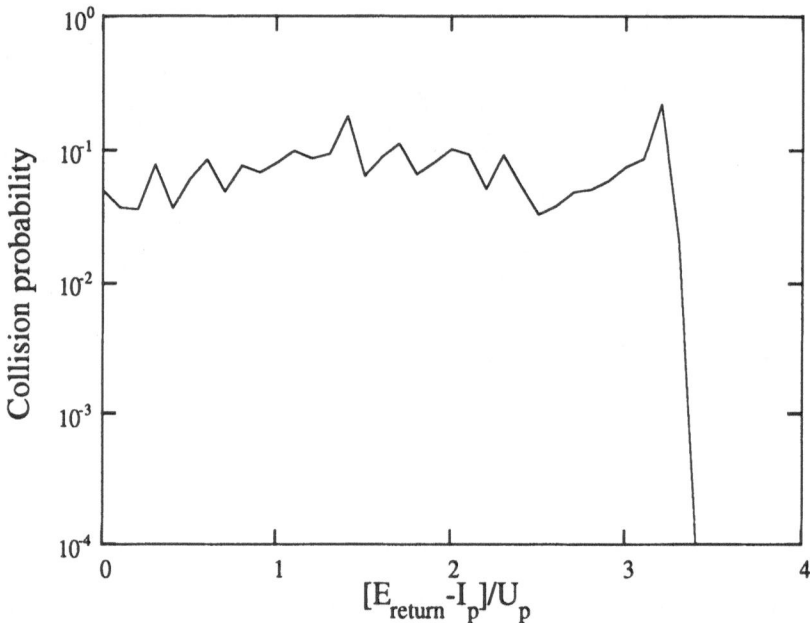

Fig. 8 Energy histogram for electron trajectories returning to the nucleus for hydrogenic, soft-Coulomb potential: $\hbar\omega = 0.025$ au and $I = 1 \times 10^{14}$ W/cm^2.

The harmonic plateau has a distinct cutoff because the field can transfer a limited amount of energy to the electron before it returns to the nucleus. Because of this the harmonics beyond the plateau, coming only from the short-range distortion of the charge density, should behave perturbatively. Although it is beyond the scope of this contribution to consider the phase matching effects on the harmonic signals, we should mention that the angular distributions of the harmonics depend on the effective order, p, of the single-atom signal.[4,15] In the example shown in Fig. 4b we assumed the effective order was 5 which is representative of what we find in our calculations. This says, for example that the 41st harmonic, if it is within the plateau, scales with incident intensity as I^5 rather than the I^{41} one would expect from a perturbative response. The width of the angular distribution scales roughly as $p^{-1/2}$ which means the spatial distribution of the plateau harmonics should be *much* broader than for those beyond the cutoff. In a separate contribution to this meeting the group from Imperial College[19] showed that the angular distributions of harmonics generated in helium exhibited this expected dramatic narrowing as the cutoff in the plateau is approached. This picture of two distinct sources for high-order harmonics is also reinforced by the observed intensity dependences of the individual harmonics. Both the Stanford group[3a] and the Saclay/Lund collaboration[20] have observed that the harmonics initially rise very rapidly as the intensity increases then flatten out very abruptly as they join the plateau. In this narrow transition region the source of the harmonics is changing from the short-range dynamics to that governed by the acceleration of the electron far from the atom by the oscillating field.

Finally we wish to consider the consequences of this picture of two-step ionization dynamics on the other emission process occurring at the same time as harmonic generation. This is the production of very high energy electrons (ATI). If we compare the xenon results presented above for photon and electron emission rates we can see that there is a distinct difference in that the cutoff in the harmonic spectrum is not present in the ATI. In Fig. 9 we compare the single-atom emission strengths for the two processes at four different intensities. The harmonic emission strength is given by ω^3 times $|d_q|^2$. We have also noted in this figure the predicted cutoffs, $I_p + 3U_p$, for the harmonics. As discussed above, without additional collisions with the nucleus the electrons would not have energies larger than $3U_p$, but they clearly do. Also by examining Fig. 7 we see that the electrons released near the maximum of the field during the cycle, which will have the lowest initial drift velocity, are those responsible for the highest harmonic photons. Therefore, this model shows there is not a simple, direct connection between the ATI and harmonic spectra. That is, the portion of the wave function responsible for highest energy photons does not necessarily correspond to that which ends up in the ATI peaks of the same order. However below the harmonic cutoff the harmonic and ATI strengths are clearly very similar. The reason the overall strengths scale similarly with intensity is that these are governed by the initiation step which is the rate at which electrons are released to the continuum. This quantum mechanical first step, tunnelling through the barrier, defines the number of electrons available to either produce photons or be measured in the ATI spectrum. It is this aspect of the ionization process which depends sensitively on the details of the atomic potential. The subsequent evolution of the electrons in the continuum should be essentially the same for all systems, depending only on the ionization potential and the ponderomotive energy. This explains our earlier observations[11,21] of virtually identical plateau distributions for several model potentials all of which were constructed to have the same I_p. The models included the hydrogen atom and molecule, a short range potential which had no excited states and a one-dimensional soft-Coulomb potential. The overall strengths of the calculated plateaus for these systems were quite different, reflecting the different rates at which electrons were promoted to the continuum, but the shape of the spectrum was very similar with the cutoff in all cases being

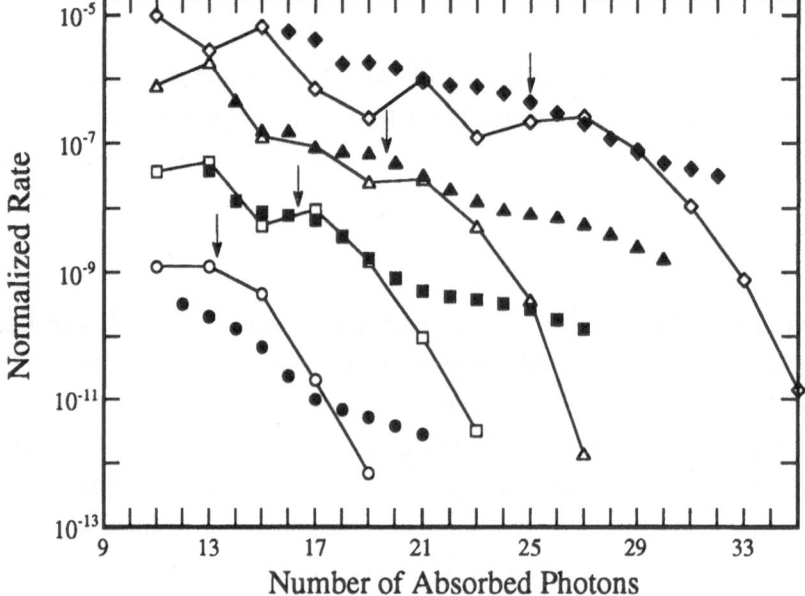

Fig. 9 Photon (open symbols) and electron (filled symbols) emission rates for xenon at 1.06 μm and intensities of 1 (circles), 2 (squares), 3 (triangles) and 5 (diamonds) x10[14] W/cm[2]. The harmonic rates have been scaled to agree with the electron rate for the 21st order peaks for the highest intensity shown.

accurately predicted by $I_p + 3U_p$. The harmonic spectrum probes the distribution of velocities near the nucleus while the electron spectrum measures the distribution of drift velocities after the electrons have completely escaped from the effects of the potential. One signature of this difference is that the additional collisions necessary for the higher energy electrons should lead to increasing angular distributions of the ATI peaks. In Fig.10 we show calculated[2] halfwidths of the angular distributions for hydrogen ATI peaks which do show minima near $3U_p$ followed by increasing breadth for higher orders. These results agree well with the observations of Wolff et al.[22] for the same conditions.

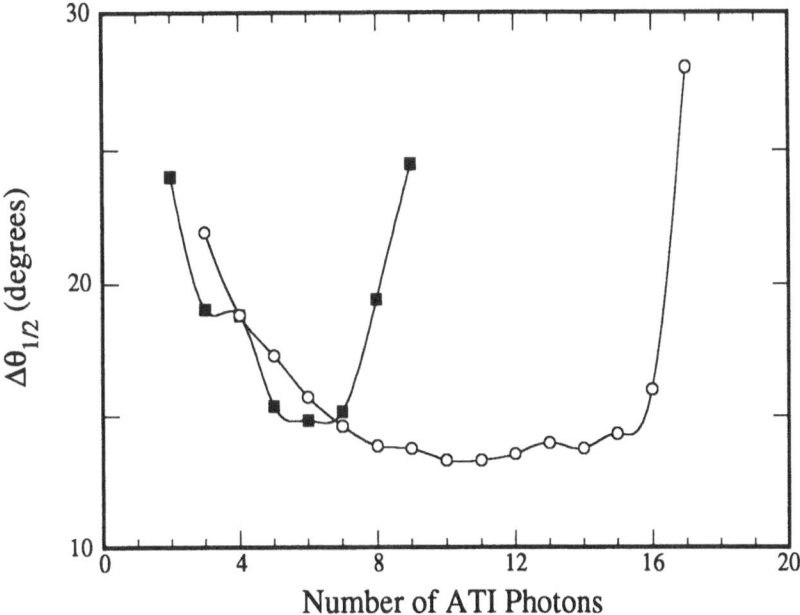

Fig. 10 Half heights of ATI peaks calculated for hydrogen at 2 (squares) and 3 (circles) $\times 10^{13}$ W/cm^2 and a wavelength of 1.06 μm.

The calculations we have presented in this paper illustrate the application of our single-active-electron model of intense-field, multiphoton processes. We have found that it is very successful in modelling very high order, non-perturbative processes in rare gas atoms. The simple two-step, quasi-classical model which accounts for many of the detailed characteristics of the photon and electron emission processes provides significant new insight into the excitation dynamics under intense field conditions.[23]

ACKNOWLEDGEMENT

We wish to acknowledge the very significant contributions to this work by our collaborators Anne L'Huillier and Phillipe Balcou from Saclay for the results presented on harmonics and Louis Franklin DiMauro and Baorui Yang from Brookhaven for the ATI results. This work has been carried out under the auspices of the U. S. Department of Energy at the Lawrence Livermore National Laboratory under contract number W-7405-ENG-48.

REFERENCES

*   Present address: Department of Chemistry, University of California at San Diego

1.   G. Petit, P. Agostini and F. Yergeau, J. Opt. Soc. Am. B 4:765 (1987).

2. K. J. Schafer and K. C. Kulander, Phys. Rev. A 42:5794 (1990); K. J. Schafer, Corr Phys. Comm. 63:427 (1991).

3. (a) J. Macklin, J. D. Kmetec and C. L. Gordon III, Phys. Rev. Lett. 70:766 (1993); (b) L'Huillier and P. Balcou, Phys. Rev. Lett. 70:774 (1993).

4. A. L'Huillier, K. J. Schafer and K. C. Kulander, J. Phys. B 24:3315 (1991).

5. K. C. Kulander, K. J. Schafer and J. L. Krause, Time-dependent studies of multiphot processes, in: "Atoms in Intense Radiation Fields," M. Gavrila, ed. (Advances in Atom Molecular and Optical Physics, Supplement 1) Academic Press, New York (1992)

6. K. C. Kulander and T. N. Rescigno, Comp. Phys. Comm. 63:523 (1991).

7. K. C. Kulander, J. L. Krause, K. J. Schafer, S. W. Allendorf, J. K. Crane, K. S. Bu' and M. D. Perry, in Time Dependent Quantum Molecular Dynamics: Experiment a Theory, J. Broekhove and L. Lathouwers, Eds. (Plenum, New York, 1992).

8. S. W. Allendorf, J. K. Crane, K. S. Budil and M. D. Perry, to be published.

9. K. C. Kulander, Phys. Rev. A 38:778 (1988).

10. K. C. Kulander and B. W. Shore, J. Opt. Soc. B 7:502 (1990).

11. J. L. Krause, K. J. Schafer and K. C. Kulander, Phys. Rev. A 45:4998 (1992).

12. K. Burnett, V. C. Reed, J. Cooper and P. L. Knight, Phys. Rev. A 45:3347 (1992).

13. K. J. Schafer, B. Yang, L. F. DiMauro and K. C. Kulander, Phys. Rev. Lett. 70:xx (1993).

14. A. L'Huillier, P. Balcou, C. Candel, K. J. Schafer and K. C. Kulander, Phys. Rev. 46:2778 (1992).

15. B. W. Shore and K. C. Kulander, J. Mod. Opt. 36:857 (1989).

16. P. B. Corkum, N. H. Burnett and F. Brunel, Phys. Rev. Lett. 62:1259 (1989).

17. J. Javanainen, Q. Su and J. H. Eberly, Phys. Rev. A 38:3430 (1988). The parameter the denominator of this potential is set to 2.0 to obtain a binding energy of 13.6 eV.

18. H. A. Bethe and E. E. Salpeter. "Quantum Mechanics of One- and Two-Electron Atoms Plenum, New York (1977).

19. R. A. Smith, J. W. G. Tisch, M. S. N. Ciarrocca, S. Augst and M. H. R. Hutchinso this workshop.

20. J. Larsson, A. Persson, S.Svanberg, C.-G. Wahlström, P. Balcou and A. L'Huillier, th workshop.

21. J. L. Krause, K. J. Schafer and K. C. Kulander, Phys. Rev. Lett. 68:3535 (1992).

22. B. Wolff, H. Rottke, D. Feldmann and K. H. Welge, Z. Phys. D 10:35 (1988).

23. We note that Paul Corkum has independently extended his quasi-static model (Ref. 16) also along these lines to consider both harmonic generation and impact ionization of the ion core by returning electrons.

ATOMS IN SUPER-INTENSE FIELDS: STABILIZATION IONIZATION AND HIGH-HARMONIC EMISSION

P.L. Knight[1], M. Protopapas[1], C. Keitel[1] and S. Vivirito[1]
K. Burnett[2], V.C. Reed[2], and S.C. Rae[2]

[1]Blackett Laboratory, Imperial College
London SW7 2BZ, UK
[2]Clarendon Laboratory, Oxford University
Oxford OX1 3PU, UK

1 INTRODUCTION

In this paper we shall describe our work on the use of the Kramers-Henneberger frame to describe the dynamics of an atomic electron in a pulsed super-intense field[1-5]. We have developed highly-efficient numerical schemes which solve the evolution equation by Crank-Nicolson methods on a very large grid[2]. We use this method to describe the total ionization yields, photoelectron spectra and the high harmonic frequencies radiated by the driven atomic source. We shall describe how the laser pulse shape influences the driven atomic wavepacket evolution in terms of a time-dependent superposition of dressed Kramers-Henneberger states[1]. We demonstrate how the stabilization phenomenon depends on the pulse rise times and peak intensities. We also use this time-evolving wavepacket to describe, through the use of the atomic acceleration[4], the generation of very high harmonics[2], and link these to stabilization and the nature of the dressed atomic potential[6].

2 PHYSICS AT ATOMIC FIELD STRENGTHS

It will help if we first address the question of orders of magnitude when we talk of super-intense fields. The laser intensity I is related to the electric field by $I = \epsilon_0 c E^2 / 2$ where ϵ_0 is the permittivity of free space , so that

$$I(W/cm^2) = 1.33 \times 10^{-3} [E(V/cm)]^2. \tag{1}$$

A laser intensity of $10^{15} W/cm^2$ thus implies an electric field strength of about $10^9 V/cm$. This should be compared with the electric field from the nucleus seen (on average) by an electron in a state with the principal quantum number n,

$$E_n = \frac{e}{4\pi\epsilon_0 a_0^2 n^4},$$ (2)

where a_0 is the Bohr radius. For the ground state, $n = 1$ and the atomic field is $E_{n=1} = 5.14 \times 10^9 V/cm$, corresponding to an effective intensity of $3.5 \times 10^{16} W/cm^2$ which is the atomic unit of intensity. Such intensities are readily achievable using short pulse chirped pulse amplification of Nd:YAG or YLF radiation, from dye lasers, Ti-Sapphire lasers and KrF systems[7]. As we will see, the laser electric field will strongly interact with the electrons with a scale determined essentially by the ponderomotive energy[8] given by

$$E_p = \frac{e^2 E^2}{4m\omega^2}$$ (3)

where m is the electron mass and ω the laser frequency; for Nd:YAG radiation with a wavelength of one micron, $E_p = 1eV$ for an intensity of $10^{13} W/cm^2$. A new high intensity regime will be defined by the onset of relativistic corrections : the ponderomotive velocity approaches the velocity of light at $10^{18} W/cm^2$ for Nd laser radiation, and at $10^{19} W/cm^2$ for KrF. Such high fields are attainable only by short pulse laser techniques; it is however worth remembering that at visible wavelengths even a six femtosecond laser pulse containing a few optical cycles is actually several hundred atomic units of time long.

What kind of new physics do we expect to encounter at atomic units of intensity ? Quite clearly the evolution is both non-perturbative and dynamic, with the nature of the field turn-on and turn-off being important in controlling the accessible states of the dressed atom. Ideas of multiphoton ionization by a well-defined minimum number of photons will be replaced firstly by the idea of tunnelling of the electron in the laser-deformed atomic potential[9]. At higher intensities we will expect to see the field-electron interaction completely dominate the Coulomb interaction (although whether the electron can in fact absorb or emit photons will depend upon the proximity of the nucleus). The free electron dressed by an intense laser field "quivers" under the influence of the ponderomotive interaction; as it oscillates (with huge amplitude) across the nuclear potential we will expect to see new regimes of ionization and high-harmonic emission. Throughout our calculations we will employ the Kramers-Henneberger (KH) frame[10] in which the quiver motion is explicitly recognized as the dominant influence. In this way we can observe directly the role of the quivering wavepacket in determining the ionization behaviour and the potential for stabilization in realistic pulses rather than in constant amplitude fields.

3 MODEL ATOMS AND MODELS OF INTERACTION

Throughout this paper we report on results obtained using the one-dimensional "Rochester potential" for a one-electron atom[11]

$$V(x) = \frac{-e^2}{(a_0^2 + x^2)^{1/2}}.$$ (4)

In atomic units (which we employ from now on), the bare atomic Hamiltonian is

$$H_0 = -\frac{1}{2}\frac{\partial^2}{\partial x^2} - \frac{1}{(1 + x^2)^{1/2}}.$$ (5)

This supports a series of bound states including a Rydberg series and a continuum, with appropriate alternating parity eigenvalues, but avoids the singularity at the nucleus which would create un-necessary problems for grid integration. We employ a box integration method with a box size of order 3000 a.u and a grid spacing of 0.1 a.u and typically including 71 bound states in our various state projections to determine ionization and occupation probabilities. The binding energy is 0.67 a.u. ($18eV$) which is unphysically high but can easily be rescaled in eq. 5.

We work directly with the KH frame[10] and transform our wavefunction from the laboratory frame $\psi_{lab}(x,t)$ by

$$\phi_{KH}(x,t) = U_1 U_2 \psi_{lab}(x,t), \qquad (6)$$

where U_1 is a contact transformation to remove the quadratic vector potential term

$$U_1 = \exp\left[\frac{i}{2}\int_{-\infty}^{t} d\tau\, A^2(\tau)\right], \qquad (7)$$

and U_2 transforms to an accelerated frame which oscillates at the laser frequency

$$U_2 = \exp\left[-\int_{-\infty}^{t} d\tau\, A(\tau)\frac{\partial}{\partial x}\right]. \qquad (8)$$

In the KH frame, the time-dependent Schrödinger equation, with the laser interaction becomes simply that of an electron evolving in a time-dependent potential, as

$$i\frac{\partial \phi_{KH}}{\partial t}(x,t) = \left[-\frac{1}{2}\frac{\partial^2}{\partial x^2} + V(x + \alpha(t))\right]\phi_{KH}(x,t). \qquad (9)$$

where $\alpha(t)$ is the classical time-dependent excursion under the influence of the ponderomotive interaction,

$$\alpha(t) = -\int_{-\infty}^{t} A(\tau)d\tau. \qquad (10)$$

The advantage of the KH frame is immediately apparent from eq. 9: once the electron is far away from the nucleus it evolves as a free electron wavepacket. The quiver motion in the laboratory frame has been transformed away by eq. 8 and need not be followed in a precise grid integration.

We follow Eberly and coworkers[11] and solve the time-dependent Schrödinger equation (but in the KH frame, we should emphasise) by a Crank-Nicolson method[3]: a box size of 3000 a.u. with a grid spacing of 0.1 a.u. and 80 timesteps per optical cycle will in the KH frame be feasible for quite realistic pulses. A $50fs$ pulse excitation under these conditions can be described using about one hour of Convex computer time. We employ either smooth pulses, with electric fields given by

$$E(t) = E_0 \sin^2\left(\frac{\pi t}{T}\right)\sin(\omega t), \qquad (11)$$

where E_0 is the peak amplitude and T the duration, with $E(t)$ set to zero outside T, or we ramp the field on with a turn-on given by eq. 11 but hold the field constant from the peak onwards. We obtain $\phi_{KH}(x,t)$ in this way from eq. 9, then identify the field-free states, transform back into the laboratory frame and make the appropriate projections to generate spectra and probabilities.

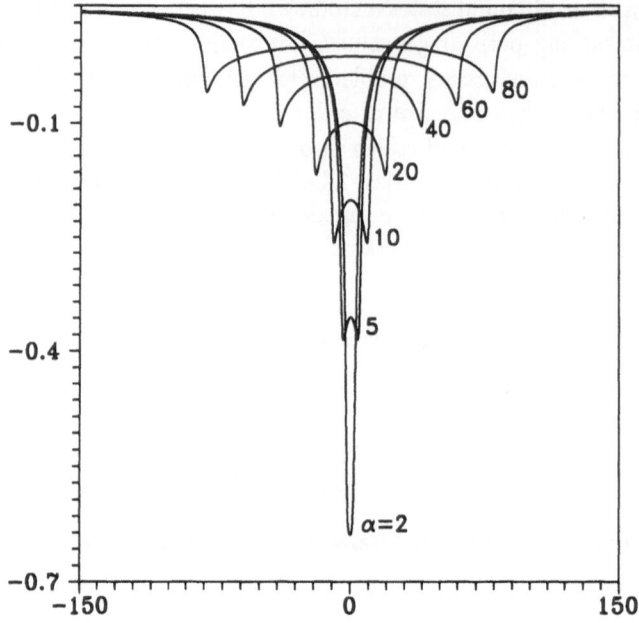

Figure 1 Time averaged KH frame potentials for different values of α.

4 KH FRAME POTENTIALS AND WAVEFUNCTIONS

It is helpful to examine the "dressed potential" of the electron in the KH frame for a <u>constant</u> amplitude laser field, to interpret our results. In eq. 9 the electron moves in a time-dependent potential reflecting the quiver of the electron in the laser field. If the laser frequency is high enough, the dominant influence will be the time-averaged potential

$$V_0(\alpha, x) = \frac{1}{T} \int_{-T/2}^{T/2} V(x + \alpha(t))dt. \tag{12}$$

where T is the period. In figure 1 we show the variation of $V_0(\alpha, x)$ with distance for increasing values of the ponderomotive amplitude α. For small α the potential still has a minimum at $x = 0$. But as α increases the dressed (time-averaged) potential bifurcates with two minima located essentially at the classical turning points $\pm\alpha$ of the quiver motion[6]. If the time-dependence of $V(x + \alpha(t))$ can be neglected (i.e if the field frequency is sufficiently high), then the dressed atomic electron behaviour will be governed by the stationary eigenstates of

$$i\frac{\partial \phi_{KH}}{\partial t}(x, t) = \left[-\frac{1}{2}\frac{\partial^2}{\partial x^2} + V_0(\alpha, x) \right] \phi_{KH}(x, t). \tag{13}$$

In figure 2 we show the dressed potential $V_0(\alpha, x)$ as a function of x for $\alpha = 18.49(E_0 = 73.96, \omega = 2)$ and the probability densities of the first few KH stationary eigenstates. Note that the ground state (fig. 1(a)) (as well as the fourth state) have <u>minima</u> in their probability densities at $x = 0$. This will become important in the dynamical evolution described in later sections, for an electron which starts in the field-free ground state will have a low overlap with the dressed ground state because of this minimum. A smooth field turn-on will access initially those KH dressed levels with maxima at the origin[1] such as the third level, shown in figure 2(b). The outer maxima of the probability densities visible in figure 2 reflect the extended time spent at the classical turning points $\pm\alpha$, so that if we could populate a dressed ground state alone, we would see <u>in the KH frame</u> a dichotomous spatial wavepacket[12]. In the laboratory frame we still need to average over the quiver and a three-peaked "trident" wavepacket will be obtained.

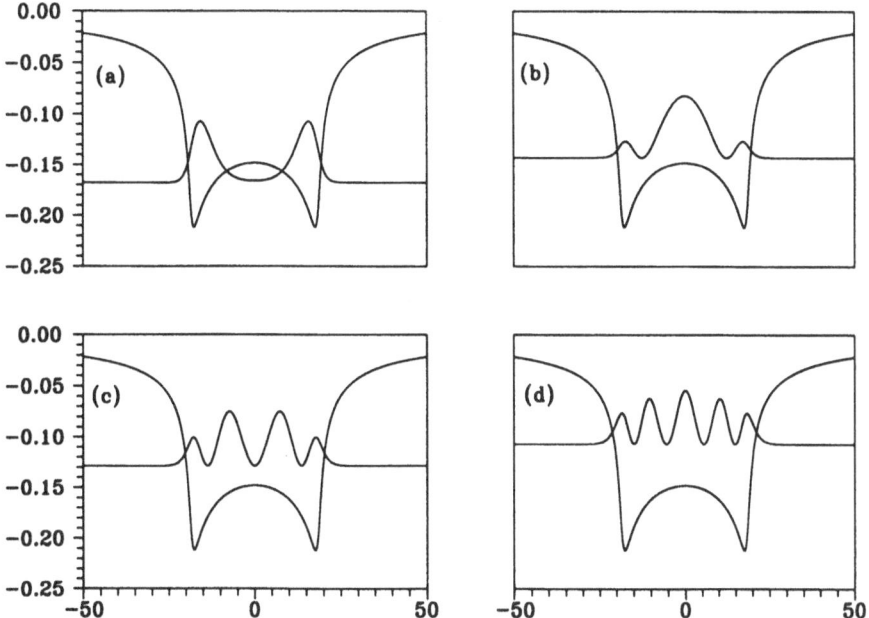

Figure 2 Modulus-squared of the KH bound eigenstates for $\alpha = 18.49$. In (a) $n_{KH} = 1$. In (b)-(d), the $n_{KH} = 3, 4$ and 5 eigenstates respectively are shown.

5 DYNAMICS OF INITIALLY PREPARED WAVEPACKETS IN SUPER-INTENSE LASER FIELDS

Eberly and coworkers have studied in detail the transition from time-independent rates to quivering wavepackets of one-dimensional atomic ionization in suitable laser fields[6,11,13]. Here we concentrate on the effects of a strong quiver amplitude ($\alpha \gg 1$) on various initially prepared wavepackets. Firstly we examine the artificial case in which at $t = 0$ we prepare the electron somehow in the lowest eigenstate of the time averaged dressed potential (eq. 13). In figure 3 we plot the probability of ionization by a suddenly turned-on field, starting from such a state, for $\alpha = 18.49, E_0 = 73.96$ and for a very <u>high</u> frequency, $\omega = 2$.

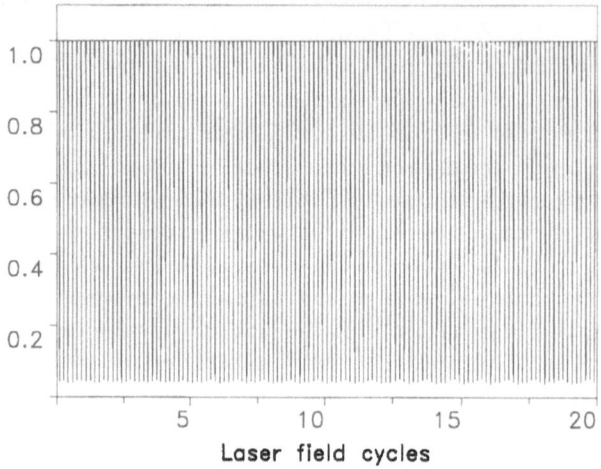

Figure 3 Probability of ionization with time, for the KH ground eigenstate with $\alpha = 18.49$ and $\omega = 2$, from a suddenly turned-on field

The first thing to note is the rapid oscillations in the probability visible from a nearly constant lower value of about 0.05 to an upper value of unity. Now a unity value suggests full ionization, but this is not the case here. What we are seeing is the effect of the quiver: the electron wavepacket oscillates to and fro across the nucleus. Whenever the electron is far from the nucleus, the apparent ionization probability rises to unity, but returns to a low value as the electron packet traverses the nucleus[1,6,11,13]. The true ionization is given by the lower levels of these oscillations and is revealed by turning off the field. For the high frequency of figure 3, we see very little true ionization: the KH ground state of the time-averaged dressed potential hardly sees the rapid oscillations as a perturbation and the atom remains stable (apart from that population lost in the turn-on transients). If we now reduce the field frequency to $\omega = 0.52$, appropriate to two photon ionization, keeping $\alpha = 18.49$, we find that an initially prepared KH eigenstate is now not stable: figure 4 reveals not only the characteristic wavepacket oscillation but also a slow rise in the lower levels reflecting ionization. The value of $\omega = 0.52$ is just too low for the dressed time-averaged potential to entirely dominate the electronic motion and the time-dependent components of $V(x+\alpha(t))$ induce transitions out into the real positive energy ionized continuum.

So far we have examined the survival of a prepared KH eigenstate in an intense laser field. Such an eigenstate tries hard to match the perturbed environment in which the electron finds itself. Of course we should examine the survival of an electron initially in a field-free eigenstate to a smoothly turned-on field. If we examine the evolution of an atom initially in its ground state exposed to a field which smoothly turns-on (with a $\sin^2(\pi t/T)$ envelope) in 5.25 cycles and then is held at a constant amplitude $E_0 = 73.96$, so that $\alpha = 18.49$ we again find the ionization or survival depends sensitively on the field frequency[1,13].

For $\omega = 0.52$ we again see a wavepacket oscillation, but of diminishing amplitude as the atom ionizes. But a higher frequency leads to a dramatically different behaviour shown in figure 5.

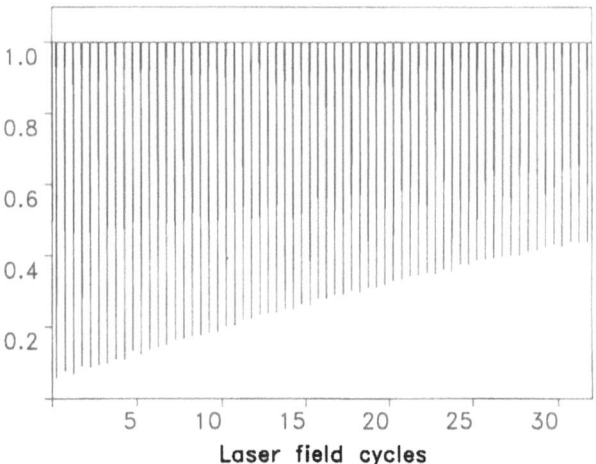

Figure 4 Probability of ionization with time, for the KH ground eigenstate with $\alpha = 18.49$ and $\omega = 0.52$, for a $\sin^2$ ramp turn-on field.

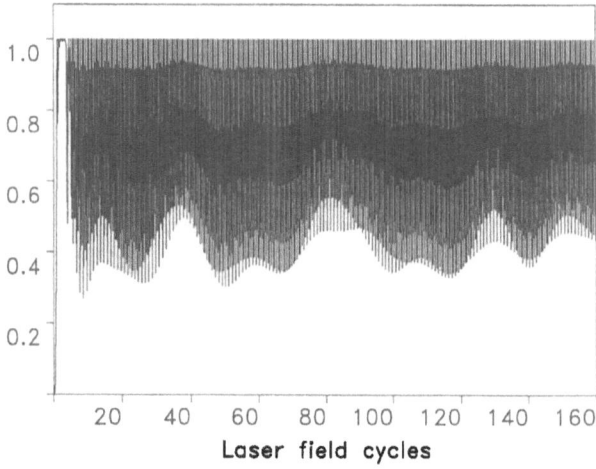

Figure 5 Probability of ionization with time, for the normal ground state, with $\alpha = 18.49$, $\omega = 2$ and a $\sin^2$ ramp turn-on field.

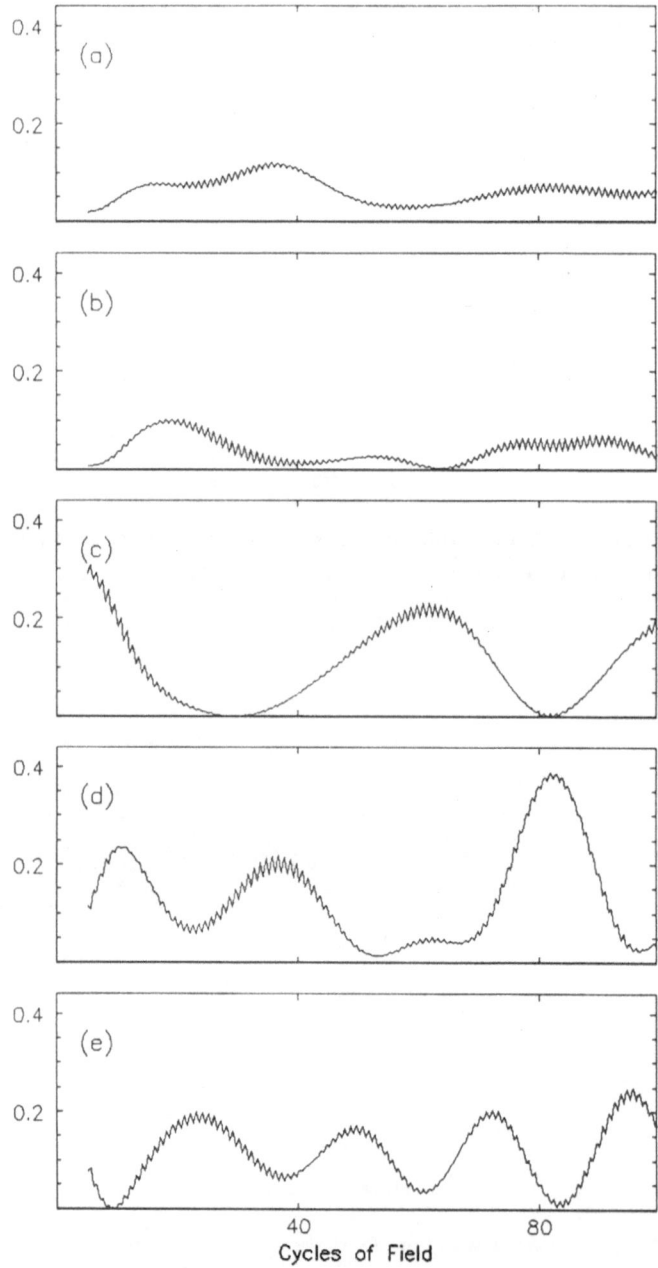

Figure 6 Population of KH bound states with time in a ramped pulse. The populations of the $n_{KH} = 1$ to 5 states are shown in (a) to (e) respectively.

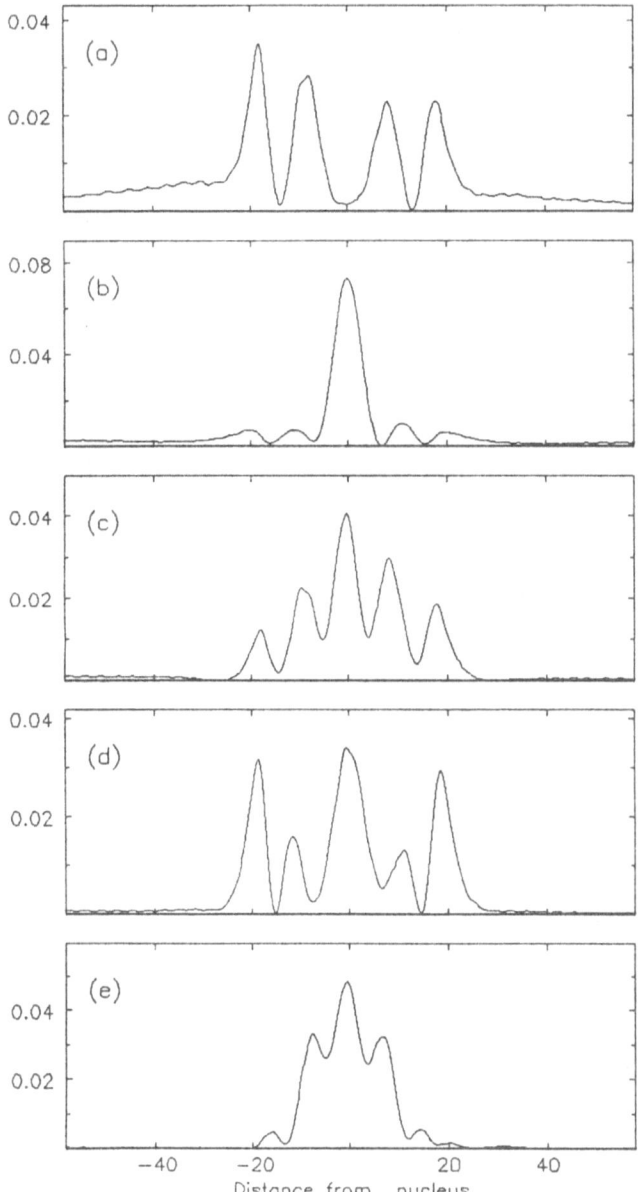

Figure 7 Modulus-squared of the wavefunction for same pulse as figure 6 after (a)30,(b)50,(c)76,(d)96 and (e)158 cycles.

Quite substantial ionization ($\sim 40\%$) takes place in the initial turn-on region, but once the electron experiences the super-intense field we see a wavepacket oscillation which is modulated by a much smaller frequency. The modulation occurs because the electronic wavepacket is made up of a superposition of KH states which beats at their frequency differences[1,14]. In figure 6 we show the time development of the population in the KH eigenstates for such a case. We have chosen again a 5.25 cycle turn-on followed by a constant amplitude field of frequency $\omega = 2$ (corresponding to a wavelength of $22.8nm$) with an intensity of $1.9 \times 10^{20} W/cm^2$. The KH state which populates the easiest at short times is the third, reflecting its greater overlap with the initial field-free state as discussed earlier. But we note that many KH states are in fact populated, and moreover their populations evolve (by Raman-like transitions[15] induced by the time-dependent potential). Figure 7 shows the resulting time-evolved wavepacket at various snapshot times following the field turn-on, showing a multipeaked rather than dichotomous structure but still essentially spread out to regions extending from $+\alpha$ to $-\alpha$, as expected.

Purely smooth pulses (without the constant amplitude and turning-off smoothly just as in eq. 11) generate very similar behaviour. We have examined the evolution of a field-free ground state in a 100 cycle pulse of wavelength $22.8nm$, $\omega = 0.52$, pulse duration of $8fs$ and peak intensity $1.9 \times 10^{20} W/cm^2 (\alpha = 18.49)$: for such an unrealistic pulse less than 20% ionization is generated, almost entirely by the initial transients. More realistically we see almost the same behaviour for an atom exposed to a $175fs$ pulse of $530nm$ radiation of peak intensity $10^{14} W/cm^2$ provided the electron is initially prepared in the $n = 20$ Rydberg level.

6 HIGH HARMONIC GENERATION

It is straightforward to show that substantial continuum excitation in above-threshold ionization (ATI) will lead to the emission of high harmonics[16] and that in the "modestly high" intensity region there is a close relationship[17] between harmonics and ATI. If we write the electronic state vector at time t in terms of discrete states $|n\rangle$ and continuum states $|E\rangle$

$$|\psi(t)\rangle = \sum_n a_n(t)|n\rangle + \int dE \, a_E(t) |E\rangle \qquad (14)$$

then the dipole moment responsible for the radiation of high harmonics

$$d(t) = \langle \psi(t)| \, \hat{d} \, |\psi(t)\rangle \qquad (15)$$

contains bound-bound terms $a_n(t)a_m^*(t)$ and bound-free terms $a_n(t)a_E^*(t)$. We might expect the highest frequencies to involve transitions from high-lying continuum states back to the discrete ground state. Their intensities will be governed by $|d(t)|^2$ which by the above will be proportional to $|a_E(t)|^2$ which of course governs the electron spectrum in ATI. This highly qualitative argument should not be taken too literally, and fails completely in the super-strong field regime.

The harmonics radiated by long pulses are certainly governed by the dipole evolution. The harmonic spectrum may be computed[18] by calculating $|\psi(x,t)\rangle$ as above, forming $d(t)$ as defined by eq. 15, taking a fast Fourier transform to give $d(\omega)$ and plotting $|d(\omega)|^2$. For short pulses this procedure will fail. What really radiates is the dipole acceleration[4] $\ddot{d}(t)$, which for the one-dimensional potential, eq. 5, can be evaluated very easily using Ehrenfest's theorem.

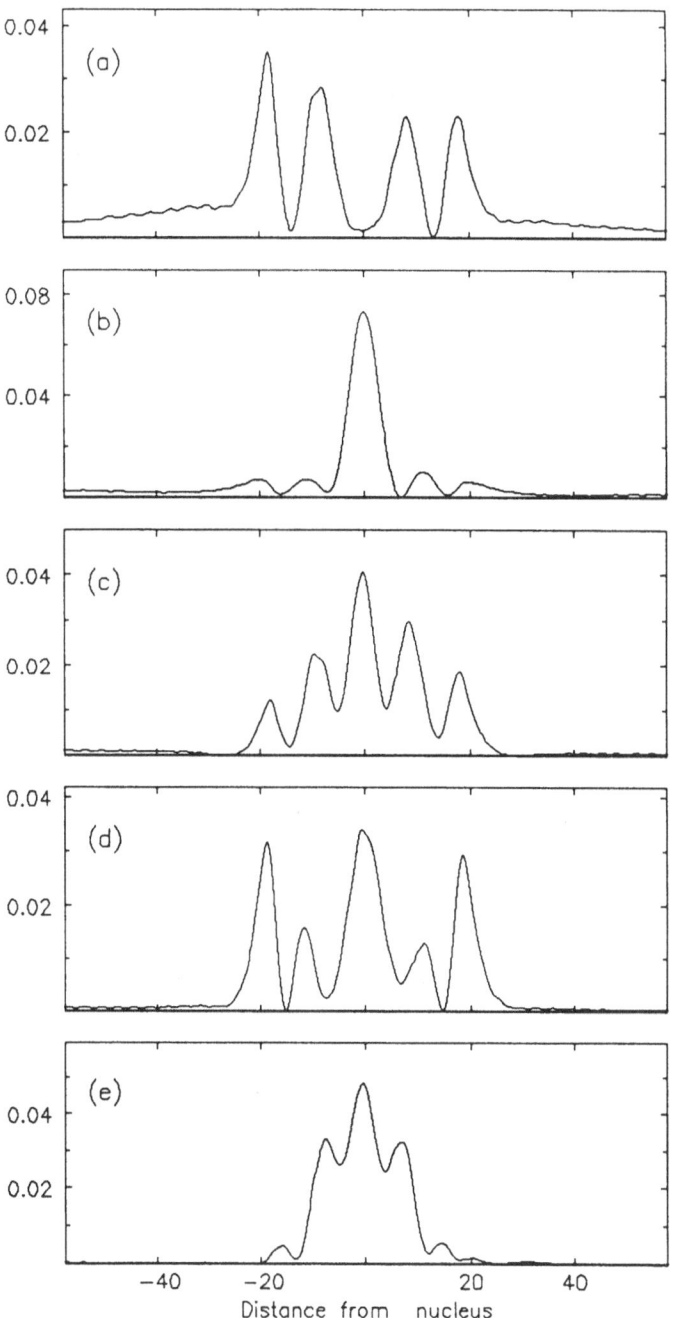

Figure 8 Harmonic spectra for a $\sin^2$ ramped turn-on pulse with $\omega = 0.52$ for increasing peak intensity.

121

Once this acceleration is Fourier transformed, one may expect only a factor of ω^4 difference from the dipole spectrum. For example we have computed the spectrum for a $50fs$ pulse of $300nm$ radiation of peak intensity $2 \times 10^{14} W/cm^2$. This value of field strength is not sufficient to cause rapid ionization and the dipole and acceleration forms of the spectrum (apart from the ω^4 factor) do not differ substantially. If, however, the field peak intensity is increased to $3.2 \times 10^{15} W/cm^2$, we see rapid and substantial ionization. The electronic wavepacket not only oscillates with the familiar quiver motion but also drifts away from the nucleus rapidly in a direction determined by the phase of the field turn-on. A linearly increasing drift motion leads to an increasing dipole moment and a quite artificially large background which swamps the harmonics. Yet an examination of the dipole acceleration (ignoring the simple quiver) reveals that it terminates as soon as the electronic wavepacket moves substantially away from the nucleus: a free electron is <u>not</u> accelerated by the laser field; the true acceleration spectrum reveals harmonics (although noisy) and <u>not</u> a featureless continuum background[4].

So far we have considered high harmonic generation by modest field intensities and by field intensities close to the atomic unit where over-the-barrier (OTB) ionization proceeds. We have noted that in the OTB region the harmonics are not very clearly defined but are noticeable. If, however, the field amplitude is further increased into the stabilization region, the well-resolved harmonics return[2], out to very high orders. In figure 8 we illustrate these various regimes for a field frequency of $\omega = 0.52$, turned-on in 5.25 cycles with the peak laser field amplitude (in atomic units) increasing from 0.05 to 5; the spectra are computed from the acceleration by Fourier transforming after the initial transients of the turn-on have died away. In the first three figures we are working in the modest intensity region where there are clearly resolved harmonics (and as it happens, well resolved ATI peaks). The next two figures are in the OTB region where the harmonics become increasingly noisy reflecting the rapid ionization. But the final frame shows the restoration of clear harmonics when the laser field is sufficient to create a stabilized wavepacket: the harmonics extend to very high orders and are quite broad. One might ask why they are so broad, implying they originate in short time intervals of the evolution.

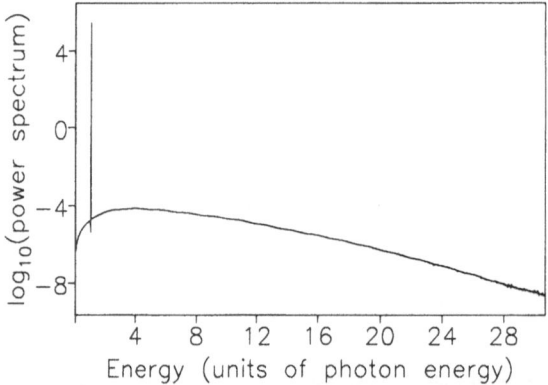

Figure 9 Harmonic spectrum from the ground state of the time-averaged KH potential, evolved under the full Hamiltonian with $\omega = 0.52$ and electric field 5 a.u.

The connection between very high harmonics and stabilization is quite subtle. If we imagine that we prepare an initial wavepacket in the KH ground state and turn on the intense field, we can possess <u>no</u> dipole (or dipole acceleration) and the electron cannot radiate harmonics. Nevertheless, this is true only in the very high frequency limit in which such perfect stabilization occurs. At lower frequencies, the time-averaged part of the potential is not enough to describe the entire electron dynamics and the full time-dependent potential is needed. In figure 9 we show the power spectrum of radiation emitted by an electron prepared in the KH ground state evolving under the influence of a field of wavelength $88nm$ and intensity $8.8 \times 10^{17} W/cm^2$. We observe radiation principally at the fundamental, followed by a broad and smooth background lacking harmonic content. To see harmonics in the stabilization region we need to create a more modulated wavepacket. But this is automatically achieved merely by starting in the field-free ground state and exciting a superposition of KH states as we saw earlier. In figure 10 we divide the harmonic spectrum emitted by an electron initially in the field-free ground state excited by a high frequency field $\omega = 2$ which is turned on in 5 cycles and then held constant with amplitude $E_0 = 15$, into two parts.

The first part is due to the bound part of the KH wavepacket and the second due to the continuum part. By this we mean that at the end of the field <u>turn-on</u> we examine

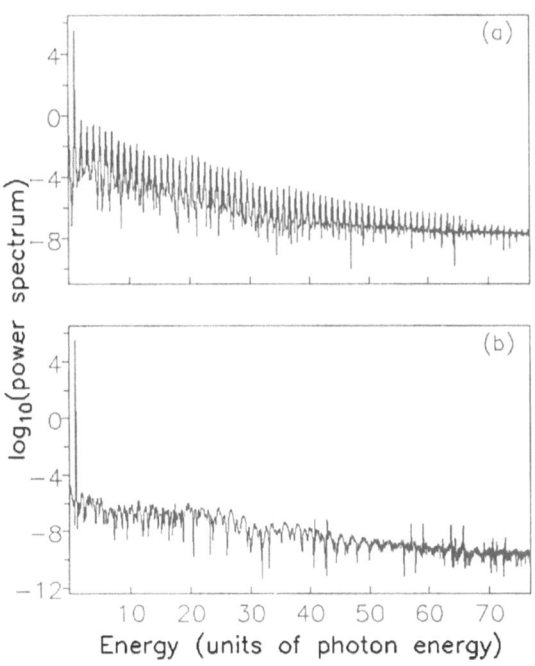

Figure 10 Harmonic spectrum from the KH frame (a) bound and (b) continuum states after a 5-cycle pulse turn-on with $\omega = 2$ and $E_0 = 15$. The harmonics are taken from the 5th to 50th cycle of the field.

the amplitudes in the KH bound and continuum states; then we evolve the bound part of the packet under the influence of the full time-dependent potential in the constant amplitude stabilization region. We do the same for the continuum part. Now of course in the stabilization region, what is bound stays essentially bound, albeit in a breathing, evolving wavepacket. Figure 10 shows that the comb of very high harmonics originates from the stabilized wavepacket motion and not from high-lying continuum states as expected from the modest-intensity argument given after eq. 15. The continuum here generates just background radiation and elastic, fundamental radiation. The evolving wavepacket motion for the parameters of figure 10 is shown in figure 11, where we plot the acceleration (omitting the quiver electric field term) as a function of time. Note both the slow modulation and the rapidly varying peaks; these can be interpreted in terms of the oscillating electron wavepacket moving in the KH dressed potential. This dressed potential has steep barriers which reflect the wavepacket.

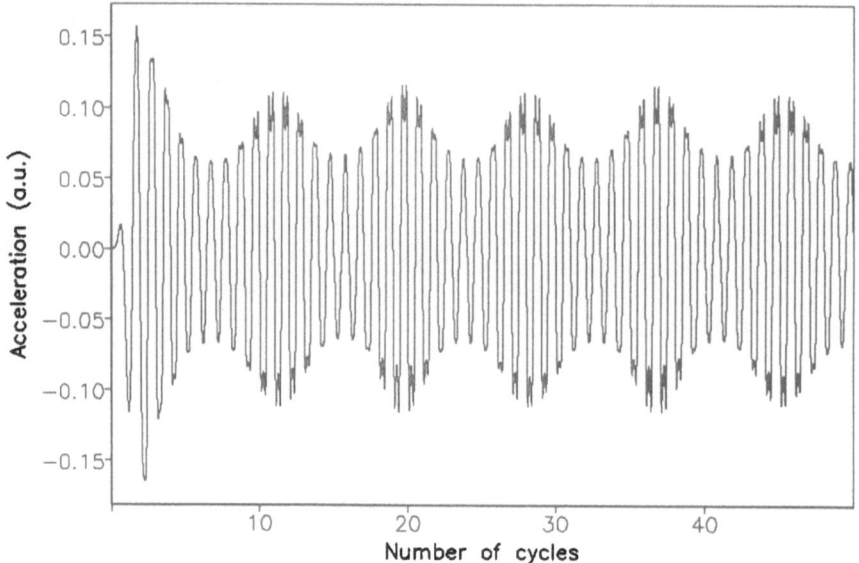

Figure 11 Acceleration as a function of time for same pulse as in figure 10. The electric term in the acceleration has been omitted for clarity.

7 RELATIVISTIC REGION

A rather unexplored region opens when we increase the electric field strength further such that the ponderomotive energy reaches the order of the rest mass of the electron:

$$\frac{e^2 E^2}{4m\omega^2} \approx mc^2 \tag{16}$$

In atomic units, this means $E/2\omega \approx (137)$. In other words, one half of the maximal classical oscillatory velocity gets close to the speed of light which, in atomic units, is equal to the inverse of the fine structure constant. Quite obviously, a fully relativistic treatment becomes necessary in this region. As we are here talking about laser field intensities of the order of 10^{18} to $10^{20} W/cm^2$ for optical radiation, the electron interacts with a huge number of photons at the same time. This makes a classical treatment quite realistic as long as one averages over suitable initial conditions via a standard Monte Carlo method. Several approaches have been given in the literature[19,20], though none of these has gone into the relativistic regime by exploring field strengths big enough to accelerate the electron to speeds close to c in an atomic potential.

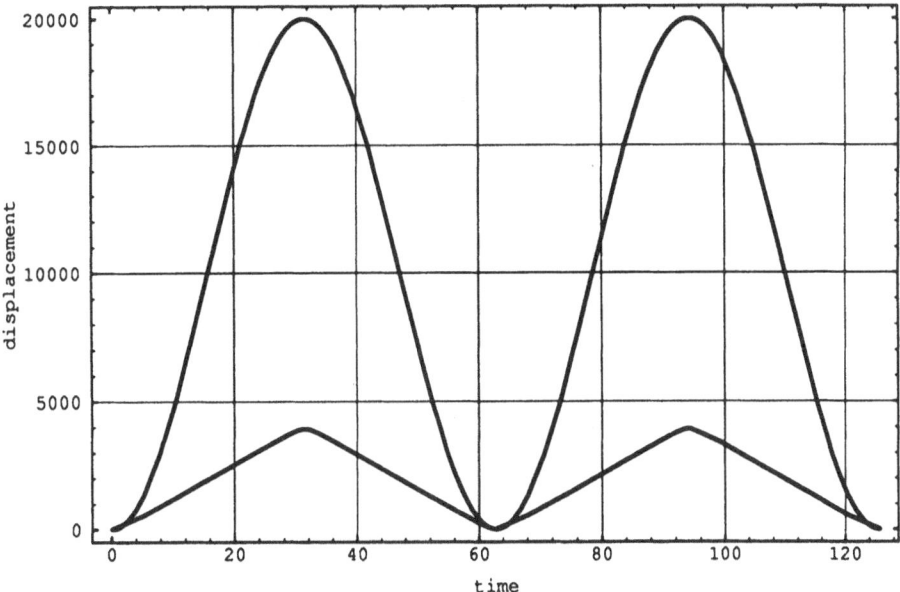

Figure 12 Relativistic and nonrelativistic trajectories for an electron in the model potential and super intense field for $\omega = 0.1$ a.u. and $E = 100$ a.u. .

We here solved the classical relativistic Lorentz equation for an electron with velocity $\vec{v}$ and mass m in an electric field $\vec{E}$ due to the laser and Coulomb field and in a magnetic field $\vec{H}$ due to the laser field:

$$\frac{d(m\vec{v})}{dt} = e\vec{E} + \frac{e}{c}\vec{v} \times \vec{H} \tag{17}$$

This equation can be solved numerically. In figure 12 we have displayed the 1D trajectories of an electron in the "Rochester potential" of a proton and a suddenly turned on ultra-strong laser field. This plot shows the first two ponderomotive oscillations for the electron in the relativistic and nonrelativistic case. The latter shows as expected a sinusoidal oscillation with maximal velocity E/w here larger than the speed of light, whereas the true relativistic trajectory reveals that the electron first accelerates quite fast to close to the speed of light then almost keeps its velocity constant until it suddenly slows down at the classical turning point in phase with the nonrelativistic motion. We have also investigated the two-dimensional case, in which the field polarization lies in the initial orbital plane and found a quite substantial sensitivity to the inclusion of the magnetic field. From the acceleration of the relativistic electron we have moreover evaluated the harmonic spectra in the relativistic domain, which demonstrates the two contributing sources of harmonic radiation: the acceleration of the electron across the nucleus and the anharmonic excursions shown in figure 12 caused by the relativistic dressing of the electron. This will be described in more detail in a future publication.

8 CONCLUSIONS

We have studied the processes of stabilization, ionization and high harmonic generation in the interaction of atoms with super-intense fields. We have introduced the one-dimensional hydrogen model and have described the numerical and intuitive advantages of solving the corresponding Schrödinger equation in the Kramers-Henneberger frame. For high enough field frequencies we were able to average over the time dependence of the KH-potential and thus evaluate the lowest energy KH stationary eigenstates. As to be expected, we find from these wavefunctions, that the electron has its highest probability close to the classical turning points and a minimum at the centre ot the potential. We have followed the dynamics of such a KH state and also of superpositions of KH states generated by the pulsed laser turn-on, and have been able to investigate clearly the ionization process without the usual veiling influence of the quiver. This has enabled us to have a close look at the importance of the pulse shape and in particular the turn-on and turn-off phases of the laser field.

For the evaluation of the harmonic spectra, we have emphasized on its determination via the acceleration of the electron. The drift motion of the dipole evolution gives rise to an artificial background and may hide the harmonics. In the regime of over-the-barrier ionization we often find the harmonics dissappear due to the fast ionization of the electron. Increasing the intensity furthermore to the stabilization regime, however, we see the broad harmonics reappear up to very high order. By separating the contributions of the spectrum into bound and continuum parts, we found that high harmonics originate from the bound stabilized motion of the electron. This we explain via strong reflection of KH-wavepackets at the steep barriers of the dressed potential. Our present interest lies in the rather unexplored relativistic domain of harmonic generation. The sudden acceleration of the relativistic trajectories, as seen in figure 12, may be related back to analogous steep barriers in the corresponding KH-potential.

ACKNOWLEDGEMENTS

This work was supported in part by the UK Science and Engineering Research Council and by the Science Plan of the European Community.

REFERENCES

1. V.C.Reed, P.L.Knight and K.Burnett, *Phys.Rev.Lett* 67:1415 (1991).
2. V.C.Reed, K.Burnett and P.L.Knight, *Phys.Rev.* A 47:R34 (1993).
3. V.C.Reed and K.Burnett, *Phys.Rev.* A 42:3151 (1990);V.C.Reed and K.Burnett, *Phys.Rev.* A 43:6217 (1991).
4. K.Burnett, V.C.Reed, J.Cooper and P.L.Knight, *Phys.Rev A* 45:3347 (1992).
5. K.Burnett, V.C.Reed and P.L.Knight, *J.Phys.* B 26 (1993).
6. For comprehensive reviews of stabilization, see articles by Eberly,Grobe,Law and Su, by Kulander, Schafer, and Krause and by Gavrila in "Atoms in Intense Laser Fields," M.Gavrila, ed., Academic Press, London 1992, which also contain references to the original pioneering work of Mittleman,Gavrila and coworkers.
7. see e.g M.H.R.Hutchinson, *Contemp. Phys.* 30:355 (1989).
8. H.G.Muller, A.Tip, and M.J. van der Wiel, *J.Phys.B* 16:L679 (1983);
   R.R.Freeman, P.H.Bucksbaum, W.E.Cooke, G.Gibson, T.J.McIlrath and L.D. van der Woerkom, in "Atoms in Intense Laser Fields" M.Gavrila, ed., Academic Press, London 1992, and references therein.
9. L.V.Keldysh, *JETP* 20:1307 (1965); S.Augst, D.Strickland, D.D.Meyerhofer, S.L.Chin and J.H.Eberly *Phys.Rev.Lett.* 63:2212 (1989).

10. H.A.Kramers, "Collected Scientific Papers," North Holland, Amsterdam 1956; W.C.Henneberger *Phys.Rev.Lett.* 21:838 (1968).

11. J.Javanainen, J.H.Eberly and Q.Su *Phys.Rev.* A 38:3430 (1988); Q.Su and J.H.Eberly *Phys.Rev.* A 44:5997 (1991).

12. M.Pont, N.R.Walet, M.Gavrila and C.W.McCurdy *Phys.Rev.Lett* 61:939 (1988).

13. Q.Su and J.H.Eberly, *J.Opt.Soc.Am* B 7:564 (1990); Q.Su, J.H.Eberly and J.Javanainen *Phys.Rev.Lett.* 64:862 (1990).

14. K.C.Kulander, K.J.Schafer and J.L.Krause *Phys.Rev.Lett.* 66:2601 (1991).

15. K.Burnett, P.L.Knight, B.R.M.Piraux and V.C.Reed *Phys.Rev.Lett* 66:301 (1991).

16. We will not address here the problems of many atom emission and phase-matching which are necessary to explain the pioneering observations of high harmonic emission by A.McPherson, G.Gibson, H.Jara, U.Johann, T.S.Luk, I.A.McIntyre, K.Boyer and C.K.Rhodes [*J.Opt.Soc.Am.* B 4:595 (1987)] and M.Ferray, A.L'Huillier, X.F.Li, L.A.Lompré, G.Mainfray and C.Manus [*J.Phys.* B 21:L31 (1989)].

17. B.W.Shore and P.L.Knight *J.Phys.* B 24:325 (1987).

18. K.C.Kulander and B.W.Shore *Phys.Rev.Lett* 62:524 (1989); K.C.Kulander and B.W.Shore *J.Opt.Soc.Am* B 7:502 (1990); B.W.Shore and K.C.Kulander *J.Mod.Opt* 36:857 (1989); J.H.Eberly, Q.Su, and J.Javanainen *Phys.Rev.Lett.* 62:881 (1989).

19. G.A. Kyrala, *J. Opt. Soc. Am.* B 4:731 (1987).

20. G. Bandarage, A. Maquet, T. Ménis, R. Taïeb, V. Véniard, and J. Cooper *Phys. Rev. A.* to be published.

MULTIPHOTON IONIZATION OF ATOMIC HYDROGEN IN INTENSE LASER PULSES: THE ROLE OF RYDBERG STATES

D. Feldmann, H. Rottke, K.H. Welge, and B. Wolff-Rottke

Fakultät für Physik
Universität Bielefeld
D 4800 Bielefeld 1
Germany

INTRODUCTION

Multiphoton ionization (MPI) of atomic hydrogen by intense laser radiation can serve as the simplest test for comparison between experiment and theory. We wish to describe the present state of knowledge.

In general the interaction between an atomic electron and intense laser fields may be devided into three regimes depending upon the relative strengths of the two interactions to which the electron is exposed. There are the two limiting situations in which either the Coulomb potential or the radiation field dominates and the intermediate case where they are of comparable strength. The MPI processes which we shall discuss here cover a regime of intensities and wavelengths in which all three situations are involved during the process: MPI starts from the initial groundstate which can be described as dominated by the Coulomb interaction between proton and electron with the radiation field being a perturbation only. In the final state an electron is leaving the proton in the radiation field which now dominates. The third and most complicated regime lies in between, in the region of excited states of the atom where static Coulomb interaction and the influence of the oscillating field are of comparable strength. In our case this happens for excited states which are characterized by the prinzipal quantum numbers $n = 2$ and $n = 3$ in the limit of no radiation field. It is evident that such MPI processes cannot be described by a perturbative approach.

A major complication arises from the experimental restriction, that high radiation intensities can only be provided by pulsed focused laser beams. This leads to the well known problem that the spatio-temporal intensity profile must be taken into account when experimental and theoretical results are to be compared.

Below, we shall first describe the model of intensity induces resonant ionization processes, which was first proposed by Freeman et al.[1]. Here we present a simple version related to our experimental conditions.

Super-Intense Laser-Atom Physics, Edited by
B. Piraux *et al.*, Plenum Press, New York, 1993

The experimental energy spectra of photoelectron will be given and interpreted in the framework of this model and then they will be compared with results of theoretical calculations.[2,3,4] Angular distributions of photoelectrons will be presented and finally an extra series of peaks in the electron energy spectrum will be discussed, for which we have three tentative explanation.

THE MODEL OF AC-STARK-SHIFT INDUCED RESONANCES APPLIED TO HYDROGEN ATOMS.

Figure 1 shows an energy level diagram of atomic hydrogen and its intensity dependence for laser radiation at a wavelength of 608 nm which corresponds to a photon energy of $\hbar\omega = 2.04$ eV. The energies of the groundstate dressed by n photons are indicated by the dashed lines labeled $(1s + n\hbar\omega)$. The ac-Stark shifts of the bound levels have been calculated in the perturbative regime by Gontier and Trahin[4] and nonperturbatively by Dörr et al.[2] using a Floquetansatz. The $1s$-groundstate shifts perturbatively down by a small amount of 10 meV every 10^{13} Wcm^{-2}. The $n = 2$ and $n = 3$ levels are strongly coupled by the radiation field because the photon energy is close to their energy separation. They are strongly mixed with increasing intensity and due to strong coupling to the continuum they also broaden considerably. Therefore it is not reasonable to label them by quantum numbers at higher intensities. Their increasing width is indicated by dashed lines which cease when their width exceeds a few tenths of an eV. The ionization limit (indicated by $n = \infty$) shifts ponderomotively which means by $W_p \approx 0.344$ eV per 10^{13} Wcm^{-2} which corresponds to the classical mean quiver energy of a free electron in the oscillating electric field of the light. Within the resolution of this figure, high Rydberg states shift by the same amount, parallel to the ionization limit. For $n = 4$ and $n = 5$ the shifts deviate from ponderomotive behaviour. Levels with different angular momentum exhibit different shifts, as indicated for $4p$ and $4f$. (All shifts are drawn for $m = 0$ levels which are the only ones accessible by linearly polarized light from the $1s$-groundstate.)

Resonantly enhanced ionization processes can occur at intensities at which a shifted level "intersects" a dressed ground state of the same parity. Such an "intersection" can be a crossing or an avoided crossing depending upon the relative magnitudes of the multiphoton Rabi frequency and the ionization width. Details have been discussed in theoretical papers[5]. A resonant ionization process via the $4f$ state is indicated along the line of arrows. At this "resonance intensity" $I_R(4f)$ the shifted $4f$ state is resonantly coupled to the ground state by seven photons. After absorption of at least one more photon the electron can be released into the continuum with a drift energy of

$$E_K(4f) = \hbar\omega - \big(IP(I_R) - E_{4f}(I_R)\big)$$

or at the corresponding higher ATI-energies

$$E_K^S(4f) = E_K(4f) + S \cdot \hbar\omega$$

where S indicates the number of excess photons absorbed in the continuum. $IP(I_R)$ and $E_{4f}(I_R)$ denote the ionization potential and the energy of the $4f$-state at this resonance intensity I_R. Under short pulse conditions, which we have in our experiment, the quiver energy of the photoelectron is transfered back to the radiation field when the laser pulse intensity drops.

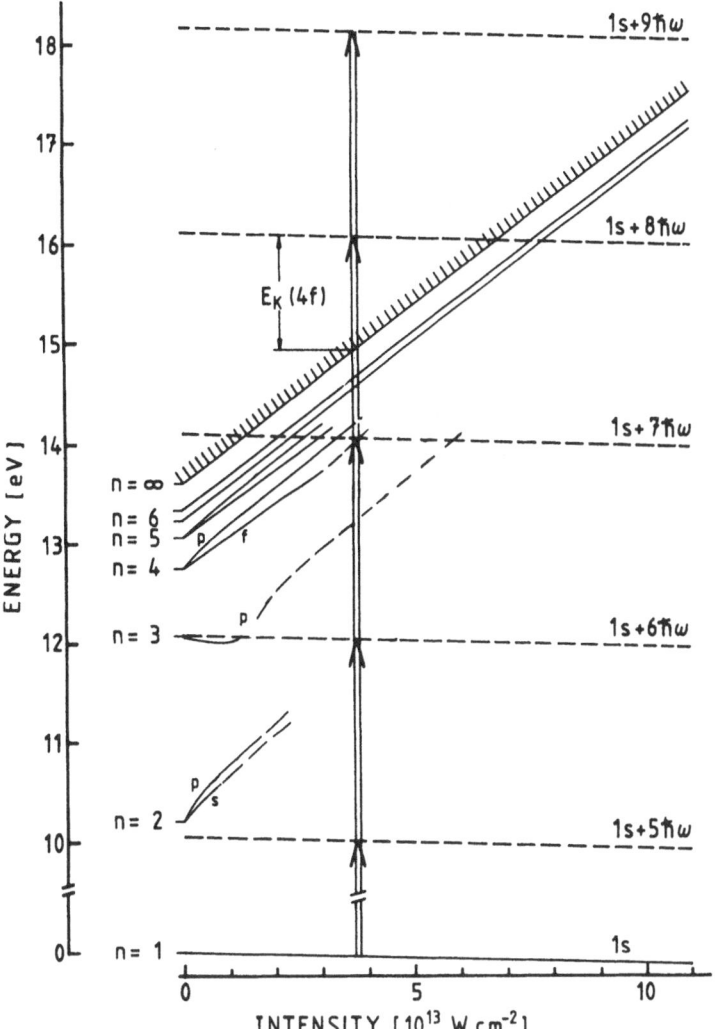

Figure 1. Energy diagram of atomic hydrogen and intensity dependent shifts. The ground state dressed by n photons $(1s + n\hbar\omega)$ is indicated by dashed, almost horizontal lines. A resonantly enhanced MPI-process via $4f$ is shown along the line of arrows. For more details see text.

All the higher Rydberg states $n > 4$ are tuned through seven photon resonance at somewhat lower intensities.

It can also be seen in this figure 1 that seven photon ionization is possible only at low intensities up to $1.9 \cdot 10^{13}$ Wcm^{-2}. At this intensity the ionization limit reaches the energy of the groundstate dressed by seven photons ($1s + 7\hbar\omega$). Ionization by a minimum number of eight photons is possible in the subsequent intensity interval between $1.9 \cdot 10^{13}$ Wcm^{-2} and $7.8 \cdot 10^{13}$ Wcm^{-2}. The kinetic drift energy of the

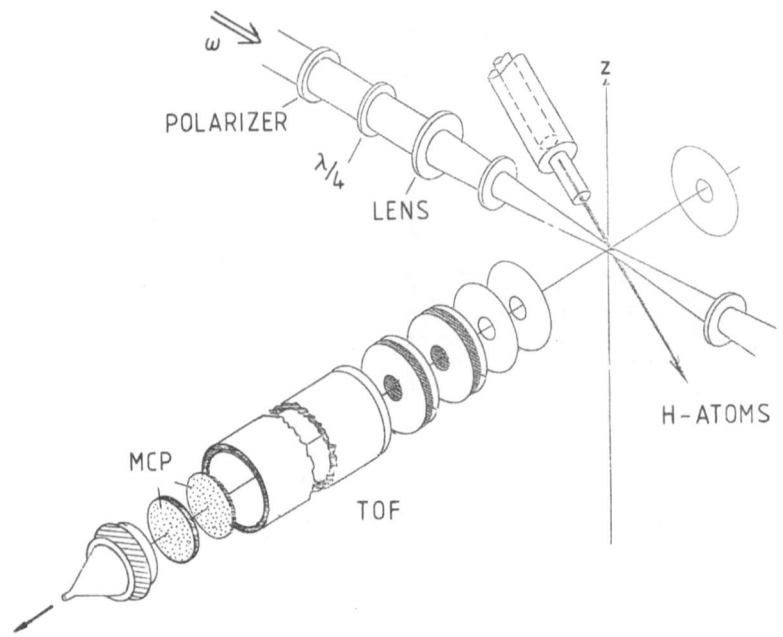

Figure 2. A schematic view at the experimental interaction region.

photoelectrons produced by eight photon absorption is a function of the intensity at which they are produced $E_K(I) = E(1s + 8\hbar\omega) - IP(I)$ where also $E(1s + 8\hbar\omega)$ weekly depends upon intensity as mentioned above.

In principle one could also expect ionization processes at intensities above $6.8 \cdot 10^{13}$ Wcm^{-2} for which at least nine photons would be necessary. However, calculations[2] show that saturation, which means depletion of target atoms, is approached in this intensity regime – for the pulse duration and wavelengths used in our experiments.

EXPERIMENT

The experimental arrangement is schematically shown in figure 2.

A beam of hydrogen atoms produced in a microwave discharge is crossed by the focused ($f = 100$ mm) laser beam. Photoelectrons are detected in a time-of-flight (TOF) spectrometer perpendicular to the laser beam. Laser wavelengths between 590 nm and 640 nm could be used, the laser pulse duration was about 0.5 ps. The direction of linear polarization could be varied by a half-wave plate ($\lambda/2$) for measurements of angular distributions of photoelectrons. Up to approximately 10 % of the unstructured background signal of electrons was due to ionization of molecules, mainly H_2.

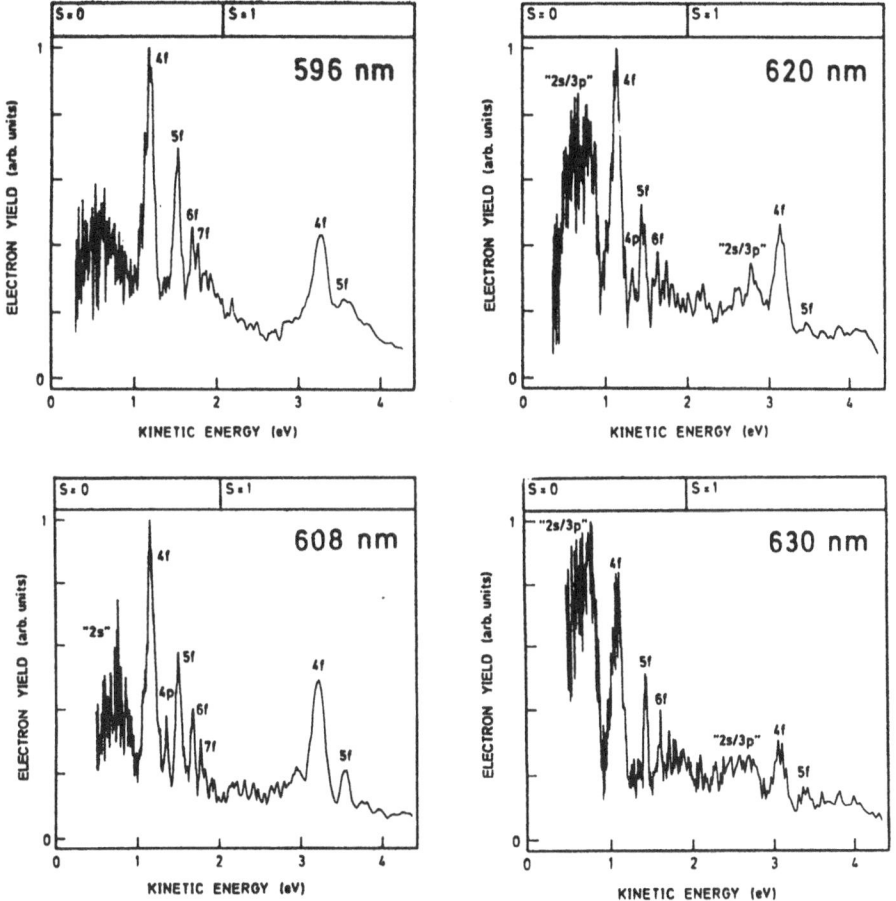

Figure 3. Energy spectra of photoelectrons for different wave lengths. The pulse peak intensity was about $(6 \pm 1) \cdot 10^{13}$ Wcm^{-2}.

RESULTS

The Energy Spectrum of Photoelectrons

Figure 3 shows typical energy spectra of photoelectrons for different wavelengths. The electrons were detected in the direction of the laser polarization. The spectra show a rich substructure of peaks between 1 eV and 2 eV. These peaks can be attributed to resonantly enhanced eight-photon ionization processes.

The participating nf states can be identified, assuming that they shift parallel with the ionization limit. Within experimental uncertainty this is correct for $5f$, $6f$ and $7f$ however, $4f$ shifts less and $4p$ more – as indicated already in figure 1. The $4p$-peak could only be identified with the help of calculations[2]. By the way this seems to be the first observation of an ac-Stark splitting of l-degenerate levels in such an intense radiation field.

The strong signal fluctuations at energies below 1 eV are an experimental artifact. The maximum below 1 eV can be attributed to the influence of $n = 2$ and $n = 3$ states, which are heavily mixed and broadended. The quantum numbers in quotation marks are only used for identification purposes.

At higher energies above 2 eV$\approx \hbar\omega$ in the next ATI-order the $4f$ and $5f$ peaks can be identified and at longer wavelengths also contributions from "$2s/3p$". The experimental energy resolution decreases at higher energies.

With increasing wavelength the Rydberg series of peaks with $n \geq 4$ becomes less important and the broad "$2s/3p$" structure begins to dominate. The reduced signal from higher n resonances can easily be understood. At longer wavelengths which means smaller photon energies the "resonance intensities" of these states becomes lower, and because they are higly nonlinearly coupled to the groundstate by seven photons the corresponding MPI processes are largely reduced.

As atomic hydrogen is the simplest real atom it has been used as a test case for many theoretical calculations ranging from perturbative calculations[4], classical simulations[6] to "single diabatic Floquet state" approximation[2,3] and solutions of the time dependent Schrödinger equation[7]. For the simulation of a realistic experimental laser pulse it is necessary to determine ionization probabilities for "all" pulse intensities present in a laser focus up to a maximum pulse peak intensity. This has up to now been prohibitive for an application of the last methode mentioned above.

As a consequence of the large ac-Start shifts which can be larger than $\hbar\omega$ it is evident that calculations which have used an undisturbed atom give no realistic picture.

In the following, we shall consider only results of those calculations which were performed for the wavelengths and intensities corresponding to our experimental conditions. An all-order multiphoton resonant theory has been developed and applied by Gontier and Trahin[4]. They were able to calculate an electron energy spectrum which is in good agreement with our experiment at $\lambda = 608\ nm$ for electron energies below 1.8 eV with respect to the nf peaks for $n = 4, 5$ and 6. They also consider very carfully the influence of the laser pulse peak intensity and pulse duration on the shape of the electron spectrum.

An even more detailed comparison is possible with results of calculations performed by Dörr et al.[2,3]. The theoretical details may be found somewhere else.[2,3]. Figure 4 shows an example of a comparison for a wavelength of $\lambda = 608\ nm$ and two different laser pulse peak intensities. The uncertainly of the experimental peak intensity is of the order of ± 20 %. The vertical scales have been adjusted at 1.2 eV in the lower spectrum.

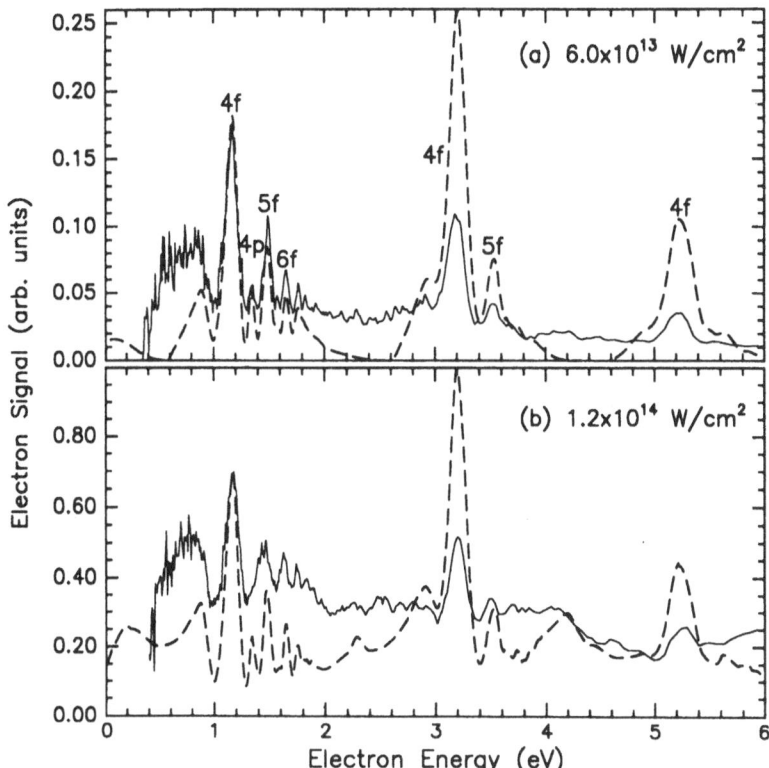

Figure 4. Photoelectron spectra for two different intensities. Full curve: experimental result; broken curve: theoretical result. The wavelength is $608 nm$ and the pulse duration $0.4\ ps$.

The general structure of sharp peaks coincides nicely. The experimental spectra show a higher unstructured background signal for $6 \cdot 10^{13}$ Wcm^{-2}. There is a reasonable agreement in the relative peak heights between 1 eV and 2 eV. The height of the experimental signal below 0.7 eV is not reliable because the transmission of low energy electrons through the time-of-flight spectrometer degrades.

A significant difference exists with respect to the relative ATI-contributions of the $4f$ and $5f$ resonances. The calculations predict higher contributions at higher orders than those in the experiments. If the model picture of resonant ionization processes occuring mainly around resonance intensities is correct, this discrepancy cannot arise from the experimental uncertainly or fluctuations with respect to the spatio-temporal intensity distribution in the focal volume. The relative ATI-distribution should only be determined by the branching ratio at the "resonance intensity". It is obviously this branching ratio which differs between experiment and theory. Presently we have no convincing explanation for this discrepancy.

In the context of these nf resonant ionization processes we wish to address the question of population being transfered into these states and its possible survival. De Boer and Muller[8] have recently observed population in f-states of xenon after an intense 100 fs pulse. They have proposed a two step model in which these excited states are populated when the intensity during the laser pulse is around the resonance intensity and that ionization from these excited states can proceed from then on during the rest of the laser pulse. This opens the possibility for some excites state population to survive.

We find no evidence supporting the applicability of this model for the $4f$ resonance of atomic hydrogen. The calculations[3] predict a survival time of a few fs for the $4f$ state at resonance intensity. Two experimental observations even contradict implications of this model. We do not see a broadening of the $4f$ peak with increasing laser peak intensity, and also the ratio of $4f$ peaks belonging to different ATI-orders does not change with peak intensity – both within the margins of experimental uncertainly.

In addition we should mention a significant experimental difference. We use five times longer laser pulses. The survival chance of excited state population should increase for shorter laser duration – see also below.

Before we finish this section on energy spectra of photoelectrons we wish to point out that the use of short intense laser pulses provides a methode to do spectroscopy of all excited levels of an atom in less than $10^{12} s$. In the focus volume of such a laser pulse resonance intensities for all n-states of atomic hydrogen are present. This type of ac-Stark spectroscopy applies one often applied methode of spectroscopy: "Tune the atom not the wavelength". This ac-Stark spectroscopy is certainly no high resolution methode because the signature of excited states is obtained in the energy spectrum of photo electrons.

The close correspondence between this ac-Stark spectroscopy and wavelength spectroscopy may be seen in figure 5. The upper part shows an energy spectrum of the photoelectrons, and because of the correspondence between electron energy and the intensity at which they are created we can also use an intensity scale as indicated. The lower part shows the wavelength dependence of the generalized seven photon ionization cross section calculated by Karule[9] using perturbation theory. For n between 4 and 7 the spectra look similar. The $p - f$ splitting of the $n = 4$ state was not included in the calculations and the special situation for $n = 2$ and $n = 3$ has already been adressed above.

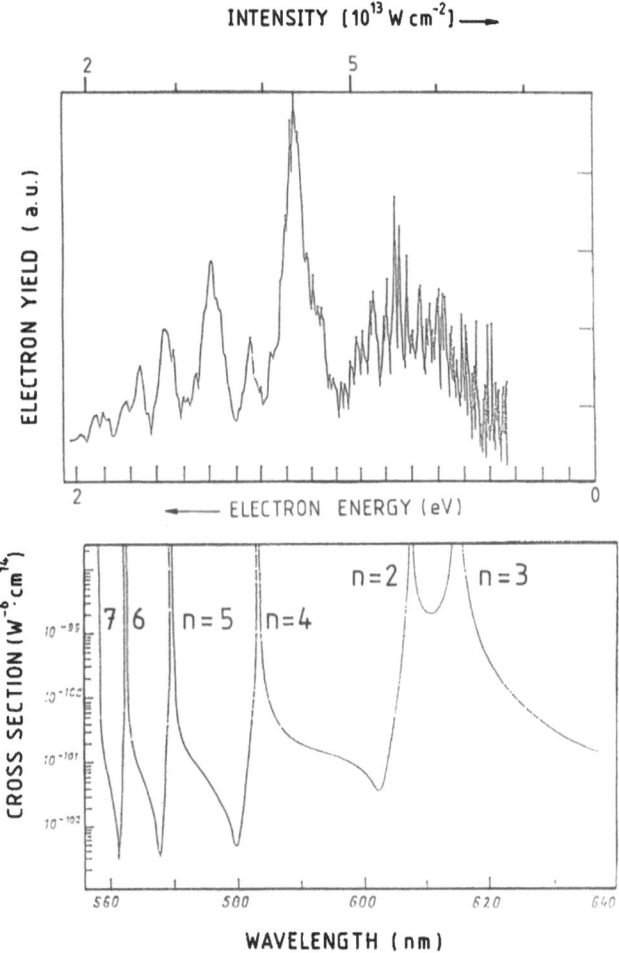

Figure 5. Comparison of a spectrum obtained by ac-Stark scan of the atomic levels at a fixed wavelength (upper part), and the wavelength dependence of the MPI-cross section, calculated by Karule[9] (lower part). For more details see text.

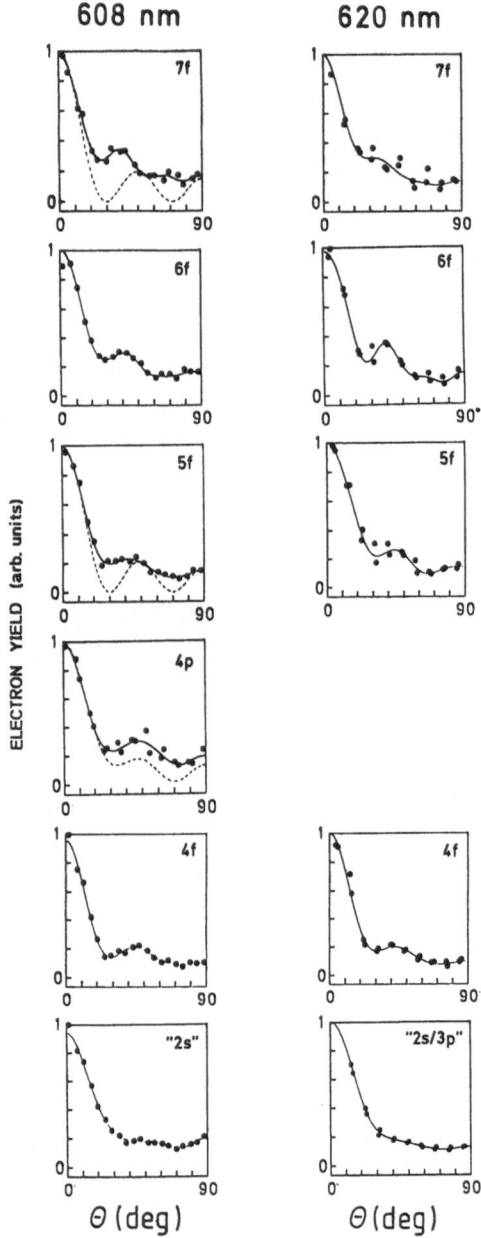

608 nm **620 nm**

ELECTRON YIELD (arb. units)

Θ (deg) Θ (deg)

Figure 6. Angular distribution of photoelectrons. The solid line through the experimental points (dots) is a fit to Legendre polynomials. The broken lines are results of calculations by Gontier and Trahin[11].

Angular Distributions of Photoelectrons

In general differential cross sections like angular distributions provide data for a more detailed comparison with theory or models. The angular distribution of photoelectrons from MPI by linearly polarized light can be described as a super-position of partial waves with respect to the angular momentum of the final continuum states. For linearly polarized light in dipol approximation the following selection rules are applicable for each photon step: $\Delta l = \pm 1$; $\Delta m = 0$. According to a propensity rule by Fano[10] excitation steps with $\Delta l = +1$ use to be stronger than those with $\Delta l = -1$. In several simple cases it has been found that angular distribution of outgoing electrons can be used to asign the angular momentum of intermediate resonances contributing to the MPI-process.

Figure 6 shows experimental angular distributions of energy selected electrons corresponding to peaks in the energy spectrum figure 3. The dots represent experimental points. The full lines are fits to these points by sums of Legendre polynomials. For three distributions we have included the results of calculations by Gontier and Trahin[11], shown as dashed lines.

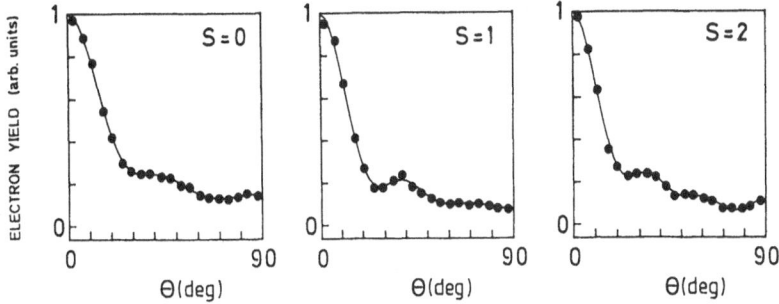

Figure 7. Angular distributions for different ATI-orders. Otherwise like figure 6.

The $4p$ distributions agree quite well. The undulatory character reminds of a g-wave. This contradicts the simple expectation that adding one unit of angular momentum by the ionizing last photon to a p-state should result in an outgoing d-electron. Instead it seems to indicate that the state which we label by $4p$ has admixtures of other (higher) angular momentum when it is in the radiation field at its "resonance intensity". The $5f$ distributions show similarity with respect to the undulations and the $7f$ distributions differ significantly. However, it should be kept in mind that in all these three cases the experimental peaks sit on a remarkably high unstructured background signal.

In figure 7 we show angular distributions for different ATI- orders. For this purpose the signals were integrated over the energy interval of one ATI-order. Although the undulatory structure of the distributions is not very pronounced, it may be seen, that the number of wiggles increases with S. This is an indication of accumulation of angular momentum in accordance with Fano's[10] propensity rule.

We hope that more detailed comparison with theory will be possible in the near future[12].

An Extra Series of ATI-Peaks

In this subsection we wish to discuss an additional series of peaks in the energy spectrum of the photoelectrons. Figure 8 shows extended experimental spectra for three different laser pulse peak intensities. The features we wish to discuss here are the extra peaks marked by the arrows which show up at around multiples of the photon energy. There is no significant contribution at $E_K \approx \hbar\omega \approx 2$ eV in the lowest ATI-order.

The relative contribution of these peaks is not visible in the spectrum at the lowest intensity of $6 \cdot 10^{13}$ Wcm^{-2} and it dominates the high energy part of the

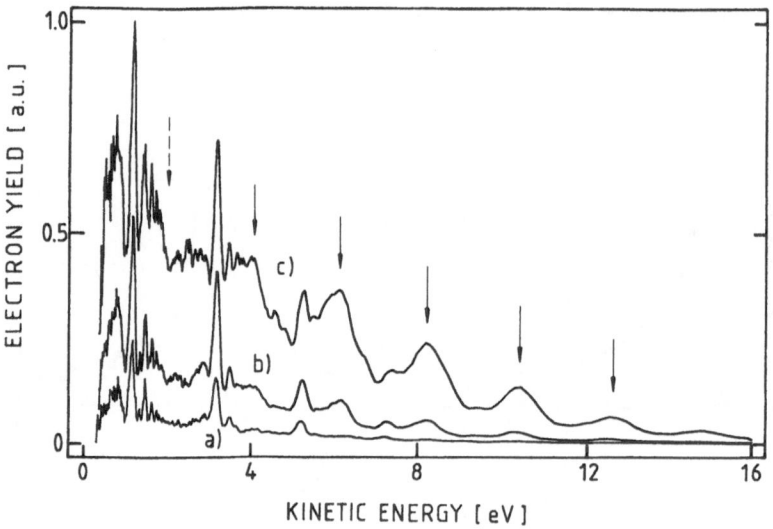

Figure 8. The spectrum of photoelectrons for three different intensities. The series of extra peaks is marked by the arrows. The intensities are: a) $6 \cdot 10^{13}$ Wcm^{-2}, b) $8 \cdot 10^{13}$ Wcm^{-2}, c) $1.2 \cdot 10^{14}$ Wcm^{-2}.

spectrum at $1.2 \cdot 10^{14}$ Wcm^{-2}. At longer wavelengths the extra peak series becomes less important.

Two main conclusions can be derived:

First, the high intensity threshold for the appearance of the peaks and the high energies (ATI-orders) at which the electrons show up, clearly indicate that they are produced by ionization processes which proceed at high intensities above $6 \cdot 10^{13}$ Wcm^{-2} only.

Second, the electron energies around multiples of the photon energy restrict the possible participation of bound state resonances to states which are either very loosely bound near the ionization limit or to states which lie one or two photons below the ionization limit at "resonance intensity".

We want to discuss three tentative explanations for this extra series of peaks. Two of them are based on theoretical calculations and the last one is a qualitative model based on Rydberg wave packets.

1. Recent calculations for MPI of H^--ions by Faisal and Scanzano[13] predict a series of extra peaks at multiples of the photon energy. They show up at high intensities only and do not shift with intensity. They are related to the opening of new reaction channels for stimulated multiphoton bremsstrahlung processes and they are a type of Wigner cusps. Two arguments make this explanation somewhat doubtful for our case of MPI of neutral atomic hydrogen. First, Wigner cusps usually are abrupt changes of underlying cross sections. We do not observe an underlying signal at higher ATI orders. Second, for a neutral atom there exists a smooth transition from ionization continuum down into the region of high Rydberg states, which do not exist in H^--ions. Therefore in atoms, stimulated bremsstrahlung can smoothly turn into stimulated recombination below the continuum limit. This might wash out cusp structures which are related to the opening of new channels.

2. The calculations by Dörr et al.[2] show a state which exhibits a peculiar ac-Stark shift. It is included in figure 1 and starts from $3p$ at low intensities. As mentioned, $3p$ is strongly coupled to $2s$ and the low lying continuum – at low intensities. At intensities around 10^{13} Wcm^{-2} this coupling to the continuum changes into a coupling with high Rydberg states. Reliable calculations were not possible in this range and the further evolution of this state with increasing intensity is undetermined. At somewhat higher intensities a new state appears which shifts slightly more than the ionization limit. As shown in figure 1 this state intersects the $(1s + 7\hbar\omega)$ energy at an intensity above $6 \cdot 10^{13}$ Wcm^{-2}. This high "resonance intensity" and its energy, being almost one photon below the ionization limit at this intensity, make it a good condidate for serving as resonance in an MPI process leading to the extra series of ATI peaks. Presently there exist no calculated electron energy spectra for energies above 6 eV.

3. The third model picture may be characterized as time-delayed ionization of electrons parked in Rydberg states as wavepackets.

Conventionally, Rydberg wavepackets are prepared by excitation of a superposition of Rydberg states with broad band ps-laser radiation. In a classical picture: This preparation occurs at small distances from the ion core. Then the wave packet evolves and explores the outer region of the Coulomb potential, it is reflected and approaches the core region again. During this excursion, at large distances from the core photoabsorption is inefficient – it more or less behaves like a free electron. Its classical round trip time is $\tau(\bar{n}) \approx 2\pi\bar{n}^3$ where $\bar{n}$ is the mean principal quantum number of the superposition. Efficient photo absorbtion is possible when the electron is near the ion core again. More details may be found in the literature[14].

In our model wavepackets can be excited by "resonant" multiphoton excitation during the rising part of the laser pulse when the Rydberg states are rapidly ac-Stark tuned through resonance. Figure 9 shows the typical temporal laser intensity profile for a point in the focal volume which is exposed to a pulse peak intensity of 10^{14} Wcm^{-2}.

The Rydberg states from $n = \infty$ to $n = 10$ are tuned through resonance within 25 fs, as indicated in figure 9a) and on a larger scale in figure 9b). The picture of "resonant" excitation is certainly a simplification for this short time scale.

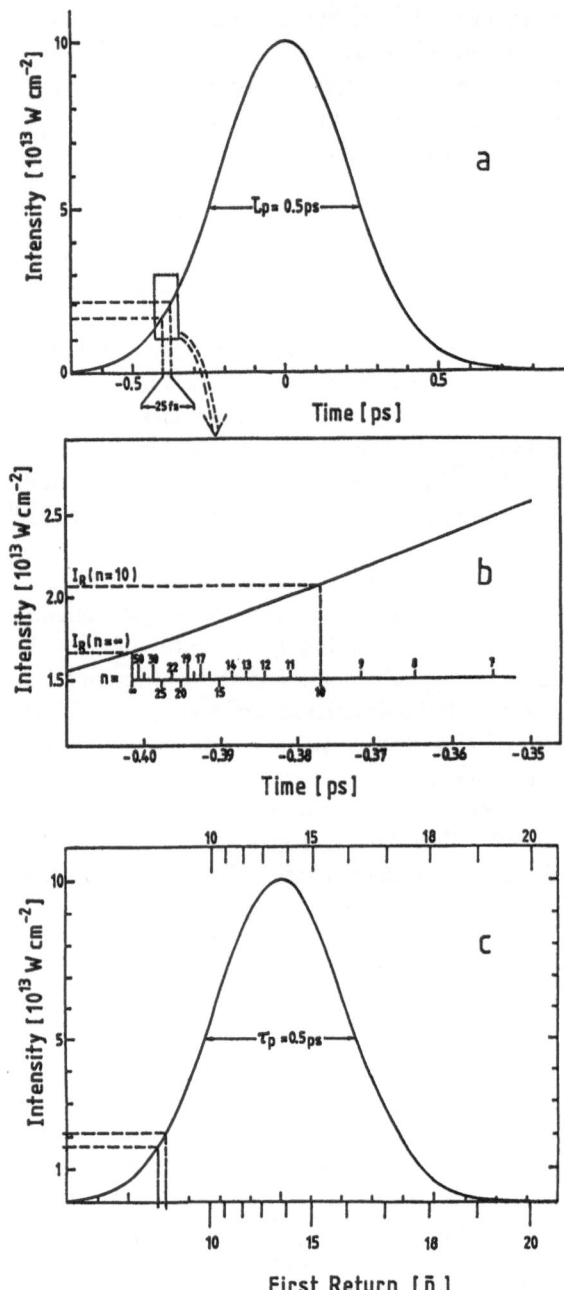

Figure 9. a) A typical temporal laser pulse intensity profile. The "resonance intensities" for high Rydberg levels are indicated. b) The corresponding part and "resonance intensities" during the pulse rise is shown on a larger scale. c) The same profile as a) with the return times of packets with different ñ-values.

Nevertheless, a coherent preparation of Rydberg states and creation of a wave-packet should be possible. The detailed composition of the packets and their $\bar{n}$-value may differ largely depending upon the location in the focal volume and its corresponding intensity profile. After excitation the packet moves away from the ion core and can hardly be photoionized until it returns. Figure 9c shows the return times for wave packets (or parts of them) belonging to different $\bar{n}$-values. It can be seen that contributions with $\bar{n}$ between $\bar{n} \approx 10$ and $\bar{n} \approx 16$ return during the highest intensities of the laser pulse. Therefore they can now be ionized more efficiently than during their excitation at lower intensities. Ionization at these high intensities can lead to absorption of many excess photons in the continuum giving photoelectrons at high energies – as is observed. In this model picture also the other observations can be explained. At longer wavelengths the excitation of packets is less efficient, because high Rydberg states are tuned through resonance at lower intensities, and even small changes of the "resonance intensities" can make a large effect because the excitation is a highly non-linear seven-photon process. High Rydberg states are energetically very close to the ionization limit and they shift parallel to it with intensity. This can explain the peak positions at multiples of the photon energy.

One consequence of this model is the possibility of survival of electrons in high Rydberg states ($\bar{n} \geq 18$ in our case) because the corresponding wavepackets do not return to the core during the laser pulse. This effect should be even more pronounced for shorter laser pulses.

We presently favour the last two explanations and believe that further experiments and (or) calculations will enable us to decide which process dominates.

SUMMARY

Resonant ionization processes play a dominant role in MPI of atomic hydrogen in short laser pulses. using linearly polarized light at wavelengths around $600\ nm$ and intensities below $10^{14}\ Wcm^{-2}$. The energy spectrum of the photoelectrons reveals that almost all excited states are ac-Stark tuned through resonance. In principle one can obtain a spectrum in less than $10^{-12}s$. A comparison with recent theoretical calculations shows good agreement with respect to peak positions and relative peak heights in the low energy range. The relative contributions of resonant processes to different ATI-channels disagree.

We find all kinds of ac-Stark shifts: perturbative behaviour of the ground state for which the static Coulomb field dominates, perturbative behaviour for the high lying Rydberg states and the continuum limit which show free electron behaviour, and irregular shifts for $n = 2$ and $n = 3$ and $n = 4$ states.

The intense field ac-Stark splitting of the $n = 4$ is observed.

Only a few of the experimental angular distributions can presently be compared with theory.

An extra series of high energy electron peaks was observed and waits for a definite explanation.

In conclusion we wish to point out that quantitative agreement between experiment and theory begins to show up for some details like energy spectra of photoelectrons for MPI processes in a regime where neither the atomic potential nor the radiation field can be treated perturbatively. This does not mean that everything is clearly understood.

ACKNOWLEDGEMENTS

Support from the Deutsche Forschungsgemeinschaft SFB 216 is gratefully acknowledged.

REFERENCES

1. R.R. Freeman, P.H. Bucksbaum, H. Milchberg, S. Darack, D. Schumacher, and M.E. Geusic, Phys.Rev.Lett. 59, 1092 (1987)

2. M. Dörr, R.M. Potvliege, and R. Shakeshaft, Phys.Rev. A41, 558 (1990)

3. M. Dörr, D. Feldmann, R.M. Potvliege, H. Rottke, R. Shakeshaft, K.H. Welge, and B. Wolff-Rottke, J.Phys. B 25, L 275 (1992)

4. Y. Gontier, and M. Trahin, Phys.Rev. A46, 1488 (1992)

5. for example see: M. Dörr, R.M. Potvliege, and R. Shakeshaft, J.Opt.Soc.Am. B7, 443 (1990)

6. see contributions in this book

7. K.C. Kulander, K.J. Schafer, and J.L. Krause in "Atoms in Intense Laser Fields" edited by M. Gavrila (Academic Press, New York 1992)

8. M.P. de Boer, and H.G. Muller, Phys.Rev.Lett. 68, 2747 (1992)

9. E.M. Karule, priv. communication, and printed in Russian, Riga 1975

10. U. Fano, Phys.Rev. A32, 617 (1985)

11. Y. Gontier and M. Trahin, J.Opt.Soc.Am. B7, 463 (1990)

12. Calculations are being done by the authors of ref. 2

13. F.H.M. Faisal, and P. Scanzano, Phys.Rev.Lett. 68, 2909 (1992)

14. G. Alber, and P. Zoller, Phys. Rep. 199, 231 (1990)
    I. Sh. Averbukh and N.F. Perel'man, Sov.Phys.Usp. 34, 572 (1991)
    J.A. Yeazell and C.R. Stroud, Phys.Rev. A43, 5153 (1991) and references therein

STRUCTURE IN THE ABOVE THRESHOLD IONIZATION (ATI) SPECTRUM UNDER STRONG FEMTOSECOND PULSES

W. Nicklich[a,b], H. Kumpfmüller[a,b], H. Walther[a,b],
Huale Xu[c], X. Tang[c] and P. Lambropoulos[a,c]

(a) Max-Planck-Institut für Quantenoptik, D-8046 Garching, Germany
(b) Sektion Physik der Universität München, Germany
(c) Department of Physics, University of Southern California
    Los Angeles, CA 90089-0484

I. ATI SPECTRUM IN CS

Above threshold ionization (ATI) has until recently been observed mostly in the rare gases[1-4] with the exception of one experiment and related theory on hydrogen[5]. A recent experiment[6] on Cs using extremely short (around 80 fs) pulses at around 600 nm and intensities as high as 10^{14} W/cm² demonstrated the possibility of observing extensive ATI structure even though the atom can ionize with the absorption of two photons. The reason is that the extremely short pulse duration allows the atom to experience intensity sufficiently high for ATI to develop before saturation (complete ionization) by the lowest order process.

These data[6] shown in Fig. 1 exhibited a slight substructure on the side of the peaks as well as a blue shift of the main peak. The reason for this blue (towards higher photoelectron energies) shift is the AC Stark shift of the ground state which had an opposite sign and much smaller size in previous experiments. The shift of the ATI peaks (in atomic units) in first nonvanishing order is given by

$$\Delta E_g - \Delta E_{thres} = \frac{I}{4}\left(-a_g(\omega) - \frac{1}{\omega^2}\right)$$

where ΔE_{thres} is the shift of the ionization limit which is equal to the ponderomotive shift and ΔE_g is the shift of the ground state determined by its polarizability. The shift of the ionization threshold by 0.38eV at I = 10^{13}W/cm² is overcompensated by the ground state shift amounting to 0.90eV. If we include the temporal saturation ($I_{sat} = 4.2 \times 10^{12} W/cm^2$) by calculating the position of the maxima of the first two ATI-peaks, we obtain the solid lines in Fig. 2. The shifts in the low intensity region are in good agreement with Eq. (1). At higher intensities the shifts become constant, since we reach saturation. The mismatch between the theoretical and experimental saturated values

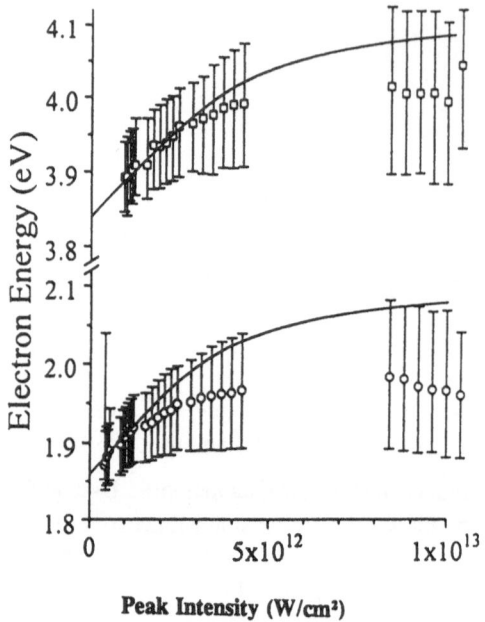

Peak Intensity (W/cm²)

Figure 1. (a) Experimental electron energy spectrum measured after multiphoton ionization of Cs with laser pulses of about 90fs duration and a peak intensity of about 10^{14}W/cm² al a wavelength of 621 nm in the direction of the laser polarization. (b) TD calculation at an intensity of 5 x 10^{12}W/cm². The pulse shape is trapezoidal with 6 optical cycles turning the field on or off and 30 cycles at maximum. The peaks below threshold correspond to population trapped in excited states. (a) and (b) are separated slightly in the vertical direction for visual convenience.

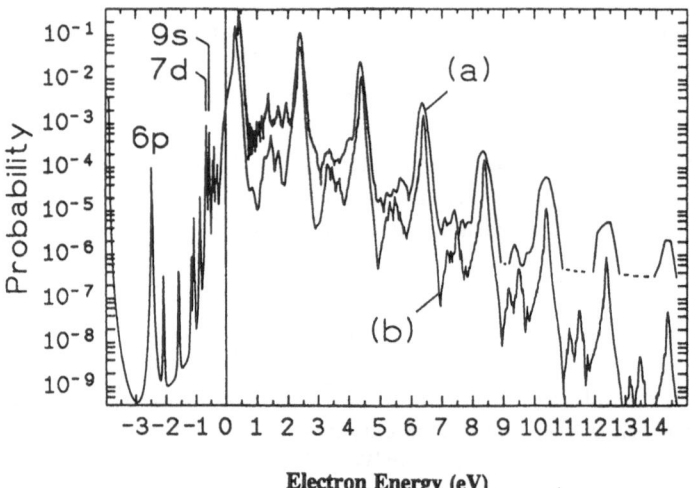

Electron Energy (eV)

Figure 2. Shift of the first and the second ATI-peak versus laser peak intensity. The dots show the mean energy, the error bars indicate the width of the peaks (i.e. the region that contains 68% of all counts). The solid lines indicate the shift expected from PT including saturation. The pulse duration is 70fs.

(Fig. 2) can be explained if we assume that our theoretical saturation intensity is too large by about a factor of two.

The substructures appearing in Fig. 1 have an origin entirely different from that observed in earlier experiments in which the lowest order process of ionization was more than six. In these cases, excited atomic states could come into resonance by shifting upward by an amount practically equal to the ponderomotive shift. These are states coupled to the ground state by a few-photon Rabi frequency which is weak compared to the single-photon ionization from that state. Thus whatever substructure was observed in those cases, it was due to excited states populated (sometime during the pulse) by a "weak" Rabi frequency and depending on the situation (angular momentum of the state etc.), ionized either instantly or somewhat later, with the possibility that part of the population is retained after the end of the pulse. Here we have the diametrically opposite situation. Since the energy of one photon is not too far above the 6s $\rightarrow$ 6p transition and the matrix element is fairly large, we have an enormous Rabi frequency (equal in fact to $1.0^6 \times 10^8 \sqrt{I}$ rad/s, where the intensity I is to be inserted in W/cm^2). This makes the Rabi frequency larger than the detuning even below 10^{12} W/cm^2. As a result, we have an immediate deposition of population to the 6 p state. The absorption of one more photon from there reaches excited states of the type 9d, 10s etc, from where the atom ionizes by the absorption of one more photon. We have thus at the same time a non-resonant 2-photon ionization from the 6s and a 3-photon three-step ionization through the 6p and subsequent s and d states. Of course, as explained above, the 6s shifts upwards. The substructure is therefore due to the population in excited states which is initiated by the sudden population of the 6p. This can also be viewed (in a different language) as an AC Stark splitting of a rapidly varying magnitude as the intensity rises and falls. Although only one laser is involved, the energy analysis of the photoelectrons serves as a probe of that splitting, as well as of the shift of the ground state. Since the field is detuned considerably from the strong transition 6s $\leftrightarrow$ 6p, the peak heights of the two components are asymmetric; one being the main (tall) ATI peak and the other being the small side peak with its satellites due to the further population of higher states.

The laser pulses used in the experiment have a pulse duration of 70 to 35 fs and a peak intensity of 10^{14}W/cm^2. The wavelength of the laser pulse is about 635 nm. Details about the experimental setup can be found elsewhere[6].

The ionization potential of Cesium is 3.89eV, while the photon energy about 1.96eV. We herefore expect two photon ionization very close to the threshold. The one photon step is away from any resonance and about halfway between the 6P and 7P states. In order to explain the measured results we have to assume a small contact potential of about 0.2eV in the electron spectrometer. Since the height and position of the lowest peak is strongly affected by the closely lying threshold and the contact potentials in the time of flight setup, it is omitted in the discussion.

The measured electron spectrum in Fig. 1a, a result of 5×10^5 counted electrons, shows ATI-peaks up tu 14eV, the highest one corresponding to a nine photon process. The peaks are separated by the photon energy. The height of the peaks decreases with energy above threshold.

The spectrum calculated through TD theory[7,8] is shown in Fig. 1b. the general structure is remarkably similar to that of the experiment including the substructures (sidepeaks). One reason for the difference in the ratio of the ATI peak heights between theory and experiment may be the fact that in the experiment the measured electron signal is in the direction of the polarization, while the theory produces angle integrated spectra.

II. ATI SPECTRUM AND SUBSTRUCTURES IN Na AND Li

In the previous section, we presented experimental evidence for substructure in the ATI peaks of Cs under a very short pulse, and provided a theoretical interpretation based on non-perturbative time-dependent (TD) theory matching its basic features. We also provided a physical description of the underlying processes and their difference from previous experiments. This physical description and picture suggest that Na and Li, under the same conditions, should exhibit similar but more pronounced and robust features for two reasons: (a) The lowest order ionization is 3 instead of 2, which means that it will require a bit higher intensity to saturate. (b) The detuning of the photon from the single-photon resonance transition is considerably smaller, which should entail more pronounced Rabi oscillations and hence substructure. We consider these two atoms in this section.

The absorption of two photons in Na and Li are far from resonance (by at least 800 cm^{-1}) with any bound states. The only excited states that could conceivably come into 2-photon resonance (as they shift upwards during the pulse) are the 3d in Na and the 3d and 4s in Li. But these states become strongly coupled (albeit nonresonant) to the respective first excited states (nP) and through those to the ground state. Thus their shifts are not expected to be simple functions of the intensity. Moreover the Rabi frequency between the ground and first excited states is enormous at intensities of the order of $10^{12}W/cm^2$. One would therefore expect new and hitherto unexplored effects, provided the pulse is sufficiently short for the atom to experience the necessary intensity as demonstrated in the previous section on Cs where extensive Above Threshold Ionization (ATI) was reported.

In order to understand first the expected dynamical behavior, we examine photoelectron spectra of Na and Li under pulses of photon energy 2eV for which the resonant transitions (at zero field) are detuned by 800cm^{-1} for Na and 1200cm^{-1} for Li, with the photon energy being smaller than the energy of the resonance transition 3s-3p in Na and larger than the energy of the resonance transition 2s-2p in Li.

In Fig. 3 we show photoelectron spectra in Na calculated through TD theory for pulses of trapezoidal pulse shape and peak intensities (a) 10^{12} W/cm^2 and (b) 2 x 10^{12} W/cm^2. First, we note that the substructure of the peaks is enhanced by almost two orders of magnitude as compared to that obtained in Cs under identical conditions (Fig. 1). Second, as also shown in Fig. 3, substantial amount of population is trapped in excited states even at the end of the pulse. Note that, although the first excited state 3p can never shift into resonance during the pulse, and is well outside the pulse bandwidth, it still does get populated because of the large Rabi frequency. Third, almost every subpeak except the central one, can be associated with direct ionization from some excited state, as indicated in Fig. 3. Closer examination shows that the central peak shifts with the peak intensity, and that the energy difference between the central peak and the subpeaks due to direct ionization from 4d and 5s are at 0.32eV and 044eV for peak intensities 10^{12} and 2 x 10^{12} W/cm^2, respectively. This can be traced to Rabi-like splitting due to the off-resonance strong coupling of the ground state 3s to the first excited state 3p. Rabi-like splittings must be correlated with oscillations in the populations of the states involved in the coupling. This we examine below.

In Fig. 4a we plot the populations in the 3s, 3p, and 4 d states as a function of time for peak intensity $I=10^{12}W/cm^2$. They clearly show a very strong Rabilike

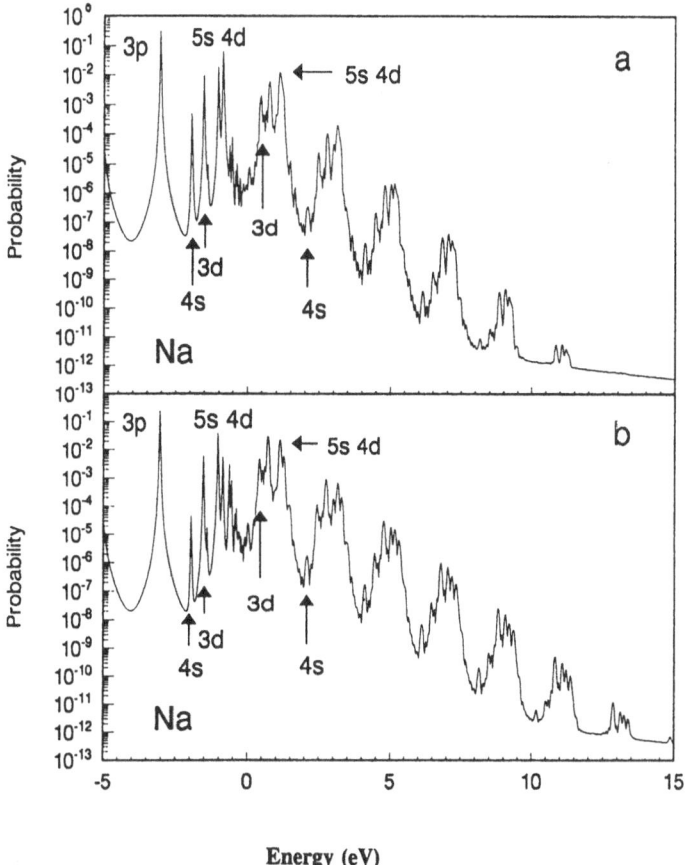

Figure 3. The calculated ATI spectra for the Na with photon energy = 2eV, the pulse shape is trapezoidal with 6 optical cycles turning the field on or off and total 32 optical cycles (67 fsec). The peaks below threshold correspond to population trapped in excited states. (a) Intensity $I = 10^{12}$W/cm². (b) $I = 2 \times 10^{12}$W/cm².

oscillation between the 3s and 3p, i.e. whenever there is a peak in 3s one can find a dip corresponding to the 3p. For Na one more photon from the state 3p can also reach the neighborhood of the excited states 4d and 5s, while a third photon will ionize the atom. Thus we have also calculated the population in the 4d but the coupling here is much weaker compared to the coupling between the 3s and 3p, so that the two-photon relatively weak ionization from the strongly coupled quasi-two-level system can serve as a probe of the splitting. The behavior in Fig. 6a does indicate that this is not a pure two-level system with decay, but with some perturbers. That is why we can see a kind of population revival[11] in the 3p and 4d. In Fig. 4b we have shown the populations in the 3s and 3p states for higher peak intensity ($I = 2 \times 10^{12}$W/cm²) at which, the population revival becomes much more pronounced. This is simply because at higher intensity, more states are coupled. The population revival here is different from the collapse and revival of a Rydberg wavepacket[9]. In the Rydberg wavepacket, one excites a superposition of Rydberg states and then observes the evolution of the wavepacket in a field-free

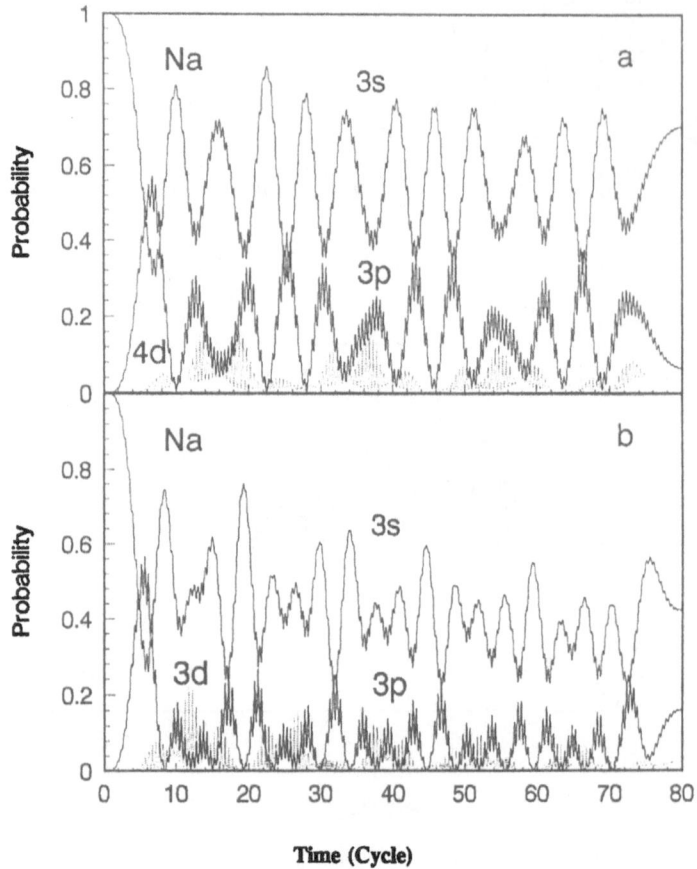

Figure 4. Populations in 3s, 3p and 4d states of Na as function of time with the same condition as in figure 1, but total 80 optical cycles. (a) $I = 10^{12}W/cm^2$ (b) $I = 2 \times 10^{12}W/cm^2$

condition. The revival we have shown here for Na, is in the presence of strong fields and can be viewed as the revival of a state after the electron has made long excursion (has spread out) into the continuum. If we assume that only the 3s and 3p are coupled, the Rabi-like splittings are 0.32eV for $I = 10^{12}W/cm^2$ and 0.6eV for $I = 2 \times 10^{12}W/cm^2$, and they are quite close to the values of the separation between two subpeaks in the ATI spectra as we discussed earlier.

As we mentioned above, Li would be another candidate for strong coupling behavior. The conditions for Li are a little bit different. First, one 2eV photon will be above the excited state 2p. Second, the detuning of the one photon transition between the 2p and 3d is much smaller than in the analogous case of Na, and also the Rabi frequencies are almost equal for those two couplings, ($\Omega_{sp} = 8.23 \times 10^{14}$ rad/sec and $\Omega_{pd} = 8.27 \times 10^{14}$ rad/sec for $I = 2.5 \times 10^{12}$ W/cm^2). Thus Li is expected to behave more like a three-level system or even multilevel system when the laser intensity is high.

Fig. 5 shows spectra of Li at the intensity of 2.5 x 10¹²W/cm² with photon energy equal to 2eV. We can see very clear structures in the ATI peaks some of which can be assigned to the direct ionization frome excited states. Some of them cannot at this time be assigned to any state, which may due to Rabi splitting and those subpeaks also shift as the intensity changes. In Fig. 6 we plot populations of 2s, 2p and 3d as a function of time, because due to the strong coupling of those states, we expect to see large Rabi-like oscillation as indeed is the case in Fig. 6. But we also note unusual behavior for the 2p state in that it hardly oscillates. Its population, after some initial transient oscillation settles into a relatively constant value which can be understood as population stabilization due to coherent transitions between a few levels, despite the fact that the electron is driven by the strong field far out into the continuum. This suggests that Li should withstand unusually large intensity before ionizing completely.

We have in fact observed this behavior. Under conditions identical to those of the ATI experiment in Cs, where Cs was found to saturate at intensities around 4 x 10¹² W/cm², we find that Li has not saturated even at 10¹⁴W/cm² and pulse duration 70fs.

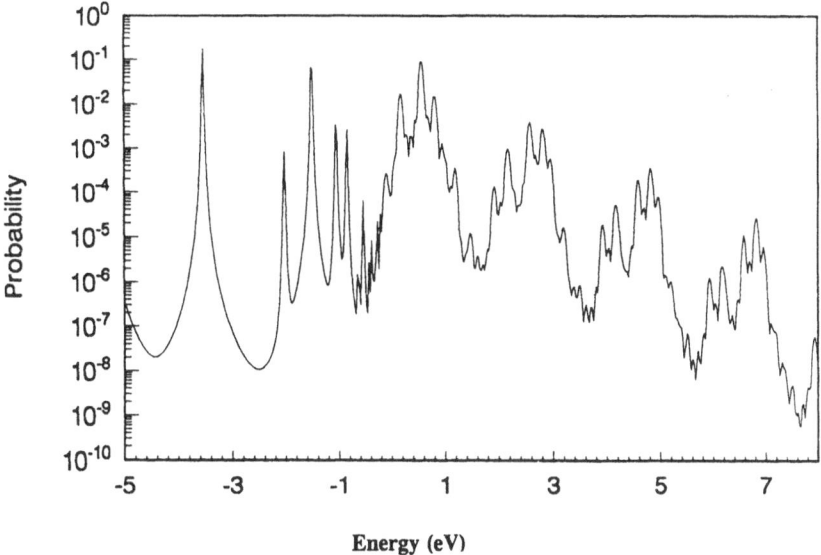

Figure 5. The calculated ATI spectrum for Li with peak intensity of pulse of 2.5 x 10¹⁴W/cm². Pulse shape and the photon energy are the same as in Fig. 3.

We have also observed very pronounced structure in the ATI spectrum as shown in Fig. 7. Given that for computational reasons, we do not yet have theoretical spectra integrated over the spatial distribution of the radiation, we do not expect a detailed matching between theory and experiment, but only a correspondence between the most important (robust) features of the spectrum. These should be the energy positions of the most prominent photoelectron substructures. Indeed as seen in Fig. 7, those substructures

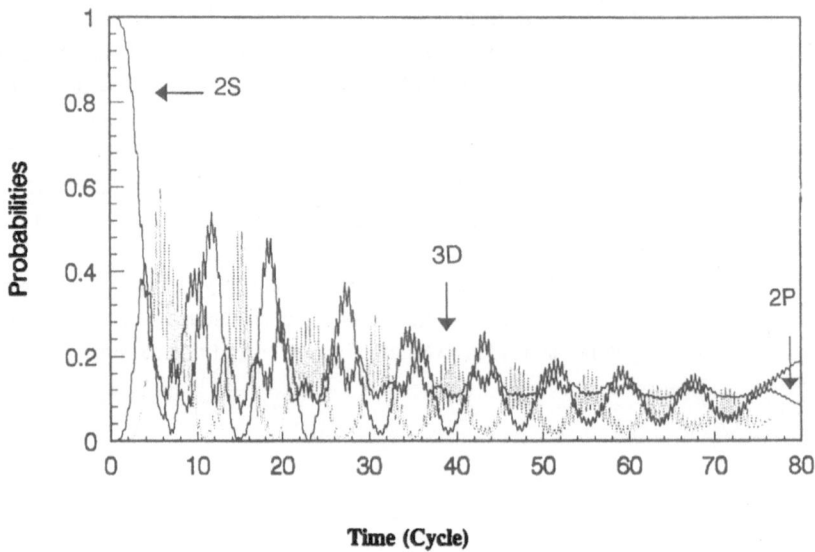

Figure 6. Populations in 2s, 2p and 3d states of Li as function of time with the same conditions in Fig. 5.

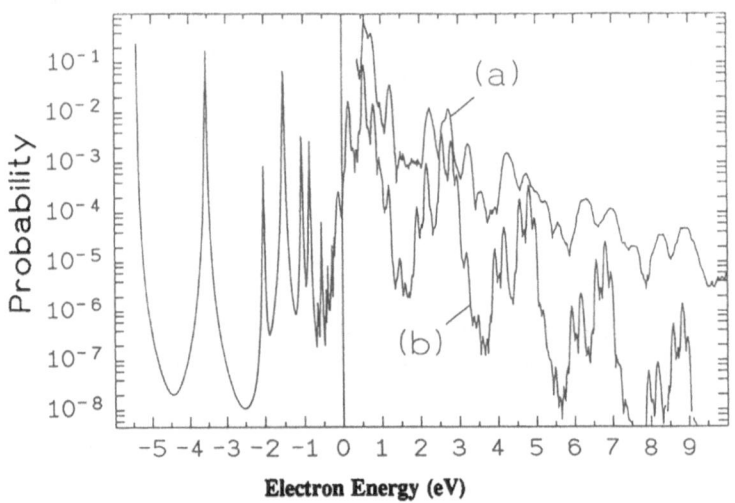

Figure 7. ATI spectra for Li. (a) Experimental data with pulse duration of 35fs . (b) Same as Fig. 5.

of the theory that are sufficiently pronounced to survive the background due to stray electrons are mirrored in the experimental photoelectron spectrum. As expected, the minima between successive ATI peaks are much deeper in the theory owing to the absence of space averaging and of course of stray electrons. In fact all subpeaks in the theoretical photoelectron spectrum can be correlated with ionization from some of the excited states whose population at the end of the pulse is also shown in the part of Fig. 7 that corresponds to negative energies.

ACKNOWLEDGEMENTS

The theoretical part of this work was supported by NSF under Grant N°. PHY-9013434 and DOE under Grant N° DE-FG03-87ER60504.

REFERENCES

1. P. Agostini, F. Fabre, G. Mainfray, G. Petite and N.K. Rahman, Phys. Rev. Lett. 42, 1127 (1979)

2. P. Kruit, J. Kimman, H.G. Muller and M.J. van der Wiel, Phys. Rev. A 28, 249 (1983)

3. T.J. McIlrath, P. H. Bucksbaum, R.R. Freeman and M. Bashkansky, Phys. Rev. A 35, 4611 (1987); R.R. Freeman, P.H. Bucksbaum, H. Milchberg, S. Darack, D. Schumacher and M.E. Geusic, Phys. Rev. Lett. 59, 1092 (1987)

4. H.G. Muller, H.B. van Linden van den Heuvell, P. Agostini, A. Antonetti, M. Franco and A. Migus, Phys. Rev. Lett. 60, 565 (1988); P. Agostini, P. Breger, A.L'uillier, H.G. Muller, G. Petite, A. Antonetti and A. Migus. Phys. Rev. Lett. 63, 2208 (1990)

5. H. Rottke, B. Wolff, M. Brickwedde, D. Feldmann and K.H. Welge, Phys. Rev. Lett.64, 404 (1990). M. Dörr, R.M. Potvliege and R. Shakeshaft, Phys. Rev. Lett. 23, 2003 (1990)

6. W. Nicklich, H. Kumpfmüller, H. Walther, X. Tang, Huale Xu and P. Lambropoulos, Phys. Rev. Lett. 69, 3455 (1992)

7. K. C. Kulander, Phys. Rev. A 35, 445 (1987)

8. Huale Xu, X. Tang and P. Lambropoulos, Phys. Rev. A 46, R2225 (1992)

9. D.R. Meacher, P.E. Meyler, I.G. Hughes and P. Ewart, J. Phys. B: At. Mol. Opt. Phys. 24, L63 (1991)

IONIZATION IN STRONG FIELDS

P. Agostini[1], E. Mevel[1], P. Breger[1], A. Migus[2] and L. F. DiMauro[3]

[1]Service de Recherches sur les Surfaces et l'Irradiation de la Matière
CE Saclay
91191 Gif Sur Yvette (France)
[2]Laboratoire d'Optique Appliquée, ENSTA, Ecole Polytechnique
91120 Palaiseau, (France)
[3]Chemistry Department, Brookhaven National Laboratory
Upton, 11973 New York (USA)

INTRODUCTION

The process of atomic ionization by intense electromagnetic field is still the subject of active investigations. Three aspects of this process have recently attracted attention. The first is the progressive loss of the multiphoton character as the field becomes progressively stronger. The second is related to the trapping of population in excited states during transient resonances. The third concerns the process of multiple electron ejection.

What may be called the multiphoton character of ionization stems essentially from the framework of perturbation theory. To lowest non-vanishing order, for instance, the ionization rate should scale as I^N (where I is the intensity and N the number of photons necessary to reach the continuum) and deviation from this power law is a sign of the influence of higher order terms in the perturbation expansion. Physically this usually corresponds to including the atomic states ac-Stark shifts which play a dominant role either in the neibourghood of intermediate resonances or in the kinetic energy of the outgoing electron. One way to handle these shifts is to work within the dressed-atom picture and use the Floquet ansatz. All atomic states acquire intensity-dependent complex energies and intermediate resonances appear as level crossing, or anti-crossing at certain intensities for which the corresponding dynamic detunings become zero. In the limit of low frequency fields, due to the huge number of photons involved in the process, none of these view points is practical and it is much more efficient to think in terms of a classical field. When the frequency is low enough, it becomes effective to describe the process as a dc-field tunnel ionization or even over-the barrier ionization (OBI) when the field becomes strong enough. Theoretically, it is possible, up to a certain point, to extend the Floquet calculation to the tunneling limit. Experimentally, this is more difficult because depletion of the ground state severely limits the intensity range over which it is possible to study the dynamics of the ionization process. Investigations of the transition from one regime to the other in ionization of rare gases are presented in the first section of this paper.

The question of population trapping in states excited through transient resonances during the pulse has recently attracted a lot of attention This question arises in the problem of adiabatic transfer to excited states, the physics of Rydberg wavepackets in strong short pulses, the stabilization of atoms in strong fields. Furthermore, recent apparently contradictory experiments have spurred interest in this problem. In the second section, this question is addressed through the description of a single-state transient resonance in a case study using the resolvent operator method. It is shown that the space and time localization of ionization, which was believed to be contradictory with population trapping, is in fact the key to understand that residual population may be left in the excited state. The two viewpoints recently opposed in the literature [1] are therefore shown to be compatible.

Finally the last section summarizes recent advances in the problem of two-electron ejection in strong fields. One of the long-standing questions is to evaluate the role of electron correlations in such a process since, contrary to double photoionization (one photon, two electrons), multiphoton double ionization does not require such correlations. In the language of perturbation theory, this question is the same as asking the role of intermediate two-electron states in a multiphoton transition coupling the ground state to the double ionization continuum. In a non-perturbative language, this has been viewed as a possible shake-off process or even as a tunnel or over-the-barrier ionization of the first electron followed by an e-2e collision of this electron with its parent ion. Recent findings in double ionization of xenon and helium atoms are briefly reported and discussed in this section.

FROM THE MULTIPHOTON TO THE DC-FIELD IONIZATION REGIMES

It is well known that atoms can be ionized by photons with an energy much smaller than the ionization potential. To the lowest non-vanishing order of time-dependent perturbation theory, the probability of such a process is proportional to the power $2N$ of the field, (or the power N of intensity) if N photons are necessary to reach the ionization threshold (Fig.1). The LOPT framework is sufficient only if all dynamic detunings remain much larger than the radiation bandwidth. Such detunings are defined as the energy difference between the energy of an integer number of photons m and the atomic states perturbed energies:

$$\Delta_j = m\,\hbar\omega - \widetilde{E}_j \tag{1}$$

$$\widetilde{E}_j = E_j + S_j \tag{2}$$

where S_j denotes the ac-Stark shift of the intermediate state j. It is only to this approximation that the distinction between a resonant and a non-resonant transition is meaningful. Under these conditions, the kinetic energy of the outgoing electron is (E_I ionization potential):

$$E_0 = N\,\hbar\omega - E_I \tag{3}$$

$$E_0 = N\,\hbar\omega - E_I$$

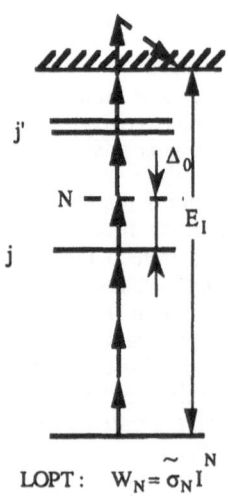

$$\text{LOPT}: \quad W_N = \widetilde{\sigma}_N I^N$$

Fig.1 Energy diagram for MPI.

For N larger than a few units, however, the intensity at which MPI is observable, the LOPT framework is insufficient. The Stark shifts of most atomic states become of the order of the photon energy with two important consequenses:

(i) The effective ionization potential is:

$$\tilde{E}_I \cong E_I + U_p \tag{4}$$

where U_p is the average classical kinetic energy of a free electron quivering in the electromagnetic field, also called ponderomotive energy. This is due to the difference between the Stark shifts of the ground state and Rydberg states.

(ii) There is no longer a clear distinction between resonant and non-resonant MPI since transient resonances are bound to occur during the rise of the pulse as the various resonance intensities are reached in the focal volume.

Another important characteristic of MPI in intense fields is that the atom absorbs many more photons than the minimum N (ATI).

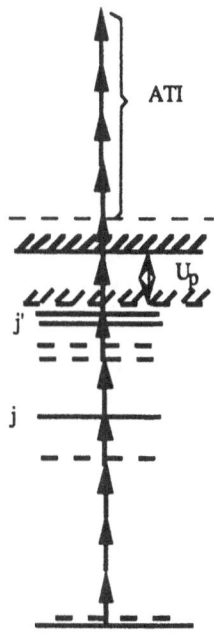

Fig.2 MPI in strong field: Stark-shifted states and effective ionization potential.

In the case of long wavelength, a crude estimate of the maximum electron kinetic energy may be obtained from a classical calculation known as the "Simple Man Theory" by van Linden van den Heuvell and Muller [2]: let us assume that an electron is released at t' by the ionization process with negligible kinetic energy in a classical field E cosωt. Its instantaneous kinetic energy is:

$$T = \frac{e^2 E^2}{2m\omega^2} (\sin^2\omega t + \sin^2\omega t' - 2 \sin\omega t \sin\omega t') \tag{5}$$

and the average:

$$\langle T \rangle = \frac{e^2 E^2}{2m\omega^2} (\frac{1}{2} + \sin^2\omega t') \tag{6}$$

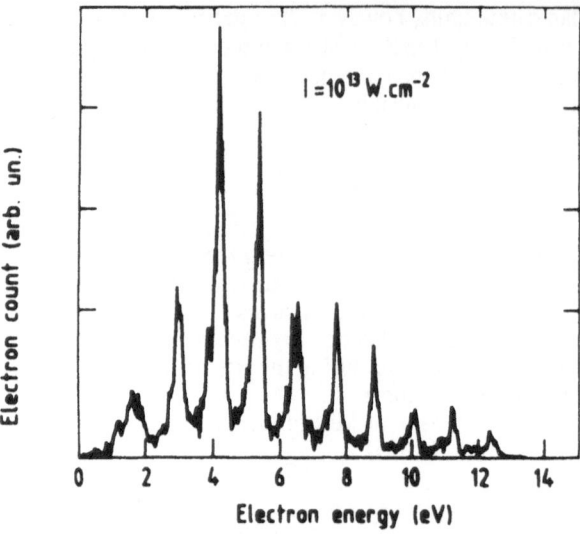

Fig.3 Electron energy spectrum from MPI of xenon showing a case where S ℏω >> Up.

The first term in (6) is the ponderomotive energy while the second represents the energy of a drift motion and depends of the phase at the time t' the electron was "born" in the field (in linear polarization).

T peaks at 8 U_P while $<T>$ peaks at $3U_P$ $U_P/\hbar\omega$ is therefore an estimate of the number of photons absorbed in excess of N. Actually, how much kinetic energy can be retained outside the laser beam depends on the pulse duration. If the pulse is very short, then the ponderomotive energy is lost and $<T>$ peaks at $2U_p$. If the pulse is long then the quiver motion which gives rise to the ponderomotive energy is converted into translation motion as the electron travels through the field gradient and $<T>$ can peak at $3U_P$. In the case of circular polarization, the drift velocity is constant since the field amplitude is constant. It can be shown [3] that in this case, the drift velocity is normal to the field at the time of the electron release. The cases for which the Simple Man Theory holds are those where U_p is much larger than the photon energy (essentially far infrared or microwave regions of the spectrum). It must be stressed however that it is very crude for the visible or near infrared. For instance at 1064 nm and 10^{13} W.cm^{-2}, $U_p = 1$ eV while the number of excess photons absorbed in MPI of xenon (N=11) is 10 (Fig. 3).

At still higher intensities, the ionization regime tends towards a dc-field ionization. It is easy to understand that, if the field is strong enough, long enough then the electron can be released either through tunneling or over-the-barrier processes.

The width of the barrier is of the order of E_I / eE. The probability for an electron to tunnel through the barrier is then exponentially small according the well-known Landau formula. For the critical value of the field (in a.u.):

$$E_{cr} = \frac{E_I^2}{4Z}$$

(7)

the saddle point is lowered to the ground state level and the electron can flow over the barrier. Now, for an ac-field the same formalism can be applied if the radiation frequency is sufficiently low to consider that ionization occurs through an adiabatic succession of static processes.

In the regime of strong fields, multiphoton (MPI) and tunnel ionization (TI) or over-the barrier ionization (OBI) are the two limiting cases of the ionization process. The ratio γ of the tunneling time (i.e. the width of the barrier divided by the electron velocity) to the optical period is known as the adiabaticity or Keldysh parameter and is generally used to separate the two regimes:

$$\gamma = \sqrt{\frac{E_I}{2\,U_p}}$$

(8)

In the low frequency limit, ($\gamma \ll 1$), TI is a good description of the transition dynamics. This is typical of rare gases ionized by a CO_2 laser in which case the electron tunnels out in a time less than half the field period while its energy and momentum are subsequently determined by the Lorentz force.

In the other limit, ($\gamma \gg 1$), the multiphoton character is evident in the photoelectron spectra as a structure repeated with the photon energy period (Above Threshold Ionization (ATI)) and whose rate scales as I^N. This is typically the case for visible or UV ionization of an alkali atom.

In fact, recent theoretical calculations show that tunnel ionization domain is reached when γ roughly decreases below 1 which is a less drastic condition. So, it is interesting to explore the evolution of a typical atomic response over this intermediate regime ($\gamma \approx 1$).

According to the definition (8), γ seems to be an increasing function of E_I. However, γ is an intensity dependent parameter through U_P since:

$$U_p (eV) = \frac{e^2 E^2}{4m\omega^2} = 9.34\ 10^{-14}\lambda^2\ (\mu m)\ I\ (W.cm^{-2})$$

(9)

Therefore the effective value of γ depends on the intensity at which the ionization is actually observed. For a given atom, this intensity ranges between the appearance and saturation intensities. The appearance intensity depends on the ionization potential and experimental conditions and corresponds to the minimum value for which an electron signal is effectively detected. Furthermore, it is not possible to submit a given atom to an arbitrary high intensity because it will be ionized before sensing the peak of the pulse. The maximum practical intensity, the saturation intensity, as defined by Lambropoulos [4], that can be applied to a neutral atom is determined by its ionization probability and the pulse characteristics. The upper limit for the saturation can be reached by using state-of-the-art intense 70 fs pulses. The highest saturation intensities are obtained, under given laser conditions, with atoms having the highest ionization potential. A qualitative idea of the atomic behavior over a large range of intensities can nevertheless be obtained by studying similar atoms with increasing ionization potentials. The rare gases are well suited to this purpose and conveniently provide targets which span a factor of two in ionization potential and more than ten in saturation intensities for irradiation by 617 nm light. Then ionization of rare gases spans a range of γ's large enough to encompass both limits (from $\gamma > 1$ for xenon to $\gamma < 1$ for helium).

To study the dynamics of ionization, the energy spectra of the photoelectrons resulting from the interaction between a laser pulse and an atom are known to carry information that is not clearly present in total yield measurements. The evolution of the photoelectron energy spectra from xenon to helium [5] is summarized in Fig 4. The spectra of the lowest I.P. gases, xenon ($E_I =12.13$ eV), krypton ($E_I =13.99$ eV) and argon ($E_I =15.76$ eV), show series of Stark-induced resonances which are reproduced in the different ATI orders. The resonances obviously dominate the transitions and, for xenon and argon, the spectra are relatively robust against intensity changes. As the peak intensity is increased beyond the appearance intensity, more resonances appear towards low energies but the spectra maintain their general structure. Therefore, from xenon to argon the ionization process keeps a clear multiphoton character in the sense that both the atomic structure and the photon energy play the major role and are clearly apparent on the spectra.

In xenon, the now well known 6-photon resonances of the nf states are observed (n=4,8). The appearance intensity is about $4\ 10^{12}$ W.cm^{-2} corresponding to the 8f resonance. A small contribution from the p-states is also detected. Another resonance appears at $3.9\ 10^{13}$ W.cm^{-2} (Fig. 1) due to the 7p state. As the peak intensity is increased beyond $6\ 10^{13}$ W.cm^{-2}, the spectra show essentially no change in structure for a given ATI order but the relative amplitudes of the different orders seem to change. The resonant processes are clearly dominant over the nonresonant ones as seen from the low background between the resonant structures.

In krypton, the appearance intensity is $1.5\ 10^{13}$ W.cm^{-2} and, again, the resonances dominate. However, the intensity behavior of the spectra is in strong contrast with the previous ones: the resonances merge into a continuum up to 9 10^13 W.cm^-2 and a strong resonance, identified as the 4f state eventually emerges for intensities beyond this value. This corresponds to an exceptionally large Stark shift, as discussed in detail by Mevel et al. [6] due to a combination of favorable factors (large static detuning: 2.9 eV and high saturation intensity). In spite of this unusual behavior, the ionization process remains undoubtedly multiphoton.

In argon, the spectra show again outstanding resonances and behave similarly as intensity is increased. Quantitatively, the intensity necessary to obtain spectra is 10^{14} W.cm^{-2}, but the appearance intensity is 2.5 10^{13} W.cm^{-2}. Saturation intensity is about 2 10^{14} W.cm^{-2}.

In contrast, the two lightest rare gases present spectra that drastically differ from the previous ones. Let us consider neon first. Its I.P. is significantly higher (21.56 eV) and so is the appearance intensity (1.6 10^{14} W.cm^{-2}). The electron energy spectrum (Fig. 4) is observed up to an energy of 50 eV and presents a series of broad peaks whose separation is the photon energy. Some substructures seem still to be present in the spectra but they have not been identified. In conclusion, the multiphoton character is still obvious.

Finally in helium, the appearance intensity is about 60 times larger than for xenon, (2.5 10^{14} W.cm^{-2}). As shown in Fig. 4 the electron spectrum, which extends to 90 eV does not show structure above 30 eV.

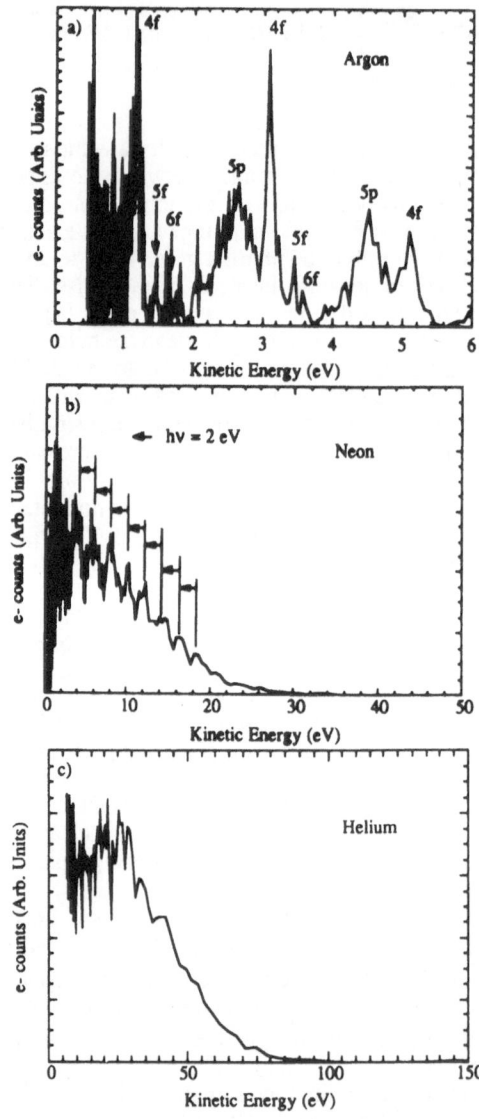

Fig.4 Electron energy spectra showing progressive disappearance of structures. The spectra a), b), c) have been recorded at 10^{14}, 6 10^{14} and 1.5 10^{15} W.cm^{-2} respectively.

The disappearance of all structures in the spectrum is not, per se, a proof of the tunneling regime. Actually, a Floquet calculation in the case of hydrogen shows that resonant structures may still be visible as the tunneling regime is approached [7]. However, at intensities above the critical intensity (at which the saddle-point potential becomes equal to the electron binding energy), the ionization occurs in times of the order of $2\pi/\omega_{at}$ where ω_{at} is the atomic-orbital frequency. The Floquet calculation becomes inadequate then but still gives an indication of the widths. As they are much larger than the photon energy, it implies that all structures are washed out. It is interesting to check whether the spectra correlate to the variation of the adiabaticity parameter. Using the experimental intensities, one gets the γ values shown in Fig.5.The upper and lower values correspond to the experimental appearance and saturation intensities respectively. It is clear that there is some correlation since values for xenon to krypton are above unity while for neon and helium they reach below one where the tunneling regime begins. However the γ values can be only indicative since they remain close to one contrary to the case of CO_2 laser experiments. On the other hand the electron spectra are significantly different.

In summary, this experimental study of rare gases ionization by intense 100 fs, 617 nm laser pulses by electron spectrometry shows the following trend: at lowest intensity, the ionization process is unambiguously of multiphoton character and the Stark-induced resonances are dominant. States with Stark shifts as high as 2.9 eV are identified and the corresponding resonances are still outstanding. At intensity about 10 times larger, the resonances are lost but the ATI structure is still apparent. Eventually, even this structure is lost as expected when the shifts and widths become comparable to the atomic-orbital frequencies and at the onset of the tunneling regime, when the ionization process concentrates in times of the order of the optical period. The ionization process is more easily interpreted as a two-step process: tunnel ionization followed by classical motion of the free electron in the field.

SPACE LOCALIZATION AND BOUND STATE POPULATION IN SHORT PULSE RESONANT MULTIPHOTON IONIZATION

Picosecond and subpicosecond pulses have revealed [8] the resonant nature of multiphoton ionization (MPI). In short, the optical field required to observe MPI is usually strong enough to induce Stark resonances which temporarily increase the ionization rate and dominate, in general, over the nonresonant process. Due to averaging over the pulse space-time intensity distribution, the resonances are normally unobservable in the total ionization yield but easily appear as sharp structures (typically 30 to 50 meV wide) in the photoelectron energy spectrum produced by short pulses. Due to ponderomotive shift of the ionization limit, the energy at which the structure appears is uniquely determined by the corresponding state's binding energy and the photon energy [8]. From the uncertainty principle, it may be argued that the sharp structures indicate that the atomic state remains in resonance for a time comparable to the inverse of its width (100 fs). It was therefore asserted [9] that the resonant structures originate only from the region of the interaction volume where the intensity (I_r) needed to bring

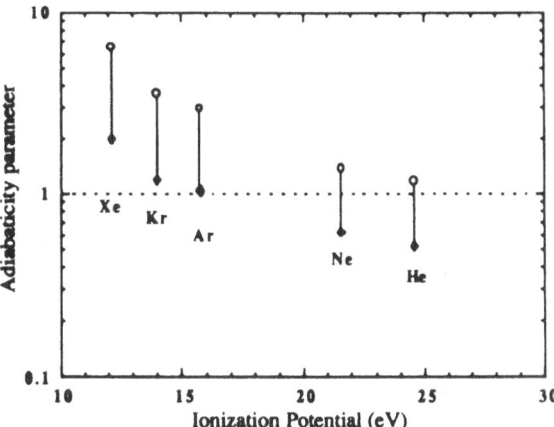

Fig.5 γ values corresponding to appearance ($\circ$) and saturation ($\Diamond$) intensities. The solid line indicates the range of values accessible to experiment for each gas.

the state into resonance is reached at the *top* of the pulse's temporal envelop where the shift rate is close to zero. This scenario results in a long enough time scale for adiabatic passage to occur from the ground state to the excited state which is then immediately ionized at a constant intensity, namely Ir. This model, which is based on a generalized Landau-Zener (LZ) level-crossing theory, led to the "shell" model of ionization owing to the spatial shape of such a region in a gaussian pulse. While the model was consistent with the photoelectron results, the FOM group made the most interesting discovery that population may survive in the Stark-shifted excited state after the end of the strong pulse [10]. From this observation, it was alternatively proposed that the excited state may indeed be populated when the resonance occurs during the rise of the pulse and subsequently ionized at various intensities during the rest of the pulse. The resulting photoelectron energy remains unchanged *provided* that the state has the *same* rate of shift as the ionization limit, i.e. the ponderomotive shift. An obvious characteristic of this scenario is that it contradicts the shell model since now the resonance intensity may be reached at any time during the pulse and therefore *anywhere* in the interaction volume. However, a recent experiment provided reconfirmation of the "shell" model by observing ionization through a state which has a shift rate *different* from the ponderomotively shifted threshold. The resulting photoelectron spectrum showed a sharp resonance [1] which apparently refutes the FOM scenario and raises questions about the origin of the observed excited state population.

In fact, the shell model is perfectly compatible with a residual excited state population. To demonstrate this, it is sufficient to consider a model which uses an isolated resonance [11]. To compute the residual population in the excited state, as well as the ionization probability or the electron energy spectra, it is convenient to use the well known resolvent-operator technique. [12]. When applied to a two-level atom, it has proved extremely successful in accounting for all the low intensity resonant MPI experiments [12]. It does provide some advantages in the present discussion since it is completely analytical and therefore is not computer intensive. A general objection would be that it is basically a time-independent model and therefore inadequate to describe the situation at hand. However, this limitation is the same for the Floquet theory which has nevertheless proven effective in treating the case of even very short pulses [7]. The single excited state involved in this calculation should be adequate to describe the situation in which the resonant state is sufficiently isolated from adjacent states. The specific case studied here is the (7+1)-photonionization of H atoms having a 7-photon dynamical resonance with the 4f state. The photon energy used is 0.0738 a.u. (618 nm) and is choosen to closely match the experiment [10]. In our two-state model, the ground state at zero energy has a Stark coefficient α_g and the resonant state has energy E_r and Stark coefficient α_g. The relevant parameters of the problem in atomic units are: the static detuning from the resonance $\Delta_0 = m \Omega_p - E_r$ (where Ω_p is the photon energy and m the number of photons coupling the two states); the resonance intensity $I_r = \Delta_0/(\alpha_r + \alpha_g)$; the generalized Rabi frequency $R_{gr} I^{m/2}$; and the ionization width of the excited state $\gamma_r I$, where I is the laser intensity and a one photon coupling γ_r of the excited state to the continuum is assumed. The probabilities that the system be ionized (P_i) or left in the excited state (P_r) at time t, are given by (for a complete derivation see for example Gontier and Trahin [12]):

$$P_i(I,t) = 1 - \frac{1}{\left|Z^+ - Z^-\right|^2}\left[\left|Z_r^+ e^{-iZ^+t} - Z_r^- e^{-iZ^-t}\right|^2 + R_{gr}^2 I^m\left|e^{-iZ^+t} - e^{-iZ^-t}\right|^2\right]$$

$$P_r(I,t) = \frac{R_{gr}^2 I^m}{\left|Z^+ - Z^-\right|^2}\left[\left|e^{-iZ^+t} - e^{-iZ^-t}\right|^2\right]$$

(10)

where Z^+ and Z^- denote the two poles of the resolvent operator:

$$Z^\pm = \frac{1}{2}\left[\tilde{E}_r + \tilde{E}_g - i\gamma_r I \pm \sqrt{\left[\Delta_0 - (\alpha_r + \alpha_g)I + i\gamma_r I\right]^2 + 4 R_{gr}^2 I^m}\right]$$

(11)

and

$$\widetilde{E}_{r,g} = E_{r,g} + \alpha_{r,g} I \qquad (12)$$

The direct coupling of the ground state to the continuum can normally be neglected and, in the present case, $\alpha_g \ll \alpha_r$. The values of the atomic parameters are [13]: $R_{gr} = 0.65 \; 10^9$, $\gamma_r = 1.1$, $\alpha_r = 184.421$, $\alpha_g = 4.62552$, $E_r = 0.46875$ a.u. and the intensity unit is $1.4038 \; 10^{17}$ W.cm^{-2}. Figure 6 shows the real part of the poles as a function of intensity in the region of the level crossing (actually a small avoided crossing) at this photon energy. All the physics of the resonance is contained in this crossing diagram and in the pulse duration. Figure 6 displays P_i and P_r in the same intensity range for a time $t = 2500$ a.u. which is long enough for the resonance to build up (see below). Note, that the ionization and residual bound state population is localized at the resonant intensity defined by the crossing point. Figure 6 shows the probabilities for the system to end up in the continuun, excited state, and ground state after a square pulse of intensity I_r. It is clear that, if the pulse is not too long, some population may subsist in the excited state (up to 17.5% for a 120 fs pulse). However, the amount that is left strongly depends on the static detuning (through the resonance intensity), the ionization rate $\gamma_r I$ of the excited state, and the pulse duration τ. The space localization of the probabilities (basis of the shell model) has obviously its origin in their sharp dependences on intensity, shown above. For a gaussian beam, the intensity is given by:

$$I(r,z,t) = \frac{\exp\left[-\dfrac{r^2}{1+z^2}\right]}{1+z^2} \exp\left[-\dfrac{\left(t - \frac{z}{c}\right)^2}{\tau^2}\right] \qquad (13)$$

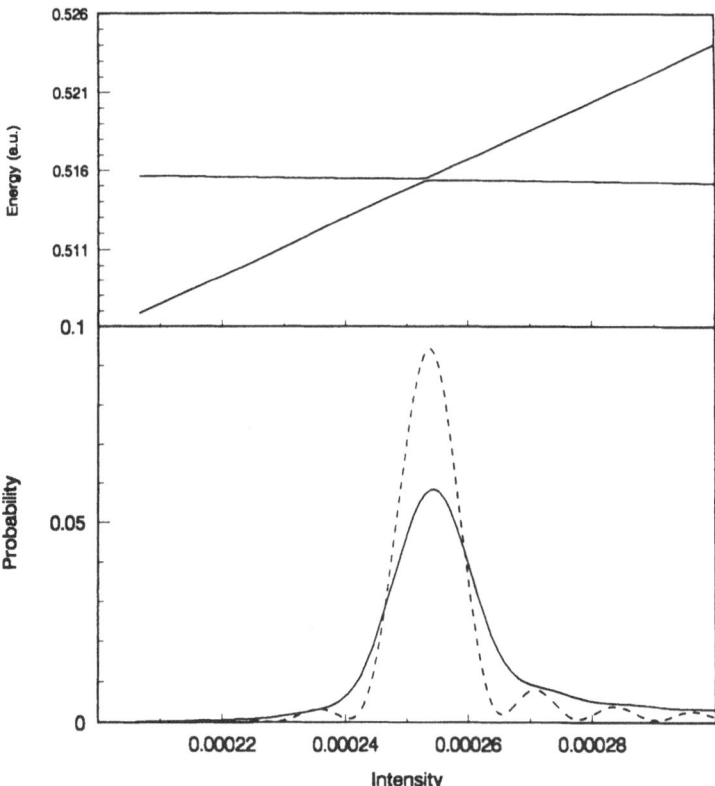

Fig.6 Real parts of the poles as a function of intensity and Pi (solid line) and Pr (dashed line) vs intensity.

As $I(r,z,t)$ evolves in time along the pulse, the resonance intensity is reached in differen regions of the beam. However, the effective time τ_{eff} it remains on resonance also varies along the pulse. It will stay on resonance longer, if I_r is reached at the top rather than during the rise or fall of the pulse. The populations in the groung state (P_g), excited state (P_r) and continuum (P_i) are displayed in Fig.7 as a function of the pulse duration for the resonance intensity. It is clear that in order to have $P_r \neq 0$ the pulse should not be too long. Contour plots o $P_r[I(r,z,t),\tau_{eff}]$ (where r and z are the space coordinates) for a pulse with a peak intensity $2I_r$ show the space localization in Fig.8. They provide snapshots of the populations as the pulse evolves in time. In the rising edge of the pulse, some population is non-resonantly driven at the center of the beam, but at a virtually undetectable level. When the pulse reaches its maximum a t = 0, most of the population transfer occurs in the spatial regions of the beam where the intensity is I_r in full agreement with the LZ analysis by McIllrath et al. [9]. Note, we observe identical localization and time dependence of the ionization probability, P_i. The significan addition of this calculation is that it allows evaluation of the residual excited state populatior which is observed to be *appreciable and localized* after the pulse. The basic reason for this occurence is that even at the top of the pulse τ_{eff} may be too short to saturate the transition since $\tau_{eff} < \tau$. However, an increase of the resonance intensity results in both an increase of the peak population and a rapid decrease of its lifetime. For instance, a 2% increase in the photor energy results in a maximum excited state population of 30% (to be compared to the 17.5% in Fig.6) but it is reduced to zero after a time of only 10^4 a.u.. For a 5% photon energy increase, this time is reduced to 5000 a.u.. If approximately 10% of the total population is left in the 4f state at 618 nm, the model predicts that this value would be reduced to zero for 590 nm photons, in agreement with the experiment [14].

Finally, we have examined the photoelectron spectrum that arises when a resonant state has a shift different from the continuum limit. First, it must be realized that ionization always occurs over some intensity range, albeit small. Therefore, the question of the structure broadening is always present but our analysis demonstrates that the energy shift is not expected to play a major role. This is confirmed by a calculation where the electron energy spectra (the ionization probability as a function of the energy mapping of the intensity) are simulated for three values (1.5, 1, .5) of the ratio α_p/α_r (α_p ponderomotive shift coefficient) by linearly mapping the energy K to intensity as $K = 8\,\Omega_p - E_I - \alpha_p I$. In spite of an obvious broadening and a trivial shift, the structures remain relatively narrow. At the same time, the amount of population left in the excited state remains constant and spatially localized.

In conclusion, both the LZ and the present calculation predict a sharp dependence of the ionization probability around the resonance intensity. Both time and space localizations of ionization follow. However, the LZ level crossing theory implies that all the ground state

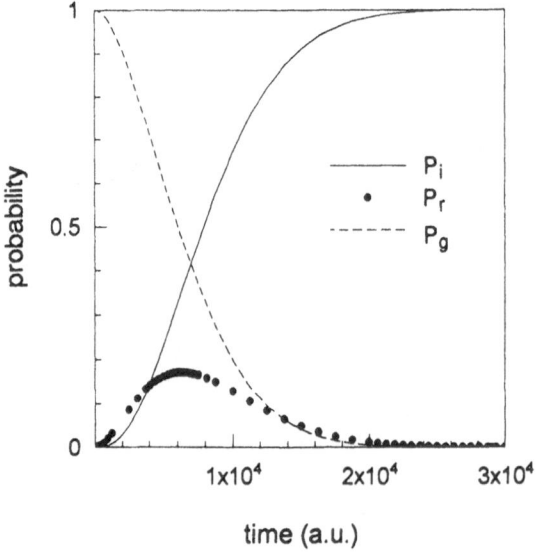

Fig.7 Ionization and excitation probabilities vs pulse duration.

population is transferred to the excited state (most efficiently in the regions of the interaction volume where the resonance intensity is reached at the top of the pulse) and that any residual excited population will be returned to the ground state during the fall of the pulse. This may be a very good approximation for high Rydberg states with very low ionizing rates but it does not seem to be adequate for realistic couplings in the case of the 4f state. The dynamics of the system is more accurately described by the present calculation in which the amount of population left in the excited state clearly depends on the excited state ionization rate and the time. Furthermore, this localization produces sharp energy structures even in the case of a state shifting differently from the continuum limit. However, the above calculation clearly shows that the FOM result is rather atypical and fragile, since a small increase in I_r or in the pulse duration will in general be sufficient to remove all excited state population. Finally, it should be stressed that the present calculation is only a first approximation to describe situations where such large shifts are observed. In particular, all the couplings are described to lowest order. The agreement between the experiment and the present prediction must be ascribed to the fact that the 4f state is very weakly coupled to other states. More complete treatments using a larger basis or time-dependent solution of the Schrödinger equation are required to make accurate predictions in the general case.

MULTIPHOTON DOUBLE IONIZATION

Direct vs sequential double ionization

Suran and Zapesochnyi [15] made the first observation of a multiphoton double ionization process. Irradiation of strontium atoms by 1064 nm pulses resulted in Sr^+ and Sr^{++} ions. At first glance, this was not a very exciting observation: strontium is ionized by five-photon absorption to produce an alkali-like ion which can, in turn, be ionized by absorbing eleven photons. The double ionization process seems to be just a sequence of single ionizations. However, a very straightforward calculation reveals that this picture is oversimplified: typical cross-sections for 5 and 11-order processes are $\sigma_5 = 10^{-140}$ cm^{10} s^4 and $\sigma_{11}=10^{-336}$ cm^{22} s^{10} respectively. Using a flux of 10^{30} photons.cm^{-2} s^{-1} yields a ratio of 10^{-16} for the corresponding rates w_{11}/w_5 while the experimental value was about 10^{-2}. This 14-order-of-

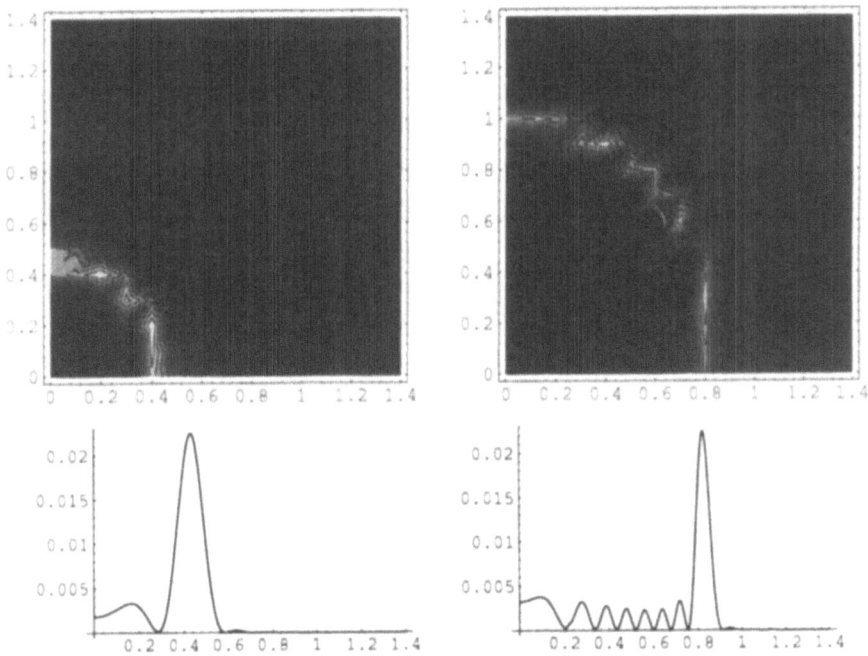

Fig.8 Spatial localization of the probabilities at two instants in the gaussian pulse.

magnitude discrepancy was at the origin of the idea of a "direct" multiphoton double ionization process in which the two electrons are simultaneously driven to the double continuum:

$$A + N\hbar\omega \Rightarrow A^{++} + 2\,e^-$$ (14)

Delone et al. proposed solution to this puzzle [16] as a two-step process: the two electrons would first be driven above the first ionization threshold through two-electron intermediate states; then they would diffuse through the dense quasi-continuum of doubly-excited states up to the second threshold with a very large probability, a process they coined as "direct" by contrast with the "sequential" discussed above.

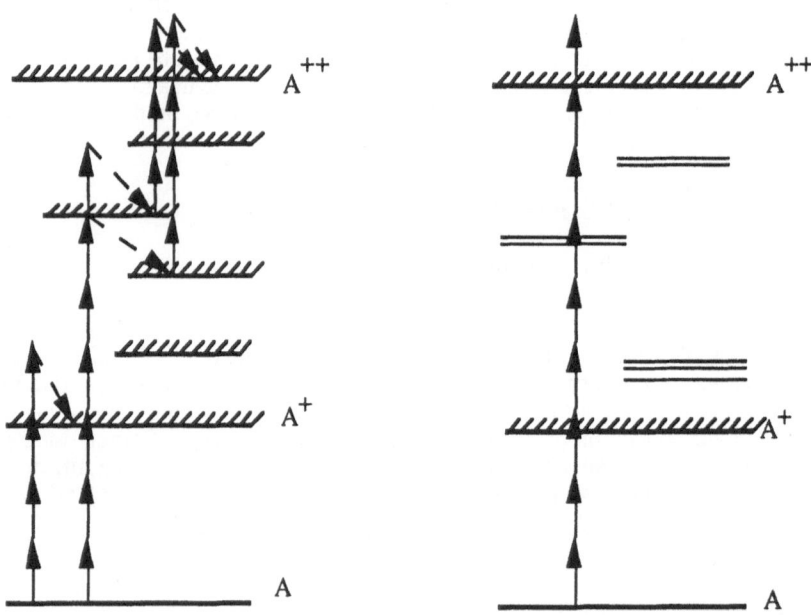

Fig. 9 Possible sequential double ionization processes in an atom. The broken arrows indicate the photoelectrons energies.

Fig.10. The "direct" multiphoton double ionzation process seen as a transition between the ground state and the double continuum. All intermediate states are *atomic* states.

Experiments using tunable lasers as well as electron spectroscopy revealed that, at least in the case of alkaline-earth atoms, the most likely process is the following: the atom absorbs one or two excess photons above the first continuum and ionizes leaving the ion in the ground state or in one of the first excited states. The ion makes then multiphoton (possibly resonant) transition to the second ionization threshold. The resonances as well as the number of possible sequential channels make the probability of the process much larger than the "typical" value used in the argument above. It was then realized that the "direct" process (Fig. 10) has a very small branching ratio and is therefore quite difficult to detect. Actually, to our knowledge it has not been demonstrated so far.

The situation is similar for the multiple ionization of rare gases for which charge states of 8+ (complete stripping of the outer shell) are now currently reported. There is a general consensus that this occurs sequentially, the electrons being peeled off one at a time.

Why is the non-sequential ionization probability larger than the sequential one?

There are now several evidences for what may be coined as non-sequential double ionization processes. Historically, the first one [17] refers to multiphoton double ionization of xenon by 532 nm, 50 ps pulses. The signature for a nonsequential process is provided by a characteristic hump on the doubly charged ion yield as a function of intensity. The point is that the Xe^{++} yield *cannot* be fitted by a single rate. The corresponding cross-sections are obtained from the fit to the data and compared to "standard" values (for instance values based on scaling laws derived from perturbation theory) and the non-sequential cross-section found anormally large. This result has been recently reproduced at a slightly different wavelength [18]. The laser characteristics (527 nm, 50 ps, 1.5 W, 10^{13} W.cm^{-2}) are very close to the one used by L'Huillier et al. [17] with, however, a repetition rate of 1 KHz. This high rate allows to use counting techniques and accumulate a good statistics in a reasonable time. The spectrometer is a time-of-flight with a parabolic mirror collector (Fig.11). The same device can be used to record ion charge spectra.

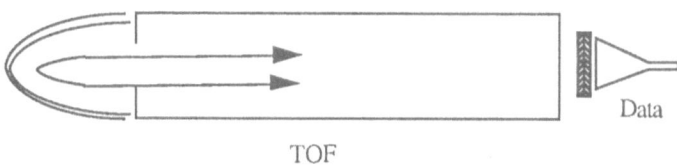

Fig.11 Schematic of the tof spectrometer used for the electron-electron coincidence and ion yield measurement.

The result is shown in Fig.12. First, with linear polarization, the Xe^{++} yield reproduces the hump observed by L'Huillier et al. [17]. The appearance intensity is lower than the saturation intensity for the Xe$^+$ production and is about 6 10^{12} W.cm^{-2}. At this intensity, the ponderomotive energy at 527 nm is about 150 meV, that is much smaller than the photon energy. The ionization is therefore expected to be well in the perturbative regime.

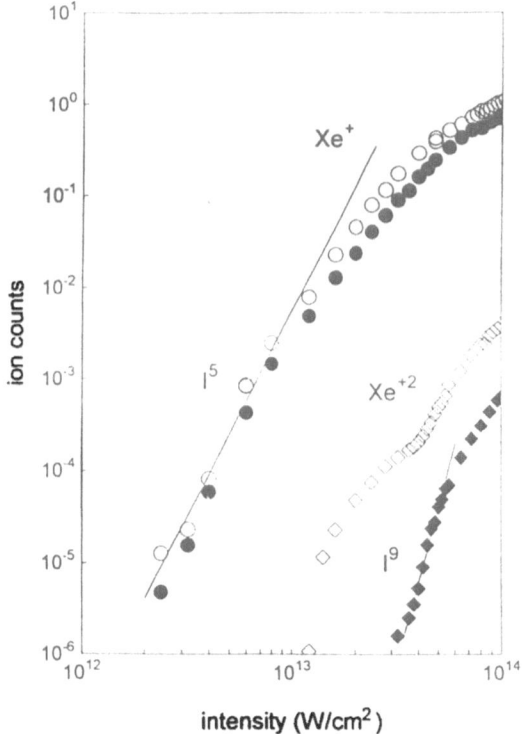

Fig.12 Ion yields in MPI of xenon by 527 nm, 50 ps pulses. The open circles and diamonds are taken with linearly polarized light and the closed circles and diamonds with circularly polarized light.

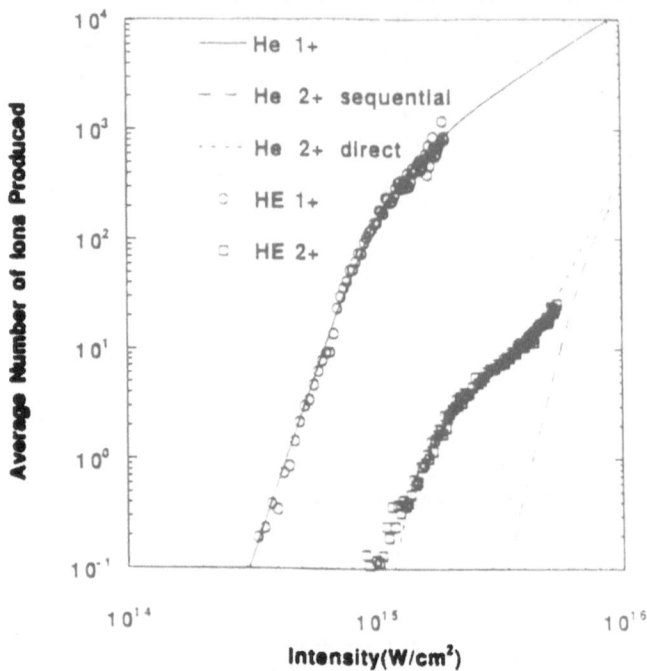

Fig.13 Ion yield data compared to sequential and nonsequential double ejection rates. (From Fittinghoff et al., 1992)

Below saturation, the Xe^+ data is fitted by a I^5 power law, in good agreement with the LOPT prediction. In order to test the origin of the hump we have measured the same yields with circular polarization. Fig.12 clearly shows a suppression of the hump in this case. One possible interpretation of the suppression is that the hump is due to a transient resonance with a low angular momentum two-electron state [4]. However, the resonant state cannot be easily identified and this conjecture, although very probable, is not yet proved. On the other hand, if the large rate corresponding to the hump was due to the direct double ionization process (in the sense of Delone), the electron-electron coincidence measurement should confirm it. At saturation intensity, the percentage of doubly charged ions is about .5%, a rather small value but still reasonable to attempt the measurement. A *negative* result was obtained, indicating that this process is negligible under the above conditions.

Similar evidence was produced by two observations [18], [19] in MPI of helium by 100 fs, 617 nm pulses. In this case, single ionization is better described in terms of dc-field ionization, or more precisely in terms of tunneling rates calculated according to the Amosov-Delone-Krainov theory. The experimental result (Fig.12) shows the same behavior for the He^{++} yield as xenon. Using the cycle averaged rates Γ given by the tunneling theory gives a good prediction for the He^+ yield and the upper part of the He^{++} data. However there is a strong discrepancy at appearance intensity for He^{++}. Here again, one must use two rates to fit the data. This time however, resonances cannot be invoked since the ionization of He (and *a fortiori* He^+) is in the tunneling regime as asserted by the electron energy spectra (see above) and the good agreement with the tunneling rates. Fittinghoff et al. [19] used the idea of a shake-off ionization of the second electron by the first one to justify that the double ejection cross-section is larger than predicted by the ADK theory [20]. In the case of helium it can be safely asserted that, if the double ejection is more probable than predicted for a sequence one one-electron ejection, it means that the two electron interact. In this sense, this result qualifies to be the first evidence of a nonsequential double ionization.

A similar experiment has been carried out with the femtosecond laser source in Palaiseau [18]. Briefly, the experiment consisted of an amplified colliding pulse mode-locked dye laser system producing 2 mJ/pulse of 617 nm, 100 fs radiation at 20 Hz repetition rate. Laser energy was controlled with an external power attenuator, and the polarization was varied with a quarter-wave plate. The absolute intensity and polarization were calibrated using the well-known short pulse Stark resonances of xenon. The light was focussed by f/4 optics into an ultra-high vacuum chamber with a base pressure of 10^{-9} torr. An effusive source of helium was regulated to produce a target pressure of 10^{-6} - 10^{-8} torr (50-100 total ions per shot maximum) depending upon the intensity. A parabolic mirror spectrometer with a one meter time-of-flight tube was used either in an ion or electron collection mode with proper biasing of the electrostatic optics. The ion signal was digitized by a LeCroy 9450 oscilloscope, and the counts were binned according to the laser energy to within ± 2 %. Care was taken to insure reduction of artifacts due to space-charge, contact potentials, and background ions, specifically H_2^+.

Figure 14 illustrates the intensity dependent yield curves for singly and doubly ionized helium with linearly and circularly polarized 617 nm, 100 fs radiation. Each data point (bin) represents a single laser intensity averaged over approximately 1,000 to 2,000 laser shots. The shape of the ion yield curves produced with LP light (Fig. 14a) accurately reproduces the structure seen by Fittinghoff et al. Also, there is excellent agreement between their shifted intensity scale and that shown in Fig.14 which was independently calibrated. Examination of ion curves clearly reveals the presence of discontinuity for LP light. The He^{2+} production in the low intensity (0.9 – 1.3 10^{15} W.cm^{-2}) range can be identified as the nonsequential regime of two-electron ionization, while above 3 10^{15} W.cm^{-2} ionization occurs sequentially and at a rate consistent with an ADK tunnel model. Again, circular polarization suppresses the hump (Fig. 14b). This ensemble of results *in helium* is well interpreted within a model recently proposed by Corkum [21] which can be oulined as follows: double ionization is a three-step process: (1) one electron is emitted by TI or OBI; (2) the "free" electron acquires a quiver energy in the field and, in linear polarization essentially no drift motion [3]; (3) it makes an ionizing collision with its parent ion which removes the second electron. In circular polarization, however, the drift velocity is such that the electron never comes back to the ion and the only possibility to remove the second electron is through a TI or OBI of He^+. There is apparently an excellent agreement with the observed data. However, this scenario cannot explain the xenon case since there the quiver energy is much smaller than the Xe^+ ionization potential.

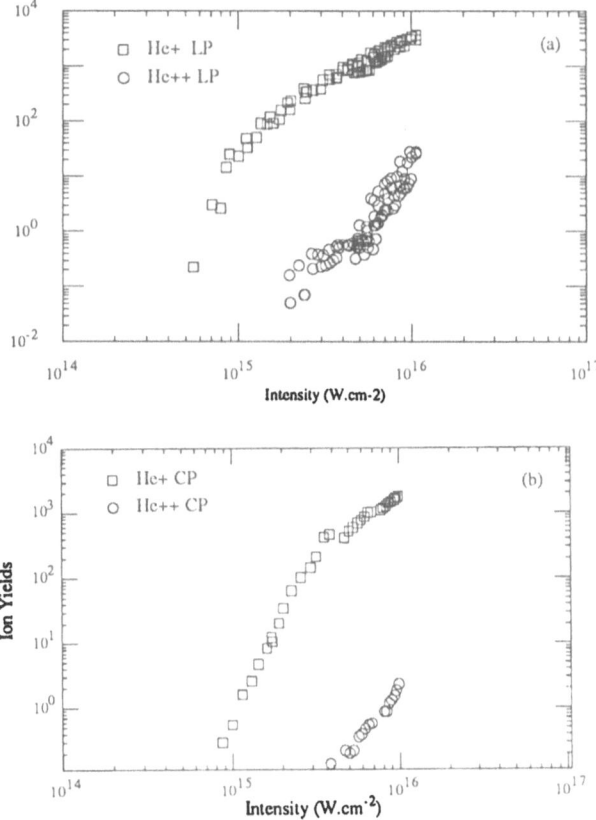

Fig.14 Ion Yields for He^+ and He^{++} for linear (a) and circular (b) polarizations. (FromWalker et al., 1993)

CONCLUSION

In gradually stronger electromagnetic fields, ionization of atoms is less and less influenced by the atomic structure. This is clearly seen in the photoelectron electron energy spectra in which first, the Freeman resonances, second the ATI structure are progressivelly lost. The perturbative description (even the Floquet ansatz) of the ionization process becomes inadequate when the ionization time becomes of the order of the atomic orbital period.

In the regime where the dynamic resonances play the major role, one question arose recently about the amount of population left in the excited states after the ionizing pulse. A Floquet analysis of a model case leads to the conclusion that this amount crucially depends on the static detuning, the excited state ionization cross-section and the pulse duration. As a consequence, cases where a substantial population is left in the excited state may be found. Even in those cases, the ionization is still strongly localized in space as predicted by the level-crossing theory. However, this "trapping" is rather fragile and is easily destroyed, for instance, by increasing the laser fluence.

When two electrons are removed from the atom under moderate intensity, the process is, in most cases, just a sequence of one-electron ejections. However, the ion yields reveal that double ionization rates can be much larger than the sequential ones both at relatively low intensity ($5 \ 10^{12}$ W.cm^{-2}) and much higher ones ($5 \ 10^{15}$ W.cm^{-2}). While in the first case it is possible to invoke an accidental transient resonance (with a doubly excited state) this is ruled out in the second one when the ionization regime is essentially over-the-barrier ionization (OBI). An alternate model of OBI of the first electron followed by an e-2e collision of the quivering electron with it's parent ion, recently proposed [21] seems to provide a very satisfactoty explanation of the observed yields. However it cannot possibly explain the first case in which the quiver energy is much smaller than the ion ionization energy. Interestingly enough, in both cases the rate is drastically reduced in circular polarization, but for certainly different reasons [18].

ACKNOWLEDGMENTS

This work was partially supported by the EEC under the Science Program (contract Sc1-0103C) and NATO, contract SA2.05(RG910678).

REFERENCES

1. G. N. Gibson, R. R. Freeman, T. J. McIlrath, Verification of the dominant role of enhancement in short pulse multiphoton ionization, *Phys. Rev. Lett.* 69:1904 (1992); G. N. Gibson, R. R. Freeman, T. J. McIlrath, *Optics and Photonics News* 3:22 (1992).
2. H. B. van Linden van den Heuvell and H. G. Muller, Limiting cases of excess-photon ionization in "Multiphoton Processes", S. J. Smith and P. L. Knight eds., Cambridge University Press, Cambridge (1988).
3. P. B. Corkum, N. H. Burnett and F. Brunel, Multiphoton ionization in large ponderomotive potentials, in "Atoms in Intense Laser Fields", M. Gavrila ed., Advances in Atomic, Molecular and Optical Physics (Suplement I), Academic Press, New York (1992).
4. P. Lambropoulos, Mechanisms for multiple ionization of atoms by strong pulsed lasers, *Phys. Rev. Lett.* 55:2141 (1985).
5. E. Mevel, P. Breger, R. Trainham, G.Petite, P. Agostini, A. Migus, J. P. Chambaret and A. Antonetti, Atoms in strong optical fields: evolution from multiphoton to tunnel ionization, *Phys. Rev. Lett.* 70:406 (1993).
6. E. Mevel, P. Breger, R. Trainham, G.Petite, P. Agostini, J. P. Chambaret, A. Migus and A. Antonetti, Contrasted behaviour of Stark-induced resonances in multiphoton ionization of krypton, *J. Phys. B: At. Mol. Opt. Phys.* 25:L401 (1992).
7. M. Dörr, R. M. Potvliege and R. Shakeshaft, Tunneling ionization of atomic hydrogen by an intense low-frequency field, *Phys. Rev. Lett.* 64:2003 (1990).
8. R. R. Freeman, P. H. Bucksbaum, H. Milchberg, S. Darack, D. Schumacher, M. E. Geusic, Above-threshold ionization with subpicosecond laser pulses, *Phys. Rev. Lett.* 59:1092 (1987).
9. T. J. McIllrath, R. R. Freeman, W. E. Cooke and L. D. van Woerkom, Complex spatial structure of the ion yield arising from high-intensity multiphoton ionization, *Phys. Rev.A* 40:2770 (1989).

10. M. P. De Boer and H. G. Muller, Observation of large populations in excited states after short-pulse multiphoton ionization, *Phys. Rev. Lett.* 68:2747 (1992).
11. P. Agostini and L. F. DiMauro, Space localization and bound state population in short pulse resonant multiphoton ionization, *Phys. Rev. A: Rapid Comm.* (1993) to be published.
12. Y. Gontier and M. Trahin, in "Multiphoton Ionization of Atoms", S. L. Chin and P. Lambropoulos eds., Academic Press, New York (1984).
13. M. Trahin (1992) unpublished.
14. H. G. Muller (1992) private communication.
15. V. V. Suran and I. P. Zapesnochnyi, *Sov. Tech. Phys. Lett.* 1:420 (1975).
16. N. B. Delone, V. V. Suran and A. Zon, "Multiphoton Ionization of Atoms" Chin S. L. and Lambropoulos P. eds., Academic Press, Orlando, Fla (1984).
17. A. L'Huillier, L. A. Lompré, G. Mainfray and C. Manus, Multiply-charged ions induced by multiphoton absorption in rare gases at 0.53 μm, *Phys. Rev. A* 27:2503 (1983).
18. B. Walker, E. Mevel, Y. Baorui, P. Breger, J. P. Chambaret, A. Antonetti, L. F. Dimauro, P. Agostini, Double ionization in the perturbative and tunneling regimes, (1993) to be published.
19. D. N. Fittinghoff, P. R. Bolton, B. Chang, K. C. Kulander, Observation of nonsequential double ionization of helium with optical tunneling, *Phys. Rev. Lett.* 69:2642 (1992).
20. M. V. Ammosov, N. B. Delone, and V. P. Krainov, Tunnel ionization of complex atoms and of atomic ions in an alternating electromagnetic field, *Sov. Phys. JETP* 64:1191 (1986).
21. P. B. Corkum, A plasma perspective on strong field multiphoton ionization, *Phys. Rev. Lett.* (1993) to be published.

STABILIZATION OF EXCITED STATES AND HARMONIC GENERATION : RECENT THEORETICAL RESULTS IN THE STURMIAN-FLOQUET APPROACH

R. M. Potvliege and Philip H. G. Smith

Physics Department
University of Durham
Durham DH1 3LE
England

INTRODUCTION

This paper essentially consists of a description of two unrelated sets of numerical results. On the one hand, we present the results of recent calculations of non-linear polarizabilities of atomic hydrogen, in which their variations and the variations of the dipole susceptibility at harmonic frequencies have been resolved on a fine mesh of intensity of the incident laser field. This will make a detailed analysis of high-order harmonic generation in this gas possible. On the other hand, we also study the intensity dependence of the lifetime of excited (circular) states of hydrogen. We show that adiabatic stabilization should be observable in experimental conditions which are realisable with existing laser systems, as has been suggested previously. An excellent guide to an experimental study is provided by a simple empirical law predicting, in good approximation, the intensity at which the lifetime passes by a minimum value.

The Floquet approach has been extensively employed, these last few years, in the theoretical study of photoionization (or photodetachment) and harmonic generation in simple atomic systems exposed to intense laser radiation.[1-4] The theory can be described in a few words, without entering into the details. We assume first that the laser can be represented by a classical, single-mode monochromatic electric field of angular frequency ω; the generalization to the case of a multicolor field is straightforward and interesting.[3,5] Hence, the time-dependent Schrödinger equation for the system possesses stationary solutions, whose wave vectors have the form

$$|\Psi(t)\rangle = \mathrm{e}^{-iEt/\hbar} \sum_{N=-\infty}^{\infty} \mathrm{e}^{-iN\omega t}|\psi_N\rangle. \tag{1}$$

The harmonic components $|\psi_N\rangle$ satisfy a system of time-independent coupled equations,

$$(E + N\hbar\omega - H_0)|\psi_N\rangle = V_+|\psi_{N-1}\rangle + V_-|\psi_{N+1}\rangle, \qquad N = 0, \pm 1, \pm 2, \ldots, \tag{2}$$

where H_0 is the field-free Hamiltonian and the operators V_+ and V_- are the two components of the dipole operator $D(t) = V_+ \exp(-i\omega t) + V_- \exp(i\omega t)$. We work in the velocity

Super-Intense Laser-Atom Physics, Edited by
B. Piraux *et al.*, Plenum Press, New York, 1993

gauge, with the contribution to the total Hamiltonian that is quadratic in the vector potential transformed out by a unitary transformation. Since the system is initially in a bound state before the laser field is turned on, in the cases we are interested in, this system of equations should be solved subject to Siegert boundary conditions. That is, in position space and asymptotically far from the nucleus, the harmonic components should be a superposition of free waves coupled by the dipole interaction and distorted by the long range Coulomb potential — with outgoing spherical waves in the open channels, since there is no incident wave, or exponentially damped waves in the closed channels. Therefore, the quasienergy E is an eigenvalue of the system of equations, and is complex: $E = E_0 + \Delta_{ac} - i\Gamma/2$. The real part of E is the unperturbed energy of the initial state, E_0, shifted by the ac Stark shift Δ_{ac}. The total rate of multiphoton ionization (or multiphoton detachment) is simply $\Gamma/\hbar$.

In our calculations, the eigenvalue problem was solved numerically in position space by expanding the harmonic components on a discrete basis set consisting of spherical harmonics $Y_{lm}(\hat{\mathbf{r}})$ and complex radial "Sturmian functions" $S_{nl}^{\kappa}(r)$,

$$S_{nl}^{\kappa}(r) = N_{nl}^{\kappa} (-2i\kappa r)^{l+1} e^{i\kappa r} \, {}_1F_1(l+1-n, 2l+2, -2i\kappa r). \tag{3}$$

The complex parameter κ must be chosen with $0 < \arg(\kappa) < \pi/2$, otherwise the basis set would not be appropriate for representing the wave function with the required asymptotic behaviour.[6] The final results (ionization rates or polarizabilities) are independent of the choice of κ if $\arg(\kappa)$ is in the correct angular range and enough basis functions are included in the expansion. The numerical results presented here are fully converged, in this respect as well as in respect to the number of harmonic components and partial waves which were retained in equation (2).[7] The coefficients of the expansion

$$\langle \mathbf{r} | \psi_N \rangle = \sum_{nlm} c_{nlm}^{(N)} \frac{1}{r} S_{nl}^{\kappa}(r) Y_{lm}(\hat{\mathbf{r}}) \tag{4}$$

and the quasienergy are found by solving the non-hermitian system of homogeneous linear equations which results from projecting equation (2) onto each basis function.[8]

Clearly, the Floquet approach is adequate only to the extent that any intrinsically time-dependent effect, arising from the finite duration of the pulses used in actual experiments, is negligible. In other words, to the extent that the system can be represented, at all time, by a single stationary quasienergy state. This approximation may break down, because of the temporal variation of the field strength, at intensities where the complex quasienergies of several quasienergy states are sufficiently close to each other. This situation is unlikely in the case of adiabatic stabilization of circular states which is discussed later on in this paper. The case of H(1s) at long (1064 nm) wavelength, which is studied in next section, is different. There we assume that the only time-dependent effect is the diabatic passage of anti-crossings, the population being transferred from one adiabatic quasienergy state to the other, without transfer to other dressed states either off-resonance or as Stark-shift-induced resonances are passed. This assumption seems to be reasonable, considering that the gap at anti-crossings between the real part of the quasienergy of the diabatic dressed 1s states and the real part of the quasienergy of any diabatic dressed excited state is extremely narrow at 1064 nm, in the intensity range studied.[9]

HARMONIC GENERATION

Having obtained the Floquet wave vector of the dressed ground state, one can calculate the dipole moment $d(t)$ of the atom exposed to the incident intense laser field (which is supposed to be polarized linearly along the z-direction):

$$d(t) = \langle \Psi(t) | ez | \Psi(t) \rangle = e^{-\Gamma t/\hbar} \sum_{q=1}^{\infty} \left[d(q\omega) e^{-i\omega t} + \text{c.c.} \right]. \tag{5}$$

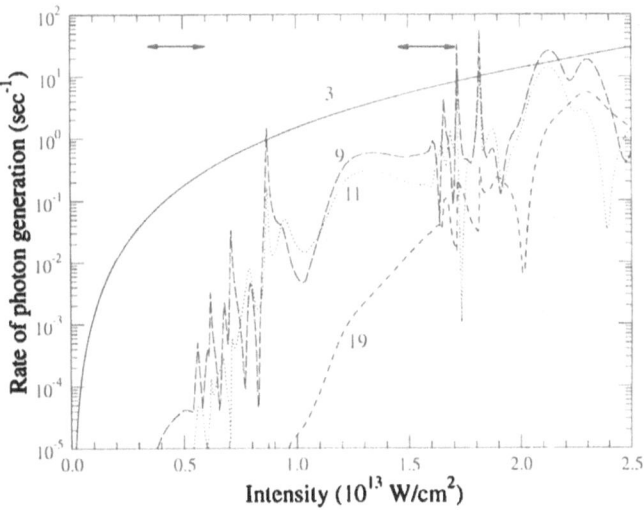

Figure 1. Total rate of harmonic generation by a hydrogen atom, at $q = 3, 9, 11$, or 19, versus the intensity of the incident 1064 nm laser field. The arrows indicate the ranges of intensity where the results are less reliable.

In terms of the individual harmonic components, we have

$$d(q\omega) = \sum_N \langle \psi_{N-q} | ez | \psi_N \rangle. \tag{6}$$

Note that both the angular and the radial parts of the bra-vector are complex-conjugated, when forming the scalar product. The normalization of the Floquet wave vector and the technical difficulties encountered in evaluating $d(q\omega)$ on the Sturmian basis have been discussed in Ref. 10.

When the rate of multiphoton ionization is much smaller than the frequency of the laser beam, and the variation of the intensity is slow enough to be neglected, the polarization of the medium $\mathcal{P}(t)$ oscillates in time with the same period as the incident field :

$$\mathcal{P}(t) = \frac{1}{2} \sum_{q=1}^{\infty} \left[\mathcal{P}_q \, e^{-i\omega t} + \text{c.c.} \right]. \tag{7}$$

This quantity depends on the response of the atoms and of the photoelectrons to the incident field and to the (much weaker) fields at harmonic frequencies which are generated in the medium. Taking into account only the atomic response to the incident field, the polarization of the medium at frequency $q\omega$, $\mathcal{P}_q$, is simply $2\mathcal{N}d(q\omega)$, where $\mathcal{N}$ is the local atomic density ($\mathcal{N}$ decreases slowly in time under the effect of multiphoton ionization). Both the modulus and the argument of $\mathcal{P}_q$ are important for describing the propagation of the harmonic fields in the medium. However, the total rate of harmonic generation by an *isolated* atom, Γ_q, does not depend on the argument of $d(q\omega)$:

$$\Gamma_q = \frac{4q^3\omega^3}{3\hbar c^3} |d(q\omega)|^2. \tag{8}$$

The essential features of the variation of $|d(q\omega)|$ with the intensity I of the incident field are well known : these quantities are enhanced strongly by Stark-shift-induced resonances (unless $q\hbar\omega$ is much smaller than the excitation energy of the lowest excited state) and their mean values increase less and less rapidly with I and q than the perturbative I^q power law; moreover, as q increases for a fixed (high) intensity, the value of $|d(q\omega)|$ first decreases, then stays roughly level in a "plateau", before decreasing rapidly.[11] These features are striking in the results of Figure 1 for photon generation from the dressed ground state of H in a linearly polarized 1064 nm field, which are more detailed but otherwise in excellent agreement with those obtained recently by J.L. Krause and coworkers for a larger range of intensity.[12] Since the number of radial functions in expansion (4) is necessarily small, typically about 30 for each symmetry and photon number, it has not been possible to obtain accurate results in the ranges of intensity where multiphoton resonances with dressed Rydberg states can occur. These ranges are indicated by arrows in the figure. The variation of Γ_3 is close to the result predicted by lowest order perturbation theory; there is, however, some departure at high I caused by the the next-to-leading order term in the perturbation series. (For example, lowest order perturbation theory underestimates Γ_3 by 26% at 2×10^{13} W/cm^2.[10]) In contrast, there is a profusion of narrow structures in Γ_q for $q > 5$ which can be attributed to the non-perturbative interaction of the dressed ground state with dressed excited states, in the same ranges of intensity where resonances manifest in the rate of multiphoton ionization.[10] At some intensities, the rate of photon generation can actually become larger in the 11th harmonic than in the 3rd harmonic. That Γ_{19}, is larger than Γ_9 at 2.5×10^{13} W/cm^2 is a manifestation of the expected "plateau" in the variation of $|d(q\omega)|$ with q at high intensity. However, it is clear that the relative strengths of the harmonics in the plateau vary rapidly with I. The deep, broad minima of Γ_9, Γ_{11} and Γ_{19} between 2.0 and 2.5×10^{13} W/cm^2 are also worth noting; they do not appear to be associated with any Stark-shift-induced resonance. Not only does the modulus of $d(q\omega)$ oscillate markedly with I, but also the argument. One sees that both the real and the imaginary parts of $\mathcal{P}_q$ are far from being gently-varying functions of I.

If we also take into account the contribution to $\mathcal{P}(t)$ of the atomic response to the harmonic fields that are generated in the medium, we can write, with $\mathcal{E}_1$ and $\mathcal{E}_q$ denoting, respectively, the electric field strengths of the incident field and of the q-th harmonic,

$$\mathcal{P}_q = \mathcal{P}_q^d(\mathcal{E}_1) + \mathcal{N}\chi^{(1)}(-q\omega, q\omega; \mathcal{E}_1)\,\mathcal{E}_q + \cdots \tag{9}$$

The terms that are neglected in the right-hand-side of this equation are either of higher order in $\mathcal{E}_q$, or correspond to wave-mixing between harmonic fields. The first term, $\mathcal{P}_q^d$ is simply the atomic response at frequency $q\omega$ to the monochromatic field $\mathcal{E}_1$. The second term is proportional to $\chi^{(1)}(-q\omega, q\omega; \mathcal{E}_1)$, the "linear" susceptibility at frequency $q\omega$ of the atom *dressed non-perturbatively* by the incident field. Hence, $\mathcal{P}_q^d$ is the driving term for the generation of the q-th harmonic field, while the propagation of this field throughout the gas depends on $\chi^{(1)}(-q\omega, q\omega; \mathcal{E}_1)$. It is easy to see that $\chi^{(1)}(-q\omega, q\omega; \mathcal{E}_1)\mathcal{E}_q/2$ is the contribution linear in $\mathcal{E}_q$ to the dipole moment

$$d^{(2c)}(q\omega) = \sum_N \langle \psi_{N-q}^{(2c)} | ez | \psi_N^{(2c)} \rangle \tag{10}$$

where $|\psi_N^{(2c)}\rangle$ are harmonic components of the Floquet wave vector of the atom submitted to the two-colour electric field $\mathcal{E}_1\hat{\mathbf{z}}\cos\omega t + \mathcal{E}_q\hat{\mathbf{z}}\cos q\omega t$.[3,13] Provided the reduced mass effects are neglected, this susceptibility scales with the charge of the nucleus:

$$\chi^{(1)}(-q\omega, q\omega; \mathcal{E}_1)_{Z=1} = Z^4\chi^{(1)}(-q\omega Z^2, q\omega Z^2; \mathcal{E}_1 Z^3)_Z. \tag{11}$$

Like $\mathcal{P}_q^d$, $\chi^{(1)}(-q\omega, q\omega; \mathcal{E}_1)$ is a complex quantity which can be quite sensitive to Stark-shift-induced resonances. Although its imaginary part vanishes when $\mathcal{E}_1 = 0$ and $q\hbar\omega < E_0$, it is in general non-zero. The importance of the non-perturbative effects is illustrated by the results for the phase mismatch

$$\Delta k_q = \frac{2\pi}{c} q\omega\mathcal{N}[\chi^{(1)}(-q\omega, q\omega; \mathcal{E}_1) - \chi^{(1)}(-\omega, \omega)], \tag{12}$$

Table 1. Real ($\Re$) and imaginary ($\Im$) parts of the phase mismatch Δk_q, in cm^{-1}, for a density $\mathcal{N} = 5 \times 10^{17}$ atoms per cm^3. The wavelength of the incident beam (at fundamental frequency) is 1064 nm. The results for Xe were calculated in perturbation theory by L'Huillier *et al* (the question mark indicates that the result is somewhat uncertain).[14] The imaginary part is denoted by ϵ where smaller than 0.01 cm^{-1}. The numbers in the last column are the (real) phase mismatches for a population of free electrons at density $\mathcal{N}$.

q	Xe (perturb. th.) $\Re$	$\Im$	H (1×10^{11} W/cm^2) $\Re$	$\Im$	H (2×10^{13} W/cm^2) $\Re$	$\Im$	Free electrons
3	0.14	0	0.43	ϵ	0.65	ϵ	43
5	0.84	0	2.7	ϵ	4.1	ϵ	77
7	3.3(?)	0	14	ϵ	18	13	110
9	22(?)	0	−140	0.01	−1	100	142
11	−5.7	19	−77	6.3	−42	35	174
13	−14	17	−34	27	−32	25	207
19	−23	6.7	−39	9.7	−39	9.7	303

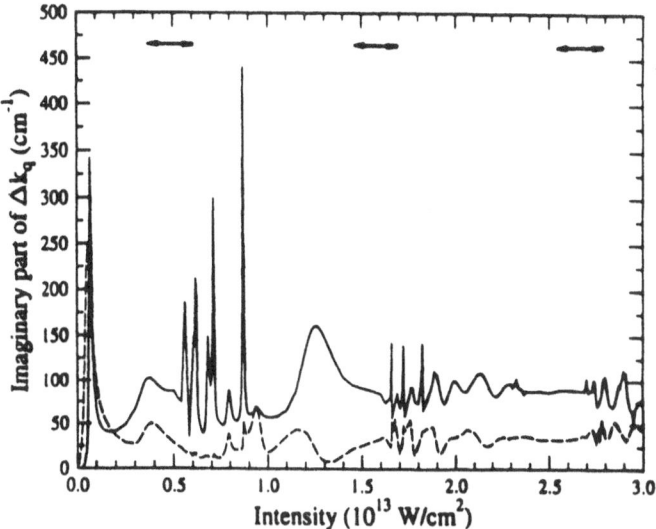

Figure 2. Imaginary part of the phase mismatch Δk_q for $q = 9$ (solid curve) or 11 (dashed curve), versus the intensity of the incident 1064 nm laser field. The arrows indicate the ranges of intensity where the results are less reliable.

where $\chi^{(1)}(-\omega,\omega)$ is the linear susceptibility of H(1s) at the fundamental frequency, which are presented in Table 1 and in Figure 2. In Table 1, the phase mismatch for atomic hydrogen is compared, for two different intensities, with corresponding results for xenon (in the weak field limit) and for a population of free electrons, the latter being intensity-independent. The trends shown by the results for $q \not\approx |E_0|/\hbar\omega$ are typical: as $\mathcal{E}_1$ increases, Δk_q remains close to its zero-$\mathcal{E}_1$ value, or changes slowly as predicted by perturbation theory. As expected, it is of the same order of magnitude in atomic hydrogen as in xenon and the atomic contribution to the phase mismatch is rapidly swamped by the contribution of the photoelectrons as their population builds up at high intensity. On the other hand, well marked enhancements associated with Stark-shift-induced resonances can be observed for harmonic orders $q \approx |E_0|/\hbar\omega$ (in particular for $q=9$). For example, the $1s$ dressed state is brought into resonance with the dressed $4p$ and $4f$ states at about 1×10^{12} W/cm^2; the total ionization rate is too small to be calculated at such a low intensity,[10] but as seen from Figure 2 this intermediate-state resonance manifests dramatically in the susceptibility. The strongest enhancements of Δk_9 between 0.6 and 0.9×10^{13} W/cm^2 are due to resonances with dressed states with g-character. The imaginary part of the susceptibility is appreciable for these harmonics, and, presumably, cannot be neglected when analysing how these fields are generated.

ADIABATIC STABILIZATION

It is well known that at high frequency the rate of ionization of an atom initially in the ground state increases with the intensity of the incident beam, passes by a maximum value, and eventually decreases as the field distorts the atom; however, the ionization rate becomes so large before decreasing that the atom has no chance to survive more than a few optical cycles. It has been suspected for a long time that highly excited states, circular states in particular, should also become stable against ionization at high intensity, if the energy of the photons is larger than the zero-field binding energy, however, with the added advantage that the ionization rate remains small enough at all intensities that the atom could survive a pulse of realistic duration. This prediction was verified in a time-dependent calculation by Pont and Shakeshaft[15], and in an approximate Floquet calculation by Vos and Gavrila[16]. We have recalculated the lifetime of the state studied by Vos and Gavrila, without making their "Born approximation". Our results, which are shown in Figure 3, are in qualitative agreement with theirs, the main differences being that the lifetime of the atom is longer and that the onset of stabilization occurs at lower intensity than they found. The full Floquet results are supported by results of time-dependent calculations.[17,18]

Numerical results for a few other initial states are presented in Table 2. Comparison of these results, and of similar results for a number of initial states and frequencies, suggests a simple scaling law for the case of linear polarization: the intensity I_{min} at which the lifetime of the atom reaches its minimum value is given in good approximation (in W/cm^2) by $\omega^3 (l + 1 - m)! \, m \, I_0$, where l and m are the angular momentum and magnetic quantum number of the initial (unperturbed) state, the angular momentum being quantized along the polarization axis of the field, ω is the angular frequency of the laser expressed in atomic units, and $I_0 = 3.51 \times 10^{16}$ W/cm^2. This relationship, which might be coincidental to some extent,[19] is verified for $0.04 \leq \omega \leq 0.1$, with $m = 4, 5$ or 6, $l = m$, $m+1$ or (with some disagreement) $m+2$, and principal quantum numbers $n=l+1$, $l+2$ or $l+3$ (I_{min} depends but weakly on n). It does not extend to states whose initial l-value is greater than $m+1$ or $m+2$, since they often undergo avoided crossing with other states as the intensity varies. However, it remains a good guide for estimating I_{min} for circular Rydberg states. For example, the empirical scaling law predicts $I_{min} \approx 7.9 \times 10^9$ W/cm^2 for $n=20$, $l=m=19$, $\lambda = 20$ μm, while the "exact" I_{min} is 5.02×10^9 W/cm^2.

We cannot confirm the power-law increase with intensity of the lifetime in the adiabatic stabilization regime found by Vos and Gavrila, since Floquet calculations for super-intense fields are too demanding. As is illustrated by the results shown in Figure 4, it may be that the rate of increase of the lifetime depends more on the angular momentum of the initial state than on the frequency or the magnetic quantum number. However, more work would be necessary to establish this trend. It should be noted that the increase of the lifetime with laser intensity, for $I > I_{min}$, is not rapid, and may be so

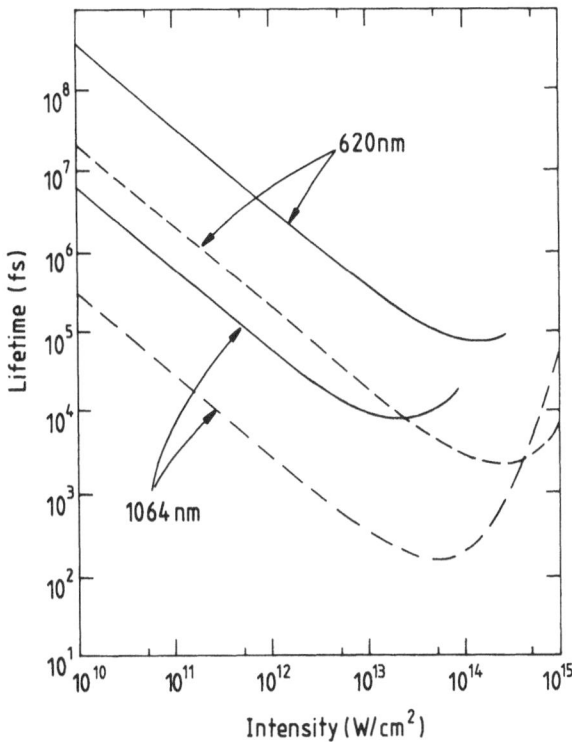

Figure 3. Lifetime ($\hbar/\Gamma$) of a hydrogen atom which is initially in the state $n = 7$, $l = 6$, $m = 5$, versus the intensity of the incident laser field, for two different wavelengths. The polarization is linear. Solid curves: present results; dashed curves: results obtained in "Born approximation".[16]

Table 2. Half-life, $T_{1/2} = \hbar \log 2/\Gamma$, of a few excited states of atomic hydrogen for two different values of the wavelength of the incident laser beam, at the intensity $I_{\min}$ (in units of 10^{13} W/cm^2) where it reaches its minimum.

| | | | 1064 nm | | | | 620 nm | | | |
| | | | Linear pol. | | Circular pol. | | Linear pol. | | Circular pol. | |
n	l	m	$T_{1/2}$	$I_{\min}$	$T_{1/2}$	$I_{\min}$	$T_{1/2}$	$I_{\min}$	$T_{1/2}$	$I_{\min}$
5	4	4	146 fs	0.9	5.31 fs	6.1	571 fs	5.5	10.0 fs	42
8	4	4	322 fs	1.0	21.9 fs	5.6	1.09 ps	5.9	39.7 fs	36
8	5	4	502 fs	1.8	173 fs	8.1	2.58 ps	12	536 fs	60
8	6	4	1.71 ps	4.0			12.9 ps	35		
7	5	5	1.30 ps	1.25	21.3 fs	9.0	7.19 ps	7.5		
7	6	5	5.59 ps	2.3	337 fs	26	51.5 ps	15		
8	6	5	4.24 ps	2.3	402 fs	17	36.3 ps	15		

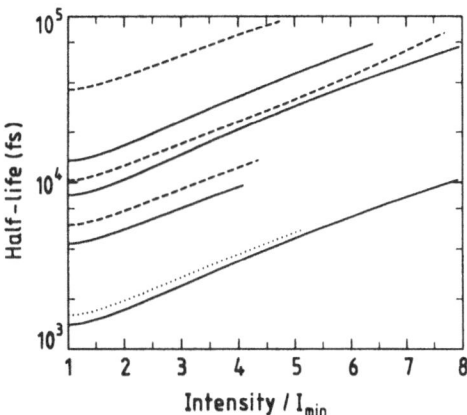

Figure 4. Half-life ($\hbar \log 2/\Gamma$) of a hydrogen atom versus the intensity of the incident laser field. The polarization is linear. On the horizontal scale, 1 corresponds to the intensity where the lifetime reaches its minimum value. The initial states (n,l,m) and incident wavelengths are, from top to bottom, (8,6,5) 620 nm; (8,5,5) 532 nm; (8,6,5) 842 nm; (6,5,5) 620 nm; (7,6,5) 1064 nm; (8,5,4) 532 nm; (8,4,4) 532 nm; and (6,5,5) 1064 nm. The different styles of curve correspond to different angular momenta of the initial state.

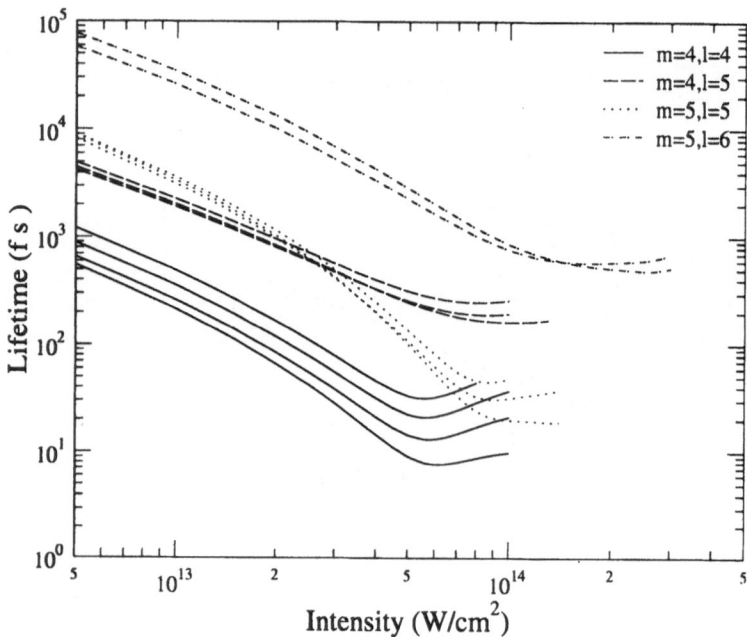

Figure 5. Lifetime ($\hbar/\Gamma$) of a hydrogen atom versus the intensity of the incident 1064 nm laser field. The polarization is circular. Results for the first few lowest lying states are shown for each symmetry.

slow that an experimental detection of stabilization would not be possible if the target excited atoms are not produced in a very small region of the laser focus.

Results for circular polarization are also given in Table 2, as well as in Figure 5, for the sense of rotation of light such that stimulated emission does not bring the dressed initial state in resonance with dressed states belonging to lower lying manifolds. The lifetime also passes by a minimum value, but this occurs at much higher intensity (and much larger rate) than in the case of linear polarization.

ACKNOWLEDGMENTS

This work greatly benefited from enjoyable collaborations, with M. Pont on adiabatic stabilization, and with A. L'Huillier on harmonic generation (in particular during RMP's stay at CEN Saclay in 1989). It is supported by the UK Science and Engineering Research Council, and by a RIC grant from the University of Durham.

REFERENCES

1. For a review of the literature published before the middle of 1991, see e.g. R.M. Potvliege and R. Shakeshaft, Nonperturbative treatment of multiphoton ionization within the Floquet framework, *in*: "Atoms in Intense Laser Fields", M. Gavrila, ed., *Adv. At. Mol. Opt. Phys.* Supplement 1 : 373 (1992).
2. M. Pont, R.M. Potvliege, R. Shakeshaft, and Zh.-j. Teng, Low-frequency theory of multiphoton ionization II, General formulation and further results for ionization of H(1s), *Phys. Rev. A* 45:8235 (1992).
3. R.M. Potvliege and P.H.G. Smith, Two-colour multiphoton ionization of hydrogen by an intense laser field and one of its harmonics, *J. Phys. B* 25:2501 (1992).
4. The high frequency form of the theory is described by M. Gavrila, Atomic structure and decay in high-frequency fields, *in*: "Atoms in Intense Laser Fields", M. Gavrila, ed., *Adv. At. Mol. Opt. Phys.* Supplement 1 : 435 (1992).
5. S.-I Chu, Generalized Floquet theoretical approaches to intense-field multiphoton and nonlinear optical processes, *Adv. Chem. Phys.* 73:739 (1989); M. Dörr, R.M. Potvliege, D. Proulx, and R. Shakeshaft, Multiphoton processes in an intense laser field VI: Two-color ionization with incommensurable frequencies, *Phys. Rev. A* 44:574 (1991).
6. See e.g. R.M. Potvliege and R. Shakeshaft, Nonperturbative calculation of partial differential rates for multiphoton ionization of a hydrogen atom in a strong laser field, *Phys. Rev. A* 38:1098 (1988). The method is formally equivalent to working with a real basis set but a complex-dilatated Hamiltonian, and, in many respects, is very similar to the one previously developed by Chu, Reinhardt, and coworkers [S.-I Chu and W.P. Reinhardt, Intense field multiphoton ionization via complex dressed states: Application to the H atom, *Phys. Rev. Lett.* 39:1195 (1977); A. Maquet, S.-I Chu, and W.P. Reinhardt, Stark ionization in dc and ac fields: an L^2 complex-coordinate approach, *Phys. Rev. A* 27:2946 (1983).] Our method and theirs differ in the choice of basis functions, the choice of gauge, the partitioning of the matrix, and in that we prefer working with the true real Hamiltonian.
7. The results presented by L. Plaja and L. Roso at this workshop suggest that it is sufficient to retain but a few partial waves in time-dependent calculations of rates or harmonic generation, since the wave function does not need to be accurate at large distance, this process being mostly due to bound-bound transitions near the nucleus. More partial waves are certainly required in time-independent Floquet calculations of the same quantitites. This may be because the Floquet wave function must satisfy an eigenvalue problem in which one cannot disentangle short and large distances.
8. The equation is multiplied, before integration over space, by each basis function in turn, the angular part (*not* the radial part) of the latter being complex-conjugated. The

method provides a variational stationary principle for the quasienergy [B.R. Junker, Recent computational developments in the use of complex scaling in resonance phenomena, *Adv. At. Mol. Phys.* 18:207 (1982)]. More details on the numerical techniques can be found in the appendix of Ref. 1.

9. However, it would be interesting to analyse the interaction between dressed states thoroughly, in the light of recent experimental results [M.P. de Boer and H.G. Muller, Observation of large populations in excited states after short-pulse multiphoton ionization, *Phys. Rev. Lett.* 68:2747 (1992)] and the residual discrepancy between the experimental and theoretical ATI spectra in hydrogen at 608 nm [M. Dörr, D. Feldmann, R.M. Potvliege, H. Rottke, R. Shakeshaft, K.H. Welge, and B. Wolff-Rottke, The energy spectrum of photoelectrons produced by multiphoton ionization of atomic hydrogen, *J. Phys. B* 25:L275 (1992)].

10. R.M. Potvliege and R. Shakeshaft, Multiphoton processes in an intense laser field: Harmonic generation and total ionization rates for atomic hydrogen, *Phys. Rev. A* 40:3061 (1989). An excellent check of the correctness of our procedure for obtaining the dipole moments is that the Floquet results are in excellent agreement with the results of time-dependent calculations reported in Ref. 12.

11. The fact that the effective order of non-linearity of the dipole moment is lower than the harmonic order is of great importance to understand the experimental high intensity harmonic spectra; see A. L'Huillier, K.J. Schafer, and K.C. Kulander, High-order harmonic generation in xenon at 1064 nm: The role of phase matching, *Phys. Rev. Lett.* 66:2200. The plateau in the dipole moment of an isolated atom scales linearly with the ponderomotive energy P (see e.g. Ref. 10) and extends to harmonic orders of the order of $(|E_0| + 3P)/(\hbar\omega)$ [J.L. Krause, K.J. Schafer, and K.C. Kulander, High-order harmonic generation from atoms and ions in the high-intensity regime, *Phys. Rev. Lett.* 68:3535]. This is supported by our results, although we have not been able to obtain converged results for very high intensity.

12. J.L. Krause, K.J. Schafer, and K.C. Kulander, Calculation of photoemission from atoms subject to intense laser fields, *Phys. Rev. A* 45:4998 (1992).

13. The linear term can be extracted from $d^{(2c)}(q\omega)$ by numerical differentiation, with $\mathcal{E}_q$ being set to very small values (small values of $\mathcal{E}_q$ make it possible to treat the coupling to this field by a fast iterative method when solving the eigenvalue problem).

14. A. L'Huillier, X.F. Li, and L.A. Lompré, Propagation effects in high-order harmonic generation in rare gases, *J. Opt. Soc. Am. B* 7:527; *ibid.* 7:2137.

15. M. Pont and R. Shakeshaft, Observability of atomic stabilization in an intense short pulse of radiation, *Phys. Rev. A* 44:R4110 (1991). The results of some of the time-dependent calculations reported in this article for ionization of H($4f_3$) have been found to be in good agreement with the results of Floquet calculations [R. Potvliege and R. Shakeshaft, Multiphoton processes in atomic hydrogen, *in*: "Electronic and Atomic Collisions", W.R. Standage, I.E. McCarthy, and M.C. Standage eds., Adam Hilger, Bristol (1992)].

16. R.J. Vos and M. Gavrila, Effective stabilization of Rydberg states at current laser performances, *Phys. Rev. Lett.* 68:170 (1992). These authors carried out a Floquet calculation of the lifetimes, in the Kramers-Henneberger frame, but they neglected the nucleus-electron Coulomb interaction in the final state of the photoelectron and described it by a plane wave. The difference between their results and ours can be attributed entirely to this "Born approximation".

17. M. Pont, private communication (January 1992).

18. The results of Figure 3 are in excellent agreement with the results of an independent full Floquet calculation by A. Buchleitner and D. Delande (manuscript, 1993). These authors also show that for circular states one has stabilization without dichotomy; this fact was also pointed out, independently, by M. Pont, at this workshop.

19. The scaling in $m\omega^3$ has a dynamical origin [e.g. R. Shakeshaft, Atoms in strong laser fields: From tunneling to stabilization, *Comm. At. Mol. Opt. Phys.* 28:179 (1992)].

THE USE OF THE TIME CORRELATION FUNCTION IN FLOQUET THEORY

Thomas Millack

Laboratoire Chimie Physique, Université P. et M. Curie
11, rue P. et M. Curie, 75231 Paris CEDEX 05, France

INTRODUCTION

In the description of the interaction of short and strong laser pulses with atoms or molecules the Floquet picture [1] provides in general the best basis for an analysis even down to pulse lengths of the order of 10 or 20 laser cycles. This is so because the Floquet states are the equivalents of the stationary states for the case of a constant laser amplitude and make it possible to separate the effects of the laser frequency from the effects of the time variation of the laser amplitude. The diagonalization of the Floquet Hamiltonian, however, is very difficult if the system contains a continuum and there is essentially only one method to do this, the complex dilatation method [2]. On the other hand, the direct, numerical solution of the time dependent Schrödinger equation is much easier and there are a number of efficient approaches, mostly based on real coordinates and energies [3,4]. Therefore the complex dilatation method can not be directly used to analyse a wavefunction calculated with such a real algorithm. It is the purpose of this contribution to describe an alternative method introduced into Floquet theory recently [5] to obtain informations about the Floquet content of a given wavefunction. This method is based on the calculation of the (local) time correlation function which can be calculated by solving the time dependent Schrödinger equation and which uses real energies and coordinates. It can therefore very easily incorporated into a given solution algorithm of the time dependent Schrödinger equation.

FOUNDATIONS

By the (local) time correlation function we understand the quantity

$$P(\tau) = <\Psi(t_0 \mid \Psi(t_0+\tau)> \tag{1}$$

where $\Psi(t_0)$ is the wavefunction of the system propagated in the pulse up to the time t_0 and $\Psi(t_0+\tau)$ the wavefunction obtained by propagating in a constant amplitude field with $\Psi(t_0)$ as initial condition. The point of propagating in a constant amplitude field is that the Floquet content does not change any more and that the wavefunction can be expanded into a complete set of Floquet states

$$\Psi(t_0+\tau) = \sum_\alpha \int_0^\omega d\varepsilon \, c_\alpha(\varepsilon) \, e^{-i\varepsilon(t_0+\tau)} \, \phi_{\varepsilon,\alpha}(t_0+\tau) \tag{2}$$

where α is a general index counting all possible degeneracies of the quasienergy ε. Note that because ε stays real in (2) the spectrum of the Floquet Hamiltonian is continuous and we have to integrate over one Brillouin zone. The $c_\alpha(\varepsilon)$ (by definition periodic in ε) are the occupations of the different degenerate quasienergy states and contain the informations about the Floquet content of the wavefunction. Initially, at $t_0=0$, they are just a delta function pointing to the initial condition of $\Psi(t)$. Later in the pulse, the population moves with the adiabatically connected Floquet state and so the $c_\alpha(\varepsilon)$ are expected to show a pronounced maximum at the position of the corresponding energy. As the system starts to ionize this maximum should be broadend according to the decay rate and further peaks should appear if other Floquet states get occupied. As we will show below this is indeed true, but only the first half of the story. The $c_\alpha(\varepsilon)$ also contain informations about the past of the wavepacket, i.e. the parts of the wavepacket which have been ionized earlier in the pulse, are still present at time t_0. The second half of this paper will deal with a method to separate these two contributions.

The structure of the $c_\alpha(\varepsilon)$ can be made visible with the help of (1). Note that because $P(\tau)$ does not depend on α, the degeneracies can not be resolved with the present method. Inserting (2) into (1) yields

$$P(\tau) = \sum_{k=-\infty}^\infty \int_0^\omega d\varepsilon \, d_k(\varepsilon) \, e^{-i\tau(\varepsilon+k\omega)} \tag{3}$$

with

$$d_k(\varepsilon) = \sum_{\alpha,\alpha'} \sum_{k'=-\infty}^\infty \int_0^\omega d\varepsilon' \, c_{\alpha'}^*(\varepsilon') \, c_\alpha(\varepsilon) \, e^{i(\varepsilon'-\varepsilon+k'\omega-k\omega)t_0} \langle \phi_{\varepsilon',\alpha'}^{k'} | \phi_{\varepsilon,\alpha}^{k} \rangle \tag{4}$$

Here $\phi_{\varepsilon,\alpha}^k$ denote the Fourier coefficients of $\phi_{\varepsilon,\alpha}(t)$. The reason why (4) is a quite complex expression lies in the fact that different Floquet states are not orthogonal for different times (modulo $2\pi/\omega$). Using the equal time orthogonality relation for Floquet states we obtain

$$\sum_{k=-\infty}^\infty d_k(\varepsilon) = \sum_\alpha |c_\alpha(\varepsilon)|^2 := C(\varepsilon) \tag{5}$$

which is the total occupation probability of a given quasienergy. Because the $d_k(\varepsilon)$ are proportional to the sum of all $c_\alpha(\varepsilon)$, although with in general unknown coefficients, the structure of the $c_\alpha(\varepsilon)$ is reflected in the $d_k(\varepsilon)$. We will discuss this point in the next section. But before, we should mention one more point. Taking $\tau = 1\,2\pi/\omega = lT$ in (3) with l equal to an integer we can use (5) and obtain

$$P(lT) = \int_0^\omega d\varepsilon\ C(\varepsilon)\ e^{-il\varepsilon T}$$

$$. \tag{6}$$

So P(lT) is the l'th Fourier coefficient of $C(\varepsilon)$ and can be used to reconstruct $C(\varepsilon)$ via its Fourier series.

THE FOURIER TRANSFORMED TIME CORRELATION FUNCTION

By Fourier transforming (3) we can already obtain a number of qualitative informations about the $d_k(\varepsilon)$ and therefore about the $c_\alpha(\varepsilon)$. If the $c_\alpha(\varepsilon)$ have a peak at a specific ε, the $d_k(\varepsilon)$ will also have a peak at this position, although with a different amplitude, which is repeated in each Brillouin zone counted by k. If such a peak can be assigned to a specific Floquet resonance, the number of corresponding peaks visible will give a lower bound to the number of Fourier coefficients of this state as is clear from (4). Further, the relative peak heights of different classes of peaks will give an indication about the relative fraction of population contained in these peaks. This, however, is a very loose statement, because we found that the Fourier transformed time correlation function is in general much more sensitive to the presence of different resonant states than indicated by their relative occupation. And finally, if t_0 approaches the end of the pulse, the picture obtained should approach the ionization spectrum after the pulse. This follows from the work of Feit, Fleck and Steiger [6] and was applied by Horbatsch [4] who used this method for the case of a time independent Hamiltonian.

In order to increase the signal to noise ratio of the Fourier transform and to avoid the aliasing problem it is of advantage to use a window for the calculation of the Fourier transform. So we take the Fourier transform of (3) as

$$\widetilde{P}(E) = \frac{1}{T_P} \int_0^{T_P} dt\ P(t)\ (1 - \cos(\frac{\pi t}{T_P}))\ e^{iEt}$$

$$\tag{7}$$

where T_P is the length of the time interval taken for the constant amplitude propagation.

Let us now turn to an example. We have solved the time dependent Schrödinger equation for the one-dimensional soft Coulomb model potential $V(x) = -1/\sqrt{1+x^2}$ [7] with a standard Crank-Nicholson algorithm. The interaction with the laser field was taken in Dipol approximation and the calculation was performed in the Kramers-Henneberger frame. Starting from the ground state we applied a pulse of the form $E(t) = E_0\ \sin^2(\pi t/t_p)\ \sin(\omega t)$ where t_p is the pulse length. As parameters we choose $\omega = 0.152$ a.u., $E_0 = 0.075$ a.u. and

t_p=50T. From the investigations of Reed and Burnett [8] we expect that with these parameters the system stays essentially in a single Floquet state with no significant intermediate resonance, but that the final ionization spectrum shows a pronounced interference structure due to parts of the wavepacket ionized at different times. Figure 1 shows the Fourier transformed correlation function at t_0=5T, the beginning of the pulse.

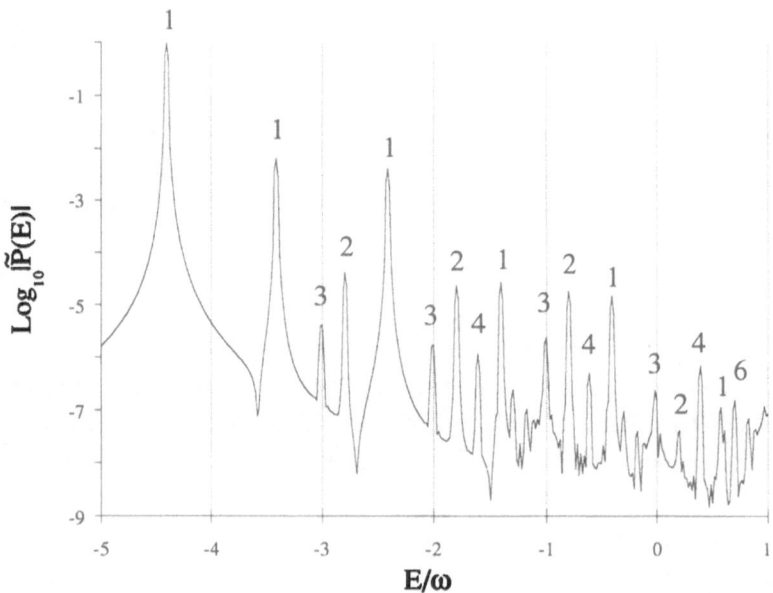

Figure 1. The Fourier transform of the time correlation function at t_0=5T for a sin-square pulse of the length of 50T at a frequency of ω=0.152a.u. and a maximum field strength of E_0=0.075 a.u. The numbers indicate to which bound state of the unperturbed Hamiltonian the corresponding peaks belong.

Shown are 6 Brillouin zones indicated by the vertical lines starting from the zone -5 because initially 5 photons are needed to ionize the system. As the field strength is still very small the ac Stark shift is neglegible and most of the peaks can easily be identified. The numbers above each peak show the number of the corresponding level of the soft Coulomb potential it represents. It should be mentioned here that the n=5 state is nearly degenerate with the n=1 state and can not be resolved. For this low field strength it is possible to determine the populations of the states and in fact the occupation of the ground state is 0.999999 and except for the states n=2 and n=3 the occupation of the other states is less than the accuracy of the calculation (10^{-7}). The presence of these states is due to the fact that numerically we have to turn on the field non-adiabatically, however small. This shows that the Fourier transform is indeed very sensitive. If now the field is turned on further the odd numbered states shift to the left and the even numbered to the right, but because we perform our calculation in the Kramers-Henneberger frame the shift is very small for high lying levels.

The ground state shifts strongest, but does not cross the ionization threshold, at maximum field strength it is at $E/\omega=0.12$ (modulo ω). At $t_0=14T$ it has a noticable avoided crossing with the n=2 state, but most of the population follows the diabatic curve and only about 10^{-3} of the population is transfered to the other branch. There is an equivalent avoided crossing at $t_0=39T$ when the field is turned off and in between these two times the ionization is strong. The build up of ionized population can be seen in the spectrum of the time correlation function by the presence of broad features which no longer shift if the amplitude is varied further. In Figure 2 the spectrum at $t_0=45T$ is shown which is the same field strength as in

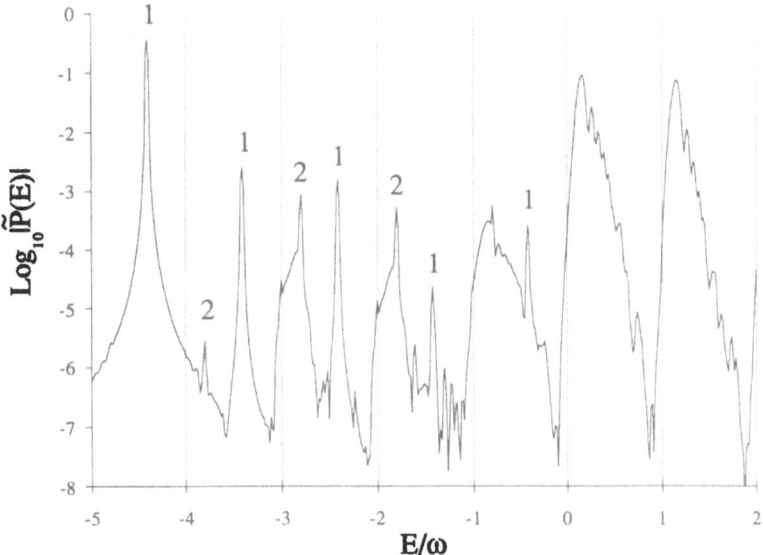

Figure2. The Fourier transform of the time correlation function at $t_0=45T$ for a sin-square pulse of the length of 50T at a frequency of $\omega=0.152$ a.u. and a maximum field strength $E_0=0.075$ a.u. The numbers indicate to which bound state of the unperturbed Hamiltonian the corresponding peaks belong. The broad features at the right hand side are connected to the ATI photoelectron spectrum.

Figure 1. The peaks on top of the broad structures on the right hand side are not connected to any resonance, but are in fact the interferences or rainbow structures discussed by Reed and Burnett [8]. Indeed these peaks appeared only after the energy position of the ground state connected Floquet state had passed the corresponding position at times belonging to the turn off of the pulse. If we now compare these structures with the ATI spectrum calculated after the pulse by projecting onto unperturbed continuum states (Figure 3) we see that Figure 2 is already a very good representation of the photoelectron spectrum. The differences are coming mainly from the window used to calculate Figure 2. So the time correlation function makes it possible to study the development of the ionization spectrum and to answer the question whether the peaks visible in the spectrum are due to resonantly enhanced ionization or due to

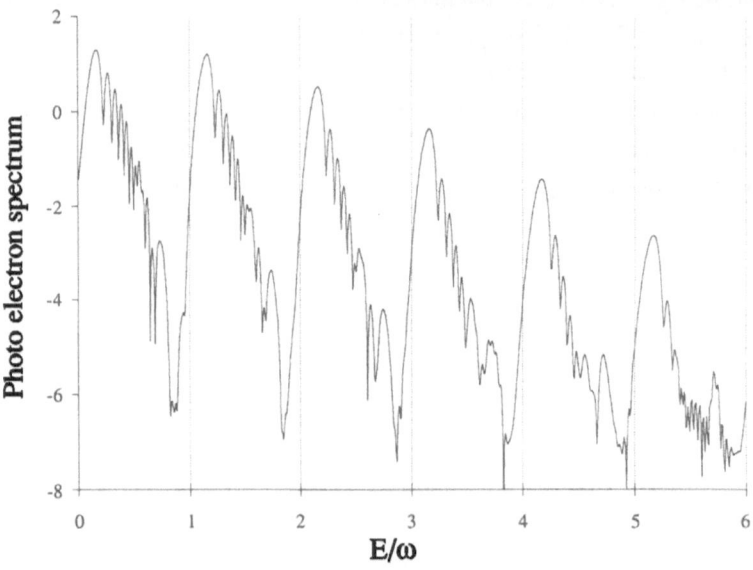

Figure3. The photo electron spectrum after a pulse of length t_p=50T, frequency ω=0.152 a.u. and maximum field strength E_0=0.075 a.u. with the ground state as initial condition. This picture was calculated by projecting onto the unperturbed states after the pulse.

interference effects. A possible larger split of the population at an avoided crossing can also be observed, although, as will be discussed below, the exact determination of the population of single Floquet states is not easily possible.

RESONANCE REPRESENTATION

Up to now we did not touch the question of how to exactly determine the positions and widths of the resonant Floquet states. The reason for this is that this it not possible as long as the already ionized portions of the wavepacket are present in the correlation function. The determination of positions and widths beyond the accuracy of the Fourier transform requires a fit of the line shape of the spectrum to a known functional form. Such a form is in general unknown for the ionized parts of the wavepacket. There is, however, a very easy and general known trick to get rid of these parts. Instead of calculating the correlation function in (1) by an integration over a (numerically) infinite space region, restrict the integration over a finite volume. As soon as the wavepacket reaches the boundary of the volume the corresponding correlation function starts to decrease exponentially. Furthermore, in the Kramers Henneberger frame the interaction has the general form $V(x+\alpha(t))$ where $\alpha(t)$ oscillates with a maximal amplitude α_0. The already ionized portions of the wavepacket have their main support at large x where the α_0 can be neglected so that these portions no longer interact with the field. A restriction to a finite interior volume will exclude exactly these parts. Figure 4 shows the equivalent to Figure 2, but now the integration volume for the time correlation function is restricted to 180 a.u. Two things are immediately obvious. First the large rainbow

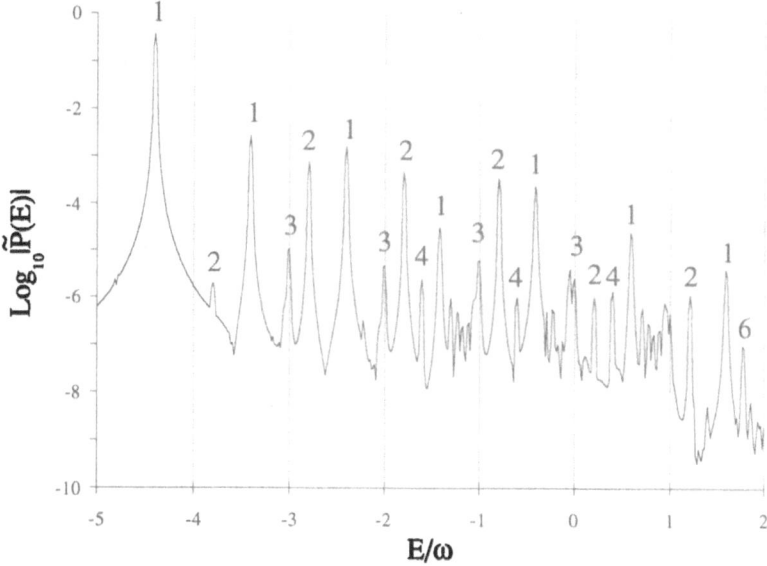

Figure4. The Fourier transform of the time correlation function at $t_0=45T$ for a sin-square pulse of the length of 50T at a frequency of $\omega=0.152$ a.u. and maximum field strength $E_0=0.075$ a.u. The integration volume for the calculation of the time correlation function was restricted to 180 a.u. The numbers indicate to which state of the unperturbed Hamiltonian the peaks belong.

structures have disappeared and a large number of peaks are now visible which were hidden before. Second the appearance of corresponding peaks is now much more uniform in their structure. This is confirmed by a line shape fitting which determines the positions and widths of the peaks. Such a fitting technique was developed by Feit, Fleck and Steiger [6] and can easily be adapted for the present case. The positions and widths of corresponding peaks are essentially equal as they should if they were to represent a single Floquet resonance state. So the representation (3) of the time correlation function can be replaced by

$$P(\tau) = \sum_n \sum_{k=-\infty}^{\infty} d_{k,n} \, e^{-i\tau(\varepsilon_n - i\Gamma_n + k\omega)}$$

(8)

where ε_n and Γ_n are the positions and widths of the different resonances. It is also possible to give an expression for the $d_{k,n}$, but the orthogonality relations for the Floquet functions are destroyed by the introduction of the finite volume and it is very difficult to replace it with some other relation. A more detailed discussion will be given elsewhere.

ANALYSIS VIA THE FOURIER SERIES

There are two questions remaining which we want to discuss in this section. The first is how

191

to improve the accuracy of the calculation and second whether it is possible to determine the population contained in a specific resonance or ionization structure. For reasons of space we will only sketch the most important points, a more detailed account will be given elsewhere. Concerning the first question the fundamental problem is that the resolution obtained with the Fourier transform is intimately connected to the time interval T_p (see (7)). So if the propagation interval used for the time correlation function is 64 laser cycles we will have 64 points in each Brillouin zone of the Fourier transform. This was the case for the Figures 1,2 and 4 and is in general enough to see most features. But if two peaks are very close this may not be enough to separate them. A possible improvement can be given with the help of equations (5) and (6). The function $C(\varepsilon)$ defined in (5) is the total occupation probability density and can be calculated by

$$ C(\varepsilon) = \frac{1}{\omega} \sum_{l=-\infty}^{\infty} P(lT)\, e^{il\varepsilon T} $$

(9)

Clearly one knows only a finite number of Fourier coefficients and this is in general not enough to obtain a converged series. It is therefore necessary to use acceleration techniques to calculate this series. The simplest and most effective algorithm to use is the well known epsilon algorithm [9]. If it converges ε can be chosen at will and the resolution is only limited by the accuracy of the determination of $P(\tau)$. The question whether and when it will converge is, however, highly non trivial. It is always necessary to check the results of the calculation for consistency for instance with the help of (5) by summing over all possible $d_k(\varepsilon)$. It is also possible to add an additional converging factor of the form $e^{-\Gamma |l| T}$. The interpretation of such a converging factor is the artificial introduction of an extra width Γ (see (8)) which ensures the decay of the time correlation function fast enough to enable the acceleration algorithm to converge.

The interpretation of the function $C(\varepsilon)$ as a occupation density opens the possibility to estimate the amount of population contained in a given feature. For this one has to integrate $C(\varepsilon)$ over a range which is dominated by this feature. This works very well if the peaks are isolated as for instance in a situation as in Figure 1. The peaks corresponding to the states n=1,2 and 3 are the only peaks visible in $C(\varepsilon)$ and are well separated. But it is clear that in general one can only get an estimate. It is not possible to exactly fix the integration range and other peaks can considerably overlap with the feature under consideration. So the only way to arrive at a better estimate would be to know the functional form of the line shape and to fit the data accordingly. But this is very difficult even for the case of a resonant representation. In the latter case the requirement would be that the sum $\Sigma_k\, d_{n,k}$ is a real number representing the ocupation of the resonance n. But in general this number is complex and such an interpretation is not possible.

APPLICATION TO HIGH HARMONIC GENERATION

The standard technique to calculate the light spectrum emitted by a single atom is to propagate the wavefunction with a constant amplitude field and to calculate the time dependend dipol or acceleration which is then Fourier transformed to give the spectrum [10]. If the unperturbed Hamiltonian has even and odd states the spectrum will normally only show odd harmonics which have the well known plateau structure the origin of which is being discussed intensly. But this is only true if a single Floquet state is dominating the wavefunction during this propagation. If two or more Floquet states are occupied one has to expect additional Hyper Raman lines above and below the odd (or even) harmonic positions where the distance from the harmonic position is given by the difference of the Floquet energies. The technique described above can now be used to determine these differences and to answer the question whether there is a strong enough avoided crossing to populate other Floquet states. This will clearly depend on the pulse parameters and one has to identify the relevant parameter regions to guide possible experiments. Work in this direction is in progress and will be published in a separate paper.

CONCLUSIONS

In conclusion it was shown how to use the (local) time correlation function to analyse atoms interacting with a short and strong laser pulse. The analysis of either the Fourier transform of this function or the total probability density makes it possible to follow the adiabatic/diabatic movement of the system along the pulse, to determine the Floquet state energy positions and widths, to detect the positions of important avoided crossings and to study the creation of ionization population.

ACKNOWLEDGEMENT

The author would like to thank Valérie Véniard and Joachim Henkel for their considerable help and support during the first stages of this work and Prof. Alfred Maquet for very helpful discussions. This work was supported by the European Community in the SCIENCE program. The Laboratoire de Chimie Physique de l'Université Pierre et Marie Curie (URA 176) is an "Unité de Recherche Associé au CNRS"

REFERENCES

[1] J.H. Shirley, Phys. Rev. 138, B979, (1965).

S.I. Chu, Adv. At. Mol. Phys. 21, 197, (1985).

[2] A. Maquet, S.I. Chu and W.P. Reinhardt, Phys. Rev. A27, 2946, (1983).

R.M. Potvliege and R. Shakeshaft, Phys. Rev. A40, 3061, (1989).

[3] K.C. Kulander, Phys. Rev. A35, 445, (1987).

J. Javanainen, J.H. Eberly and Q. Su, Phys. Rev. A38, 3430, (1988).

X. Tang, H. Rudolph and P. Lambropoulos, Phys. Rev. Lett. 65, 3269, (1990).`

special issue of Comput. Phys. Comm. 63, K.C. Kulander eds., (1991).

special issue of J. Opt. Soc. Am. B7, K.C. Kulander and A. L'Huillier eds., (1991).

M. Pont, D. Proulx and R. Shakeshaft, Phys. Rev. A44, 4486, (1991).

U.L. Pen and T.F. Jiang, Phys. Rev. A46, 4297, (1992).

S. Chelkowski, T. Zuo and A.D. Bandrauk, Phys. Rev. A46, R5342, (1992).

[4] M. Horbatsch, J. Phys. B24, 4919, (1991).

[5] T. Millack, V. Véniard and J. Henkel, submitted to Physics Lett.

[6] M.D. Feit, J.A. Fleck Jr. and A. Steiger, J. Comp. Phys. 47, 412, (1982).

M.R. Hermann and J.A. Fleck Jr., Phys. Rev. A38, 6000, (1988).

[7] Q. Su and J.H. Eberly, Phys. Rev. A44, 5997, (1991).

[8] V.C. Reed and K. Burnett, Phys. Rev. A43, 6217, (1991).

[9] C. Brezinski and M. Redvio Zaglia, "Extrapolation Methods, Theory And Practice", Elsevier Science, Amsterdam, (1991)

[10] J.H. Eberly, Q. Su and J. Javanainen, Phys. Rev. Lett. 62, 881, (1989).

K. Burnett, V.C. Reed, J. Cooper and P.L. Knight, Phys. Rev. A45, 3347, (1992).

J.L. Krause, K.J. Schafer and K.C. Kulander, Phys. Rev. A45, 4998, (1992).

Some recent developments in multiphoton ionization of atoms in intense fields

M. Pont[1], A. Bugacov[1] and S.R. Atlas[2]

[1]Physics Department
University of Southern California
Los Angeles, CA 90089-0484
[2]Thinking Machines Corporation
245 First Street
Cambridge, MA 02142-1214

1 Introduction

In this paper we would like to discuss three different topics that we have worked on recently. In the second section we would like to discuss some aspects of (adiabatic) stabilization, addressing what we believe its mechanism to be, as well as some recent computations that we have carried out in order to examine whether the breakdown of the dipole approximation in strong fields could perhaps overcome stabilization. In the second section we would like to give a short account of a formalism that we have developed in order to be able to include correlation effects within the Floquet approximation treatment of two-electron atoms that ionize in strong, low frequency fields. In the final section, we would like to summarize our efforts to numerically integrate the time-dependent Schrödinger equation for a two-electron system subject to an intense pulse of laser light. These efforts are taking place on the massively parallel 'Connection Machine'. Since massive parallelism, a new concept in supercomputing, has only made its entrance in atomic physics quite recently, we deem it worthwhile, without going into too much detail, to give a brief overview of some particular features of this machine and how one goes about developing efficient algorithms on it.

2 Stabilization

2.1 Introduction

As the year of birth of 'stabilization' one could perhaps take 1990[1, 2, 3, 4, 5]. This means that this discovery is actually very recent, yet it has already given rise to lots of activity by many contributors [6]. Perhaps it is too early to describe its history, but

Super-Intense Laser-Atom Physics, Edited by
B. Piraux *et al.*, Plenum Press, New York, 1993

it can nevertheless be refreshing in view of some more recent developments.

In the past there had been various attempts to develop a theory to describe in some approximation the multiphoton ionization of atoms in superintense fields, most notably by Mittleman[7, 8]. Mittleman was the first to recognize one of the major catches[8]. Whereas others had thought along lines of treating the binding potential as a perturbation all together and taking the free-particle propagation in the field as the lowest-order approximation[9], he recognized that this was not the right thing to do, since the initial channel is bound and hence the neglect of the binding potential in the initial channel is questionable. Now we know that the right thing to do is to 'dress' the initial state in the KH-frame. The residual interaction can then be considered 'weak' provided the frequency is sufficiently high[7, 10, 11].

2.2   The mechanism: stabilization without distortion

In 1984 Gavrila and Kaminski (GK) developed a scheme for free-free scattering in high-frequency fields[10] that was later extended to treat multiphoton ionization[11]. But the latter version turned out to be a reformulation of a, perhaps, forgotten theory of Mittleman[7], except for a subtle point that later turned out to be quite crucial.[1] The region of validity of the GK-theory reads:

$$\omega \gg I_p(\alpha_0), \tag{1}$$

where I_p stands for ionization potential of the atom *in the field* and ω for the circular frequency of the light. Mittleman took the more restrictive field-free ionization potential in Eq.(1). He also seems to have overlooked that for high magnetic quantum number states (and linear polarization) Eq.(1) is rather easily satisfied, what later turned out to be crucial for the observability of stabilization[12, 13, 14]. The GK-theory claims that the atom would ionize at 'moderate', yet unknown rate, depending on the degree to which this condition is satisfied. Its prediction is intuitively clear: If the proton in the KH-frame oscillates more rapidly, than the typical round-trip time of the electron about the nucleus *in the field*, the electron would basically feel the time-averaged electrostatic potential exerted by the proton and hence the atom would be stable. Hence the atomic electron cloud could be distorted while oscillating as a free electron in the laboratory frame, yet it would be relatively stable against multiphoton ionization. This type of stabilization we call 'high-frequency stabilization'. In the GK-theory stability is obtained if the parameter α_0 is kept fixed while the frequency is increased. This is something different from 'high-intensity' stabilization, namely a decrease of the ionization rate for *fixed frequency and increasing intensity*[2, 3, 5].

The GK-theory was shortly followed by a computation by Pont and Gavrila [12] showing a dramatic decrease of the ionization potential with α_0, a feature that is accompanied by the splitting of the electron cloud into two disjoint parts, which they titled 'dichotomy'. This suddenly 'upgraded' the *high-frequency* theory to a *high-intensity* theory and suggested the possibility of stabilized hydrogen atoms in very strong fields of probably not unrealistically high frequency, although the problem how to ever get the atom in such a strong field remained. The prediction of Pont and Gavrila was confirmed shortly after by Eberly and Grobe [4], who indeed observed stabilized 'dichotomous' states, though in 1 dimension (see also later work by Kulander, Schafer and Krause [6]).

---

[1]There is another point. Whereas Mittleman only wrote down an expression for the total width, the GK-theory also gives an expression for the (angular dependent) partial rates.

What was missing still was the calculation of decay rates. Meanwhile, Pont [2] had managed to simplify the GK expression (which is still considered to be too difficult to evaluate without further approximations) for the ionization amplitude to the extent that it allowed computation and analysis of the decay rates:

$$f_n = \frac{1}{\pi k_n^2} \int_{-\pi}^{+\pi} e^{in\phi} e^{i\alpha\,(\phi)\cdot k_n}\,\Phi(-\alpha\,(\phi))d\phi. \tag{2}$$

Here Φ is the (possibly distorted) wavefunction in the KH-frame and k_n is the momentum of the photoelectron after absorption of n photons by the atom. Eq.(2) predicts high-intensity stabilization, but more importantly Eq.(2) predicts that *stabilization occurs even when there is no distortion at all*. This was a totally new prediction. Up to that time, stabilization was considered to be due to the decrease of the ionization potential making the frequency high w.r.t. the ionization potential in the field in accordance with Eq.(1) or alternatively due to the distortion of the electron cloud which decreases the probabilty of the electron hitting the nucleus and hence ionization. These latter two effects are of course related to each other.

Equation (2) provides the following picture of *the mechanism of high-intensity stabilization*. In the KH-frame we have a stationary (possibly distorted) electron cloud. Outgoing electron waves (representing multiphoton ionization) are created in the vicinity of the momentary position of the proton as it passes to and fro (see Fig.1). Equation (2) contains the *probability amplitude* $\Phi(-\alpha\,(\phi))$ to find the electron at the position of the proton and a *phase factor* $\exp\{i\alpha\,(\phi)\cdot k_n\}$ associated with the displacement of the proton from the origin of the Kramers reference frame by $-\alpha\,(\phi)$. These two notions formed the key to what was according to Pont the physical mechanism underlying the high-intensity stabilization of an atom in very strong fields: The first stabilizing effect arises from the phenomenon of *'radiative distortion'* of the electron cloud, as argued already.

But Eq.(2) shows a totally new aspect that has nothing to do with distortion: In very intense fields we have $\alpha_0 k_n \gg 1$ hence the de Broglie wavelengths of the photoelectrons are small with respect to the spatial extension ($2\alpha_0$) of the source from which they are emitted (the distance over which the proton oscillates), see Fig.1. It thus follows that the phase factor $\exp\{i\alpha\,(\phi)\cdot k_n\}$ varies rapidly with the phase ϕ of the field. Consequently, waves emitted from different positions of the proton in the KH-frame tend to cancel each other due to *'destructive interference'* except in directions nearly orthogonal to the polarization axis where $\alpha\,(\phi)\cdot k_n$ can be small. This suppresses the outgoing flux of electrons with increasing α_0 (or at fixed frequency, with increasing intensity) and hence we have high-intensity stabilization without distortion. Whereas the distortion sets in gradually with α_0 (and may actually be insignificant for Rydberg states to be discussed below), it is the latter effect that we think is causing the sharp turnover of the Floquet rate when plotted as a function of the intensity [2, 3, 5].

An independent analysis by Dörr et al. [5] arrived at the same condition: When the ponderomotive energy gets larger than the photon energy:[2]

$$P = \hbar\omega \tag{3}$$

the ionization rate attains a maximum. These latter authors had however an apparently totally different point of view and claimed the stabilization effect was kinematical in origin, whereas the explanation that Pont had given (see above) suggested a quantum origin. These two pictures were shown to be equivalent in a later paper by Pont and Shakeshaft [13] and in fact these authors were able to heuristicaly rederive Eq.(2). It is of course the kinematical picture that is the most appealing. Equation (3) was

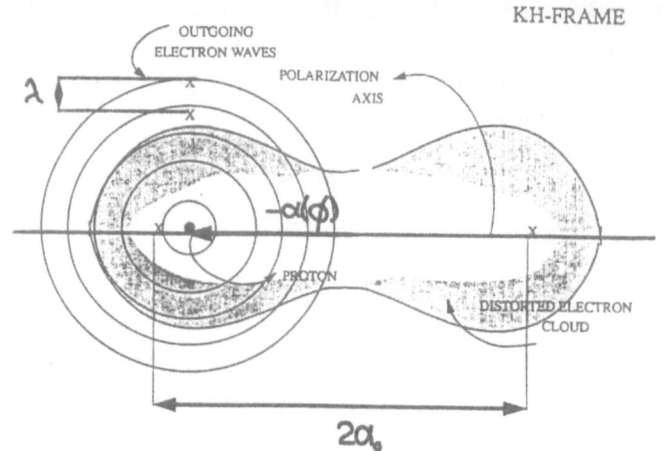

Figure 1. Outgoing electron waves are formed in the vicinity of the proton. In the stabilization regime, the de Broglie wavelength is small with respect to $2\alpha_0$ and hence waves emitted from different positions of the proton will interfere 'destructively'.

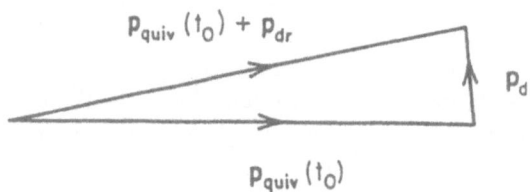

Figure 2. The electron driven by the electric field, impinges on the nucleus at time t_0 with the quiver momentum $p_{quiver}(t_0) = k_q$, and it scatters practically elastically with momentum transfer $p_{dr} = k_f$. The two sides of the triangle are practically equal.

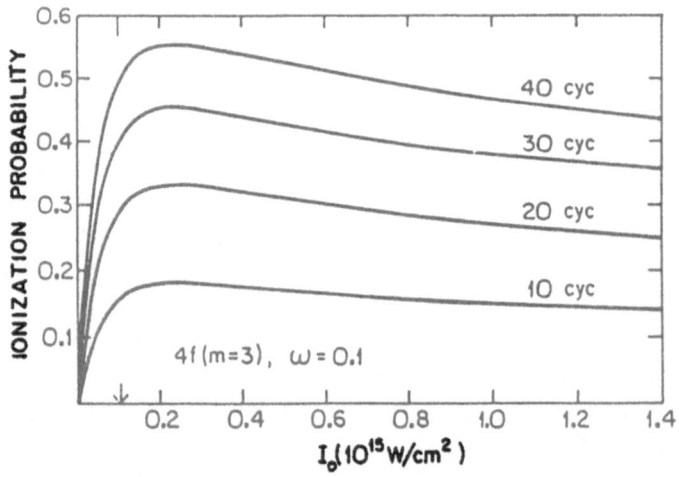

Figure 3. Ionization yield from the $4f(m = 3)$ state of hydrogen at the end of a Gaussian pulse of frequency 0.1 a.u. as a function of the peak intensity, for various durations.

supported by all the up to then available data. The evidence for Eq.(3) has meanwhile gotten quite overwhelming with recent extensive Floquet calculations by Potvliege and Smith (see elsewhere in this book), who empirically confirmed Eq.(3) but showed that in general there is a nontrivial factor involved – presumably of geometrical origin – that depends on the azimuthal and magnetic quantum number of the initial state and which we had neglected in Eq.(3).

Let me sketch here our kinematical mechanism of stabilization. We envisage that at some instance the electron impinges on the nucleus with a momentum equal to the quiver momentum $\mathbf{k_q}$. If it is to leave the atom with the drift momentum $\mathbf{k_f}$, it must have collided with the nucleus to give a final momentum $\mathbf{k_q} + \mathbf{k_f}$. (Under the circumstances of intense, high-frequency fields this collision happens 'suddenly'.) See Fig.2. Yet, this process can only take place provided there is but a small change in the energy of the electron, for otherwise the likelyhood would be very small. But when P is larger than ω, $|\mathbf{k_f}|$ is small with respect to $|\mathbf{k_q}|$ ($\frac{1}{2}|\mathbf{k_f}|^2 \propto \omega$ and $\frac{1}{2}|\mathbf{k_q}|^2 \propto P$). Hence $\mathbf{k_f}$ has to be nearly orthogonal to $\mathbf{k_q}$. In other words: the photoelectron is kinematically restricted to emerge in directions nearly orthogonal to the polarization axis. This finding is consistent with calculations of angular distributions in the stabilization regime carried out by Pont[2]. Thus, the restriction in phase space implies stabilization. Therefore the 'destructive interference' discussed above is simply the quantum manifestation of the same effect, the origin of which is kinematical.

To proof our claim that dichotomy or distortion is not necessary for stabilization, let us turn to Fig.3 - which by the way we believe was the first demonstration of the experimental feasability of a test of the prediction of stabilization namely for Rydberg states of high magnetic quantum number[13, 14]. Here we have presented the total yield obtained by solving the time-dependent Schrödinger equation for a hydrogen atom initially in a circular $[4f(m = 3)]$ state that is exposed to a Gaussian pulse with different peak intensities. The turnover of the yield marks the onset intensity of stabilization, in accordance with our formula Eq.(3). At an intensity of $1.0 \, 10^{15}$ W/cm^2 - which is well in the stabilization regime as one can easily convince oneself - we plotted the charge cloud as a function of time (see Fig.4). It shows no sign of significant distortion while it oscillates almost perfectly like a free electron in the field.

2.3 The breakdown of stabilization

In textbooks one may find the statement that the dipole approximation is valid as long as the wavelength of the light is large with respect to a typical atomic size. This, however, is in general wrong; it is only true for weak fields and single photon ionization. To see this, we rather should compare the total photon momentum (which comes in packages of $\hbar\omega/c$ per photon with c the speed of light) with the available atomic momentum, say $\hbar/a_n$, with $a_n = a_0 * n$ the Bohr radius and n the principal quantum number. If the number of photons would equal one, it would yield the text book statement. But in strong fields there are actually many photons, say N, contributing and this number should be multiplied by $\hbar\omega/c$ to get the total momentum imparted by the field to the atom. This number N turns out to equal $P/(\hbar\omega)$ as one could have guessed. To show this, we note that the initial state, in the high-frequency regime, will basically shift along with the ionization threshold (the electron cloud will oscillate like a

---

[2]This is the same as saying: 'When the de Broglie wavelength of the photoelectron gets smaller than the size of the source...' or $\alpha_0 k_1 = 1$. This is easily shown from the formulae $\alpha_0 = I^{(1/2)}/\omega^2$ and $k_1 \approx \sqrt{2\omega}$.

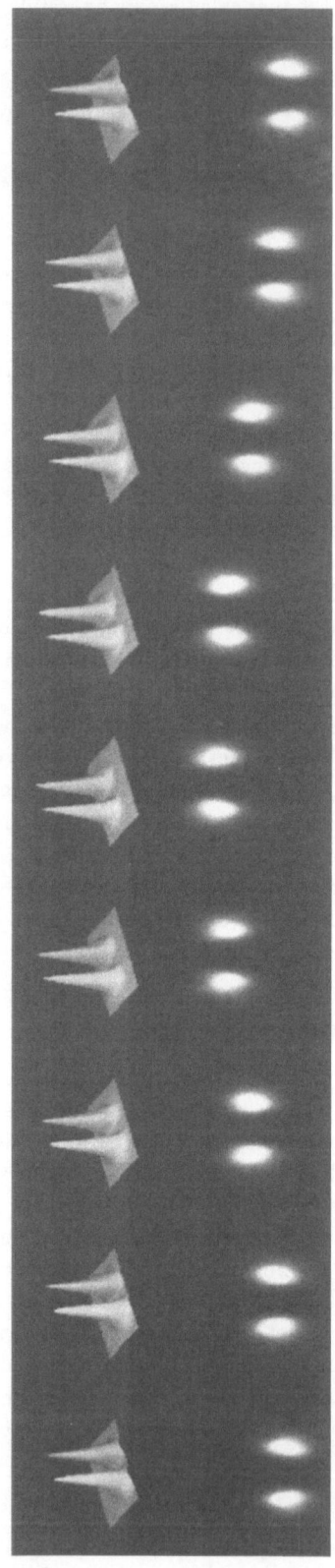

Figure 4. Time evolution of the probability distribution for the electron cloud in a hydrogen atom during exposure to a short (30fs) but very intense (10^{15} Watt/cm²) laser pulse of blue light (455nm). The atom was prepared in an excited state (4f, m=3). The lower half of each figure is a 2D projection of the probability distribution shown in the upper half. The snapshots were taken at equally spaced instants in time during one full cycle of the light wave, starting from the instant at which the intensity reached its maximum value. The nucleus of the atom (not shown) remains stationary at the center of the base plane of each snapshot, while the electron cloud oscillates about the nucleus, following the field in a coherent fashion. This sequence displays one of the remarkable features of 'stabilization': although the electron clouds makes large excursions about the nucleus - the amplitude of oscillation shown here is approximately 17 Bohr radii - it nevertheless conserves its shape as time evolves, and the probability of the electron actually leaving the atom is surprizingly small.

free electron, which adds P to its free-field energy).[3] The total imparted momentum by the field is therefore P/c and this proves that N is indeed given by $P/(\hbar\omega)$. Therefore, we expect the dipole approximation in strong fields to break down about when $P \approx c/n$ (in atomic units). On the other hand, we expect relativistic effects to become important when the electron speed becomes comparable to the speed of light, i.e., when $P \approx c^2$. Hence it is retardation that we think we have to worry about first. (This does not mean that there may not still be other effects, such as dipole radiation that could be important at high intensities, and that have been neglected in the computations up to now.)

One can envisage the breakdown of the dipole approximation (to lowest order) as due to the addition of a spatially homogenous magnetic field that complements the dipole electric field so as to give the correct Poynting vector. However, this magnetic field does *not* observe conservation of the magnetic quantum number (we lost the axial symmetry) and hence Rydberg states with high magnetic quantum number [13, 14] get coupled to lower-lying states. For these lower-lying states the frequency is in general not high and so they will not stabilize, but rather rapidly ionize through tunneling. This leads to destabilization of the state. (We have already encountered a similar form of 'destabilization' in a different context. Within the dipole approximation this happens e.g. for $m = 0$-states for which the frequency is high with respect to their binding energy, but not w.r.t. the ground state binding energy, see in particular Fig.2 in Ref.[13].) In fact, the Rydberg state could even be resonantly coupled to lower-lying states through the action of the magnetic field.

We express here our concern about the observability of dichotomous Rydberg states as proposed by Vos and Gavrila[14]. It appears that the intensity to attain dichotomy for Rydberg states is so high that the ponderomotive energy may no longer be small with respect to c/n. Whereas the ionization by the dipole electric field is known to stabilize, the ionization occuring due to the presence of the magnetic field appears to rise indefinitely with increasing intensity and hence *at sufficiently high intensities stabilization ceases to exist*. In their paper, Vos and Gavrila consider $n = 7$ states, where according to us the dipole approximation breaks down when $P \approx 20$. According to them, dichotomy sets in between $\alpha_0 = 500$ and 1000. For the values of the frequency $\omega = 0.0735$ and 0.0428 that they considered, this implies $P = 337 - 1351$ and $P = 114 - 458$, respectively. Hence we think that, for the cases they presented, the states would be destabilized before dichotomy would have set in.

To illustrate our claim we have solved the Schrödinger equation bringing into account the retardation by expanding the vector potential through first order in $1/c$. This we have done in order to simplify the computational complexity that arises as a consequence of loss of axial symmetry. A preliminary result is shown in Fig.5. It shows that the effect is significant and can amount to about a 18% increase in the yield. In Fig.5 we had chosen a very modest intensity (in comparison to what is needed to induce dichotomy) and a relatively high frequency and low principal quantum number [2p(m=1)]; the numerical value of nP/c is roughly 0.1. In order to observe dichotomy for Rydberg states one needs either substantially higher intensity, or substantially lower frequency which implies a higher principal quantum number. It follows from the discussion above that in these cases the destabilization due to non-dipole terms should be quite significant[15].

---

[3]Provided that we can neglect distortion of the electron cloud, which is a reasonable assumption for high Rydberg states that we have in mind here.

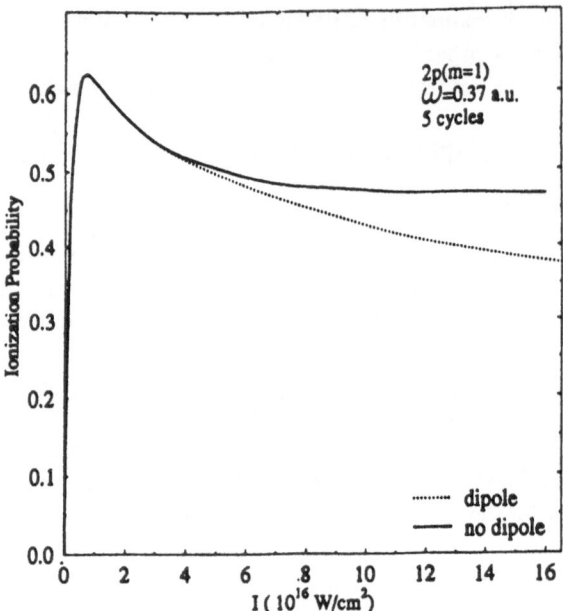

Figure 5. The influence of non-dipole terms on the stabilization of a hydrogen atom exposed to a five cycle pulse of frequency 0.37 a.u.. Even for this case of low principal quantum number, high frequency and relatively moderate intensity, the influence of non-dipole terms is significant. This could imply that in the example studied by Vos and Gavrila, the dichotomous limit cannot be reached.

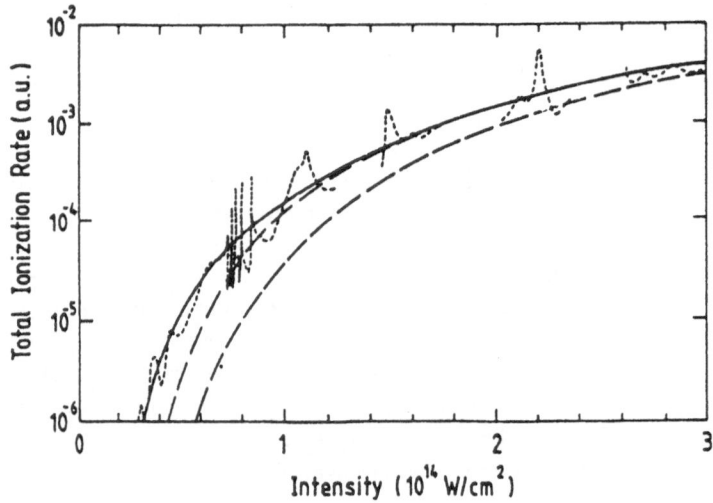

Figure 6. Width Γ_{ac} vs intensity for H(1s) irradiated by linearly polarized light of wavelength 616 nm. The short dashes are 'exact' results and the lower long dashed line, the upper long dashed line, and the solid line are those given by the first term (cycle-averaged dc), the first two terms, and the first four terms respectively of the ω^2-expansion.

3 A low-frequency theory for multiphoton ionization in strong fields

The treatment of multiphoton ionization of two-electron systems is difficult if one wants to take into account the fact that the two electrons are not moving independently. Accurate calculations taking into account correlation in weak fields (by perturbation theory) have only recently become available[16]. (See also the contribution by Shakeshaft elsewhere in this book.)

In perturbation theory one expands e.g. the quasi-energy in powers of the field strength:

$$E_{ac}(F,\omega) = \sum_n E_{2n}(\omega)F^{2n}. \tag{4}$$

However, perturbation theory is restricted to fairly weak fields. From perturbation theory we cannot learn anything about the role of correlation when two-electron atoms are ionized by intense, pulsed lasers. In addition, mostly many photons are needed to surpass the high ionization potentials of atoms. This is especially true when the atoms are exposed to strong fields of low frequency, since we know that under these circumstances the ionization potential is increased by the ponderomotive energy. To set up a 'complete' theory one would have to do a full Floquet calculation, a formidable task for a two-electron system. However, if we would average over all 'structures' due to multiphoton resonances and thresholds one could hope to be able to construct a theory that is much simpler and yet includes correlation and nonpertubative effects at the same time.

Recently, we have achieved this goal. The expansion turns out to be the counterpart of perturbation theory, i.e., we expand in powers of the frequency with expansion coefficients depending on the field strength:

$$E_{ac}(F,\omega) = \sum_m E^{(2m)}(F)\omega^{2m}. \tag{5}$$

The first term turns out to be equal to the cycle-average of the instanteneous dc quasi-energy, in agreement with tunneling theories (Keldysh[17], Perelomov and Popov[18]). As well-known, the quasi-energy approaches the cycle-averaged dc quasi-energy as the Keldysh parameter:

$$\gamma = \sqrt{I_p/2P} \tag{6}$$

becomes smaller than unity. A recent numerical demonstration of this feature can be found in Ref.[19]. In Fig.6 we illustrate this for ionization from the hydrogen 1s-state by 616 nm light.

As one can see, when the Keldysh parameter is still larger than unity the discrepancy between the cycle-averaged dc width and the exact Floquet width can be quite significant. Our expansion in powers of the frequency however interpolates quite accurately through the exact Floquet results and only a few terms are needed in the expansion to get good convergence. It does of course not reproduce the resonance structures that arise if Rydberg states shift in and out of resonance when varying the intensity. Our expansion Eq.(5) is an 'adiabatic' one, and hence cannot reproduce features that are characteristic of the discrete nature of the photon.

The keypoint of our low-frequency expansion is a judiciously chosen Ansatz for the Floquet wavefunction. We first write the usual Floquet Ansatz in the form:

$$\Psi(\tau,F,\omega) = \exp\left[-\frac{i}{\omega}\int^\tau \tilde{E}(\tau',F,\omega)d\tau'\right]\tilde{\Psi}(\tau,F,\omega). \tag{7}$$

Here $\tilde{E}$ and $\tilde{\Psi}$ are functions that depend on the phase τ of the field periodically.

Our low-frequency Ansatz consists of assuming that $\tilde{E}(\tau, F, \omega)$ and $\tilde{\Psi}(\tau, F, \omega)$ can be expressed in a power series in ω. Substitution of our Ansatz in the Schrödinger equation allows us to solve the Schrödinger equation in increasing powers of the frequency. To lowest order equation Eq.(7) is just the adiabatic approximation.

We also have found a dispersion relation relating the real and imaginary part of the quasi-energy obtained from our formalism:

$$ReE_{ac}(F, \omega) = \frac{2}{\pi} \int \frac{ImE_{ac}(F', \omega)F'dF'}{(F'^2 - F^2)} \tag{8}$$

and we have a perturbation double series:

$$E_{ac}(F, \omega) = \sum_n \sum_m \beta_{2n}^{(2m)} F^{2m} \omega^{2n}, \tag{9}$$

which can be obtained by expanding the coefficients in either Eq.(4) or Eq.(5). It turns out that for atomic hydrogen the β coefficients are ratios of integers and we have generated them to rather large order. By combining Eq.(8) and Eq.(9) we have been able to confirm the Keldysh-Perelomov-Popov formula [17, 18] for the ac-tunneling of an electron bound by a Coulomb potential. More details can be found in Ref.[20]. We are presently working on the application of our formalism to two-electron systems as He and H$^-$.

4   The solution of the time-dependent Schrödinger equation for two-electron systems

At present there exist several methods that allow the numerical solution of the Schrödinger equation for one-electron systems.[21] The next natural step would be to consider the two-electron case. Here one encounters a formidable problem. The size of the relevant atomic state space N is much larger (perhaps three orders of magnitude) in going from two to five spatial dimensions. The computational effort can actually scale as N^3 (if, for example, a matrix diagonalization is needed) making it a computational physicist's nightmare and challenge. This is the reason that most prior work has been limited to two dimensions (see Grobe and Eberly, and Parker and Clark elsewhere in this volume).[4] We will give more details about our method below. For our computational resource, we have chosen the Connection Machine CM-200, which is built by Thinking Machines Corporation (TMC).[5] Its bigger brother, the CM-5, is presently one of the fastest computers in the world (see Fig. 7). The CM-200 is a massively parallel SIMD (single instruction multiple data) supercomputer. The introduction of massive parallelism in atomic physics is quite recent, and since we believe that the concept of massive parallelism is about to revolutionalize Computational Physics we deem it worthwhile to elaborate on some of its features.

As is well-known, serial computers such as Sun workstations basically multiply numbers two at a time; this is the oldest concept. In the seventies and eighties, vector supercomputers revolutionized the world of scientific computing. A vector supercomputer lines up the elements of a (long) vector in vector pipes. The vector elements

---

[4]Lambropoulos et al. have obtained a preliminary result in 5D using a B-spline representation of a basis constructed from Hartree-Fock orbitals, presented at this conference.

[5]The CM-200 shares the same architecture as the CM-2, but has a faster clock and improved microcode.

Figure 7. The connection machine (Thinking Machine Corporation 1991) model CM-5 supercomputer is the first with a complete parallel architecture that scales to a Teraflop performance. The system offers the range of production software and hardware capabilities that bring parallel supercomputing into the mainstream. Thinking Machine Corporation is the world's leading manufacturer of highly parallel supercomputers and the pionneer in scalable computing techniques. Photo credit: Steve Grohe.

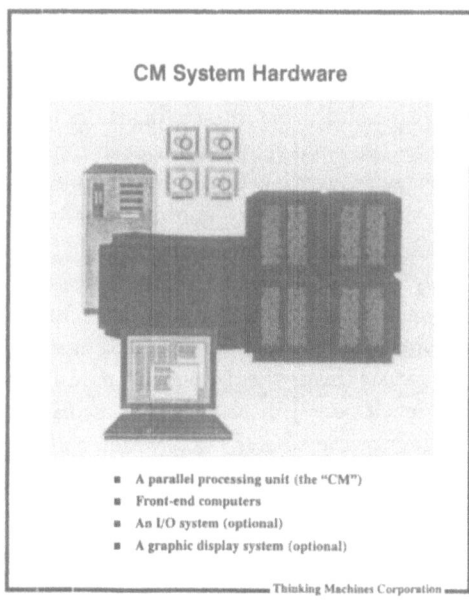

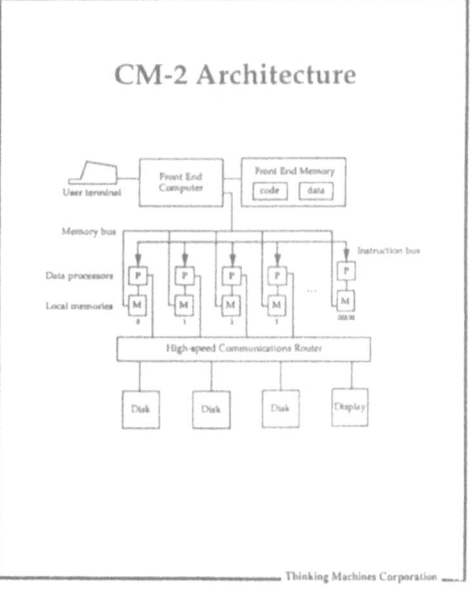

Figure 8. The CM-2 hardware. Right: The CM-2 architecture.

undergo the same operation, one by one, rapidly carried out by a single processor. Most recently, a new revolution has taken place, with the introduction of massively parallel supercomputers. In a SIMD supercomputer, a single operation is applied *simultaneously* to all the elements in a large data set. In our work, these data sets are typically very large matrices. The CM-200 contains a large number of processors,[6] and the compiler associates a single processor with each element of the matrix.[7] A front-end computer (typically a Sun workstation) downloads blocks of instructions to a sequencer inside the CM-200, which then broadcasts the instructions to all processors simultaneously. See Fig. 8. (A detailed description of the architecture and operation of the CM-200 can be found in Ref.[22]).

For example, the elementwise multiplication of two matrices A and B proceeds by first assigning each element of matrix A to a different processor, followed by the same procedure for matrix B and finally in-processor multiplication of $A_{ij} * B_{ij}$ for all i, j simultaneously. While the individual CM-200 processors are rather slow, we would like to remind the reader that the power of the machine comes from letting thousands of them operate at the same time. Were one to use Cray processors at each node of the CM-200 instead, the machine would be astronomically fast but would also have an astronomical price tag! The CM-200 therefore works best for large problems, which means that it needs substantial core memory to be efficient. The CM-200 at Los Alamos has 8Gb core memory available.

Since the benefits of massively parallel computing are most evident with very large data sets, it is not surprising that large scale I/O is an important part of working on the CM-200. In our work, we have divided our computations into two stages: a set-up stage, in which we calculate all of the matrix elements that will be needed later on; these matrix elements are subsequently fed into the program that carries out the actual time integration (propagation stage). In between the set-up and propagation stages the large matrices are stored on a Data Vault (also built by TMC), which is a parallel disk array capable of transferring data to and from the CM-200 at speeds on the order of Gbytes per minute!

The example of elementwise multiplication of two matrices given above is a case for which pure SIMD parallelism is an ideal approach. In practice one can also have an algorithm which would not be well-suited to this programming style. Many algorithms developed over the past fifty years consist of a strategy to carry out a series of instructions in a given (smart) prescribed order. For instance, suppose one wants to calculate the exact minimum of a function of 1,000,000 variables using the conjugate gradient algorithm[23]. The computation is dominated by the time required for each function update and the fact that the algorithm is inherently sequential. Given an example such as this, one might ask, is the SIMD concept perhaps too rigid to be of practical use?

While it is true that SIMD programming has certain limitations, we would argue that the answer to this question is emphatically: NO. Many physics codes, for example, consist of a body of inner products, matrix-vector multiplications, or matrix-matrix multiplications. By virtue of commutative properties of sums and products many of such operations are rich in parallel structure (apart from the "obvious" parallelism that is present in the two latter examples). In fact this is just one class of algorithms that parallelizes well. One could also direct the skeptic to Ref. [24], which contains

---

[6] 65,536 single bit processors, which for computational purposes are better thought of as 2048 32-bit processors with associated floating point units. The actual computational work is done by the FPUs.

[7] In practice one often has more data than processors and the compiler mimics the existence of the number of processors that the machine is short. Those software processors are called *virtual processors*.

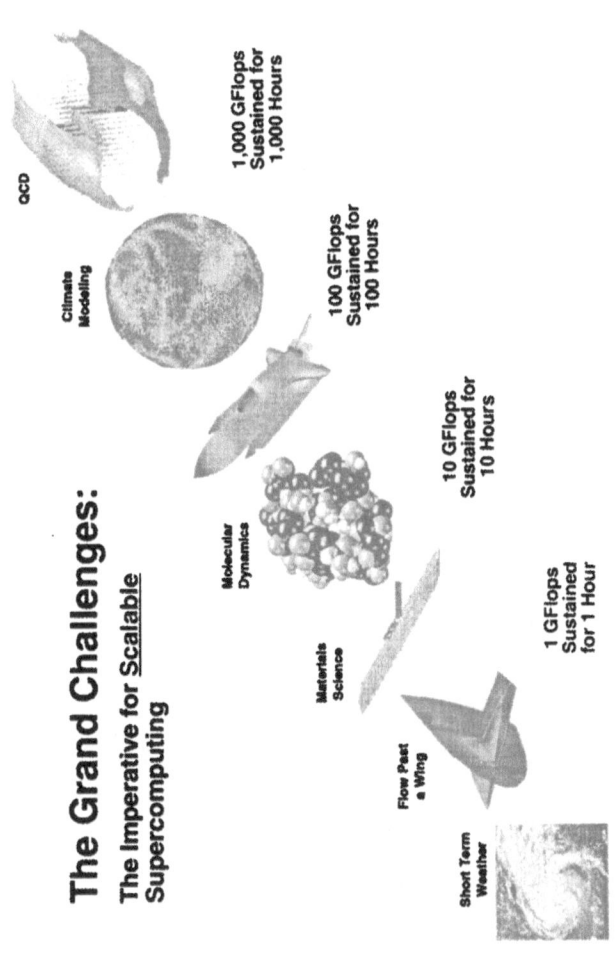

Figure 9. Example of "Grand Challenge": mathematical physics problems that require major computational resources and that are implemented successfully on Connection Machine.

a multitude of examples of various mathematical physics problems such as climate modelling, propagation of seismic waves in the earth, fluid dynamics, etc., where the SIMD concept has been applied very successfully. See also Fig. 9. The abovementioned implementations of mathematical physics problems on the CM share a number of fundamental similarities, but before discussing their basic strategy to exploit massive parallelism in more detail, let us remark on some other features of programming and porting codes to the CM.

The Connection Machine requires codes to be written in CM Fortran (CMF) rather than Fortran 77. CMF is a parallel extension of Fortran 77 which is essentially similar to Fortran 90 and allows one to very elegantly carry out parallel operations. For instance, there is a parallel IF ("WHERE"), a parallel DO ("FORALL"), and so forth; we will see some examples below. The elementwise multiplication of two matrices A and B that we encountered above simply reads in CMF, C=A*B. Of course a disadvantage of CMF/Fortran 90 is that one basically has to rewrite one's entire code in order to achieve significant benefits from running on the CM. But the pay-off can be very significant. A fortunate by-product is that CMF leads naturally to clear, compact, and well-structured code. Another disadvantage is that it is sometimes necessary to replace serial algorithms with new, parallel algorithms that are completely different but are better able to take advantage of the unique architecture of a massively parallel machine. We encountered this occassionally in our work (see at the end of this section). Computational life is made easier by the availability of a software package that provides 'ready-to-go', highly efficient linear algebra subroutines in the form of a library (the Connection Machine Scientific Software Library, CMSSL), as well as other routines such as FFTs, random number generators, etc.

Let us turn now to illustrating how one might go about mapping a mathematical physics problem onto the CM.[25] Many mathematical physics problems are described by partial differential equations, for instance the Navier-Stokes equations, the Poisson equation, the Schrödinger equation, etc. The first step in solving such problems on the CM is to discretize space, for instance by using finite difference or finite element techniques. Then with every grid point one associates a processor by prescribing a certain layout of the data ("NEWS", "SERIAL" or "SEND"[22]). The solution of the equation is then already in a form that is very suitable for SIMD. Let us give an example from electrostatics, namely the Poisson equation. In Fig. 10 we show, for an example in two dimensions, how the Laplacian can be calculated by elementary circular shifts and additions *throughout space*. A fragment of CMF/Fortran 90 code that does the job is displayed also.

In principle, we could solve our time-dependent Schrödinger equation in exactly the same way as we just described for the Poisson equation. In practice, we have chosen for an entirely different method, which is based on our good experience with projection of the Schrödinger equation on a complex Sturmian basis. Adopting the LS-coupling scheme, our basis functions for the two-electron problem are given by:

$$\Psi_{\nu',\lambda',\nu,\lambda}^{\kappa} = \frac{(1 \pm P_{12})}{\sqrt{2}} S_{\nu',\lambda'}^{\kappa}(r_1) S_{\nu,\lambda}^{\kappa}(r_2) \mathcal{Y}_{\lambda',\lambda}^{L,M}(\Omega_1, \Omega_2), \qquad (10)$$

where the Sturmians $S_{\nu,\lambda}(r)$ are defined by:

$$S_{\nu,\lambda}^{\kappa}(r) = \frac{1}{\sqrt{(\nu+1)_{2\lambda+1}}}(\kappa r)^{\lambda+1} \exp[-(\kappa/2)r] L_{\nu}^{2\lambda+1}(\kappa r). \qquad (11)$$

The parameter κ is complex. Complex Sturmians have been successfully exploited in

Local Interactions:

Solving Poisson's Equation

Maxwell's Equation $\nabla \cdot \mathbf{E} = \rho$ with $\mathbf{E} = -\nabla\phi$ leads to

$$-\nabla^2\phi(\mathbf{x}) = \rho(\mathbf{x}).$$

We wish to find ϕ inside the irregular region Ω, with ϕ specified on the boundary $\partial\Omega$.

2nd order finite difference approximation gives:

$$(4\phi_{ij} - \phi_{-1j} - \phi_{+1j} - \phi_{ij-1} - \phi_{ij+1}) = \Delta x^2 \rho_{ij}.$$

Solve for ϕ_{ij}:

$$\phi_{ij} = (\phi_{-1j} + \phi_{+1j} + \phi_{ij-1} + \phi_{ij+1})/4 + \rho'_{ij}/4$$

Local, regular communications (NEWS):
apply stencil at every point (CSHIFT).

Boundary conditions:
apply IF at every point (WHERE).

Poisson's Equation

- One processor per spatial point x
- O(1) Time

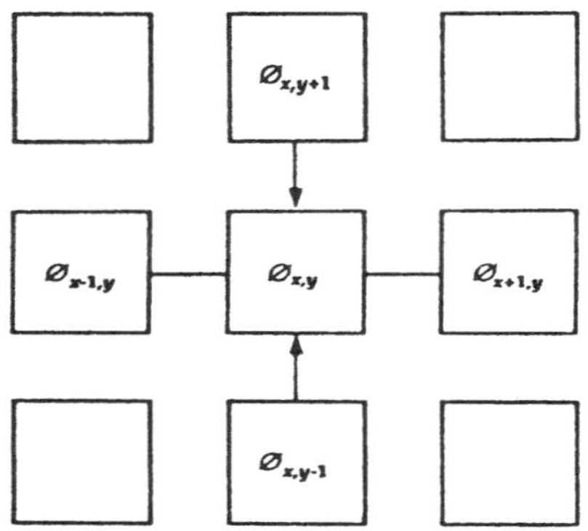

Thinking Machines Corporation

Poisson Solver— Fortran 90

```
SUBROUTINE SOLVE(PHI,RHO,INSIDE,N,ITER)
REAL PHI(N,N), RHO(N,N)
LOGICAL INSIDE(N,N)

DO I=1,ITER
  WHERE(INSIDE)
    PHI= RHO + 0.25 *
    ( CSHIFT(PHI,1,-1) +
      CSHIFT(PHI,1,+1) +
      CSHIFT(PHI,2,-1) +
      CSHIFT(PHI,2,+1) )
  END WHERE
END DO

RETURN
END
```

Figure 10. The Poisson equation on the Connection Machine is a nutshell. Left: the finite difference discretization. Middle: Scheme of the map of the grid on the processor of the Connection Machine and the communication needed to evaluate the Laplacian. Right: A fragment in FORTRAN90 to solve the Poisson equation. The Schrödinger equation could be approached in much the same way, but we have chosen for a completely different approach.

209

calculations of multiphoton rates for two-electron systems in perturbation theory[16] and in our time-dependent method for the one-electron problem of hydrogen[26] and have a well-established record in Floquet calculations for atomic hydrogen in many different circumstances.[27] Our method of time-propagation [26] consists of repeated matrix-vector (Mv) multiplications and that is where the parallelism enters. As mentioned above, Mv operations are rich in parallel structure. For example, one way to implement a matrix-vector multiply in CM Fortran would be as follows. First, perform the operation

$$\text{TEMP} = \text{SPREAD}(v, \text{DIM} = 1, \text{NCOPIES} = \text{NDIM})$$

This is makes NDIM copies of the (row) vector v, in parallel. Next, multiply the matrices M and TEMP in parallel:

$$\text{TEMP} = \text{TEMP} * \text{M}$$

Finally, carry out a parallel sum along the horizontal axis:

$$v = \text{SUM}(\text{TEMP}, \text{DIM} = 1).$$

Most of the original serial version of our code could be converted into a parallel version running on the CM in a straightforward way, although the code had to be rewritten almost completely. A major obstacle was the generation of the electron-electron repulsion matrix elements which were originally computed by recursion. The method of recursion requires sequential operation and is therefore incompatible with the efficient execution on the CM. The recursive algorithm was therefore replaced by a completely different, parallel method. Here we had an extra stumbling block to overcome: it turned out to be extremely difficult to obtain the necessary numerical accuracy. The matrix elements are computed by summation of several terms and in evaluating these sums we sometimes encountered large cancellations. Another major challenge was the efficient diagonalization of large, complex, non-Hermitian matrices on a SIMD machine. (CMSSL provides diagonalizers that can handle Hermitian matrices only). Since the computational effort of a diagonalization scales with the matrix size to the third power, finding an efficient algorithm for the CM was quite crucial. We presently have a method which is highly accurate and is quite fast: it requires roughly 30 minutes for a 2000*2000 case. It turned out that other algorithms that were recommended in textbooks as rich in parallel structure, in practice performed rather poorly on the CM if there is need for substantial data motion or communication between processors. For massively parallel machines communication can be a substantial bottleneck. "Thinking parallel" sometimes leads to amusing discoveries, e.g., all algorithms for reorthogonalization of a set of n vectors, each n long, scale in flops as n^3, for example classical Gram-Schmidt. While we found a way to parallelize the n^2 part of these textbook algorithms, we were left with a serial loop of order n, which is of course less than optimal for a parallel computer. We figured out an algorithm to do the reorthogonalization in full parallel mode in $O(1)$ sequential steps.

We are presently testing the code for relatively weak fields, proceeding onward in our parallel adventure.

Acknowledgments

The body of this work was supported by the Office of Basic Energy Sciences of the Department of Energy under contract number DE-FG03-92ER14266. The author wishes to acknowledge the Advanced Computing Laboratory of Los Alamos National Laboratory, Los Alamos, NM 87545. The Connection Machine which we utilize is located at this facility. The author is also indebted to Thinking Machines Corporation, and in particular to Dr. John Richardson, for kindly providing us with graphical material with which we have illustrated our Section 4.

References

[1] The notion of stabilization was not new, see e.g. [8]. Although predicting 'stabilization' in very intense fields, it appears however that these findings were based on theories that were seriously in error.

[2] M. Pont, Ph.D. thesis, University of Amsterdam, Amsterdam (1990); M. Pont, Phys. Rev. A **44**, 2141 (1991), ibid. 2152 (1991).

[3] M. Pont and M. Gavrila, Phys. Rev. Lett. **65**, 2362 (1990).

[4] Q. Su, J.H. Eberly and J. Javanainen, Phys. Rev. Lett. **64**, 862 (1990).

[5] M. Dörr, R.M. Potvliege, and R. Shakeshaft, Phys. Rev. Lett. **64**, 2003 (1990); M. Dörr, R.M. Potvliege, D. Proulx and R. Shakeshaft, Phys. Rev. A **43**, 3729 (1991).

[6] See for example K.C. Kulander, K.J. Schafer and J.L. Krause, Phys. Rev. Lett. **66**, 2601 (1991), and references therein.

[7] J.J. Gersten and M.H. Mittleman, J. Phys. B 9, 2561 (1976). M.H. Mittleman, *Theory of Laser-Atom Interactions* (Plenum, New York,1982), Chap.3.

[8] M. Janjusevics and M.H. Mittleman, J. Phys. B **21**, 2279 (1988).

[9] See, e.g., the book by Mittleman, Ref.[7] and Ref.[8].

[10] M. Gavrila and J.Z. Kaminski, Phys. Rev. Lett.**52**, 614 (1984).

[11] M. Gavrila, in *Atoms in Unusual Situations*, Vol.143 of NATO Advanced Study Institute Series B: Physics, ed. by J.P. Briand (Plenum, New York,1987), p.225.

[12] M. Pont, N. Walet, M. Gavrila and C.W. McCurdy, Phys. Rev. Lett. **61**, 930 (1988); M. Pont, N. Walet and M. Gavrila, Phys. Rev. **41**, 477 (1990).

[13] M. Pont and R. Shakeshaft, Phys. Rev. A **44**, R4110 (1991).

[14] R.J. Vos and M. Gavrila, Phys. Rev. Lett. **68**, 170 (1992).

[15] A full account of the work summarized in this section will be published elsewhere; A. Bugacov, M. Pont and R. Shakeshaft, submitted to PRA.

[16] D. Proulx and R. Shakeshaft, Phys. Rev. A **46**, 2221 (1992); D. Proulx and R. Shakeshaft, J. Phys. B: At. Mol. Opt. Phys. **26**, L7 (1993).

[17] L. Keldysh, Sov. Phys. JETP **20**, 1307 (1965).

[18] A.M. Perelomov and V.S. Popov, Zh. Eksp. Teor. Fiz. **53**, 331 (1967) [Sov. Phys. JETP **25**, 336 (1967).

[19] M. Dörr, R.M. Potvliege, and R. Shakeshaft, Phys. Rev. Lett. **64**, 2003 (1990); R. Shakeshaft, R.M. Potvliege, M. Dörr, and W.E. Cooke, Phys. Rev. A **42**, 1656 (1990).

[20] M. Pont, R. Shakeshaft, and R.M. Potvliege, Phys. Rev. A **42**, 6969 (1990); M. Pont, R.M. Potvliege, R. Shakeshaft and Z.-J. Teng, Phys. Rev. A **45**, 8235 (1992).

[21] For a list of references, see the special issue of the J. Opt. Soc. B **7**, 407 (1990), edited by K. Kulander and A. L'Huillier; and Comput. Phys. Commun. **63**, 1 (1991), edited by K. Kulander.

[22] *Connection Machine CM-200 Series Technical Summary*, Thinking Machines Corporation, Cambridge, Massachusetts (1991).

[23] This method can be found in many of the standardworks on numerical methods. The original reference is: M.R. Hestenes and E. Stiefel, J. Res. Nat. Bur. Stand. **49**, 409 (1952).

[24] *Los Alamos National Laboratory Advanced Computing Laboratory*, Vol. 2, Los Alamos, NM (1993).

[25] J. Bailey, *Implementing Fine-grained Scientific Algorithms on the Connection Machine Supercomputer*, Thinking Machines Technical Report Series TR90-1, Cambridge, MA (1990).

[26] M. Pont, D. Proulx and R. Shakeshaft, Phys. Rev. A **44**, 4486 (1991).

[27] Examples can be found in R.M. Potvliege and R. Shakeshaft, Phys. Rev. A **38**, 1098 (1988) and in Refs.[5, 19].

ATOMIC STATE EFFECTS IN STABILIZATION

H. R. Reiss and N. Hatzilambrou

Department of Physics
The American University
Washington, DC 20016-8058
USA

INTRODUCTION

The principal purpose of this paper is to demonstrate that the particular identity of an atom photoionized by an intense laser field has a major effect on the strong-field stabilization properties of the atom. This is done by first reviewing the origins of the method used - the Strong-Field Approximation (SFA)[1,2] - to show that it is well suited to an examination of the effects of atomic properties on multiphoton ionization. The relationship of the SFA to Kramers-Henneberger (K-H) techniques[3-5] and to the Keldysh method[6] are exhibited to establish the result that the SFA applies to all laser frequencies from the tunneling limit of the Keldysh technique to the high-frequency domain of K-H methods. Some applications of the SFA to stabilization of atoms are shown both for circular and linear polarization of the laser field over a wide range of frequencies. To emphasize the importance of the state of the atom to stabilization, some comparisons are done for several hydrogenic states and for a short range potential for a uniform binding energy. Effects of the atomic state are found to be dramatic. Finally, ionization by a circularly polarized laser of a low-lying Rydberg state ($n=10$) is studied as a function of frequency, for both s and p initial states. All cases show clear stabilization behavior, but it is more pronounced for initial p states than for initial s states.

FOUNDATIONS OF THE METHOD

The intent of the Strong-Field Approximation (SFA) is to proceed as directly as possible to a non-perturbative result for ionization rates. Towards this end, we introduce first the standard transition amplitude

$$(S-1)_{fi} = -(i/\hbar) \int dt \, (\Phi_f, H_I \Psi_i),$$ (1)

where H_I is the interaction Hamiltonian, Φ is a state vector for a system interacting only with the atomic potential, and Ψ is a fully interacting state in which effects of both the atomic potential and the laser field are included. In other words, Φ and Ψ satisfy the equations of motion

$$(i\hbar\partial/\partial t - H_0)\Phi = 0, \tag{2}$$

$$(i\hbar\partial/\partial t - H_0 - H_I)\Psi = 0, \tag{3}$$

where H_0 includes the atomic potential and H_I represents the interaction due to the laser field. Equation (1) follows from enforcing the boundary condition that measurements of the final products of the interaction are made in a region where the interaction is turned off. This corresponds to the experimental situation where a strong-field pulsed laser is used to cause ionization, so that the ions or photoelectrons are counted outside the space-time region occupied by the laser pulse. Equation (1) is exact if Φ_f, H_I, and Ψ_i are all stated exactly. The treatment of Ψ_i is the principal problem posed by Eq.(1), since it is difficult to find a general approximation suitable for a situation in which the laser field is so strong that it competes in strength with the atomic binding potential. Neither influence can be treated as a perturbation, even in the limit of extremely strong laser fields. For example, it is never suitable to treat an initially bound state as dominated by the laser field because a Coulomb potential has a singularity at the origin that can never be neglected. K-H methods have been used successfully for this problem, but only in the high frequency domain.

An alternative to Eq.(1) is the time-reversed transition amplitude

$$(S-1)_{fi} = -(i/\hbar) \int dt \, (\Psi_f, H_I\Phi_i). \tag{4}$$

This is also exact if Φ_f, H_I, and Ψ_i are exact. Here, however, the fully interacting state is the final state. This is possible to approximate quite rationally as a state that will be dominated by the laser field if the field is sufficiently intense, because Ψ_f represents an ionized electron which is in a continuum state, not a bound state. If the laser field is sufficiently strong, it will be the dominant influence on the detached electron, and Ψ_f can then be approximated by a Gordon-Volkov state, an exact solution for a free electron in a plane wave field. The approximation so made requires only that the ponderomotive potential of the electron in the laser field be greater than the binding potential of the atom[2]. There is no constraint on the field frequency. This is the genesis of the SFA, which can be presented[1,2] in full formality, including correction terms.

An important salutary result of employing Eq.(4) instead of (1) is that the atomic properties in Eq.(4) are predominantly in one place - in the initial state Φ_i. This is a state that can be expressed with great accuracy. Thus a theory based on Eq.(4) is suited to convenient exploration of the effects of atomic properties on strong-field photoionization.

Equation (4) can be regarded also as the basis of the Keldysh approximation[6], originally presented as a physically motivated *ad hoc* theory. Some authors (for example, Refs.7

and 8 and Keldysh[6] also) note that the Keldysh theory lacks field interaction in the initial state and regard this as a drawback. Thus there is the temptation to insert laser field interaction[7,8] into Φ_i, even though Eq.(4) expressly shows that this is inappropriate. A theory resulting from such a step is supported neither by Eq.(1) nor by Eq.(4).

CONNECTION OF THE SFA TO K-H AND KELDYSH METHODS

As pointed out in the preceding section, The direct-time transition amplitude of Eq.(1) and the reversed-time transition amplitude of Eq.(4) are identical in their physical content. Since one's physical intuition is more immediate in the direct-time case, it is logical to ask for the direct-time equivalent of the SFA, which is based on the time-reversed Eq.(4). One can prove directly[9] that the SFA matrix element is equivalent to a type of K-H approximation in the strong-field case where the non-interacting state Φ_f of Eq.(1) is approximated by a free-particle state, neglecting the Coulomb interaction of the ionized electron. That is, one has the correspondence

$$(\Psi_f^{GV}, H_I \Phi_i) \equiv (\Phi^{free}, H_I \Psi_i^{K-H}) \tag{5}$$

between the SFA matrix element in Eq.(4) and the strong-field K-H approximation for the matrix element in Eq.(1). In this instance, the state Ψ_i^{K-H} is used in the sense of Faisal[10], in which, after the K-H transformation is made, the resulting time-dependent Coulomb potential is averaged to the ordinary Coulomb potential. The result in Eq.(5) explains why the high-frequency approximation presented by Faisal[10] yields the same analytical form as given in Ref.1. Both are closely related to the strong-field case of the work of Gavrila and collaborators[5,11,12], as shown in Ref.13.

However, although the K-H approach is limited to high frequencies, the SFA is not. In particular, the SFA is known[1,2] to yield the correct tunneling result in the low frequency limit. In fact, the matrix element on the left-hand side of Eq.(5) is just the same as that of the Keldysh approximation except for two things. For one, the work of Keldysh[6] is done in the length gauge, whereas the SFA is in the velocity gauge. For another, the Keldysh result, though it can nominally be stated more generally, is given in the Keldysh paper in the tunneling limit, and is normally employed only in that form. There is nothing in the derivation of the SFA that implies a frequency limitation, and the correspondence with both high and low frequency alternative methods confirms that property.

STABILIZATION IN GROUND-STATE HYDROGEN

Monochromatic results for stabilization of ground-state hydrogen in strong laser fields as calculated by the SFA are given in a recent paper[13]. Transition rates as a function of intensity for a wide range of frequencies are calculated in this paper, with extensions of these results for circular polarization given here in Fig.1, and for linear polarization in Fig.2. Especially to be noted is that if state lifetimes

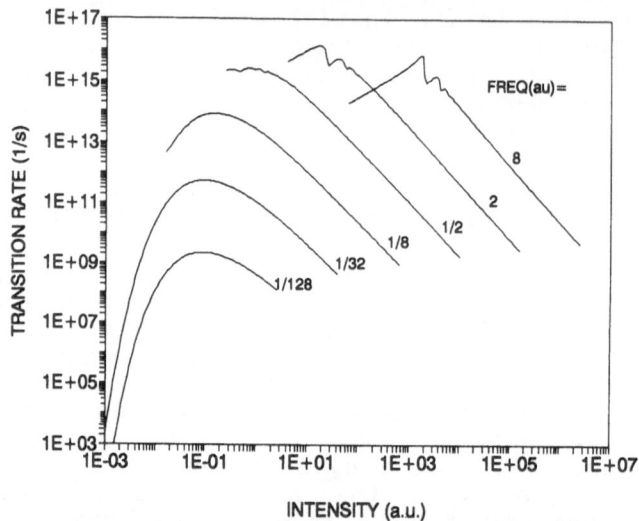

Figure 1. Transition rate for photoionization of ground state atomic hydrogen by a circularly polarized laser as a function of laser intensity for a range of laser frequencies from low (1/128 a.u.) to very high (32 a.u.). Where the terminus of each curve remains on the figure, the low intensity end is where twice the ponderomotive potential is equal to the binding energy (where SFA validity begins) and the high intensity end is where twice the ponderomotive potential is equal to mc^2 (where relativistic effects occur). These results agree closely with those of Pont and Gavrila[12].

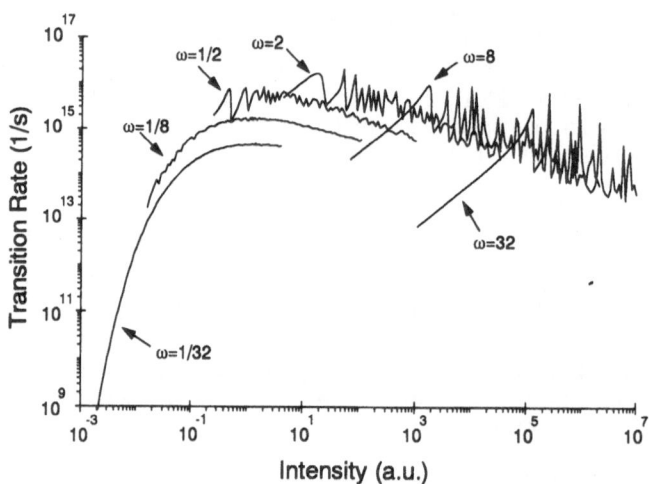

Figure 2. Transition rate for photoionization of ground state atomic hydrogen by a linearly polarized laser as a function of laser intensity for a range of laser frequencies from moderately low (1/32 a.u.) to very high (32 a.u.). Where the terminus of each curve remains on the figure, the low intensity end is where twice the ponderomotive potential is equal to the binding energy (where SFA validity begins) and the high intensity end is where twice the ponderomotive potential is equal to $0.1(mc^2)$.

in the presence of the laser field are to be calculated (as in Ref.13) instead of transition rates as in Fig.1, the results are virtually identical to those calculated by Pont and Gavrila[12]. This remains true even if the SFA calculations of Ref.13 are extended to intensities not only as high, but also as low as those in Ref.12 (although low intensities are normally eschewed in the SFA).

Figure 1 shows clear evidence of stabilization for monochromatic fields. As is well known, such behavior is difficult to detect in practice, in large part because of the temporal behavior of the pulse. All atoms in the focal region can become ionized before the stabilization region is ever reached. Figure 1 then suggests that lower frequencies are likely to be more successful to demonstrate stabilization because the maximum rate becomes too modest to reach saturation for a short pulse. The difficulty remains, however, that the spatial distribution of field intensity is such that a central focal region where stabilization can occur is surrounded by a much larger volume at lower (pre-stabilization) intensities which may dominate in terms of total ion production.

Linear polarization results in Fig.2 are superficially similar to Fig.1, but differ both in the rapid oscillations of the rates, especially at high frequency, and in the fact that even a rate averaged over the fluctuations shows a much slower decline after the maximum point than is true for circular polarization. Linear polarization is then a less likely candidate than circular for potential observation of the stabilization phenomenon.

Figures 3 and 4 show the intensity at which the maximum transition rate occurs for circular and linear polarization, respectively. What is shown in these figures is the way in which the intensity for maximum rate changes with frequency when the measure of intensity is the parameter z_1, defined[1,2] as $z_1 \equiv 2U_p/BE$, where U_p is the ponderomotive potential and BE is the binding energy of the atom. There are two remarkable features that appear in both Figs. 3 and 4. First is the sharp distinction between frequencies less than $\omega=1$ a.u., and those greater. This is discussed in some detail in Ref.13. The other is the accuracy of the high-frequency prediction[13] that the maximum rate should occur at $z_1=2[(\omega/BE)-1]$. The accuracy of this simple prediction holds for the calculated results of both Refs.12 and 13.

EFFECT OF SPECIFIC BOUND STATES ON STABILIZATION

All of the figures exhibited so far relate to ionization from the *1s* state of atomic hydrogen, which has been the focus of most of the theoretical work in strong-field processes. To observe the effects on strong-field ionization of alterations in the atomic state, we consider now several different hydrogenic states. To be able to exhibit purely the effects of the basic *n* and *ℓ* quantum numbers of these states, without potential confusion from very different binding energies, we employ a simple device. The SFA requires only the momentum-space wavefunction, not the configuration-space wavefunction. Therefore, we start with a pure hydrogen wavefunction in configuration space, Fourier transform it, and then substitute in each case the same binding energy, *BE=0.5*

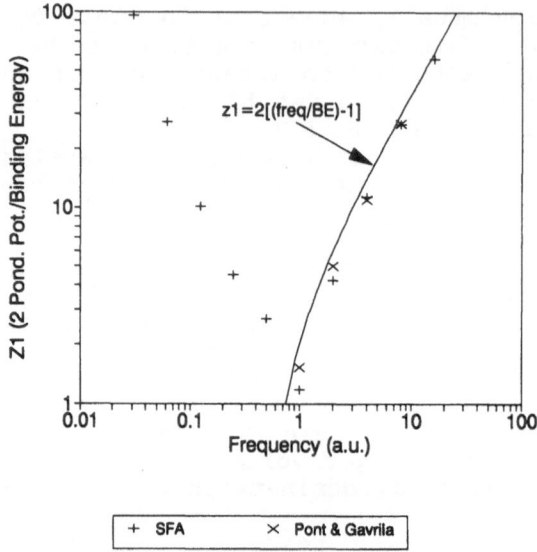

+ SFA   × Pont & Gavrila

Figure 3. Intensity at which the maximum transition rate occurs, as a function of frequency. The laser is circularly polarized and the atom undergoing ionization is atomic hydrogen in its ground state. The measure of intensity employed is the parameter z_1, defined as the ratio of twice the ponderomotive potential to the atomic binding energy. The SFA results[13] are shown by a +, and the high frequency results of Pont and Gavrila[12] are given by the × signs. The curve is the high-frequency prediction of Ref.13.

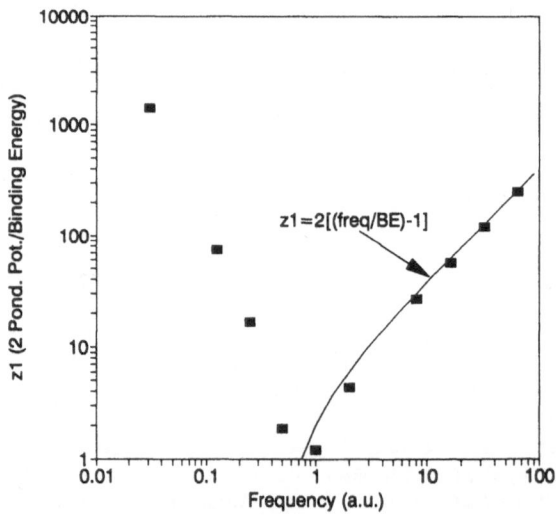

Figure 4. Intensity at which the maximum transition rate occurs, as a function of frequency. The laser is linearly polarized and the atom undergoing ionization is atomic hydrogen in its ground state. Only results for the SFA[13] are shown, since no other comparable calculations are available. See the caption for Fig.3.

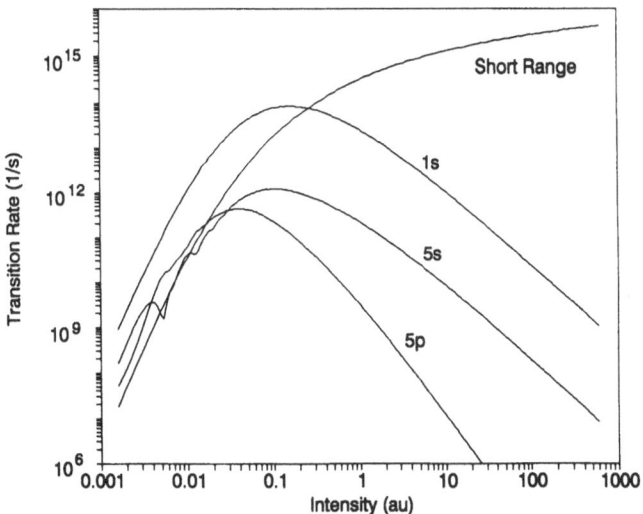

Figure 5. The effect of alteration of the atomic state on ionization transition rate as a function of laser intensity, and on stabilization behavior. The curves labeled *1s, 5s, 5p* are hydrogen atom states adjusted (in the momentum space wave function) to all have the binding energy of *0.5 a.u.* The curve for the short range potential has a configuration space wave function of Yukawa type, and is also adjusted in momentum space to have a binding energy of *0.5 a.u.* In all cases the laser is circularly polarized, with a frequency of *0.125 a.u.*

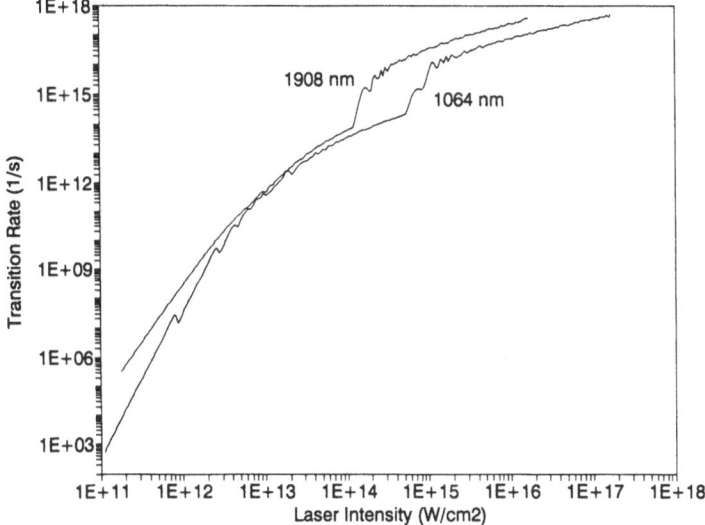

Figure 6. Transition rates as a function of laser intensity for photodetachment of an electron from a short range potential by a linearly polarized laser. For both wavelengths shown the short range potential is modeled by a potential of Yukawa type, adjusted to represent a binding energy of *3.613 eV*, as for the Cl^- negative ion.

a.u. The results so obtained with circular polarization of the laser are exhibited in Fig.5 for *1s*, *5s*, and *5p* states. The differences are striking. For example, the peak rate for *1s* is about two orders of magnitude greater than for *5s*, and even more for *5p*. Also shown in Fig.5 is the transition rate prediction for a short range potential, as for a negative ion. A wave function for a short range potential is taken to be of the form *(1/r)exp(-αr)*. For consistency in comparison with the hydrogenic results in Fig.5, the binding energy is again taken to be *0.5 a.u.* In this instance, no stabilization behavior at all is found. The short range potential yields the lowest transition rate of all the examples at low intensity, but because of the lack of a maximum rate, the short range potential eventually surpasses the other rates by a very large factor at high intensity.

The behavior shown for a short range potential in Fig.5 is for circular polarization of the laser, and for the unrealistically large binding energy of *0.5 a.u.* A more practical choice of parameters is represented in Fig.6, which shows results for the same Yukawa *(1/r)exp(-αr)* potential, but for a binding energy of *3.613 eV* (corresponding to *Cl⁻*), and for a linearly polarized laser. The ion and the wavelengths chosen are suggested by the recent experiments of Davidson et al.[14], who find the interesting result that the longer wavelength *(1908 nm)* produces more extensive photodetachment at a given intensity than does the shorter wavelength *(1064 nm)*. Figure 6 is not calculated with an accurate wavefunction, but it shows an eventual dominance of the longer wavelength (albeit at higher intensities than in the experiments).

The most significant feature of Fig.6 is that it exhibits an apparent resonance in the nominally single-bound-state potential well that gives rise to the Yukawa wavefunction. This appears to be a manifestation of the intensity-induced additional bound states in short range potentials that have been predicted to exist[15-17].

STABILIZATION OF A LOW-LYING RYDBERG STATE

It is now desired to examine the consequences for stabilization of a sharp reduction in the binding energy. For that purpose we examine ionization from a hydrogen atom state with the principal quantum number *n=10*, and a binding energy $BE=(0.5)/(10)^2$ *a.u.* This is done for circular polarization of the laser. Because of the sharp reduction of the binding energy as compared to Fig.1 for the ground state of hydrogen, the frequencies explored are reduced. Figure 7 is analogous to Fig.1 in showing transition rates for the *10s* state of hydrogen rather than the *1s* state. In both cases the laser is circularly polarized. Stabilization is again plainly manifested, and again the largest possible rate shows a sharp decline when low frequencies are considered. As before, there is the suggestion that low frequencies offer the greatest promise in terms of possible detection of the stabilization phenomenon.

A new feature to be seen in Fig.7 is the appearance of a considerable amount of structure in the curves near the peak rate and at lower intensities. This is not the same sort of structure that appeared with ionization from the ground state by a linearly polarized laser, as seen in Fig.2. The struc-

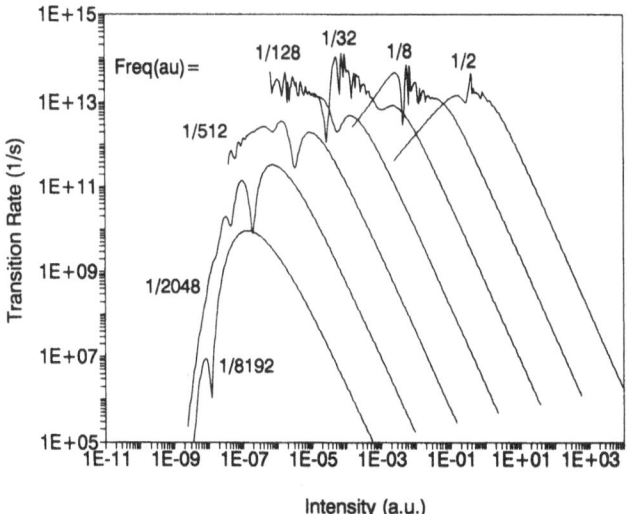

Figure 7. Transition rate for photoionization from the *10s* state of atomic hydrogen by a circularly polarized laser as a function of laser intensity for a wide range of laser frequencies. Where the terminus of each curve remains on the figure, the low intensity end marks the lower limit of the validity of the SFA approximation, and the high intensity end marks the region of specific relativistic effects, where the pondero-motive potential is $(1/2)mc^2$. This figure is to be compared with Fig.1 for ionization from the *1s* state of hydrogen

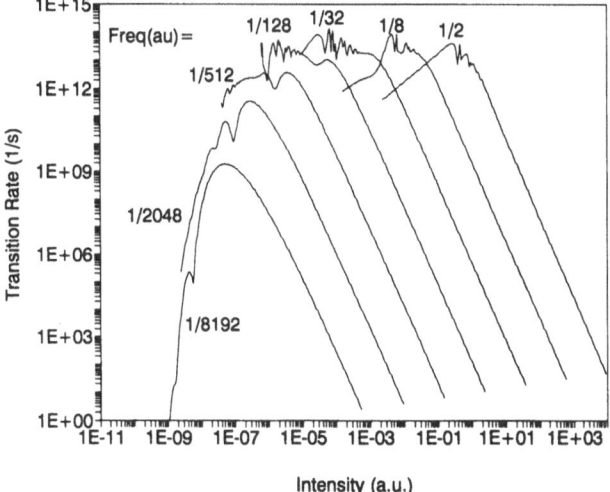

Figure 8. This figure is analogous to Fig.7, except that the initial atomic state is the *10p* state of atomic hydrogen. All other comments from Fig.7 apply here as well.

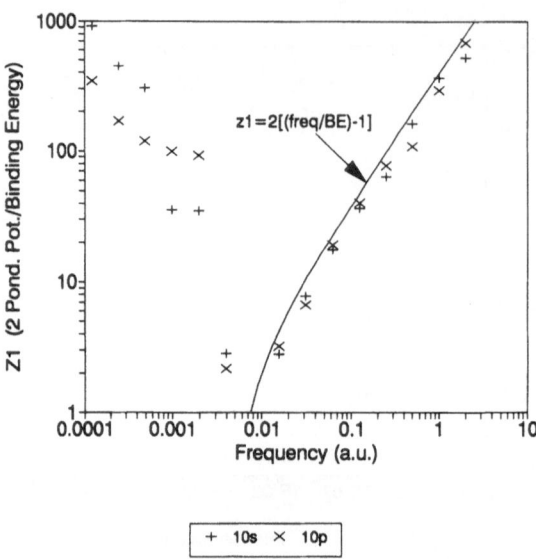

Figure 9. Intensity at which the maximum transition rate occurs, as a function of frequency. The laser is circularly polarized and the atom undergoing ionization is atomic hydrogen with principal quantum number $n=10$, and angular momentum quantum numbers $\ell=0,1$. The measure of intensity employed is the parameter z_1. It is seen that both s and p states come close to satisfying the relation expressed by the smooth curve[13] for high frequencies, i.e., frequencies greater than $1/n^2$ in a.u. For both $10s$ and $10p$ initial states, high frequency behavior is totally different than at low frequencies.

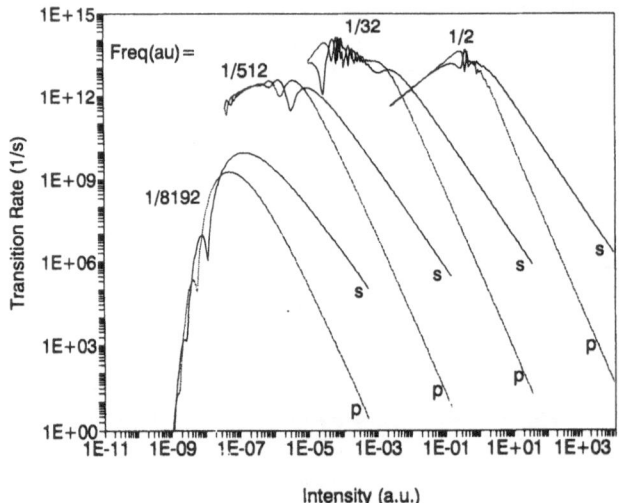

Figure 10. A direct comparison of transition rates in photoionization from initial $\ell=0$ and $\ell=1$ states of atomic hydrogen in the $n=10$ level. The s and p states, generally similar below the intensity for maximum rate, fall off at very different rates beyond the maximum.

ture in that case is mostly at intensities beyond the maximum rate. There are some well-defined features in Fig.7 even at low frequencies. There is currently no explanation for the cause of this phenomenon.

Figure 8 is the same as Fig.7 except that it represents the *10p* state of hydrogen. That is ,the binding energy is the same as for Fig.7, and the laser is circularly polarized, but the angular momentum quantum number is $\ell=1$ instead of $\ell=0$. The new structure found in Fig.7 appears again in Fig.8.

Figure 9 is analogous to Fig.3 in showing the intensity (measured in terms of z_1) at which the maximum in the transition rate occurs as a function of field frequency. As one might anticipate, energies scale as $(1/n)^2$, so the frequency at which a major qualitative change in behavior occurs is in the neighborhood of 10^{-2} *a.u.* instead of *1 a.u.* as in Fig.3. The continuous curve in Fig.9, which shows the high-frequency prediction of Ref.13, then applies to frequencies greater than 10^{-2} *a.u.*, and is seen again to be quite successful in matching the detailed calculations for both *10s* and *10p* states.

EFFECT OF ANGULAR MOMENTUM ON STABILIZATION

The results presented in the previous section for principal quantum number *n=10* make it possible to do at least a preliminary exploration of the effects of angular momentum on stabilization. Alternate curves from Figs.7 and 8 are collected in Fig.10 to permit direct comparison of the consequences of a change in initial angular momentum from an *s* state to a *p* state. For intensities lower than those at which the maximum transition rates occur, there is little difference between *s* and *p* states. Most significant is the lower maximum rate which obtains for the *p* state at the lowest frequency shown. Oscillations in the curves are quite similar, but out of phase between *s* and *p* states. The most striking difference is in the uniformly much faster fall off from the maximum rate which occurs for the *p* state.

The rapid decline from the maximum rate in the *p* state as compared to *s*, and the lower peak rate in the *p* state at low frequency suggests that an initial *p* state is more promising than an initial *s* state for a search for an observable stabilization effect. This comparison implies that initial states of even higher angular momentum might improve the prospects further, although this has not been explicitly explored to date.

ACKNOWLEDGMENT

Support for this work comes from grant number PHY-9113926 by the National Science Foundation.

REFERENCES

1. H.R. Reiss, Effect of an intense electromagnetic field on a weakly bound system, *Phys. Rev. A* 22:1786 (1980).
2. H.R. Reiss, Theoretical methods in quantum optics: S-matrix and Keldysh techniques for strong-field processes, *Prog. Quant. Electr.* 16:1 (1992).

3. H.A. Kramers, "Collected Scientific Papers", North Holland, Amsterdam, 1956.
4. W.C. Henneberger, Perturbation method for atoms in intense light beams, *Phys. Rev. Lett.* 21:838 (1968).
5. M. Pont, N.R. Walet, and M. Gavrila, Radiative distortion of the hydrogen atom in superintense, high-frequency fields of linear polarization, *Phys. Rev. A* 41:477 (1990).
6. L.V. Keldysh, Ionization in the field of a strong electromagnetic wave, *Sov. Phys. JETP* 20:1307 (1965) [*Zh. Eksp. Teor. Fiz.* 47:1945 (1964)].
7. J. Parker and C.R. Stroud, Generalization of the Keldysh theory of above-threshold ionization for the case of femtosecond pulses, *Phys. Rev. A* 40:5651 (1989).
8. M. Dörr, R.M. Potvliege, D. Proulx, and R. Shakeshaft, Multiphoton detachment of H⁻ and the applicability of the Keldysh approximation, *Phys. Rev. A* 42:4138 (1990).
9. H.R. Reiss, to be published.
10. F.H.M. Faisal, Multiple absorption of laser photons by atoms, *J. Phys. B* 6:L89 (1973).
11. M. Gavrila and J.Z. Kami ski, Free-free transitions in intense high-frequency laser fields, *Phys. Rev. Lett.* 52:613 (1984); M. Pont, N.R. Walet, M. Gavrila, and C.W. McCurdy, Dichotomy of the hydrogen atom in superintense, high-frequency laser fields, *Phys. Rev. Lett.* 61:939 (1988).
12. M. Pont and M. Gavrila, Stabilization of atomic hydrogen in superintense, high-frequency laser fields of circular polarization, *Phys. Rev. Lett.* 65:2362 (1990).
13. H.R. Reiss, Frequency and polarization effects in stabilization, *Phys. Rev. A* 46:391 (1992).
14. M.D. Davidson, D.W. Schumacher, P.H. Bucksbaum, H.G. Muller, and H.B. van den Heuvell, Longer wavelengths require lower intensity in multiphoton detachment of negative ions, *Phys. Rev. Lett.* 69:3459 (1992).
15. R. Bhatt, B. Piraux, and K. Burnett, Potential scattering in the presence of intense laser fields using the Kramers-Henneberger transformation, *Phys. Rev. Lett.* 37:98 (1988).
16. J.N. Bardsley, A. Szöke, and M.J. Comella, Multiphoton ionization from a short-range potential by short-pulse lasers, *J. Phys. B* 21:3899 (1988).
17. J.N. Bardsley and M.J. Comella, Ac Stark effect for short-range potentials with intense electromagnetic fields, *Phys. Rev. A* 39:2252 (1989).

STABILIZATION WITHIN A CLASSICAL CONTEXT

V. Véniard, A. Maquet and T. Ménis

Laboratoire de Chimie Physique - Matière et Rayonnement
Université Pierre et Marie Curie and CNRS
11 Rue Pierre et Marie Curie
F75231 Paris Cedex 05, France

INTRODUCTION

Classical mechanics has proven to be a valuable tool in the study of the response of excited hydrogen atoms in microwave fields: classical models, based on the Monte Carlo trajectory method, have been widely used to account for the ionization of Rydberg atoms (Leopold and Percival 1979, Richards *et al* 1989). According to the scaling properties of the classical hydrogen atom, the analysis of this atom-radiation interaction can in principle be extended to the low-lying atomic states, provided the field frequency and intensity are properly modified. In this context, it was shown recently that such a classical model can help to shed some light on high-order harmonic generation by atomic systems submittedto intense laser field (Bandarage *et al* 1992).

In the last few years, a number of theoretical studies have suggested that atoms could be stabilized against ionization under very strong laser pulses. Several of these studies were based on numerical simulations performed for a simplified one-dimensional model proposed a few years ago (Javanainen *et al* 1988, Eberly *et al* 1989, Su *et al* 1990). Within this framework, a one-electron atomic system is modelled through the use of a one-dimensional "soft Coulomb potential". Both quantum(Javanainen *et al* 1988, Eberly *et al* 1989, Su *et al* 1990, Reed *et al* 1991, Law *et al* 1991) and classical (Grochmalicki *et al* 1991, Grobe and Law 1991) simulations performed on this model have indicated that atoms could be stabilized at much lower frequencies than predicted by the original Gavrila-Kaminski model (Gavrila and Kaminski 1984, Pont and Gavrila 1990). Note that these results have been obtained in the case of an intense and short laser pulse, with a smooth but rapid turn-on. This leads to identify two different regimes, in which atomic stabilization can be observed. The so-called "adiabatic stabilization" occurs at very high frequency, with an extremely slow laser turn-on. By contrast, "dynamic stabilization" takes place at much lower frequency with a rapid turn-on of the laser field (a recent discussion is Gavrila 1992).

The purpose of this paper is twofold. The first part is concerned with short laser pulses and, within the classical model, the role of the Coulomb singularity, compared with the "soft Coulomb potential" is analysed. In a second section, a relatively long laser pulse is considered and a attempt is made to discuss Vos and Gavrila's results (1992) in a classical context.

SHORT PULSES

We consider first the one-dimensional atom described by the "soft Coulomb potential" :

$$V(a,x) = -(x^2+a^2)^{-1/2} \qquad (1)$$

which reproduces the long range Coulomb interaction but suppresses its singularity at the origin. The x-axis is oriented along the linear polarization of the laser field and the hamiltonian of the atom-laser system is :

$$H = p^2/2 + V(a,x) + F(t) \sin\omega t \qquad (2)$$

(atomic units are used throughout). The analysis is based on the classical Monte Carlo ensemble average method (Grochmalicki $et\ al$ 1991, Grobe and Law 1991, Bandarage $et\ al$ 1990,1992) and the ionization probability is defined as the ratio of the number of bound trajectories at the end of the pulse to the initial number of trajectories. In the following, we consider a case similar to the one discussed by Grobe and Law (1991), where the laser is smoothly turned-on over six cycles and turned-off over one cycle.

$$
\begin{aligned}
F(t) &= F_0\omega t/12\pi & & 0\leq t \leq 12\pi/\omega \\
&= F_0 & & 12\pi/\omega\leq t \leq 100\pi/\omega \\
&= F_0(102\pi-\omega t)/2\pi & & 100\pi/\omega\leq t \leq 102\pi/\omega \\
&= 0 & & t\geq 102\pi/\omega \qquad (3)
\end{aligned}
$$

where F_0 is the field amplitude and ω the frequency. We have checked that the ionization probability was quite insensitive to the detailed shape of the pulse, but a sudden turn-off can lead to an important increase of the number of ionizing trajectories.

In the case discussed by Grobe and Law (1991), the frequency was chosen to be $\omega=0.8$ a.u. and the initial energy was $E_i=-0.669$ a.u., which is the energy of the ground state in the corresponding quantal system with a=1, see Eq.(1) (Eberly $et\ al$ 1989). The results, shown in Fig.1a, clearly evidence an important decrease of the ionization probability as the laser field increases. This confirms the preceding analysis (Javanainen $et\ al$ 1988, Eberly $et\ al$ 1989, Su $et\ al$ 1990, Reed $et\ al$ 1990, Grochmalicki $et\ al$). Note that the proportion of ionizing trajectories is not given as a function of the field amplitude F_0, but as a function of the parameter $\alpha_0=F_0/\omega^2$, which is the classical excursion parameter of a free electron in a single mode laser field (Gavrila and Kaminski 1984).

In order to discuss the possible role of the Coulomb singularity, we have considered a similar system but with a deeper potential well, by choosing a smaller value of the parameter a, a=0.1, see Eq.(1). The bottom of the well is now V(0.1,0)=-10 (instead of V(1,0)=-1, in the preceding case). Two different situations have been analyzed. In the first one, the frequency ω and the initial energy are the same as in the preceding case, i.e. ω=0.8 a.u. and E_i=-0.669 a.u., the ratio of the frequency to the binding energy being $\omega/E_i \approx 1.2$. Note that this energy corresponds to some excited state in this deeper potential, the energy of the ground state of the equivalent quantal system being E_i=-3.82 a.u. The results are presented in Fig.1b : stabilization is still observed, although at much higher values of the parameter α_0 and to a much lesser extent than in the preceding case. In the second case, we have checked that if we consider initial states with enegies E_i, equal to those of the (quantum mechanical) ground state (E_i=-3.82 a.u.) and keeping constant the ratio $\omega/E_i \approx 1.2$, stabilization is also present but much less effective (see Fig.1c). In this case, stabilization occurs for the same range of values of the parameter α_0 than in Fig.1a, which corresponds in fact to higher values of the field amplitude F_0, as ω is much larger (ω=4.58 a.u.).

Our interpretation for this two sets of results is that, as the potential becomes deeper, the electron can experience much stronger accelerations. As a matter of fact, its maximum value is :

$$A_{max} = 2/(3\sqrt{3} a^2) + F_0 \qquad (4)$$

and this leads more easily to ionization as the parameter a decreases. Actually, the analysis of ionizing trajectories shows that, as long as the electron is far from the origin, the acceleration oscillates at the frequency ω between $\pm F_0$. It is only when it comes close to the origin that spikes can be seen in the acceleration, the amplitude of these spikes being much larger for a=0.1 (see Menis *et al* 1992, for a more detailed discussion).

We now come to the final point of our discussion, with a more realistic 3D-Coulomb potential. We show in Fig.1d the proportion of ionizing trajectories evolving from a microcanonical distributions of states (Leopold and Percival 1979, Richards *et al* 1989, Bandarage *et al* 1990,1992) as a function of α_0. The initial energy was chosen to be E_i=-0.5 a.u. (hydrogen atom in its ground state) and the ration ω/E_i is equal to 1.2 (ω=0.6 a.u.). Compared to the 1-D case, the ionization probability is much higher than for a=1, but quite similar to the case a=0.1. It clearly indicates that the Coulomb singularity at the origin is responsible for higher ionization probability and for the overall decrease of the stabilization process. As a conclusion, we would like to emphasize that predictions based on "soft Coulomb potentials" can lead to an overestimation of the stabilization process.

From the quantum mechanical point of vue, atomic stabilization, in the context of short laser pulses, has been ascribed to the fact that atoms are trapped in Rydberg states, which are known to be more stable in the field (Bhatt *et al* 1988). It is of interest to analyse, within the framework of classical mechanics, in which states the surviving trajectories are left. For this purpose, we consider the 3-D Coulomb problem, the initial energy being equal to E_i=-0.5 a.u., for two different values of the laser frequency. In the low frequency regime,

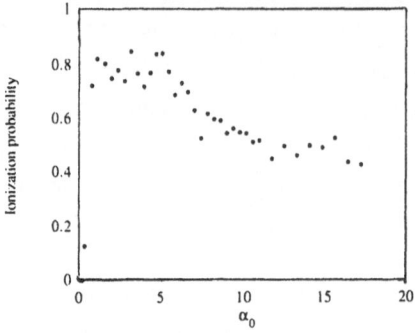

Figure 1a. Ionization probability as a function of the parameter α_0 for a=1, the laser frequency being ω=0.8 a.u. and the energy E_i=-0.669 a.u.

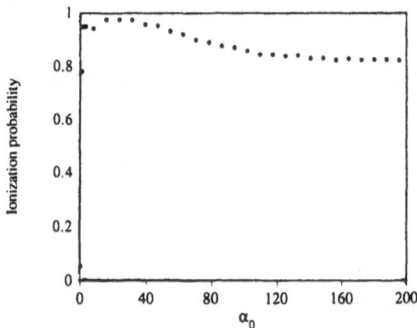

Figure 1b. Ionization probability as a function of the parameter α_0 for a=0.1, the laser frequency being ω=0.8 a.u. and the energy E_i=-0.669 a.u.

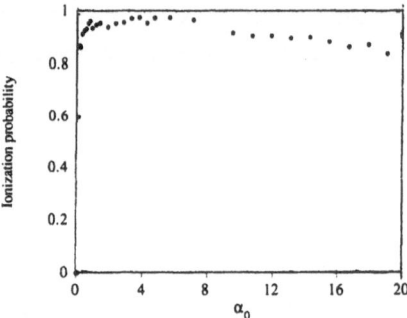

Figure 1c. Ionization probability as a function of the parameter α_0 for a=0.1, the laser frequency being ω=4.58 a.u. and the energy E_i=-3.82 a.u.

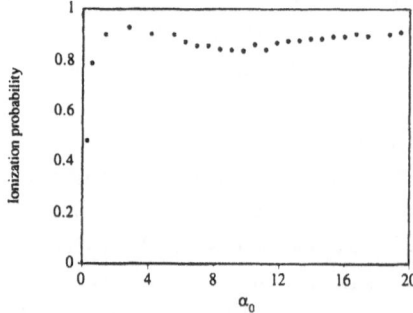

Figure 1d. Ionization probability as a function of the parameter α_0 for the 3-D Coulomb problem, the laser frequency being ω=0.6 a.u. and the initial energy is E_i=-0.5a.u.

$\omega=0.0735$ a.u., where a seven-photon ionization process is possible, almost no stabilization is observed. However, for quite a large range of values for the parameter α_0, the ionization probability does not reach its maximum value ($P_{ion}=1$). The field amplitude is chosen to be equal to $F_0=0.35$ a.u., corresponding to $\alpha_0\approx65$ a.u., which leads to 95% of ionization. Fig.2a shows the energy distribution for the non-ionizing trajectories after the end of the pulse. It appears that the electrons are left in highly excited states, the distribution being roughly centered around $E=0.0035$a.u., which corresponds to a state with $n\approx12$, n being defined as $n=(-2E)^{-1/2}$.

At a relatively higher frequency, $\omega=0.6$ a.u., we come back to the case discussed earlier, where a slight stabilization occurs. By choosing $F_0=3.5$a.u. ($\alpha\approx10$a.u.), the ionization probability is of the order of 85% and we see from the energy distribution, shown in Fig.2b, that, as in the preceding case, the electrons end in excited states. However, the distribution is now centered around $E\approx-0.03$a.u., which correspond to lower excited states, $n\approx4$. This trend would be confirmed, when looking at higher frequencies. Note that, by increasing the frequency ω while keeping the binding energy E_i constant or by increasing the ratio ω/E_i, more and more stabilization occurs.

LONG PULSES

In the original Gavrila-Kaminski model (Gavrila and Kaminski 1984, Pont and Gavrila 1990 and references therein) it was shown that atomic stabilization could be achieved in the high frequency adiabatic regime. Two conditions have to be fulfilled :

(i) the laser frequency has to be large with respect to the internal atomic frequencies

(ii) the field is monochromatic, i.e. bandwidth effects or shake-up or shake-down transitions were negligible with respect to multiphoton transitions.

It was pointed out recently (Vos and Gavrila 1992) that, starting from unperturbed Rydberg states, with high quantum number m, adiabatic stabilization could be obtained at present laser performances.

In the following, we present an attempt to discuss these results in a classical context. We consider a similar case, starting from a high Rydberg state (the initial energy is defined as $E_i=-1/2n^2$, with n=7) and the laser frequency is chosen to be $\omega=0.0735$ a.u., which satisfies condition (i). From a classical point of vue, a criterion for adiabaticity could be defined as follows: the turn-on of the field amplitude has to be sufficiently slow so that the the variation of the electric field during an atomic period is small with respect to the internal atomic field. In the present case, it leads to such a long laser pulse that the numerical calculations are too heavy to handle with. As a consequence, we have chosen a realistic picosecond laser pulse, smoothly turned-on over 50 cycles and remaining constant over 500 cycles. The ensemble of initial trajectories have a fixed eccentricity defined as

$$e=\sqrt{1-\frac{(l+1/2)^2}{n^2}} \tag{5}$$

where l is the angular momentum. The eccentricity is defined with l+1/2, instead of l, because of the Langer condition for semiclassical approaches. We show in Fig.3a the results

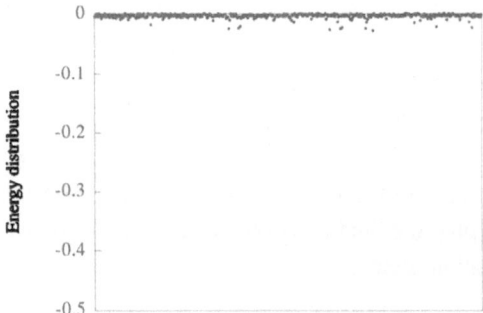

Figure 2a. Energy distribution for the non-ionizing trajectories evolving from a microcanonical ensemble for the 3-D Kepler problem. The laser frequency is $\omega=0.0735$ a.u., the field amplitude is $F_0=0.35$ a.u. and the initial energy is $E_i=-0.5$ a.u.

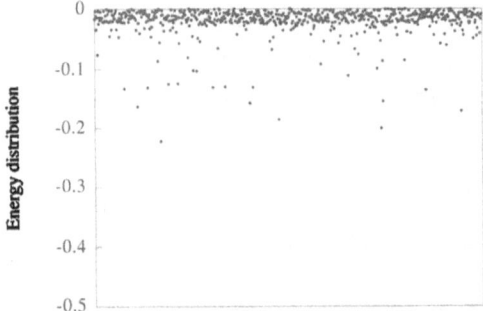

Figure 2b. Energy distribution for the non-ionizing trajectories evolving from a microcanonical ensemble for the 3-D Kepler problem. The laser frequency is $\omega=0.6$ a.u., the field amplitude is $F_0=3.5$ a.u. and the initial energy $E_i=-0.5$ a.u.

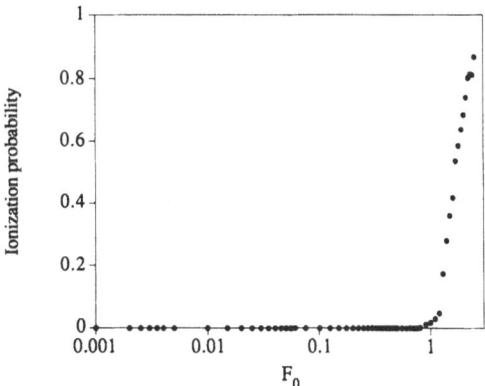

Figure 3a. Ionization probability as a function of the field amplitude F_0 for the 3-D Coulomb problem. The laser frequency is $\omega = 0.0735$ a.u. The initial state is a Rydberg state with n=7 and the angular momentum is l=6, the projection along the z-axis being m=5.

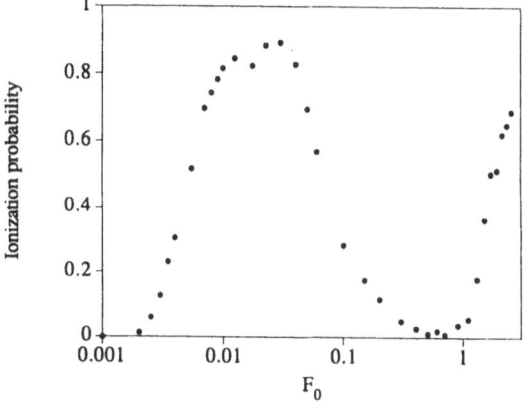

Figure 3b. Ionization probability as a function of the field amplitude F_0 for the 3-D Coulomb problem. The laser frequency is $\omega = 0.0735$ a.u. The initial state is a Rydberg state with n=7 and the angular momentum is l=5, the projection along the z-axis being m=4.

obtained for l=6 and m=5 (m is the projection of the angular momentum along the z-axis, parallel to the linear polarization of the laser). In this case, no stabilization is observed. For field amplitude smaller than 1 a.u., the ionization probability is almost equal to zero, the atomic lifetime being longer than the pulse duration. We have checked that for higher field amplitude, the ionization probability reaches its maximum value (equal to one) and then remains constant.

It turns out that the eccentricity plays a crucial role in the ionization process (as already noted by Richards, 1992). We show in Fig.3b the ionization probability as obtained when the angular momentum is set equal to l=5, which leads to a larger eccentricity than in the preceding case. The projection of the angular momentum along the z-axis is now m=4. As before, at low intensity, the atomic lifetime is longer than the pulse duration ($P_{ion} \approx 0$). But as the field amplitude increases, the ionization probability increases and then reaches a maximum value. Atomic stabilization is achieved when the field amplitude F_0 larger than 10^{-2}a.u. Note that, the turn-on being the same for all values of the field amplitude, if F_0 becomes larger than 1 a.u., the variation of the electric field during an atomic period becomes comparable to the internal atomic forces, which leads to a regime far from adiabaticity. At the present stage of the work, the sensitivity to the angular momentum (i.e. to the eccentricity) and to its projection along the laser polarization is not fully understood.

REFERENCES

Bandarage G, Maquet A and Cooper J, 1990, *Phys. Rev.* A 41:1744.
Bandarage G, Maquet A, Ménis T, Taïeb R, Véniard V and Cooper J, 1992, *Phys. Rev. A* 46:380.
Bhatt R, Piraux B and Burnett K, 1988, *Phys. Rev.* A 37:98.
Eberly J H, Su Q and Javanainen J, 1989, *Phys. Rev. Lett.* 62:881.
Gavrila M, 1992, Atomic structure and decay in high-frequency fields, *in* "Atoms in Intense Laser Fields", M. Gavrila,ed., Academic Press, Boston.
Gavrila M and Kaminski J Z, 1984, *Phys. Rev. Lett.* 52:614.
Grobe R and Law C K, 1991, *Phys. Rev.* A 44:4114.
Grochmalicki J, Lewenstein M and Rzazewski K, 1991, *Phys. Rev. Lett.* 66:1038.
Javanainen J, Eberly J H and Su Q, 1988, *Phys. Rev. A* 38:3430.
Law C K, Su Q and Eberly J H, 1991, *Phys. Rev.* A 44:7844.
Leopold J G and Percival I C, 1979, *J. Phys. B: At. Mol. Phys.* 12:709.
Menis T, Taïeb R, Véniard V and Maquet A; 1992, *J. Phys. B: At. Mol. Phys.* 25:L263.
Pont M and Gavrila M, 1990, *Phys. Rev. Lett.* 65:2362.
Reed V C, Knight P L and Burnett K, 1991, *Phys. Rev. Lett.* 67:1415.
Richards D, Leopold J G, Koch P M, Galvez E J, van Leeuwen A H, Moorman L, Sauer B E and Jensen R V, 1989, *J. Phys. B: At. Mol. Phys.* 22:1307.
Richards D, 1992, *J. Phys. B: At. Mol. Phys.* 25:1347.
Su Q, Eberly J H and Javanainen J, 1990, *Phys. Rev. Lett.* 64:862.
Vos R J and Gavrila M, 1992, *Phys. Rev. Lett.* 68:170.

DYNAMICAL STABILIZATION OF ATOMS
AT CURRENT LASER PERFORMANCES

Etienne Huens and Bernard Piraux

Institut de Physique, Unité fyam
Université Catholique de Louvain
2, chemin du Cyclotron, B-1348 Louvain-la-Neuve
Belgium

INTRODUCTION

Recent theories indicate that a one-electron system exposed to a very strong laser pulse may become stable against ionization. By stable, we mean that there is a regime of high field intensity in which the ionization probability decreases with increasing field intensity. One usually defines two kinds of stabilization: dynamical stabilization and adiabatic stabilization. Dynamical stabilization[1] results from a coherent process: following the rapid turn-on of the laser field, the system is pumped into a coherent superposition of states which is stable against ionization. This process is therefore expected to be rather sensitive to the pulse duration and to the laser frequency. By contrast, adiabatic stabilization[2] results from the adiabatic evolution of a well defined unperturbed eigenstate towards individual dressed states of the system. For photon energies higher than the unperturbed ionization potential, it turns out that the energy width of these dressed states tends to zero for very high fields, exceeding the binding field of the system.

Whether or not adiabatic stabilization is observable is still one of the most fundamental questions. Indeed, in order to undergo stabilization, the system must survive the low field intensities present in the tail of the pulse and for which rapid ionization may occur. It has been shown[3] recently that, at current laser performances, this stabilization effect might be observable when the system is initially in a state with high magnetic quantum number. Let us stress however that in this condition, the ionization rates are extremely small[4] which makes the experiment very difficult.

The results presented in this report demonstrate that with present laser technology, experimental evidences for dynamical stabilization should be relatively easier to obtain. We consider two-photon ionization of atomic hydrogen initially in its metastable state 2s by an ultrashort and intense laser pulse. Two distinct processes contribute in that case to stabilization[5]. First, the creation of a linear superposition

Super-Intense Laser-Atom Physics, Edited by
B. Piraux *et al.*, Plenum Press, New York, 1993

of the initial state and a set of p-Rydberg states and secondly, the transfer through degenerate non-resonant Raman coupling of population towards high angular momentum states. The first process which is important at all field intensities, plays a crucial role in the tail of the pulse since it is at the origin of a substantial inhibition of ionization. The second process is important only at very high field intensity where it gives rise to the creation of several spatially extended wavepackets. Since these wavepackets are characterized by a high angular momentum and have consequently to overcome an important centrifugal barrier, their overlap with the nucleus is expected to be always small. As a result, the effective interaction with the external field stays relatively weak leading to a sharp decrease of the ionization yield with increasing pulse peak intensity. This result has been obtained for pulse durations, peak field intensities and frequencies presently accessible to experiment.

This report is divided in two parts. In the first one, we give a brief outline of our theoretical approach and discuss in detail the results in the second part.

1. THEORETICAL APPROACH: A BRIEF OUTLINE

Dynamical stabilization is the result of a strong coherent excitation of many atomic states. It is therefore important to treat this problem with an approach taking the pulse time profile properly into account. In the present case, we solve the time-dependent Schrödinger equation by means of a spectral method[6].

We use the dipole approximation, the "velocity form" for the interaction Hamiltonian and assume the electric field linearly polarized. In these conditions, the time-dependent Schrödinger equation reads (in atomic units):

$$i\frac{\partial}{\partial t}\Psi(\vec{r},t) = (H_0 + \vec{A}(t).\vec{p})\Psi(\vec{r},t) \qquad (1.1)$$

where H_0 is the atomic Hamiltonian. $\vec{A}(t)$, the vector potential, is given by:

$$\vec{A}(t) = \vec{A}_0 f(t) sin(\omega t) \qquad (1.2)$$

with $f(t)$ its time profile (a Gaussian function) and ω the laser frequency. In order to solve eq (1.1), we expand the wavefunction of the system in a basis of Coulomb-Sturmian functions[7] $S_{n,l}(r)$ in the radial coordinate and spherical harmonics $Y_{l,m}(\theta,\phi)$ in the angular coordinates:

$$\Psi(\vec{r},t) = \sum_{n,l} a_{n,l}(t)\frac{1}{r}S_{n,l}(r)Y_{l,m}(\theta,\phi) \quad ; \qquad (1.3)$$

m is the azimuthal quantum number of the initial state and $a_{n,l}$ is a time-dependent complex coefficient; its initial value is obtained by projecting the initial atomic state on the Coulomb-Sturmian basis. The Coulomb-Sturmian functions are L²-integrable and form a complete set. They may be defined in term of the radial wavefunction $R_{n,l}(r/n)$ associated to the discrete spectrum of atomic hydrogen as follows:

$$S_{n,l}(r) = N_{n,l}R_{n,l}(\kappa r) \qquad (1.4)$$

where κ is a positive parameter[8] and $N_{n,l}$ a normalization coefficient[9]. The close link between the Coulomb-Sturmian functions and the hydrogen radial wavefunctions clearly suggests that the Coulomb-Sturmian basis is probably the most appropriate one for the present problem. This manifests itself by the fact that in this basis, the matrix associated to the atomic and the interaction Hamiltonian have a very simple structure. In matrix form, eq (1.1) becomes[10]:

$$i\frac{d}{dt}S\Psi = (H_0 + sin(\omega t)f(t)V)\Psi \quad ; \tag{1.5}$$

S is a tridiagonal overlap matrix (which is present because the Coulomb-Sturmian functions are not orthogonal in the usual sense). H_0 is a tridiagonal matrix and V a block tridiagonal matrix associated to the interaction, each block being bi-diagonal. Ψ is now a vector whose components are the $a_{n,l}$ coefficients.

From the numerical point of view, this formulation is very interesting because of the sparse character of system (1.5). We avoid storage problems but, everything having its bright and dark side, system (1.5) is stiff[11]: that means that if we use a usual (explicit) Runge-Kutta method to time propagate the solution Ψ, the time step needed for a given accuracy rapidly decreases with increasing size of the system. In order to overcome this problem, we use a fourth order diagonally implicit Runge-Kutta formula[12]. In that case however, one has to solve, at each time step, a large system of algebraic equations of the type Ax=b. Since in the present case, matrix A is always very sparse and banded, it is possible to use an iterative procedure, the bi-conjugate gradient algorithm[13], to solve these systems. In most of the cases treated here, the number of iterations is less than 5; that means that the total number of operations at each time step is proportional to N where N is the total number of Coulomb-Sturmian functions taken into account in the calculations.

After the time propagation, the solution $\Psi(\vec{r},t)$ of eq (1) is projected on the atomic basis. This provides a set of amplitudes $b_{E,l}$; when E, the energy, is equal to $-1/2n^2$, $\mid b_{E,l}\mid^2$ represents the probability for the system to be in a bound state whose principal and angular quantum numbers are n and l respectively. If E_i and E_{i+1} are two consecutive positive energies, $\mid b_{E_{i+1},l}\mid^2$ represents the probability for the system to be in a continuum state of angular momentum l and energy E between E_i and E_{i+1}. As a result, the electron energy spectrum $P(E)$ which represents a probability density may be calculated as follows:

$$P(E) = \frac{\mid b_{E_{i+1},l}\mid^2}{E_{i+1} - E_i}, \qquad E_i \leq E \leq E_{i+1}. \tag{1.6}$$

The number of positive energies E_i taken into account depends on the κ-parameter and the number N of Coulomb-Sturmian functions[6]. Since these functions are L^2-integrable, a truncated basis is only "covering" a finite domain (around the nucleus); this leads to an unphysical reflection of the outgoing wave from the boundaries of the domain. This reflection problem is also encountered in all methods involving space and time integration on a finite grid. We therefore expect that the number N of Coulomb-Sturmian functions has to be relatively high. It is however possible to avoid the problem of reflection by complex rotating[14] the total Hamiltonian in eq (1.1). In that case, the number of Coulomb-Sturmian functions needed is significantly reduced (≈ 100 per angular momentum). After a complex rotation of the total Hamiltonian, the corresponding time evolution operator is no longer unitary and the probability density flux which escapes at large distances, dies out rapidly. However, the probability for the system to be in a bound state after the interaction with the laser pulse is not affected. We can therefore calculate the ionization yield as follows[15]:

$$Yield = 1 - \sum_{E,l} \mid b_{E,l} \mid^2 \qquad (1.7)$$

where $E = -1/2n^2$.

2. RESULTS AND DISCUSSION

We consider two-photon ionization of atomic hydrogen initially in its metastable state 2s by an ultrashort and intense laser pulse. We present results for the ionization yield which has been calculated by means of eq (1.7).

In fig 1a and 1b, the ionization yield is studied as a function of the photon energy for various peak intensities and pulse durations. In fig 1a, the pulse duration,

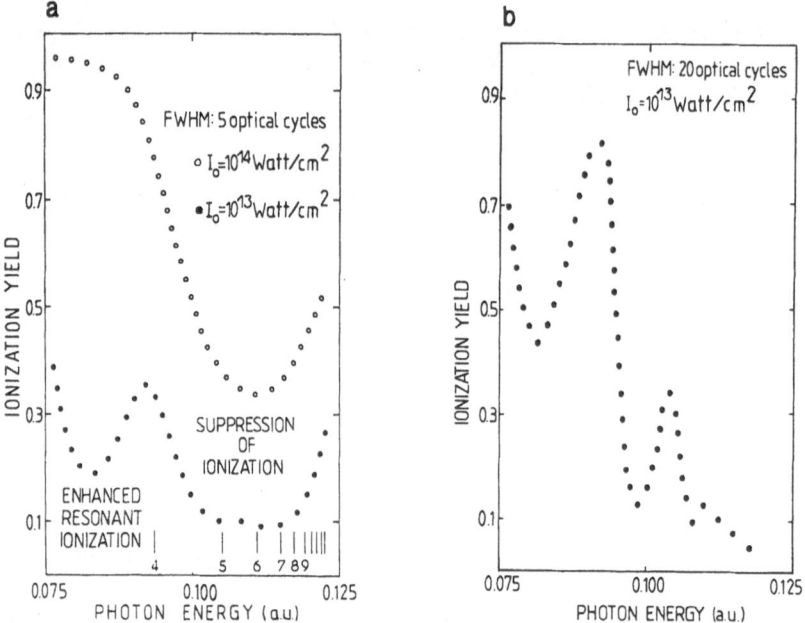

Figure 1. Ionization yield as a function of the photon energy in a.u.. in figure 1a, the pulse duration (fwhm) is equal to 5 optical cycles and two peak field intensities I_0 are considered; ● : $I_0 = 10^{13}$ Watt/cm² and ○ : $I_0 = 10^{14}$ Watt/cm². The regimes of enhanced resonant ionization and of suppression of ionization are indicated. The vertical lines with the number 4, 5, ..., refer to the (unperturbed) position of the resonances associated to the 2s-4p, 2s-5p, ..., intermediate transitions. For comparison, we consider in figure 1b the case of a 20 optical cycle pulse with a peak intensity I_0 of 10^{13} Watt/cm².

i.e the full width at half maximum (fwhm) of the pulse envelope is 5 optical cycles and two laser peak intensities I_0 are considered: $I_0=10^{13}$ Watt/cm^2 and $I_0=10^{14}$ Watt/cm^2. These peak intensities have to be compared to the intensity $(I_0=1.3 \ 10^{14}$ Watt/cm$^2)$ associated to the Coulombic binding field on the 2s "orbit". Although 5 optical cycles is still unrealistic, fig 1a is interesting because it clearly demonstates the existence of three regimes. For photon energies Ep less than 2.7 eV, ionization is resonantly enhanced as expected. The maximum presented for $I_0=10^{13}$ Watt/cm^2 and Ep$\approx$2.5 eV is the result of the 2s-4p resonant transition; we have indicated on the graph the (unperturbed) position of the first resonances. The second regime corresponds to higher photon energies between 2.7 eV and 3.25 eV. Instead of the resonant enhancement of the yield, ionization is now strongly suppressed. It is important to stress that although inhibited,

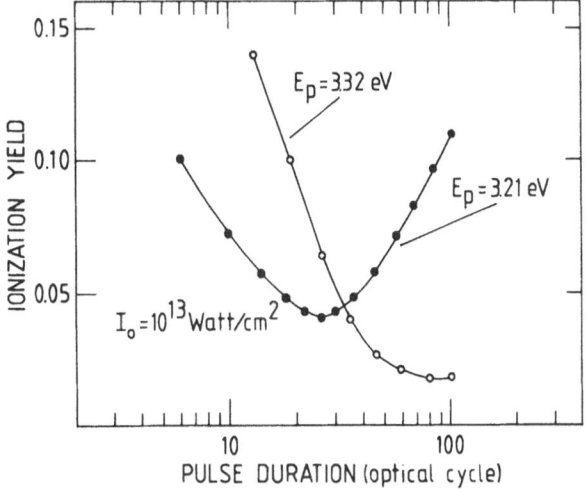

Figure 2. Ionization yield as a function of the pulse duration (fwhm) for a peak intensity I_0 of 10^{13} Watt/cm^2 and two photon energies Ep; -●- : Ep=3.21 eV and -○- : Ep=3.32 eV.

the ionization yield still increases with the laser peak intensity I_0 implying that there is no effective stabilization yet. The third regime corresponds to photon energies close to the ionization threshold. In that case, the pulse bandwidth is large enough to overlap a significant part of the continuum and produce one-photon ionization. When the fwhm is equal to 20 optical cycles and I_0 fixed to 10^{13} Watt/cm^2, enhanced ionization becomes the dominant process for photon energies less than 3.13 eV; the resonances corresponding to the 2s-4p, 2s-5p and 2s-6p transitions are now resolved. We see for instance that for Ep$\approx$2.5 eV, more than 80 percents of the population is ionized after 20 cycles. By contrast, ionization is more inhibited than for a 5 cycle pulse in the second regime: only a few percents of the total population is ionized. The behavior of the ionization yield as a function of the pulse duration is studied in fig 2 for various laser frequencies and a fixed intensity I_0 of 10^{13} Watt/cm^2.

When Ep is equal to 3.21 eV, the 2s-8p transition is quasi resonant; in this case, the ionization yield exhibits a deep minimum around 30 optical cycles. As mentioned above, the sharp increase of the ionization yield for short pulse durations is a consequence of the large spectral width of the pulse which overlaps a significant part of the continuum. Around the minimum, coherent effects are dominant; 95 percents of the population is trapped into a superposition of p-Rydberg states giving rise to a substantial suppression of ionization. For higher pulse durations, the ionization increases rapidly. In that case, the frequency bandwidth associated to the pulse is much smaller than the energy spacing between the excited p-states involved in the process. As a result, there is no more coherent effect and the population that some excited p-states (the 8p-state in particular) might acquire during the excitation process rapidly decays. When the photon energy Ep is equal to 3.32 eV, the 2s-13p transition is quasi resonant and the ionization yield reaches a minimum for a pulse duration of 100 optical cycles ($\approx$130 femtoseconds). It is also interesting to note that at its minimum, the ionization yield is much smaller than for Ep=3.21 eV: only 2 percents of the population is ionized. These results clearly indicate that for moderately intense laser pulses or in the tail of a very intense pulse, coherent effects subsist for pulse durations presently accessible to experiment. At this stage of the discussion, it is important to go back once more over the physical process at the origin of the inhibition of the ionization. Since the first 2s-np transition is quasi resonant, population is first transferred from the 2s-state towards the intermediate np-state at the very beginning of the pulse atom interaction. At higher field intensity, the np-state population is then redistributed amongst other np-states through Raman coupling or because some of these states move through resonance. As a result, a p-wavepacket is created far from the nucleus in a region of space where the effective interaction with the field is weak. In addition, this wavepacket is usually highly non-harmonic: it then collapses rapidly so that its overlap with the nucleus stays very small, ensuring a weak ionization.

Experimental results obtained recently[16] indicate that the physical mechanism discussed above is at the origin of substantial inhibition of ionization. However, effective stabilization, as defined in the introduction has never been observed. Moreover, the results discussed above clearly show that real stabilization does not occur for electric field strengths lower or equal to the atomic binding field. The situation changes drastically when much higher laser peak intensities are considered. In fig 3, we analyze the ionization yield as a function of the laser peak intensity I_0 for various pulse durations and photon energies Ep. When Ep is equal to 3.21 eV and the pulse duration to 30 optical cycles, the ionization yield increases smoothly until I_0 equals 2 10^{15} Watt/cm^2 where it reaches a maximum before decreasing sharply. In order to understand this behavior, let us analyze how, after the interaction with the pulse, the total population left in all bound states of given angular momentum, changes with I_0. The results are presented in fig 4. At moderate laser peak intensity $I_0 \leq 10^{14}$ Watt/cm^2, the population oscillates between the s-states (mainly the 2s-state) and a set of p-states; these oscillations result from Rabi type transitions between the initial state and a set of excited p-states during the interaction with the pulse. It is interesting to note that for $I_0 \leq 10^{15}$ Watt/cm^2, only three angular momenta (l=0,1 and 2) contribute to the partial wave expansion of the i total wavefunction[17]. At higher laser peak intensities, a significant part of the total population is transferred towards excited states of angular momentum l=3, l=5 while the population left in states of l=2,4 and 6 stays negligible (see fig 5). As indicated schematically in fig 6, this migration of population occurs through a degenerate non-resonant Raman process.

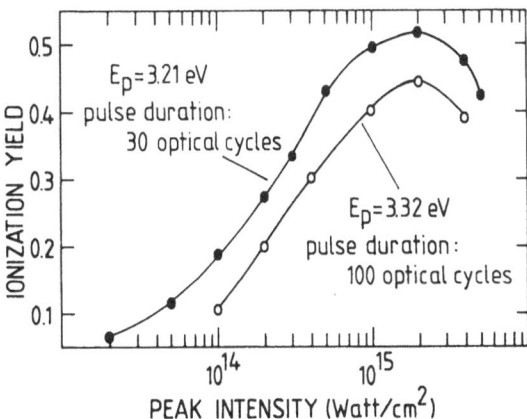

Figure 3. Ionization yield as a function of the peak intensity I_0 in Watt/cm². Two pulse durations (fwhm) and photon energies Ep are considered. -●- : Ep=3.21 eV and fwhm=30 optical cycles; -○- : Ep=3.32 eV and fwhm=100 optical cycles.

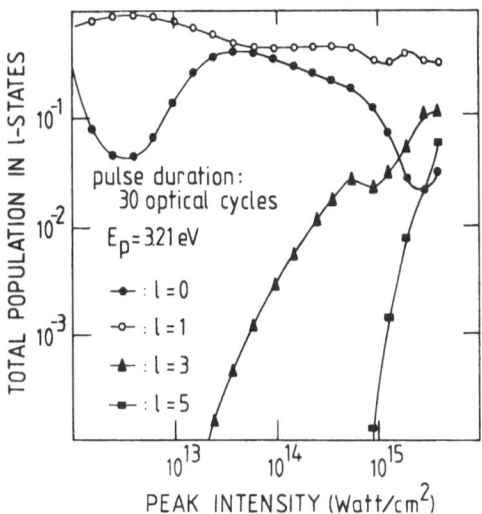

Figure 4. Total population left (at the end of the pulse) in states of fixed angular momentum l as a function of the peak intensity I_0 in Watt/cm²; -●- : $l = 0$, -○- : $l = 1$, -▲- : $l = 3$, -■- : $l = 5$. The photon energy Ep is equal to 3.21 eV and the peak intensity I_0 to 10^{13} Watt/cm². The populations in the states of $l = 2$ and $l = 4$ are too small to be represented on the graph.

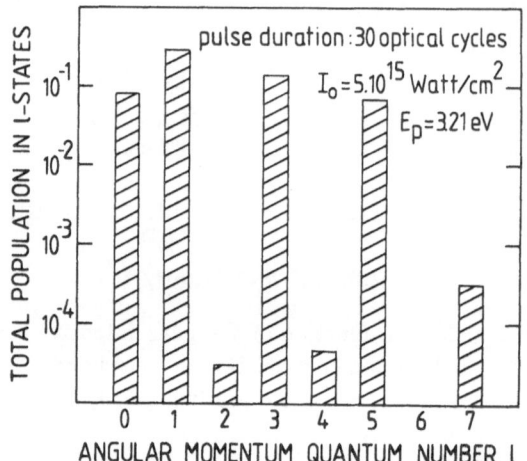

Figure 5. Total population left (at the end of the pulse) in states of fixed angular momentum l as a function of l. The peak intensity I_0 is equal to 5.10^{15} Watt/cm^2, the pulse duration to 30 optical cycles and the photon energy Ep to 3.21 eV.

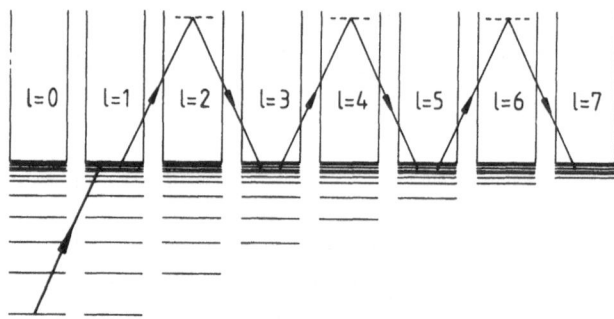

Figure 6. Schematic representation of the excitation process which leads to a migration of population towards states of high angular momentum l in the case of two-photon ionization of atomic hydrogen initially in its 2s state by an ultrashort and intense laser pulse.

This two-photon process which couples states of angular momentum l and $l + 2$ is degenerate because it involves levels characterized by the same principal quantum number n. It is non-resonant because at the frequency considered here, there is no intermediate discrete state of angular momentum $l + 1$ accessible through a resonant one-photon transition. In the present case, the transfer of population occurs via transition involving the continuum. The role of degenerate Raman coupling has been studied in detail by Grobe et al[18]. They showed that migration of population is likely to occur *only* if the process is resonant. This is not the case here, but yet, we observe a transfer

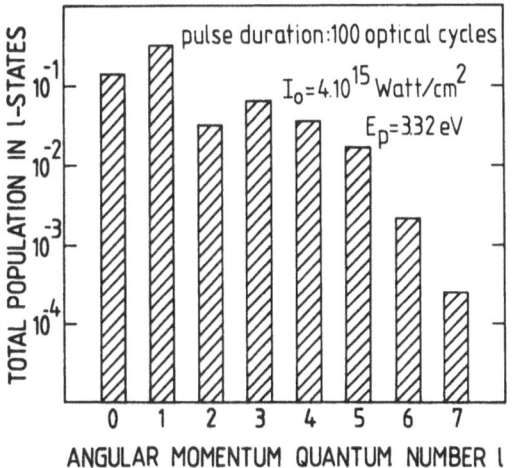

Figure 7. Total population left (at the end of the pulse) in states of fixed angular momentum l as a function of l. The peak intensity I_0 is equal to 4.10^{15} Watt/cm², the pulse duration to 100 optical cycles and the photon energy Ep to 3.32 eV.

of population. It is important to note that in their model, Grobe et al., analyze the transfer of population between *single* states of different l. Instead, we are dealing here with a coherent superpositon of states which is stable against ionization for a wide range of field intensities. As a result, the system can be exposed to laser intensites for which the degenerate non-resonant Raman coupling becomes important compared to the ionization process. For a pulse duration of 100 optical cycles and a photon energy of 3.32 eV, the behavior of ionization yield as a function of I_0 (see fig 3) is very similar to the previous case (fwhm= 30 optical cycles and Ep-3.21 eV). However, the population left after the interaction with the pulse in states of $l = 2,4,6,...$ is no longer negligible although, as shown in fig. 7, it stays lower than in states of odd angular momentum quantum number. The migration

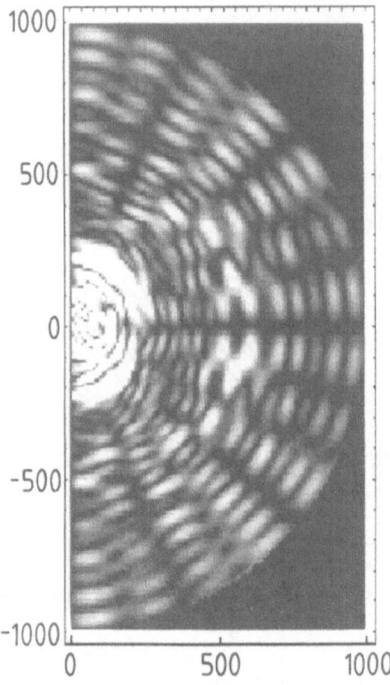

Figure 8. Cut in a 3-dimensional graph of $r^2 \mid \Psi_B(\vec{r}, t) \mid^2 sin(\theta)$ for the same case as in fig 7; Ψ_B is the projection (at the end of the pulse) of the total wavefunction on the bound states of atomic hydrogen. The vertical axis is the z-axis chosen along the direction of polarization of the electric field and the horizontal axis represents the x-axis. The numbers given in the graph are the z and x components of $\vec{r}$ in atomic units. θ is the angle between the z-axis and $\vec{r}$. Dark zones correspond to regions of very low probability density and clear zones to regions of very high probability density.

of population from a set of p-states towards states of higher angular momentum gives rise to the creation of several wavepackets far away from the nucleus, *i.e.* in a region of weak interaction with the radiation field. It is this process which is at the origin of the sharp decrease of the ionization yield as a fuction of I_0. In order to have an idea of the spatial expanse of the wavepackets, we show in fig 8, a cut in a 3-dimensional graph of $r^2 \mid \Psi_B(\vec{r},t) \mid^2 sin(\theta)$ where $\Psi_B(\vec{r},t)$ is the projection, at the end of the pulse of the full wavefunction $\Psi(\vec{r},t)$ on the bound states of atomic hydrogen. The vertical axis is the z-axis chosen along the direction of polarization of the electric field, and the horizontal axis represents the x-axis. θ is the angle between the z-axis and $\vec{r}$. Dark zones correspond to regions of very low probability density and clear zones to regions of very high probability density. We clearly observe on the graph two main regions of high probability density: the first one around the nucleus, covers a very large area, and the second one, characteristic of a p-wavepacket is at about 500 atomic units of the nucleus.

CONCLUSION

We have shown that the interaction of atomic hydrogen initially in its 2s state with an intense short laser pulse may lead to effective dynamical stabilization of the atom when the photon energy is lower than the unperturbed ionization potential. This phenomenon results from two distinct processes: at moderate field intensity, *i.e.* in the tail of the pulse, ionization is inhibited because of the creation of a coherent superposition of the initial state and a set of p-Rydberg states. For higher fields, population migrates towards states of higher angular momentum leading to the formation of several spatially extended Rydberg wavepackets which being far from the nucleus, interact weakly with the field. These processes occur at frequencies, peak field intensities and pulse durations presently accessible to experiment.

ACKNOWLEDGMENTS

One of us (B.P.) is "chercheur qualifié au Fonds National de la Recherche Scientifique". The authors are indebted to P.L. Knight, M. Pont, K. Rzazewski and R. Shakeshaft for very interesting and stimulating discussions. They would like also to express their thanks to A. Magnus and J. Meinguet for many suggestions and for mentioning to them the exitence of several powerful algorithms they used to solve the time-dependent Schrödinger equation.

REFERENCES

1. see *e.g.* M.V. Fedorov and A.M. Movsesian, J. Opt. Soc. Am. B6, 928 (1989); J. Parker and C.R. Stroud, Jr, Phys. Rev. A41, 1602 (1990); K. Burnett, P.L. Knight, B. Piraux and V. Reed, Phys. Rev. Lett. 66, 301 (1991).

2. see *e.g.* M. Pont and M. Gavrila, Phys. Rev. Lett. 65, 2362 (1990); Q. Su, J.H. Eberly and J. Javanainen, *ibid.*, 64, 862 (1990); M. Dörr, R.M. Potvliege and R. Shakeshaft, *ibid.*, 64, 2003 (1990); K.C. Kulander, K.J. Schafer and J.L. Krause, *ibid.*, 66, 2601 (1991).

3. M. Pont and R. Shakeshaft, Phys. Rev. A44, R4110 (1991); R.J. Vos and M. Gavrila, Phys. Rev. Lett. 68, 170 (1992).

4. See R.J. Vos and M. Gavrila, ref 3 and R.M. Potvliege and P.H.G. Smith in the present volume.

5. E. Huens and B. Piraux, Phys. Rev. A47, BR1568 (1993).

6. E. Huens and B. Piraux, in preparation.

7. see *e.g.* M. Rotenberg, Adv. At. Mol. Phys. 6, 233 (1970).

8. A detailed discussion about the optimal choice of κ is given in ref 6.

9. There are several ways of normalizing the Coulomb-Sturmian functions; the present convention consists in assuming the diagonal terms of the overlap matrix S equal to 1.

10. Having calculated the matrix asociated to the total Hamiltonian in the Coulomb-Sturmian basis, it is possible to switch to the atomic basis; this can be done by first diagonalizing the overlap matrix and then the matrix associated to the atomic Hamiltonian (see M. Pont, D. Proulx and R. Shakeshaft, Phys. Rev. A44, 4486 (1991)). We obtain in this way a set of energy eigenvalues; some of them are negative and close to the bound state energies of atomic hydrogen and the others are positive and associated to a discrete set of continuum states of atomic hydrogen. The distribution of the energy eigenvalues depends on the κ parameter and the number of Coulomb-Sturmian functions taken into account; this point is discussed in detail in ref 6. Some of the results presented in this report have been obtained by time propagating the total wavefunction of the system in the atomic basis.

11. E. Hairer and G. Wanner in "Solving Ordinary Differential equations II: Stiff and Differential-Algebraic problems" (Springer-Verlag, Berlin) 1991.

12. see ref 11 chap IV 6.

13. Recent advances in the field of iterative methods for solving large linear systems have been reviewed by R.W. Freund, G.H. Golub and N.M. Nachtigal, Acta Numerica *pp* 57-100 (1991).

14. C.W. McCurdy, C.K. Stroud and M.K. Wisinski, Phys. Rev. A43, 5980 (1991); M. Pont, D. Proulx and R. Shakeshaft, Phys. Rev. A44, 4486 (1991).

15. see M. Pont, D. Proulx and R. Shakeshaft, ref 14.

16. R.R. Jones and P.H. Bucksbaum, Phys. Rev. Lett 67, 3215 (1991); see also J.H. Hoogenraad and L.D. Noordam in this volume.

17. This point is also emphasized in the contribution of L. Plaja and L. Roso, this volume.

18. R. Grobe, G. Leuchs and K. Rzazewski, Phys. Rev. A34, 1188 (1986).

STRONG-FIELD INTERFERENCE STABILIZATION IN ATOMS
AND MOLECULES

M. V. Fedorov

General Physics Institute, Russian Academy of Sciences
38 Vavilov St., 117942 Moscow, Russia

Theoretical investigations of the field-induced interference stabilization of Rydberg atoms are reviewed and summarized. The main discussed results are: the strong-field dynamics of ionization (the "death valley" and stabilization), the multipeak structure of the strong-field photoelectron energy spectrum, narrowing and reconstruction of atomic quasienergy levels in the strong-field limit and so on. Two different mechanisms of the strong-field-induced stabilization in molecules are discussed. One of these mechanisms is based on the direct generalization of the ideas about interference stabilization in atoms upon the case of molecules. The second mechanism is based on the idea about the field-induced avoided crossings of electronic potential curves reconstructed in a strong resonance laser field. Competition of these two mechanisms is discussed.

1. ATOMS

The field-induced stabilization of Rydberg atoms, or the field-induced suppression of ionization are the effects studied widely now, both theoretically[1-3] and experimentally[4-6]. Let us describe briefly the existing theoretical predictions not dwelling too much upon the details of their derivation.

Stabilization

In a general case, the field-induced stabilization means that if an atom is ionized by a laser field with the electric field strength amplitude ε_0, under some conditions the ionization probability $w_i(\varepsilon_0)$, or the ionization rate $\dot{w}_i(\varepsilon_0)$ become some decreasing functions of the field strength. This rather unexpected theoretical prediction follows directly from the model of interference stabilization owing to the Λ-, or Raman-type transitions between neighboring Rydberg levels via the continuum.

Let us assume that initially (at $t = 0$) the atom is excited to some Rydberg level with the energy equal to $E_n = -1/(2n^2)$ (in atomic units used throughout the paper). This means that the initial electron wave function is given by $\Psi(\mathbf{r}, t = 0) = \psi_n(\mathbf{r})$ where $\psi_n(\mathbf{r})$ is the corresponding field-free atomic Rydberg wave function and n is the principal quantum number, n >> 1. Let such an excited atom be ionized by a light field $\mathcal{E}(t) = \mathcal{E}_0\cos(\omega t)$ with a frequency ω larger than the Rydberg electron binding energy, $\omega > |E_n|$. The Λ-, or Raman-type transitions mentioned above correspond to multiple Rydberg-continuum-Rydberg transitions shown in Fig. 1. In a monochromatic weak field such transitions between

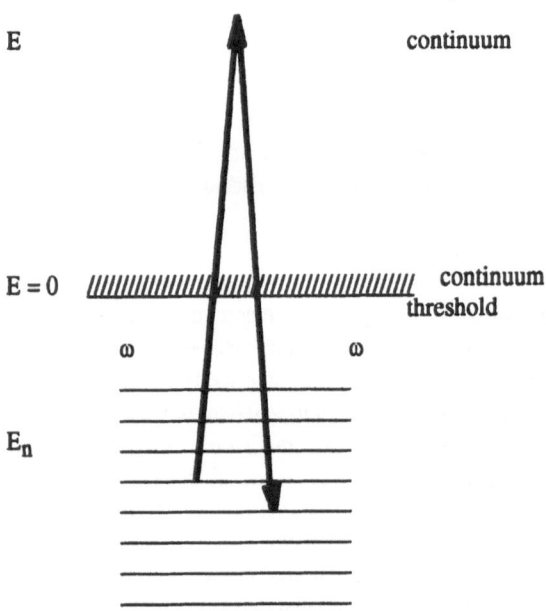

Figure 1. Λ-type Rydberg-continuum-Rydberg transitions.

different levels E_n and $E_{n'}$ ($E_n \neq E_{n'}$) are forbidden by the energy conservation law. However, if the field is strong enough, the described Λ-type transitions can provide an efficient repopulation of Rydberg levels in the process of their photoionization. Quantitatively, the strong-field criterion is determined by the condition that the weak-field Rydberg level ionization width $\Gamma_i = 2\pi |V_{nE}|^2$ becomes of the order of spacing between the neighboring Rydberg levels $E_{n+1} - E_n \approx n^{-3}$, where V_{nE} is the Rydberg-continuum dipole matrix element ($V_{nE} = - <n|d\mathcal{E}_0/2|E>$), $\mathbf{d}$ is the atomic dipole moment. In accordance with refs. 7 and 8 in the so called "quasiclassical" region (n >> 1 and E << 1) $V_{nE} \sim n^{-3/2} \times \mathcal{E}_0 \omega^{-5/3}$, and hence, the strong-field condition takes a form $\mathcal{E}_0 > \omega^{5/3}$. This is the condition under which the efficient field-induced coherent repopulation of Rydberg levels can occur, and interference of transitions from these levels to the continuum can suppress photoionization or stabilize the atom.

Both in the weak- and strong-field limits ($\mathcal{E}_0 < \omega^{5/3}$ and $\mathcal{E}_0 > \omega^{5/3}$, correspondingly) the dynamics of ionization is determined by the same simple exponential formula[1-3]:

$$w_i = \exp(-\Gamma t) \tag{1}$$

However, the decay constant Γ is different in the cases of weak and strong fields: if in the first of these cases $\Gamma = \Gamma_i = 2\pi \left| V_{nE} \right|^2 \sim n^{-3} V^2$, in the strong-field limit Γ is given by[1-3]

$$\Gamma_{str} = 2\,\pi^{-3}\,n^{-3}\,V^{-2} \tag{2}$$

In contrast with Γ_i the strong-field decay rate of equation (2) is a decreasing function of the parameter V proportional to the laser field strength amplitude $\mathcal{E}_0$. Instead of the decay rate constant Γ, the process of ionization can be characterized by the ionization time (or the Rydberg atom life-time) determined as the inverse rate constant:

$$t_i = \Gamma^{-1} = \begin{cases} \frac{1}{2\pi}\,n^3 V^{-2} & \text{for } V < 1 \\ \frac{1}{2}\pi^3\,n^3 V^2 & \text{for } V < 1 \end{cases} \tag{3}$$

The dependence $t_i(V)$ is shown in Fig. 2. The ionization time is minimal when $V = 1$, and

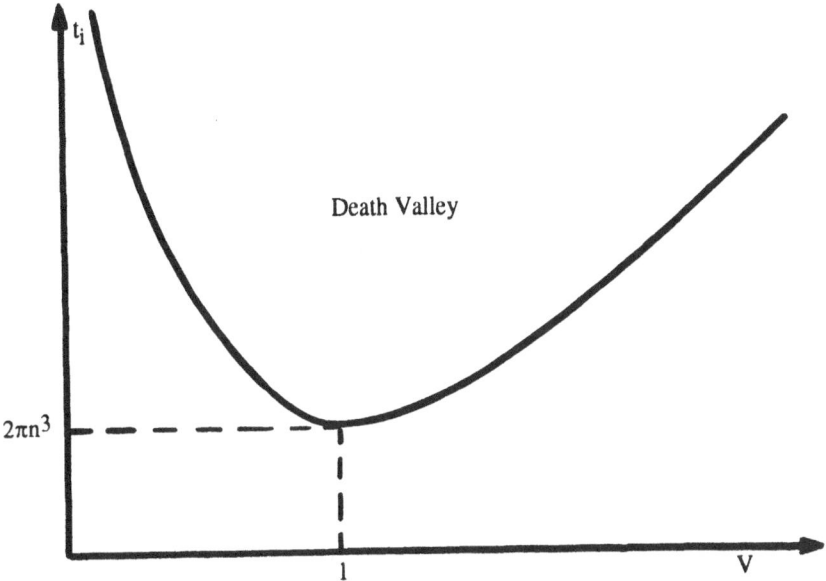

Figure 2. Ionization time vs the parameter V proportional to the field strength $\mathcal{E}_0$.

the minimal achievable ionization time appears to be equal to the classical Kepler period of motion in the Coulomb field:

$$t_{i\,min} = t_i(V=1) = 2\pi\,n^3 = 2\pi \left| E_n \right|^{-3/2} \tag{4}$$

A region around $V \sim 1$, $t_i \sim t_{i\,min}$ corresponds to the conditions of the most fast ionization and for this reason is called the death valley. The growth of $t_i(V)$ in the region $V > 1$ corresponds to the field-induced stabilization of the atom.

Multipeak photoelectron energy spectrum

The long-time limit of the weak-field photoelectron spectrum is evident: it has a single Lorenz peak centered at $E = E_n + \omega$ and having a width equal to Γ_i. The strong-field photoelectron spectrum is much more complicated. In a general case the energy distribution has been found[1-3] to be determined by the following expression applicable both in the weak- and strong-field limits:

$$w(E) = \frac{\left| V_{nE} \right|^2}{\left(E - E_n - \omega \right)^2 \left[1 + \pi^2 \left(\sum_{n'} \frac{V_{n'E}}{E - E_{n'} - \omega} \right)^2 \right]} \qquad (5)$$

The strong-field limit $(V > 1)$ of this equation in the approximation of almost equidistant levels E_n yields the photoelectron spectrum shown in Fig. 3. This is a multipeak spectrum.

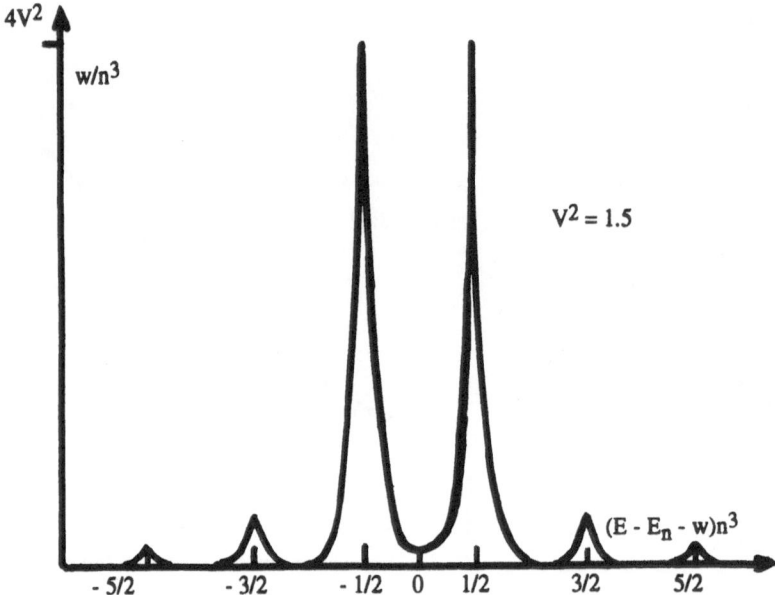

Figure 3. The strong-field photoelectron spectrum.

ts peaks are located at $E^{(n')} + \omega$ $(n' = n, n \pm 1, n \pm 2,...)$ where $E^{(n')}$ are the strong-field Rydberg quasienergies determined by equation:

$$E^{(n)} = \frac{1}{2} \left(E_{n+1} + E_n \right) \qquad (6)$$

The widths of all the peaks coincide with each other and are given by Γ_{str} of eq.(2). The heights of the two highest central peaks (closest to $E_n + \omega$) are equal to $4V^2 n^3$. All the other peaks are smaller and their heights fall as

$$w_{max}^{(n')} = w(E = E^{(n')}) = \frac{V^2 n^3}{\left(n' - n + \frac{1}{2}\right)^2} \tag{7}$$

where as previously n' = n, n ± 1, n ± 2,... The peaks of the strong-field photoelectron spectrum are narrowing and growing with a growing field strength ε_0.

Complex quasienergies

The results described above find their natural explanation in terms of such concepts as quasienergy, quasienergy wave functions, and complex quasienergy. A problem about complex quasienergies is formulated as an approximate one arising when the continuum is eliminated adiabatically and the Schrödinger equation is reduced to a set of equations for the time-dependent probability amplitudes $A_n(t)$ to find the atom on Rydberg levels. The quasienergy solutions of this set of equations are searched in the form $A_n(t) \propto \exp(-i \gamma t)$ where γ are the complex quasienergies $\gamma = \gamma' + i \gamma''$. Their real ($\gamma'$) and imaginary ($\gamma''$) parts determine positions and widths ($|\gamma''|$) of quasienergy levels (or zones). For the problem under consideration the equation for the complex quasienergy γ has been derived[1-3] and presented in the form

$$1 + i \pi \sum_n \frac{|V_{n, \gamma+\omega}|^2}{\gamma - E_n} = 0 \tag{8}$$

This equation can be solved in the weak- and strong-field limits to give the results qualitatively described in Fig. 4 where the quasienergy levels are described as zones having width $|\gamma''|$, and with their "centers of mass" located at γ'. In the limit of very strong field (V>>1) the quasienergy levels are narrowing: $\gamma'' \approx - \Gamma_{str} \to 0$ where Γ_{str} is given by equation (2). In the same limit the quasienergy levels appear to be located near $E^{(n)}$ given by equation (6). Hence, the strong-field photoelectron spectrum shown in Fig. 3 reflects directly the structure of the strong-field quasienergy levels. Also, the stabilization, or suppression of ionization described above is connected directly with the existence of narrow, or long living strong-field quasienergy levels.

Two-level model

The model derivation of the results described above has been generalized[1-3] in many ways to make a model as close to the reality as possible. On the other hand, an opposite question has been considered also: what is the simplest (though maybe not too realistic) model in which interference stabilization can occur. The answer appears to be very simple: such a system consists of two close discrete levels interacting with the continuum as it is shown in Fig. 5. If Δ_0 is the spacing between the two close levels, the condition under which the field can be considered as strong takes a form

$$|V_{\alpha E}|^2 > \Delta_0 \tag{9}$$

where $\alpha = 1, 2$. The minimal time of ionization in such a system is estimated as

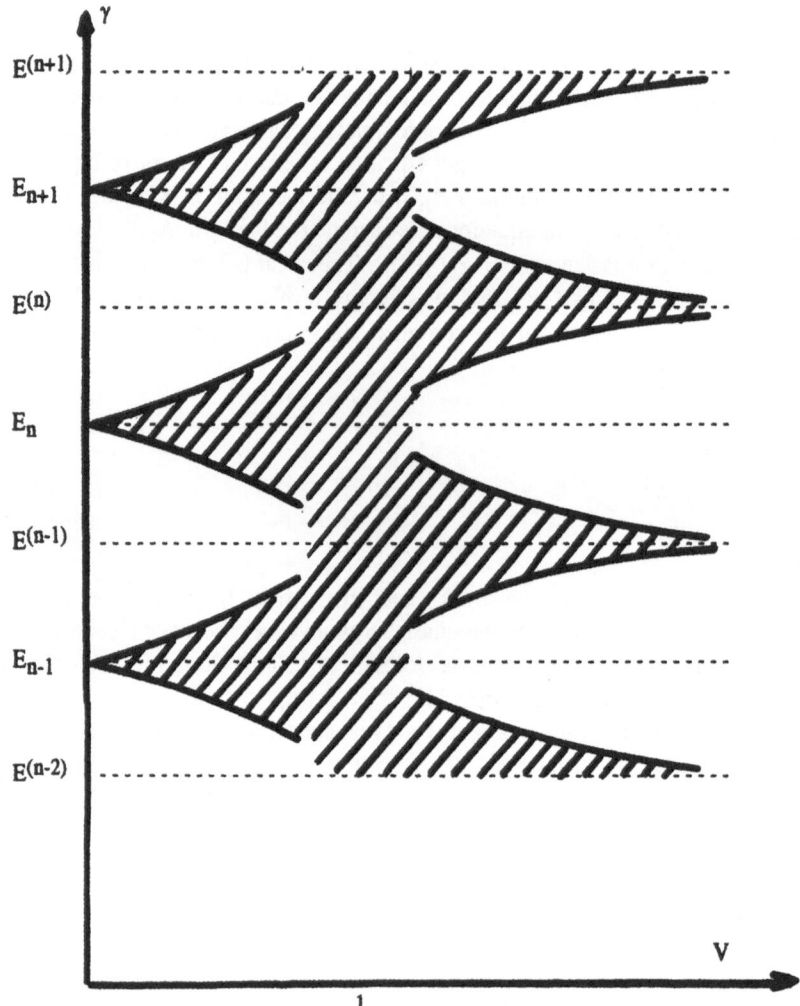

Figure 4. Rydberg quasienergy zones in a strong ionizing laser field.

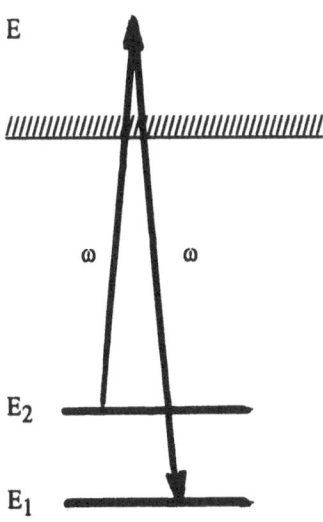

Figure 5. Two-level model in the theory of strong-field interference stabilization.

$$t_{i\ min} = 2\pi / \Delta_0 \qquad\qquad (10)$$

The main qualitative conclusion of the two-level consideration is that the results like the field-induced interference stabilization are very little sensitive to a choice of a model. For this reason the effect is expected to occur in any quantum system having close energy levels. The minimal achievable decay time is expected to be of the order of the inverse spacing between the closest levels. An addotional condition of stabilization found from the two-level considereation[1-3] consists in a requirement of close values of the bound-free matrix elements for the levels 1 and 2.

2. MOLECULES

In this Section we will try to generalize the ideas discussed above upon the case of diatomic molecules. Some alternative ideas about the field-induced stabilization will be discussed. As a whole, molecules have more degrees of freedom and are more complicated than atoms. Also the molecular energy spectrum is more complicated than the atomic spectrum. Let us begin our consideration from a short reminding of the main features of molecules.

Born-Oppenheimer approximation and Franck-Condon principle

Usually the molecular spectra are described in the framework of the Born-Oppenheimer approximation[9] This approximation exploits a large difference between the electron (m) and nucleus (M) masses, $M \gg m$. Due to this fact the electron motion is much faster than the motion of nuclei. For this reason, in fact, one can consider the nucleus kinetic energy T_R as a perturbation which can be ignored at all in the zero order approximation. In this approximation the general Schrödinger equation is reduced to a purely electronic equation, though with the potential energy, eigenfunctions ψ_n and eigenvalues E_n depending parametrically on the distance R between the nuclei, $\psi_n = \psi_n(r; R)$ and $E_n = E_n(R)$ where r denotes a set of electronic variables. By using a set of electronic functions

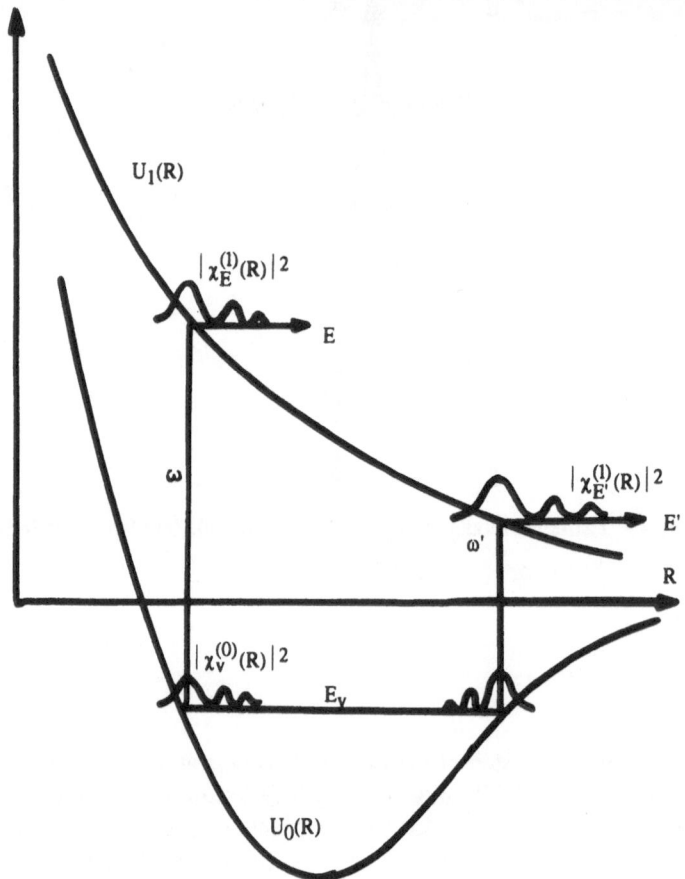

Figure 6. Potential curves $U_n(R)$, nuclear wave functions $\chi_n(R)$ and the field-induced transitions.

$\psi_n(r; R)$ as the basis we can expand now any molecular function over these functions:

$$\Psi(r, R) = \sum_n \chi_n(R)\, \psi_n(r; R) \tag{11}$$

where the expansion coefficients $\chi_n(R)$ are interpreted as the nuclear wave functions (rigorously, they have to be completed by the molecular rotational angle-dependent wave functions, but rotations are even slower than the motion over R and for this reason here are disregarded). By substituting now $\Psi(r, R)$ of eq. (10) into the original Schrödinger equation we get a set of equations for $\chi_n(R)$ which can be considered as decoupled in the Born-Oppenheimer approximation (the remaining coupling between $\chi_n(R)$ and $\chi_{n'}(R)$ is estimated to be of the order of $(m/M)^{1/4} \ll 1$). In the arising equations for $\chi_n(R)$ the electronic problem eigenenergies $E_n(R)$ play the role of effective potential energies for the nuclear motion $E_n(R) = U_n(R)$ (Fig. 6).

The same ideas about fast electron and slow nuclear motions can be used to describe approximately the field-induced transitions in molecules. The corresponding rules are formulated as the Franck-Condon principle[9]. In accordance with this principle the field-induced transitions between different electronic and nuclear states are preceding mainly under the conditions of not moving nuclei, i. e., with fixed values of R, R = const. On a map of the potential curves $U_n(R)$ (Fig. 6) this principle means that the main field-induced transitions are proceeding along the vertical lines. An additional and deeper inter-pretation

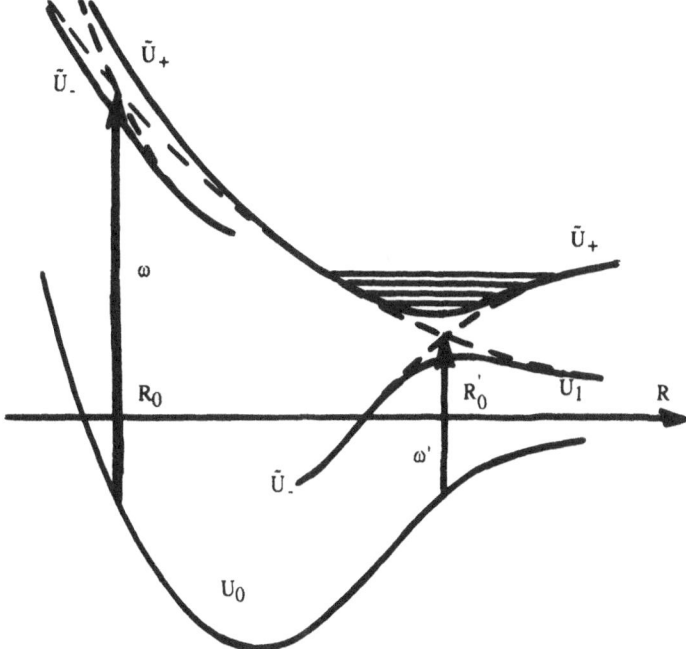

Figure 7. The field-induced avoided crossings. Two different cases corresponding to resonances on the left and right wings of the potential curves.

of the Franck-Condon principle follows from the quantum mechanical and quasiclassical analysis. The nuclear wave functions $\chi_n(R)$ of all the excited states oscillate rapidly everywhere except their classical turning points where $U_n(R) = E$, and E is the total molecular energy. For this reason the main contribution to the dipole transition matrix elements can arise from some vicinities of the classical turning points. In a complete accordance with the Frank-Condon principle the matrix elements of such transitions are expected to be maximal if the classical turning points for the initial and final states coincide with each other. In a simple geometrical picture of Fig. 6 this means precisely the same rule as formulated above: transitions along the vertical lines.

Field-induced avoided crossings

If the external laser field is strong enough and if it has a resonance between some electronic states (e. g., $U_0(R)$ and $U_1(R)$) at some point $R = R_0$ this field can give rise to a reconstruction of both potential curves and the Born-Oppenheimer variable separation procedure itself[10-15]. In the two-level approximation the reconstructed potential curves are given by

$$\tilde{U}_\pm(R) = \frac{1}{2}\{U_0(R) + U_1(R) + \omega \pm [(U_0(R) + \omega - U_1(R))^2 + (\mathbf{d}\boldsymbol{\mathcal{E}}_0)^2]^{1/2}\} \tag{12}$$

where **d** is the transition dipole moment calculated at $R = R_0$:

$$\mathbf{d} = \int dr\ \psi_1(r; R_0)\ \mathbf{d}(r, R_0)\ \psi_0(r; R_0) \tag{13}$$

and **d**(r, R) is the total electronic-nucleus molecular dipole moment. In Fig. 7 the reconstructed potential curves $\tilde{U}_\pm(R)$ are shown for two cases (i. e., for two different frequencies ω and ω') corresponding to resonances at right and left wings of the unperturbed potential curves $U_{0,1}(R)$. The arising structure of the reconstructed potential curves can be interpreted in terms of the field-induced avoided crossings.

The general equations for the nuclear wave functions $\chi_\pm(R)$ in the reconstructed electronic states ψ_+ and ψ_- (corresponding to the reconstructed potential curves $\tilde{U}_\pm(R)$ of eq. (12)) have been derived[10] and shown to have the following form (with only the most important terms retained):

$$[T_R + \tilde{U}_\pm(R) - E]\,\chi_\pm(R) = \mp\, 1/(2M)\, [\,\delta'(R)\,d/dR + (1/2)\delta''(R)]\,\chi_\mp(R) \qquad (14)$$

where $\delta(R) = \tan^{-1}[(U_1(R) - U_0(R) - \omega)/|\mathbf{d\mathcal{E}_0}|]$. The right hand side of eq. (14) describes the new "nonadiabatic" terms responsible for mixing up of the nuclear wave functions $\chi_+(R)$ and $\chi_-(R)$ belonging to different electronic states ψ_+ and ψ_-. If these nonadiabatic terms are small and mixing up of $\chi_+(R)$ and $\chi_-(R)$ is negligible, the approximation of reconstructed potential curves $\tilde{U}_\pm(R)$ can be considered as a correct one and has to be used instead of the usual approximation of the field-free potential curves $U_0(R)$. Vice versa, a large contribution from the nonadiabatic terms of eq. (13) can be considered as an indication that the best zero-order approximation is that of the field-free electronic states ψ_n and potential curves $U_n(R)$.

In the case of a resonance on the right wing of the potential curves $U_0(R)$ and $U_1(R)$ having different signs of slopes at $R = R_0'$, the potential curve $\tilde{U}_+(R)$ describes a new mode of stationary nuclear vibrations having no analog in the field-free case. Excitation of this mode can prevent the molecule from the irreversible dissociation decay, i. e., the field-induced avoided crossing can be a reason of the field-induced suppression of dissociation, or stabilization of molecules. In the case of a resonance on the right wing of the potential curves $U_0(R)$ and $U_1(R)$ one can formulate a very simple criterion justifying applicability of the approximation of reconstructed electronic wave functions and reconstructed potential curves $\tilde{U}_\pm(R)$: the gap between these two curves ($\sim |\mathbf{d\mathcal{E}_0}|$) has to be larger than the vibrational frequency in the new-formed potential $\tilde{U}_+(R)$.

If we have a resonance on the left wing of the potential curves $U_0(R)$ and $U_1(R)$ the situation is more complicated. In this case the two-level reconstruction of the electronic wave function does not give rise to a new branch of stable nuclear vibrations because the slopes of the curves $U_0(R)$ and $U_1(R)$ have the same signs at $R = R_0$. Nevertheless again in this case as previously the reconstructed potential curves can form a good weakly mixed basis if the field is strong enough. To analyze this situation let us use the approxi-mation of a linear expansion of the curves $U_0(R)$ and $U_1(R)$ near the point $R = R_0$:

$$U_{0,1}(R) = U_{0,1}(R_0) - F_{0,1}(R - R_0) \qquad (15)$$

where $F_{0,1}$ are the slopes of the unperturbed potential curves. The structure of the unperturbed and perturbed potential curves around $R = R_0$ is shown in Fig. 8 together with the corresponding nuclear wave functions $\chi_+(R)$ and $\chi_-(R)$. The symbols R_1 and R_2 denote the classical turning points for the reconstructed potential energies, i. e., the points where $\tilde{U}_\pm(R) = E$; δR_{osc} is a characteristic size of the main peaks of the wave functions $\chi_+(R)$ near their classical turning points R_1 and R_2. To estimate δR_{osc} we can use the solutions of the Schrödinger equation with the linearized potential energy expressed in terms of the Airy functions[16]. Finally we get

$$R_2 - R_1 = |\mathbf{d\mathcal{E}_0}|/\sqrt{(F_0 F_1)}, \qquad \delta R_{osc} \sim [(F_0 + F_1)/(MF_0F_1)]^{1/3} \qquad (16)$$

This is clear qualitatively that the nuclear wave functions $\chi_+(R)$ and $\chi_-(R)$ belonging to the different reconstructed electronic states ψ_+ and ψ_- do not mix with each other if $R_2 - R_1 > \delta R_{osc}$, or if

$$|d\mathcal{E}_0| > (F_0 F_1)^{1/6} [(F_0 + F_1)/M]^{1/3} \tag{17}$$

For more rigorous estimates we have to find the matrix elements from the nonadiabatic terms on the right hand side of eq. (14) between the nuclear wave functions $\chi_+(R)$ and $\chi_-(R)$ belonging to different reconstructed electronic states ψ_+ and ψ_-. This can be done with the help of a linear approximation for the potential curves $U_{0,1}(R)$ (eq. (15)). As a

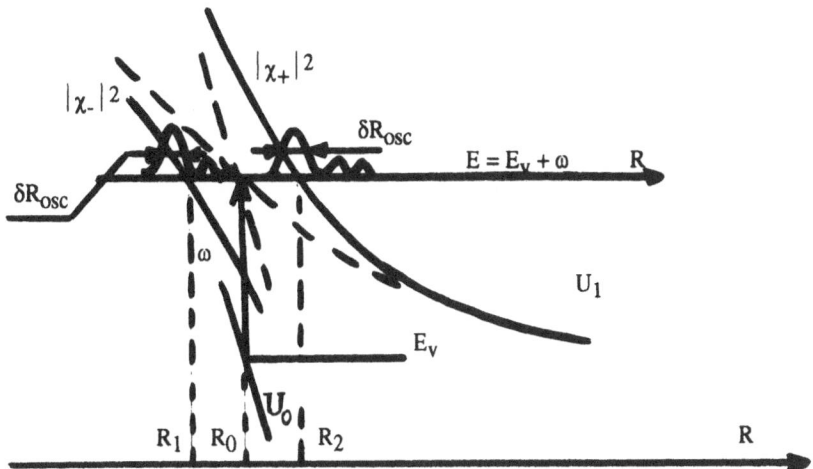

Figure 8. Potential curves and nuclear wave functions near the avoided crossing point, the left wing resonance case.

result we find that the mixing up matrix element is exponentially small if

$$|d\mathcal{E}_0| > (F_0 F_1)^{1/6} [|F_0 - F_1|/M]^{1/3} \tag{18}$$

The estimates of eqs. (17) and (18) are very similar if only $F_0 \neq F_1$. But in the case $F_0 = F_1$ the used above linear approximation for the potential curves $U_{0,1}(R)$ (eq. (15)) be-comes invalid and this case requires a separate consideration.

Though under the conditions determined by eqs. (17) and (18) the nuclear wave functions $\chi_+(R)$ and $\chi_-(R)$ belonging to the different reconstructed electronic states ψ_+ and ψ_- do not mix up with each other, this does not mean that necessarily we have the field-induced stabilization. As a whole the reconstructed potential curves are shown in Fig. 9. In addition to the avoided crossing point R_0 they have the second off-resonant crossing or avoided crossing point R_0'. It is worth to make a comment concerning similarity and difference between the pictures of Figs. 7 and 9. The first of these two pictures describes two different resonances arising in the fields of different frequencies, ω and ω', whereas the

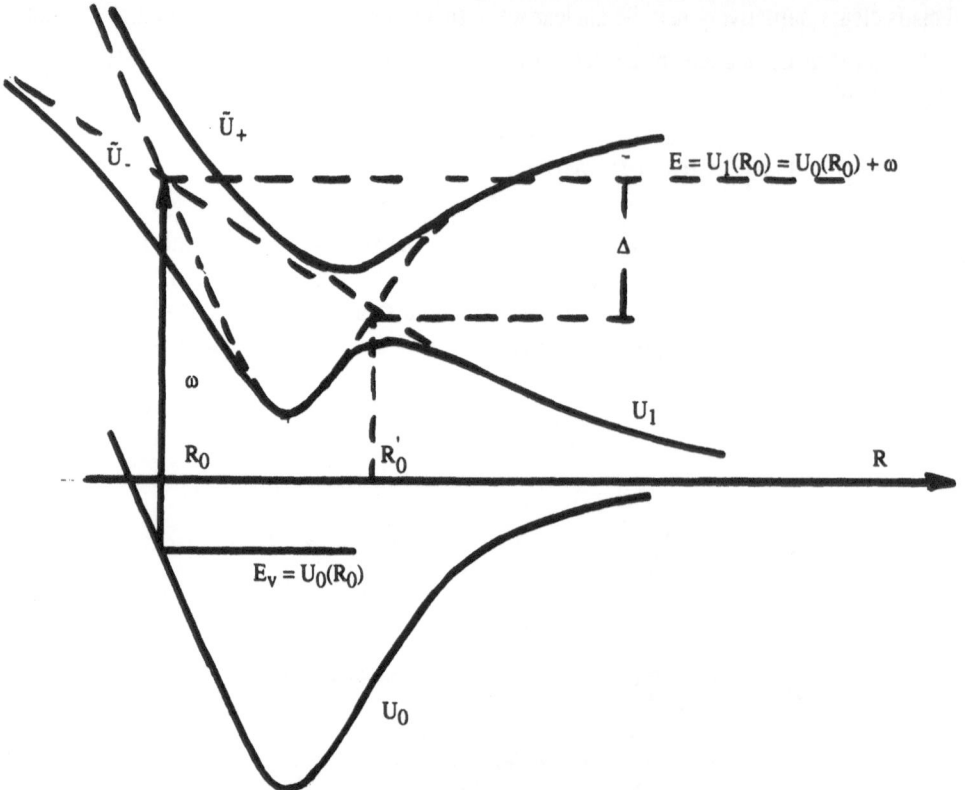

Figure 9. Two crossing points in a single frequency field.

picture of Fig 9 describes two crossings arising simultaneously and stimulated by a single-frequency field. The second crossing shown in Fig. 9 (at $R = R_0'$) is off-resonant because $U_0(R_0') + \omega - U_1(R_0') \equiv \Delta \neq 0$ (whereas $U_0(R_0) + \omega - U_1(R_0) = 0$). In difference with the point R_0 the slopes of the potential curves $U_0(R)$ and $U_0(R)$ at $R = R_0$ have different signs:

$$U_{0,1}(R) = U_{0,1}(R_0') \pm F_{0,1}' (R - R_0') \tag{19}$$

where $F_{0,1}' > 0$.

The nuclei can pass the off-resonant crossing point R_0' either along the field-free (dashed lines of Fig. 9), or along the reconstructed (solid lines) potential curves. The situation can be analyzed with the help of the well known quasiclassical approximation[16] for nuclear wave functions $\chi_\pm(R)$. Not dwelling upon any details of such an analysis let us describe only the main result. The functions $\chi_+(R)$ and $\chi_-(R)$ do not mix up with each other near the point R_0' and we have at this point the field-induced avoided crossing (solid lines on the picture of Fig. 9) if

$$| \mathbf{d'} \boldsymbol{\varepsilon}_0 | > \Delta^{1/4} | F_0' + F_1' |^{1/2} \tag{20}$$

where $\mathbf{d'}$ is determined by the same equation as $\mathbf{d}$ (eq. (13)) but with R_0 replaced by R_0'. It should be noted that Δ characterizes the nuclei kinetic energy at the point $R = R_0'$ which can be rather large, $\Delta \sim \omega$. For this reason eq.(20) characterizes a region of very strong fields, much stronger than those determined by inequalities of eqs. (17) and (18). If however the field is not so strong and

$$| \mathbf{d'} \boldsymbol{\varepsilon}_0 | < \Delta^{1/4} | F_0' + F_1' |^{1/2} \tag{21}$$

in a vicinity of the point R_0' the molecule is characterized by the field-free potential curves $U_0(R)$ and $U_1(R)$ (the dashed lines on the picture of Fig. 9). Owing to a large value of Δ inequalities given by eqs. (17) and (20) are compatible. Hence, we can make a conclusion that in the range of fields

$$(F_0 F_1)^{1/6} [| F_0 - F_1 | / M]^{1/3} < | \mathbf{d'} \boldsymbol{\varepsilon}_0 | < \Delta^{1/4} | F_0' + F_1' |^{1/2} \tag{22}$$

molecular motion is stable: being initially excited to the state $\psi_+(r; R) \chi_{E+}(R)$ near the point $R = R_0$ the molecule evolves along the reconstructed potential curve $\tilde{U}_+(R)$ near $R = R_0$ and along the field-free potential curve $U_0(R)$ near the point $R = R_0'$. As it is seen well from Fig. 9 such a motion is finite and the corresponding nuclear state is stable with respect to the dissociation. So, in the case of a resonance on the left wing of the potential curves we do not get a new vibrational branch as in the case of a resonance on the right wing. Nevertheless under proper conditions (eq. (22)) in this case we can have again the field-induced stabilization of a molecule with respect to its photodissociation owing to the field-induced reconstruction of electronic wave functions.

Interference stabilization of molecular vibrations

Let us try now to approach a problem of the field-induced stabilization in molecules from a different point of view. Let us forget for a moment about the field-induced reconstruction of the molecular electron wave functions $\psi(r; R)$ and potential curves $U(R)$. By using the field-free basis of the electron-nuclear molecular wave functions

$\psi_n(r; R) \chi_v^{(n)}(R)$ (where v is the vibrational quantum number) let us try to apply to this system the same ideas about interference stabilization as in the case of atoms (see the previous Section). This molecular-atomic analogy is expected to be most complete in the case of molecules decaying for ions in their ground states, i. e. if $U_0(R)$ has the Coulomb asymptotics: $U_0(R) \rightarrow - 1/R^2$ when $R \rightarrow \infty$. Such a molecule has a set of Rydberg-like

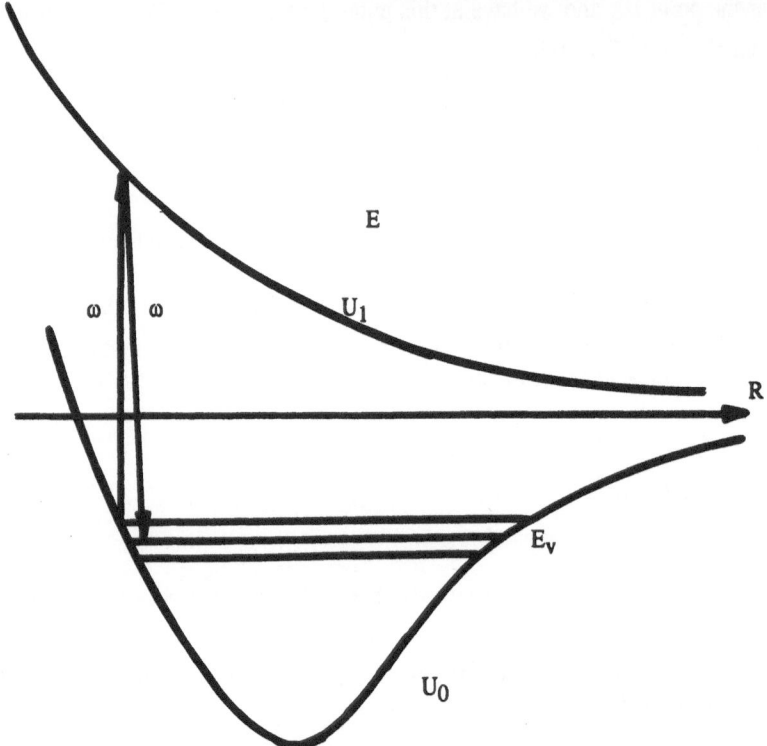

Figure 10. Λ-type electron-nucleus transitions in a molecule.

excited vibrational levels $E_v^{(0)} \approx - M/v^2$. A scheme of Λ-type transitions in such a system is shown in Fig. 10. In accordance with the general rules formulated in the previous Sections there are two main requirements under which the secondary coherent repopulati- on of neighboring levels $E_v^{(0)}$ is efficient, and subsequent transitions to the continuum from these levels interfere with each other resulting in a suppression of dissociation: i) the "dissociation" width of the level $E_v^{(0)}$ (equal to $2\pi |V_{vE}|^2$) has to be larger than spacing between the neighboring levels $E_v^{(0)}$ and $E_{v+1}^{(0)}$, and ii) the matrix elements of transitions to the continuum from the neighboring levels $E_v^{(0)}$ and $E_{v+1}^{(0)}$ have to be close to each other $V_{vE} \approx V_{v+1,E}$. To analyze the applicability conditions of these requirements we can use the mentioned above approximations of linear potential curves $U_{0,1}(R)$ (eq. (13)) and nuclear wave functions $\chi_v^{(0)}(R)$ and $\chi_E^{(1)}(R)$ expressed in terms of the Airy function[16]. In this approximations the bound-free matrix elements V_{vE} can be calculated explicitly to give the following results. The second of the two conditions formulated above (close values of matrix elements) can be fulfilled if $v \gg 1$. On the other hand, the first of the two formulated conditions (sufficiently strong field providing efficient repopulation) appears to be fulfilled under the circumstances determined by the same inequality (18) which has provided the applicability of the approximation of reconstructed potential curves around the field-induced avoided crossing point. This coincidence is very unexpected and surprising. This means that both described mechanisms of the field-induced stabilization have the same region of applicability and are hardly distinguishable. On the other hand, physically these mechanisms of stabilization seem to be essentially different. The results derived arise many questions. Has the discovered coincidence a deeper origin than a simple coincidence? Do the two stabilization mechanisms describe the same or different phenomena? If this the same phenomenon then the interference stabilization has to be simply a different language to describe the field-induced reconstruction of molecular potential curves and electronic wave functions. If nevertheless, these two mechanisms correspond to two different phenomena, the next arising question concerns the conditions under which these two phenomena and two stabilization mechanisms could be dis-tinguished. Unfortunately now we do not have convincing answers to these questions, but we hope to return to the discussed problem elsewhere.

REFERENCES

1. M.V. Fedorov and A.M. Movsesian, *J. Phys* B. 21; L155 (1988).
2. M.V. Fedorov and A.M. Movsesian, *J. Opt. Soc. Am.* B..6; 928 and 1504 (1989).
3. M.V. Fedorov and M.Yu. Ivanov, *J Opt. Soc. Am.* B. 7; 569 (1990).
4. R.R. Jones and P.H. Bucksbaum. *Phys. Rev. Lett.* 67; 3215 (1991).
5. H. Stapelfeldt, D.G. Papaioannou, L.D. Noordam and T.F. Gallagher. *Phys. Rev Lett.* 67; 3223 (1991).
6. M.P. de Boer and H.G. Muller. *Phys. Rev. Lett.* 68; 2747 (1992).
7. I.Ja. Bersons. *Phys. Lett.* A 84; 364 (1981).
8. N.B. Delone N. B., S.P. Goreslavsky and V.P Krainov. *J. Phys.* B. 16; 2369 (1983) and *J. Phys.* B. 22; 2941 (1989).
9. A.S. Davidov. "Quantum Mechanics," Pergamon Press, Oxford - New York (1976).
10. M.V. Fedorov, O.V. Kudrevatova, V.P. Makarov and A.A.Samokhin. *Opt. Commun.* 13; 299 (1975).
11. A.D. Bandrauk and M.L. Sink. *Chem. Phys. Lett.* 57; 569 (1978).
12. A.D. Bandrauk and M.L. Sink. J. *Chem. Phys.* 74; 1110 (1981).
13. A. Guisti-Suzor, X. He, O. Atabek and F.H. Mies. *Phys. Rev. Lett.* 64; 515 (1990).
14. P.H. Bucksbaum, A. Zavriyev, H.G. Muller and D.W. Schumacher. *Phys. Rev Lett.* 64; 1883 (1990).
15. A. Guisti-Suzor and F.H. Mies. *Phys. Rev. Lett.* 68; 3869 (1992).
16. L.D. Landau and E.M. Lifshitz. "Quantum Mechanics," Pergamon Press, Oxford - New York (1977).

STABILITY OF HYDROGEN ATOM IN RYDBERG STATES IN A LASER FIELD

Erna Karule

Institute of Physics
Latvian Academy of Sciences
Riga, Salaspils-1, LV-2169

INTRODUCTION

There are many experimental and theoretical investigations devoted to the multiphoton ionization (MPI) and above
threshold ionization (ATI) of atoms in a laser field. Most of
them are carried out for the ionization of atoms in ground
states. But atoms in highly excited states with large orbital
moments behave different from atoms in ground states when
exposed to a laser field. Investigations of stability of
atoms in Rydberg states in strong laser fields theoretically
are carried out using methods of classical mechanics by Vos
and Gavrila (1992) and Shepelansky. For not very intensive
laser fields perturbation theory may be used. But up to now
using perturbation theory are investigated only ionization
from ground states and excited states with main quantum
number n not higher than ten. The simplest is to investigate
hydrogen atom as for it analytical expressions for transition
matrix elements can be written. Investigating behavior of
Rydberg states one have to get asymptotic expressions for
them in the case of large main quantum number n and orbital
momentum otherwise they diverge. Investigating asymptotic
expressions in the case of states n,n-1 we got behavior of
transition matrix elements for different frequencies. Rydberg
states with large n>30 and high orbital momentum are stable
to photoionization with low frequency radiation but are not
stable to two-photon ionization.

Super-Intense Laser-Atom Physics, Edited by
B. Piraux *et al.*, Plenum Press, New York, 1993

PHOTOIONIZATION FROM RYDBERG STATES

We investigate ionization from highly excited states n with the highest angular momentum l=n-1. Ionization rate for multiphoton ionization from state n,n-1 may be written in a form

$$Q^{(N)}/I^{N-1}=4\pi^2\alpha a_0\omega I_0^{1-N}\sum_1|T^{(N)}(n,n-1;El|\omega)|^2 \tag{1}$$

where I is the radiation intensity in W cm^{-2}, α is the fine structure constant, a_0 is the Bohr radius, ω is the photon energy. We assume that all photons have the same energy. $I_0=14.038\times10^{16}$ W cm^{-2}. $Q/I^{N-1}=0.81\times10^{-17}I_0^{1-N}\omega T^{(N)}$. In the case of MPI of atoms in ground state with different number of photons N transition matrix elements usually are of the same order of magnitude. We will investigate mainly matrix elements for ionization from n,n-1 states. Considering ionization with linearly polarized light the orbital moments of initial as well of final states must be large therefore N<<n.

The simplest case we start investigation is the photoionization of hydrogen atom in a laser field. For hydrogen atom in a state n,n-1 wave function has the form

$$R_{n,n-1}=C_n r^{n-1}exp(-r/n) \tag{2}$$

where $C_n=2^n[\Gamma(2n)]^{-1/2}n^{-n-1}$
For large n and orbital moments using Stirling formula $\Gamma(m+1)=(2\pi m)^{1/2}m^m exp(-m)$ we obtain

$$C_n=\pi^{-1/4}n^{-2n-1+1/4}exp(n) \tag{3}$$

Hydrogen atom in ns state has wave function in a form

$$R_{ns}=C_0exp(-r/n)F(1-n,2,2r/n) \tag{4}$$

where for large n $C_0=2n^{-2-1/2}$

Final state continuum wave function

$$R_{El}=C_E2^{l+1}[\Gamma(2l+2)]^{-1}r^l exp(-r/q)F(l+1-q,2l+2,2r/q) \tag{5}$$

where $C_E=(2\pi)^{-1/2}k^{l+1/2}exp(-\pi/2k)|\Gamma(l+1+q)|$ $k=-i/q$

262

$$C_E = 1 \qquad \text{when} \quad k \to 0$$

For large n and orbital moments

$$C_E = [(nk)^2 + 1]^{1+1/2} \exp[-n \cdot \text{arctg}(kn)/k] \qquad (6)$$

Transition matrix elements for one-photon ionization has a form

$$T^{(1)}(nL, E1|\omega) = \int_0^{\infty} R_{E1}(r) R_{n,L}(r) r^{1+L+3} dr = C_E C_n I_B \qquad (7)$$

The radial integral I_B may be written as

$$I_B = 2^{1+1} [\Gamma(21+2)]^{-1} \int_0^{\infty} \exp(-r/q - r/p) r^{L+1+3}$$

$$\times F(L+1-p, 2L+2, 2r/p) F(1+1-q, 21+2, 2r/q) dr \qquad (8)$$

After integration over radial variable we got for I_B (Karule, 1991)

$$I_B = 2^{1+1} (21+2)_{L-1+2} / (1/p + 1/q)^{-L-1-4} [(q+p)/(q-p)]^{1+L+2-p-q}$$

$$\times (-1)^{L+1-p} \sum_{s=0}^{L-1+2} [(1+1-q)_s (1-L-2)_s] / [21+2)_s s!] (1-z)^s$$

$$\times \sum_{m=0}^{1-L+2+s} [(L-1-2-s)_m (L+1-p)_m / (2L+2)_m m!] (1+z)^m \qquad (9)$$

$$\times {}_2F_1(L+1-p+m, 1+1-q+s, 2L+2+m; 1-z^2)$$

where $z = (q+p)/(q-p)$

At the photoelectric threshold $q \to \infty$, $1-z^2 \to -4p/q$. For states with high orbital moments Eq. (8) for integral I_B simplifies and in the case when $k \to 0$ and $q \to \infty$ Gauss function reduces to $\exp(-2n)$ and in the case of one-photon ionization we have for transition matrix element expression

$$T^{(1)}(n, n-1, 01; \omega) \approx 2^n n^{3.25} \exp(-n) \qquad \omega = E_{ion} \qquad (10)$$

When $k \to 0$ transition matrix elements decreases with n. From Eq. (10) we can make conclusion that Rydberg states with highest possible orbital momentum n,n-1 are stable to one-photon ionization when the energy of a photon $\omega=E_{ion}$ where E_{ion} is the ionization potential. Let us double a photon energy $\omega_1=2E_{ion}=n^{-2}$. Then in the case of ionizatioin from n,n-1 states Gauss function reduces to exp(-πn/2) and we have

$$T^{(1)}(n,n-1;El|\omega_1)\approx 2^{n/2}exp(-\pi n/4)n^{3.25}=2^{n/2}exp(-0.8n)n^{3.25}$$

$$\omega_1=2E_{ion} \qquad (11)$$

In the case of one photon ionization from ns states Gauss function in Eq. (9) for integral I_B at large n has a form of Bessel function $J_1(4n)$ which at large n behaves as cos and sin functions. Transition matrix elements has a form

$$T^{(1)}(ns,01|\omega)\approx exp(2n)n^5\{cos[(4z-4-\pi)/4]-sin[(4z-4-\pi)/4]\}$$

where z=4n and $\omega=E_{ion}$ $\qquad (12)$

Therefore in the case of laser fields for which perturbation theory may be used states with high n and orbital momentum equal zero are unstable to photoionization. Rydberg states with highest orbital momentum n,n-1 are stable to one-photon ionization at low frequencies close to the photoionization threshold. Obviously this conclusion is true for Rydberg states which orbital moments are close to n,n-1.

TRANSITION MATRIX ELEMENTS FOR TWO-PHOTON IONIZATION

In the case of multiphoton ionization and ATI we have matrix elements in a form

$$T^{(N)}(n,n-1;El|\omega)=\int_0^\infty R_{El}(r_N)r_N^3$$

$$\times \prod_{j=1}^{N-1} \int_0^\infty G_{L_j}(r_{j+1},r_j;\Omega_j)R_{nn-1}(r_1)r_j^3 dr_j dr_N \qquad (13)$$

To get analytic expressions for transition matrix elements we used transformed Coulomb Green's function Sturmian expansion in a form

$$G_L(r,r';\Omega) = G_L^A(r,r';\Omega) + G_L^B(r,r';\Omega) \qquad (14)$$

where

$$G_L^A(r,r';\Omega)=-2^{2L+2}p^{-2L-1}\exp(-r/p-r'/p)r^L r'^L \Gamma(L+1-p)$$

$$\times [\Gamma(2L+2)]^{-1} F(L+1-p,2L+2,2r/p)\psi(L+1-p,2L+2,2r/p) \qquad (15)$$

$$p=(-2\Omega)^{-1}$$

$$G_L^B(r,r';\Omega)=\exp(-r/p)\exp(r'/p)r^L r'^{-L-2}p(L+1-p)^{-1}$$

$$\times \sum_{n=1}^{\infty} [(L+1-p)_n/(2L+2)_n n!](2r/p)^n \qquad (16)$$

$$\times \sum_{m=0}^{n-1} [(2L+2)_m/(L+2-p)_m \Gamma(-m)](-2r'/p)^{-m}$$

If $\Omega>0$ p is pure imaginary and $\mathrm{Im}G_L=\mathrm{Im}G_L^A$. G_L^B is real. G_L^B is divergent as it has $\Gamma(-m)$ in denominator. But integrating over radial variables Eq. (16) one gets convergent expressions for transition matrix elements. Using Eq. (14), (15) and (16) for Coulomb Green's function transition matrix elements we get in a form of Gauss hypergeometric functions for which exist analytic continuations. Therefore it is possible to make estimations for different frequencies in asymptotic cases when arguments and parameters are large. Using ordinary Sturmian expansion. transition matrix elements are in a form of series, which terms are polynomials. Such expressions is impossible to estimate in the case of multiphoton ionization when N>2.
Let us estimate two-photon ionization using Green's function in a form given by Eq. (14). Then the part of transition matrix elements which contains only G_L^A may be written as

$$T^{(N)}(n,n-1;E1|\omega)=\int_0^{\infty} R_{E1}(r_N)r_N^3 \qquad (17)$$

$$\times \prod_{j=1}^{N-1} \int_0^{\infty} G_{L_j}^A(r_{j+1},r_j;\Omega_j)R_{nn-1}(r_1)r_j^3 dr_j dr_N$$

If hydrogen atom wave functions are written explicitly then Eq. (17) for transition matrix elements has a form

$$T^{(N)}(n,n-1;El|\omega)=C_n C_E I_B \prod_{j=1}^{N-1} D_j I_{Aj} \qquad (18)$$

where
$$D_j=-2^{2l+2}q^{-2l-1}\Gamma(1+1-q)/\Gamma(2l+2) \qquad (19)$$

I_B is defined by Eq. (8). After integration over radial variable we have for I_B Eq. (9).

$$I_{Aj}=\int_0^\infty \exp(-r/q-r/p)r^{L+1+3}F(L+1-p,2L+2,2r/p) \qquad (20)$$
$$\times\psi(1+1-q,2l+2,2r/q)dr$$

where $p=p_{j-1}$, $p_0=n$, $q=p_j$ and $k=-i/q$
After integration over radial variable we have I_{Aj} in a convenient form to estimate the integral

$$I_{Aj}=(1/p+1/q)^{-L-1-4}\Gamma(L+1+4)\Gamma(L-1+3)/\Gamma(1+4-q)$$
$$\times \sum_{s=0}^\infty [(L+1-p)_s(L-1+3)_s(L+1+4)_s]/[s!(L+4-q)_s(2L+2)_s] \qquad (21)$$
$$\times[2pq/(p+q)]^s F(1+1-q,L+1+4+s,L+4-q+s;z^{-1}) \quad z^{-1}=(q-p)/(q+p)$$

In the case of ionization from n,n-1 states p=n and L=n-1. Then L+1-p=0 and we have

$$I_{A1}=q^{n+1+3}\Gamma(n+1+3)\Gamma(n-1+2)/\Gamma(n+3-q)$$
$$\times {}_2F_1(1+1+q,n+1+3,n+3-q;z^{-1}) \qquad (22)$$

When frequency is $\omega=E_{ion}$ energy of ejected electron is close to zero and $q \to \infty$. Transforming Gauss function to a function with argument $1-z^{-1}$ we get

$$I_{A1}\approx -n^{2n+4}\exp(-2n)[Ei(2n)+\pi ctg(\pi p)]/\Gamma(-L-p) \qquad (23)$$

From Eq. (18), (9), (19) and (23) we can get expression for transition matrix elements for two-photon ionization at the one-photon ionization threshold. In the case of two-photon ionization from n,n-1 state when p ≠ α but q=-in and integral I_B has a form

$$I_B(n,n-1) \approx 2^n \exp(-\pi n + 2n) n^{2L+10} \qquad (24)$$

Transition matrix element with $G_L{}^B$ are negligible to compare with that given by Eq. (18). Therefore we omitted it and got two-photon transition matrix element in a form

$$T^{(2)}(n,n-1;E1|\omega) \approx 2^{2n} n^{11.75} \exp(-3/4 \pi n + 2n) \qquad (25)$$

$$\times \exp(-2n)[Ei(2n) + \pi ctg(\pi p)]$$

$$= 2^{2n} n^{11.75} \exp(-0.36n) \exp(-2n)[Ei(2n) + \pi ctg(\pi p)] \qquad \omega = E_{ion}$$

where $|\exp(-2n)Ei(2n)| < (2n)^{-1}$. Stirling formula are used for

$$[\Gamma(2L+2)]^{-1} = 2^{-2L-2} (\pi)^{-1/2} n^{-2L-1-1/2} \exp(2n) \qquad L \approx n \qquad (26)$$

Below the one-photon ionization threshold due to exp(-2n) resonances are very narrow. For two-photon above threshold ionization p are pure imaginary and ctg(πp)=i.
At photoionization threshold suppressed is one-photon ionization but two-photon ionization is possible.

CONCLUSIONS

We investigated one and two-photon ionization probability for Rydberg states with high main quantum numbers n>30
and orbital moments l=n-1 in the case when and N<<n. Behavior
of states with orbital moments close to the highest will be
similar.

In laser fields for which perturbation theory may be
applied Rydberg states n,n-1 contrary to ns states are stable
to one-photon ionization at low frequencies close to ω=E_{ion}.
Stability of n,n-1 states to one-photon ionization is
increasing with n but decreasing as energy of photons are
increasing. At the photoionization threshold (frequency
ω=E_{ion}) hydrogen atom is not stable to two-photon ionization
and two-photon above threshold ionization. Therefore it would
be interesting to investigate above threshold ionization of

Rydberg states with high angular moments together with photoionization using nonperturbative methods. May be there are Rydberg states and frequencies when one-photon ionization is suppressed but two-photon ionization is possible.

Stabilization to multiphoton ionization may be reached only in intese laser fields.

It is difficult to make analytic estimation for behavior of Rydberg states in a laser field in a case of multiphoton ionization. To have more complete insight at the behavior of different Rydberg states in a laser field numerical calculations have to be performed.

REFERENCES

Vos, R. J., and Gavrila, M., 1992, Effective stabilization of Rydberg states at current laser performances, Phys. Rev. Lett. 68; 170.
Karule, E., 1990, Integrals of two confluent hypergeometric functions, J.Phys. A: Math. Gen. 23.; 1968.

STEPWISE DECAY IN THE PHOTOIONIZATION
OF RYDBERG ATOMS

J. H. Hoogenraad and L. D. Noordam

FOM Institute for Atomic and Molecular Physics
Kruislaan 407, 1098 SJ Amsterdam, the Netherlands

INTRODUCTION

Photoionization of hydrogen-like atoms with very intense laser fields has been studied intensively. Phenomena like multiphoton ionization (MPI) and excess photon ionization (EPI) have attracted considerable interest[1]. The availability of short-pulse, high-intensity lasers enables us to investigate photoionization more precisely. Most recent work, both experimentally and theoretically, is performed on the ionization from the ground state of the atom. A number of theories addressing very high intensities in either very rapidly pulsed[2] or continuous fields[3] is being developed. Experiments are performed with relatively short pulses (but often far longer than addressed by the short-pulse theories), that are not as strong as needed to verify the theoretical results. Here, we want to show calculations on one-photon ionization from high-lying excited states with pulses that could be created experimentally.

For low intensities, the photoionization yield for ionization with one photon is proportional to the intensity of the applied field and linear in the duration of the pulse. Electrons will be emitted until the end of the pulse, if the state from which they originate does not become depleted. If we do take this depletion into account, the ionization rate during the pulse decreases exponentially with time. In this paper we report circumstances in which this decay is not even exponential: The ionization rate drops in steps.

This feature is due to the use of Rydberg states. A superposition of those states (a wavepacket) can behave very much like a classical electron in a Keppler orbit: The wavepacket moves around the nucleus with a roundtrip time (τ_n) determined by the average spacing between the levels around the center state n: $\tau_n = 2\pi n^3$. A wavepacket can only be excited by a light pulse with enough bandwidth to contain several eigenstates. This is equivalent to the requirement that the length of the pulse (τ_p) needs to be short compared to the roundtrip time τ_n. We will concentrate on situations where the wavepacket-like behaviour starts to become important ($\tau_p \approx \tau_n$).

An electron can interact with light only while being close to the core: The combination of both momentum and energy conservation requires that the nucleus absorbs most of the momentum of the photon. Both excitation to a Rydberg state and ionization from such a state can only take place near the core. This behaviour can be measured directly in two-pulse experiments[4]: A wavepacket is created using a short laser pulse. Ionization with a delayed second pulse (also near the core) shows oscillations in the ionization yield with the period of the classical roundtrip time.

We now apply this reasoning to photoionization. When the system is in an eigenstate, the wavefunction has amplitude both near the core and in the remote area. Ionization will take out part of the wavefunction near the core. In this report we show that it is possible to create a hole in the wavefunction by ionizing with a short laser pulse. The pulse populates some neighbouring states, creating a wavepacket in counterphase with

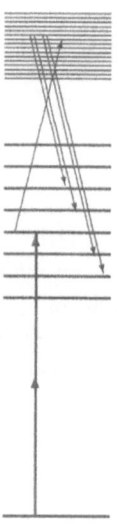

Figure 1. From the ground state, one Rydberg state is excited in a slow process. Afterwards, a short pulse is used to transfer part of the population into the continuum. By a Raman transition via this continuum, neighbouring Rydberg states will also be populated (from [6]).

the initial population. We call this hole in the wavefunction an antiwavepacket. Note that an antiwavepacket has been shown before in a computer experiment[5], although the authors did not describe it as such: A two-pulse ionization sequence was simulated, and oscillations similar to the two-pulse wavepacket experiments[4] were predicted.

An actual experiment[6] has been performed using two lasers as shown in figure 1. A single Rydberg state is excited using a nanosecond-laser. From that state, ionization with the second (picosecond) laser will also distribute wavefunction to nearby lying states. The redistributed population was subsequently probed by field ionization. Experimentally, the pulse width of the laser was not varied and therefore the effects described in this paper have not been studied experimentally. The calculations presented in this paper extend the ones presented in the experimental one.

Unlike a normal wavepacket, which is already clearly visible when the probability $r^2 |\psi(r)|^2$ is plotted, the antiwavepacket is not directly apparent this way. Subtracting the probability after ionization from the original probability $(r^2 |\psi_0(r)|^2 - r^2 |\psi(r)|^2)$, however, does show the wavepacket-like structure. In figure 2 this is shown: Immediately after the laser pulse, the difference is highest near the core. This maximum oscillates afterwards with the same roundtrip time as a normal wavepacket formed from the same states.

For short pulses, we will show that the ionization probability does not increase as rapidly with intensity as might be expected. This effect can readily be explained by the intuitive picture shown in figure 3. Assuming again that the ionization will only take place near the core, the limiting factor for ionization in medium-high fields becomes the supply of wavefunction to the core. This supply is inversely proportional to the roundtrip time of the wavepacket. We approximate the maximum ionization yield with the ratio of the pulse length to the roundtrip time

$$Y_{max} \approx \frac{\tau_p}{\tau_n}. \tag{1}$$

The yield is approaching this limit for high intensities.

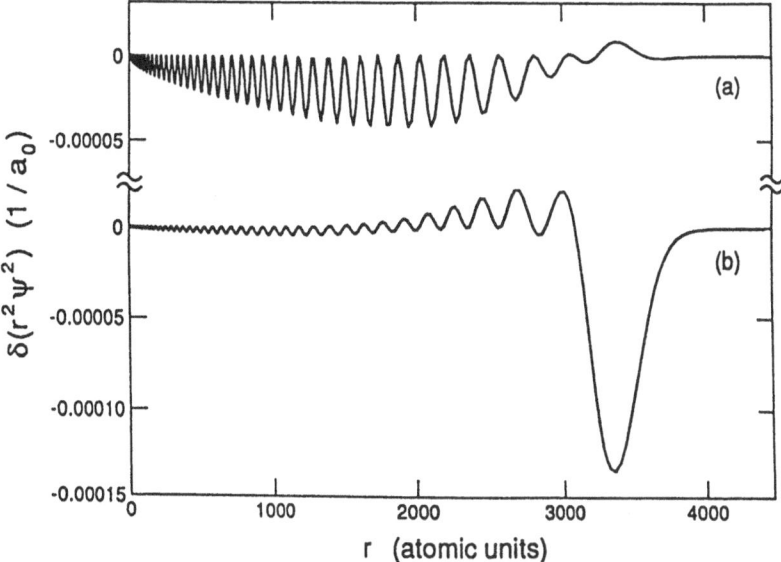

Figure 2. Antiwavepacket wavefunctions around $n = 40$. Trace (a) is calculated just after a short pulse and (b) one half roundtrip time later (from earlier calculations[6]).

METHODS

In this section we will describe the formalism used to obtain the results. We want to point out the differences and similarities with earlier calculations. Especially, we will show that Fermi's Golden Rule results are the limits for low fields.

A first approximation for the ionization rate is Fermi's Golden Rule: In this approximation, the ionization rate is supposed to be linear in both the initial population and in the transition moments. To use this approximation, the population of the original (bound) states should not change considerably during the ionization.

Within the framework of Fermi's Golden Rule, one can easily incorporate changes in the populations: By writing the ionization as a differential equation for the bound population, an exponential decay follows. For the calculations presented here, we also include transitions back to the bound states: Just after an electron is ejected after absorbing a photon, the chance exists that it is captured again by emitting a photon (stimulated emission). Due to the large bandwidth of a short pulse, the system does not have to return exactly to its original quantum state. A transfer of population can occur to neighbouring states by this Raman-like ∧-type process (See figure 1).

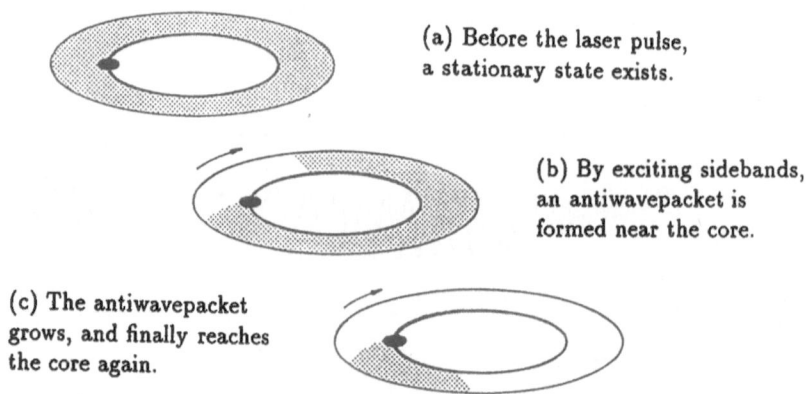

(a) Before the laser pulse, a stationary state exists.

(b) By exciting sidebands, an antiwavepacket is formed near the core.

(c) The antiwavepacket grows, and finally reaches the core again.

Figure 3. Intuitive picture of the ionization process. The ratios between the pulse length and the roundtrip time are (a) $\tau_p/\tau_n = 0$ (no pulse yet), (b) $\tau_p/\tau_n \approx \frac{1}{4}$ and (c) $\tau_p/\tau_n \approx \frac{3}{4}$.

The mathematics we use here is the same as in section III of the Parker and Stroud paper[2] and as in earlier calculations[6]. Using the dipole approximation and Rotating Wave Approximation (RWA), the Schrödinger equation in the interaction picture is:

$$
\begin{aligned}
\frac{da_E(t)}{dt} &= \sum_n V_{En} a_n(t) \exp\left(-\imath \omega_n t\right) \exp\left(\imath \omega_E t\right) \exp\left(-\imath \omega t\right) f(t) \\
\frac{da_n(t)}{dt} &= -\sum_E V_{En} a_E(t) \exp\left(\imath \omega_n t\right) \exp\left(-\imath \omega_E t\right) \exp\left(\imath \omega t\right) f^\star(t)
\end{aligned}
\tag{2}
$$

In Eq. (2), $a_n(t)$ and $a_E(t)$ are the probability amplitudes of bound and continuum states at time t. $f(t)$ prepresents the pulse shape and is proportional to the intensity of the light. $2V_{En} \sin\left(\omega t\right) f(t)$ is the matrix element of the interaction Hamiltonian connecting the states $|E\rangle$ and $|n\rangle$. ω_E and ω_n are the energies of these states and ω is the photon energy.

To solve these equations, we make two additional approximations: The first one is that the coefficients V_{En} are independent of E: The continuum is supposed to have no structure. Therefore, we will drop the index E on the matrix elements V_n. The second assumption is that the width of the continuum is infinite, essentially stating that the time the electron can be captured is infinitely small.

The continuum states $|E\rangle$ now can be eliminated from the equations (2) by formally integrating $da_E(t)/dt$ from the first line and substituting this a_E into the second line.

$$\frac{da_n(t)}{dt} = -\sum_{n'} \int_{-\infty}^{t} dt' \left(\sum_{E} V_n V_{n'} \exp\left[\imath \left(\omega_{E0} - \omega\right)\left(t' - t\right)\right] \right)$$
$$\times \exp\left(\imath \omega_n t\right) \exp\left(-\imath \omega_{n'} t'\right) \tag{3}$$
$$\times a_{n'}\left(t'\right) f\left(t'\right) f^{\star}\left(t\right),$$

with ω_{E0} the energy difference between the continuum state $|E\rangle$ and the initial bound state. The expression in the large parentheses yields $2\Gamma_n \delta\left(t' - t\right) V_{n'}/V_n$, where $2\Gamma_n$ is the Fermi Golden Rule rate of ionization from the state $|n\rangle$ into a continuum state at an energy ω_{E0} higher than its own energy. $(2\Gamma_n = 2\pi |V_n|^2)$

The equations can now be described most easily in the Schrödinger picture. The coefficients $c_n(t)$ $(= a_n(t)\exp\left(-\imath \omega_n(t)\right))$ are the probability amplitudes in this picture. To obtain the time dependent populations of the states $|n\rangle$, we solve the equations:

$$\frac{dc_n(t)}{dt} + \imath \omega_n c_n = -\Gamma_n |f(t)|^2 \sum_{n'} \frac{V_{n'}}{V_n} c_{n'}(t) \tag{4}$$

In any case where $f(t)$ is not dependent on time, equation (4) is an ordinary matrix-type differential equation. We solve it by diagonalizing this matrix. Time dependent intensities can be described with discrete steps in the intensity envelope $f(t)$.

To connect to Fermi's Golden Rule theory, we want to show that for low fields, the bound population decreases exponentially in our model as well.

When we neglect the coupling of the continuum to other bound states, equation (4) becomes a single-variable differential equation in c_n. Keeping $f(t)$ constant, the ionization rate $(-dc_n(t)/dt)$ will drop off exponentially. Even if we do include small couplings to other bound states, the right hand side transfers little wavefunction from one state into another: The neighbouring states will ionize much faster than the transfer of new population into them takes place. The only c_n that will have considerable amplitude is therefore the original state, yielding an exponential decay. It is only for high intensities that the probabilities for transfer (which grow with the square of the intensity) can compete with the ionization: The neighbouring states become populated.

Finally, we want to point out the parameters we used to obtain the results. Equation (4) depends only on the laser intensity $f(t)$, the matrix elements V_n and the energies of the bound levels. The energies are $\omega_n = \left(2n^2\right)^{-1}$. The coefficients V_n are proportional to $n^{-3/2}$, the prefactor being dependent on the wavelength of the used light and the type of states. Values for this prefactor are calculated for hydrogen s-states coupled to continuum p-states. States with other angular momenta are not included in the calculation, as the transition moments from the continuum to bound states with a higher angular momentum are an order of magnitude smaller than those back to bound states with a lower angular momentum.

We neglect that part of the wavefunction can be put into continuum levels with a higher energy by absorbing more than one photon (excess photon ionization or EPI). This wavefunction cannot be transferred back into bound states. The effect from the leak into these not included states will not make qualitative changes to the effects presented below[3].

RESULTS AND INTERPRETATION

In an experiment, the photoionization yield (the population in the continuum) is measured. Therefore, we will present our results as plots of the yield as a function of one of the simulation parameters. We calculate the photoionization yield up till time t $(Y(t))$ from the remaining bound population:

$$Y(t) = 1 - \sum_n (c_n(t))^2 \qquad (5)$$

The number of states needed to describe the system depends on the bandwidth and intensity of the laser pulse. To obtain our results, the 15 states closest to the initial state were used. Initially, only the population of the initial state is set to 1, while all other states have no probability in them. To ensure that all states that play a role are included, we verified the population in the outer 8 states to stay less than 10^{-3}.

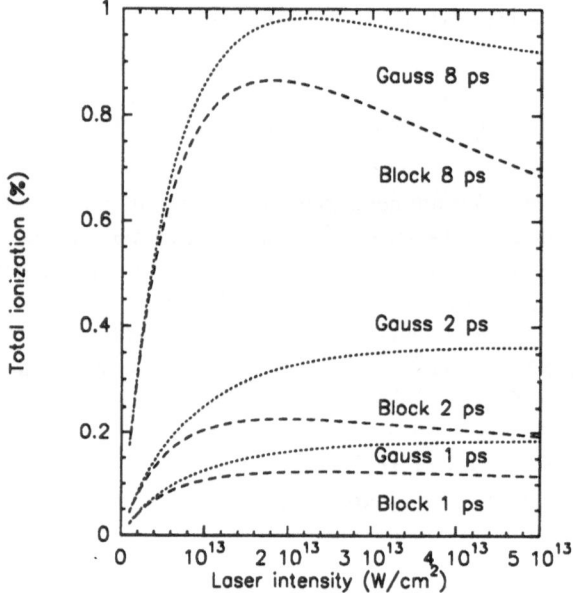

Figure 4. Total yield $(Y(\infty))$ as a function of the intensity for various pulse lengths τ_p shorter than the roundtrip time τ_n. The initial state was $n_0 = 40$ $(\tau_n \approx 10\ ps)$. The matrix elements are calculated for $\lambda = 532\ nm$.

Although we solve the matrix equation (4) for one intensity at a time, we can model any pulse shape by letting the system evolve under consecutive intensity levels. The first used shape is a "block" pulse: The light intensity is suddenly switched on and off with a variable time in between. The second shape is a Gaussian that incorporates 8 discrete levels. The width of the Gaussian is varied by elongating the duration of the levels. We verified to have a good approximation for the Gaussian by reducing the number of used intensity levels and observing convergence of the results of the calculations. Note that the fluence (number of photons in the pulse) for a Gaussian with a given peak intensity is approximately the same as the fluence of a block with the same length and peak intensity.

The first plot (figure 4) shows the total ionized fraction as a function of the peak intensity of the light pulse. We performed the calculations for both pulse shapes described above. For low intensities, the total ionization grows linearly with the intensity of the light. This is just the Fermi Golden Rule behaviour. In the initial part, the ionization is proportional to the fluence of the pulse, and thus linear in the intensity. At higher intensities, the ionization levels off to a level that depends on the pulse length. For a square one-picosecond pulse this level is 10 %, for a two-picosecond pulse it is 20 %. If we compare the results with the intuitive picture of wavefunction removed near the core, we discover that this feature can be explained easily. The total ionized part

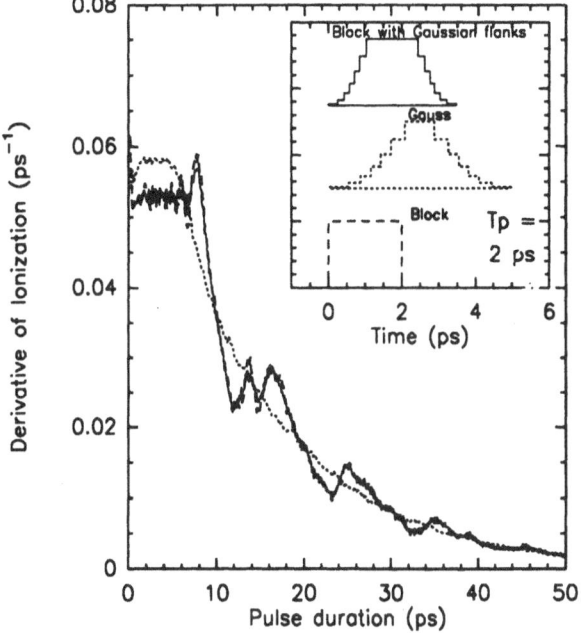

Figure 5. The differentiated ionized fraction $(\mathrm{d}Y(\infty)/\mathrm{d}\tau_p)$ as a function of the pulse length τ_p. The dashed trace is for "block" pulses, the solid trace (almost overlapping with the "block" curve in the main figure) is for the "block with Gaussian flanks" and the dotted trace is for the "Gaussian" pulse shape. $I = 5 \cdot 10^{12} W/cm^2$, $\lambda = 532 \ nm$, $n_0 = 40$. The inset shows the used pulse shapes for a 2 ps pulse.

at high peak intensities is the fraction of the wavefunction that comes near the core during the pulse, as given in equation (1). As the roundtrip time (τ_n) is approximately 10 ps, this relation is confirmed by these calculations.

For longer pulse lengths, the total ionization sometimes shows a maximum. Up to this maximum, the intuitive picture predicts the behaviour.

Although the general behaviour of the two plotted pulse shapes is similar, only the "block" line approaches the limit given by the intuitive explanation. For the Gaussian pulse, the yield exceeds this limit: The Gaussian pulse also gains "length" with the increased intensity, as the period a certain minimum intensity is present is elongated.

Next, we studied the total ionization versus the pulse length of the light. As expected, this is a rapidly increasing function (more light means more ionization). To show the underlying physics, we plotted the derivative of the ionization yield in figure 5, varying the pulse lengths for three different pulse shapes. The intensity is chosen such that 50 % ionizes during the first 10 ps.

The curve labeled "block" can also be interpreted as the time-dependent ionization rate during a block pulse at time τ_p: Taking the derivative is the effect of a small elongation of a block pulse that ends at the time τ_p, and thus the ionization during this elongation.

We observe a staircase-like decay in the "block" curve: During the first 10 ps, the ionization rate is more or less constant. This 10 ps corresponds to the roundtrip time τ_n. After this time the rate drops to approximately half the previous value. After the next roundtrip time, one could expect another plateau at approximately 25 %, but the plateaus get washed out because of the unequal spacing between Rydberg levels.

For the Gaussian pulse, the initial plateau is more or less the same, but the characteristic steps do not appear: The decay after τ_n resembles the exponential curve predicted by the Fermi's Golden Rule-type reasoning. We checked if the effect for the "block" pulses was only caused by the adiabicity of the switch-on, introducing a new pulse shape: the "block with Gaussian flanks". It consists of a 1 ps Gaussian, with a flat part inserted in the middle (see the inset of figure 5). From the calculated yield curve it is clear that the switch-on of the block is not very important: Because the traces resemble each other so closely, we conclude that the staircase-like behaviour is really caused by the formation of an antiwavepacket.

Another reason to preform calculations on this pulse shape is that when one creates a square pulse in an experiment, the flanks will never be infinitely steep. We wanted to assess how steep the edges had to be. The graphs show that this constraint is not severe.

The inset of figure 5 shows the pulse shapes for the point calculated at 2 ps. The peak intensity for the pulses is the same. The pulse length (τ_p) in the main figure is either a scaling factor (for the "Gaussian" and "block" pulses) or corresponds to the length of the flat part in the middle plus the constant rise and fall times of 1 ps (for the "block with Gaussian flanks").

We also confirmed that the step duration scaled as n^3, in accordance with the antiwavepacket explanation.

RELATED RESULTS

We want to point out some similarities with other work. The connections are based on the concept of an antiwavepacket, the methods used, and the decrease in ionization we observe in some cases.

The concept of depletion of the wavefunction near the core was introduced by Wang and Cooke[7]. They performed calculations on what they call Isolated Core Excitation to autoionizing states in two-electron atoms: A valence electron is excited to a Rydberg state. Somewhat later a second "core" electron is excited. The "core" electron can be autoionized when it interacts with the Rydberg electron. The Rydberg electron will be in a different state after the interaction, and the part of the of Rydberg-electron wavefunction that did not have this interaction yet will be decreased.

They calculate the transition moment to the autoionizing state as a function of the energy of the state to which the core electron is excited (and thus the laser frequency)[7, 8]. Taking the Fourier transform of this formula, they obtain a staircase structure very similar to figure 5.

Another result we already mentioned in the introduction, is the calculation on a two-pulse antiwavepacket experiment[5]. This calculation is based on the same physical effect as the work in the previous sections. Raman transitions as shown in figure 1 form an antiwavepacket, but in their work they use a second laser pulse to create interferences in the populations of the states that were not initially populated. They show that this population depends on the phase of the atomic wavepacket, but does not depend on the phase of the light (as would be the case for a wavepacket experiment). We were able to simulate double pulses and confirmed their findings about the populations of the states. Furthermore, the ionization yield in such an experiment shows the same oscillations in the delay between the pulses as expected from the antiwavepacket explanation.

The calculations presented in the previous section have a single Rydberg state as initial condition. Analytical results were obtained for the situation like in figure 1 already, but starting from a wavepacket[9]. The wavepacket has the same frequency envelope as the pulse that ionizes. The ionizing laser pulse can then originate from the same laser as the exciting pulse. Such a scenario is proposed as a model for multiphoton ionization[9].

Finally we want to point out some anologies to calculations on stabilization in very high electromagnetic fields, as we sometimes find a decrease in the ionization rate as a function of the laser intensity. We want to show the connections and the differences between the methods used to calculate stabilization and our work.

For extremely high fields, the wavefunction of the electrons is severely modified by the field, especially when the applied field becomes as strong as the potential felt by an electron in the state n. This distortion might decrease the amount of ionization.

Complete calculations of the effects due to this time-dependent field are extremely time-consuming. Therefore, a number of models has been proposed[10]. When the motion of the electron due to the field is larger than the field free excursion, the electrons can be taken as a fixed point in a frame, with the core oscillating rapidly. The rapid oscillations of the core potential can be averaged over, if the laser frequency is high. Stabilization then may occur due to the dichotomy of the (now semi-static) potential[11].

The states of the atom in the field can also be described as the eigenstates obtained by taking into account the coupling to the continuum and back and coupling to higher lying states in the continuum by EPI[3]. This method resembles diagonalizing the matrix in equation (4), but can be performed analytically for all the states in the atom. Very stable states are predicted above a treshold intensity. We could not reproduce this stabilization result because of our finite number of bound states.

It is still a question[12] if the approximation using first order transition moments and RWA is still valid when the fields are very intense. Parker and Stroud[2] reported considerable transfer to states which can only be reached through transitions that are forbidden for one photon. The fast-oscillating terms in the RWA contribute much to the multi-photon processes under these extremely high fields.

We have taken care to stay in the regime where the matrix elements remain small. At these low intensities we can still make the RWA and we do not expect deviations due to these very high intensity effects.

CONCLUSION

We have shown that in ionization from Rydberg levels, some structure may appear in the ionization signal as a function of the pulse length. An upper limit was given for the total ionization yield from short pulses in the medium field limit.

ACKNOWLEDGEMENTS

The authors would like to thank D. I. Duncan for the effort to create the algorithms used for the calculations and T. F. Gallagher for the support in starting this research. We especially thank H. G. Muller for his discussions on the limits of applicability of the model and the calculations of the prefactors of the matrix elements. The work described in this paper is part of the research program of the Stichting Fundamenteel Onderzoek van de Materie (Foundation for Fundamental Research on Matter) and was made possible by the financial support from the Nederlandse Organisatie voor Wetenschappelijk Onderzoek (Netherlands Organization for the Advancement of Research).

REFERENCES

1. H. G. Muller, P. Agostini, and G. Petite, Multiphoton ionization *in:* "Atoms in Intense Laser Fields," M. Gavrila, ed., Academic Press, San Diego (1992).

2. J. Parker and C. R. Stroud, Jr., Population trapping in short-pulse laser ionization, *Phys. Rev. A* 41:1602 (1990).

3. M. V. Fedorov and A. M. Movsesian, Interference suppression of photoionization of Rydberg atoms in a strong electromagnetic field, *J. Opt. Soc. Am. B* 6:928 (1989).

4. A. ten Wolde, L. D. Noordam, A. Lagendijk, and H. B. van Linden van den Heuvell, Observation of radially localized atomic electron wave packets, *Phys. Rev. Lett.* 61:2099 (1988).

5. V. P. Chebotayev and V. A. Ulybin, Synchronization of Raman transitions in highly excited hydrogen atoms: a new proposal for measuring the Rydberg constant, *Appl. Phys. B* 52:347 (1991).

6. L. D. Noordam, H. Stapelfeldt, D. I. Duncan, and T. F. Gallagher, Redistribution of Rydberg states by intense picosecond pulses, *Phys. Rev. Lett.* 68:1496 (1992).

7. X. Wang and W. E. Cooke, Wave-front autoionization: classical decay of two-electron atoms, *Phys. Rev. Lett.* 67:976 (1991).

8. X. Wang and W. E. Cooke, Wave-function shock waves, *Phys. Rev. A* 46:4347 (1992).

9. M. V. Fedorov and A. M. Movsesian, Wave packets, probabilities of transitions, and multiphoton excitation of atoms, *J. Opt. Soc. Am. B* 5:850 (1988).

10. J. H. Eberly, Atomic physics and nonlinear optics in very strong laser fields *in:* "Nonlinear Dynamics and Quantum Phenomena in Optical Systems," R. Vilasecca and R. Corbalon, ed., Springer-Verlag, Berlin (1992).

11. M. Pont, N. R. Walet, M. Gavrila, and C. W. McCurdy, Dichotomy of the hydrogen atom in superintense, high-frequency laser fields, *Phys. Rev. Lett.* 61:939 (1988).

12. H. G. Muller. private communication.

POPULATION TRAPPING IN EXCITED STATES

Harm Geert Muller and Marc Paul de Boer

FOM-Institute for Atomic and Molecular Physics
Kruislaan 407
1098 SJ Amsterdam
the Netherlands

INTRODUCTION

Resonances can play an important role in multiphoton ionisation (MPI). Especially at low and moderate intensities the presence of a resonance tends to increase the ionisation rate by many orders of magnitude. Such Resonance Enhanced MPI (REMPI) has evolved into a very useful technique for (state-selective) detection of trace atoms and molecules. For a long time it was thought that at high intensities (above 10^{13}W/cm^2) resonances were unimportant for the MPI process. This belief was based on the assumption that in this intensity range the ionization lifetime of excited states was so short that the corresponding level broadening would wipe out the resonance structure. Experiments seemed to confirm this view, since the ionisation yield did not show sharp structure as a function of wavelength.

The first indication that resonances nevertheless still played an important role at several times 10^{13}W/cm^2 was an experiment by Freeman et al.[1] on MPI of xenon. In this experiment the use of a subpicosecond laser avoided ponderomotive acceleration of the photoelectrons. In effect this made it possible to recognize the intensity at which a particular ionisation event happened, by observing the energy shift of the associated photoelectron. Such an intensity-resolved experiment shows a pronounced structure in the rate of ionisation as a function of intensity (e.g. figure 1a), with strong enhancement occurring whenever an excited state is shifted into resonance by the ac-Stark effect. The ac-Starkshift at such high intensities is of the order of an eV, which is larger than the spacing between levels of the atom, and comparable to the photon energy. Thus at any wavelength there are always regions in the focus where the intensity is such that some state is shifted into resonance, an integer number of photons above the ground state. The reason for the disappearence of the resonance structure from the wavelength scan was thus not so much the intrinsic broadening of the states, but really the broadening due to Stark shifts and the simultaneous presence of many different intensities in the focus.

STABILITY OF EXCITED STATES

Apparently excited states are much more stable than was believed initially, even at intensities where the electric force exerted on the electron by the laser exceeds the Coulomb force felt from the nucleus by a large factor. The reason for this is that·the electromagnetic field is an ac field, with a time average of zero, which has no long-time tendency to pull the electron away from the nucleus. Thus the field forces the electron into an oscillating motion (commonly referred to as quiver motion), but the center of oscillations (i.e. the short-time averaged position) is still free to move in whatever way is dictated by the comparatively weak force that binds the electron to the atom. It has been shown[2] that this separation of the electron motion into a fast quiver motion and a slow orbital motion works very well far away from the nucleus, where the third derivative of the atomic potential is negligible on the scale

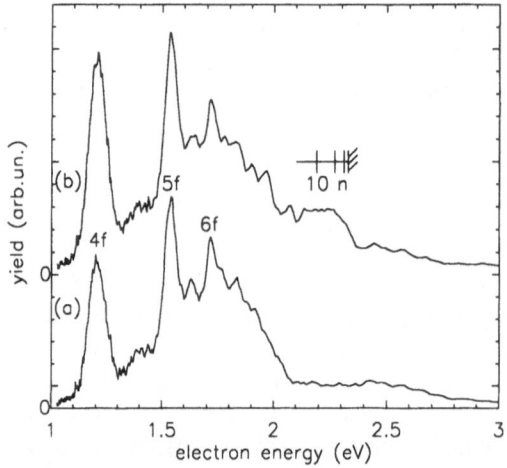

Figure 1. Electron spectra for multiphoton ionisation of xenon with linearly polarised 597 nm light in 100 fs pulses. The peaks due to several resonances with Rydberg f-states are clearly visible. In trace b an additional probe beam (532 nm, 5 ns) was present, which reveals remaining population in high (n>10) Rydberg states.

of the quiver amplitude. In order for energy to be transferred to the orbital motion (so that ionisation can occur), the electron has to approach the nucleus rather closely. Thus states that stay away from the nucleus, such as those with a sizable angular momentum, can have fairly long lifetimes.

The excited states can actually be much more stable than the ground state, and this opens the possibility for an increase of the time the electron remains bound to the atom, by exciting it to a state that has a much longer lifetime. Under conditions that would completely deplete the groundstate, this population trapping in excited states would actually cause a decrease of the ionisation yield as the resonance occurs, much in contrast to what happens in REMPI.

At this stage we want to point out that these long lifetimes have nothing to do with stabilisation, just with stability. Whether one wants to consider population trapping as a form of stabilisation is a matter of taste. In fact several different forms of stabilisation can be distinguished. It has been known for a long time that the proximity of a resonance can actually cause an ionisation rate to decrease sharply with increasing intensity, as ac-Starkshift pushes the involved state away. A similar behaviour can be expected near interference minima in between resonances[3]. Both these effects, which could be called resonance stabilisation, only occur in a very small intensity range and are critically dependent on the wavelength, in contrast to adiabatic stabilisation, in which the ionisation rate keeps consistently falling above a certain intensity[4]. It is also distinct from dynamic stabilisation, which is the temporary suppression of ionisation (by the formation of some sort of wavepacket) at the expense of larger ionisation at other times[5]. In this latter form the time dependence thus plays a crucial role. Population trapping is also sensitive to intensity and wavelength, and thus could be considered a form of resonance stabilisation.

ONIONS AND POTATOES

The exact mechanism of the ionisation through resonances in the case of subpicosecond pulses has been a subject of recent investigation. There are three times that play an important role in the REMPI process. These are the duration of the laser pulse, the lifetime of the resonant state with respect to ionisation, and the Rabi flopping time associated with the excitation of the resonance. Until recently is was thought that the Rabi time was the longest of the three, and the lifetime the shortest. Under these conditions the model proposed by Freeman et al.[1] applies, and ionisation occurs simultaneously with excitation. This excitation is strongly favoured in those regions in the focus where the resonance occurs at the peak intensity, since this makes the resonance last for a comparatively long time. Where higher peak intensities occur, the ionisation is not as efficient because the resonance occurs on the leading and trailing edges of the pulse, and lasts only for a short time. Where the peak intensity is not high enough to shift the state into resonance, there is of course no ionisation at all. Thus the ionisation mainly occurs in shells of very specific intensities, which inspired the name 'onion model' for this scenario.

Essential for the onion model are the long Rabi time and short upper-state lifetime (both compared to the pulse duration). If the Rabi rate is very fast, there is no reason for the excitation to be favoured on the peak of the pulse, and complete transfer of the ground-state population to an excited state can occur (by adiabatic passage) even on the edges of the pulse. In this case ionisation through that resonance could occur in the entire volume where the peak-intensity (in time) exceeds the intensity to shift the state into resonance. Such a distribution of the ionisation has become known under the name 'potato model'. An interesting consequence of the complete transfer of population on the edge of the pulse is the disappearence of all ionisation through states that are resonant at higher intensity. This is exactly what has been observed recently by groups in Charlottesville[6] and Amsterdam[7].

A lifetime of the upper state that is long compared to the pulse has consequences that are less obvious. If the Rabi time is long enough for the excitation to be mainly diabatic, most of the excitation will still occur in those regions where the peak intensity causes resonance. The same does not have to hold for the ionisation, though, because most of the population transferred to the upper state will survive the pulse and not be ionised. At higher peak intensities the excitation will be less efficient, but the population transferred to the upper state on the leading edge of the pulse will be exposed to much higher intensities, and thus might eventually lead to more ionisation. How much more depends on details of the pulse shape.
Population that remains behind in the upper state can be measured experimentally by probing with a laser of different colour[8], or (if the state is a high Rydberg state) by field ionisation. In the case of ionisation of xenon with a 100fs laser pulse it was shown that significant population is left behind in most resonant states, suggesting that the onion model does not apply to this case. If upper-state lifetimes are long, the population in those states is subject to

ionisation at a range of different intensities. This would lead to a broadening of the peaks in the photoelectron spectrum if the states do not shift exactly parallel to the ionisation threshold.

In the aforementioned xenon case there is apparently no problem, since the peaks indeed appear at an energy that would be predicted from the unperturbed ionisation potential of the upper state. Recently Gibson et al.[9] have demonstrated a case where the photoelectron peak does occur at a shifted position. Yet the peak is narrow, suggesting that the ionisation potential of the resonant state does not change with intensity (or that the intensity itself does not change during ionisation). No probing of excited-state population was performed in this experiment, so it could be that the lifetime of the excited state was a little shorter than the pulse duration. Under those conditions most of what is excited is also ionised even at the resonance intensity, and most of the excitation would occur at the peak of the pulse, where the intensity is stationary. The lifetime can not have been much shorter, though, since this in itself would lead to lifetime broadening of the resonant state and corresponding photoelectron peak.

Another possibility that should not be overlooked is that Starkshift of Rydberg states often behaves in a peculiar way, as nonperturbative calculations have shown[10]. After some erratic behaviour at lower intensities, during which they can move to a shifted position, they all show a tendency to shift in parallel to the ionisation limit at high intensities. This could lead to the appearence of narrow peaks at shifted positions even in the situation of long upper-state lifetime and correspondingly large populations after the pulse. The 4p-state in atomic hydrogen is an example of this behaviour[11].

To conclude this discussion on the merits of various vegetables for high-intensity laser physics we want to remark that the various possibilities that can occur are well understood, but what will occur in practice depends on atomic parameters that are not very well known (at non-perturbative intensities) and show a high variability between different atoms and even between states of the same atom. Under some conditions the dynamics becomes important, and it is no longer sufficient to describe the system by a stationary rate (as Floquet theory does) or with a simplified pulse shape (like the much beloved square pulse).

TRAPPING POPULATION

Let us return now to the topic of population trapping in resonantly excited intermediate states. This trapping can occur if the ionisation rate from the ground state exceeds that from the excited state. Even in the case that these rates are similar, the fact that the two levels couple to the same continuum will cause an interference between their respective ionisations[12]. This leads to the occurrence of a long and a short-living superposition of these states. In the limit that one of the states lives much longer than the other, the long-living superposition will resemble that state very much, and will live only slightly longer.

Let us, for definiteness, consider the case of MPI of xenon by 600nm light. In this case seven photons are required for ionisation from the ground state, while most excited states can be ionised by a single photon. In the (lowest-order) perturbative description (LOPT) of the MPI, the rate from the ground state would behave as a seventh-order power law. It might seem logical that this steep dependence on intensity might lead to a very high ionisation rate, that will become larger than any one-photon rate provided the intensity is high enough. On deeper inspection, however, this seems unlikely to be possible.

In LOPT the seven-photon process is described as a six-photon excitation to all other states, be it resonantly or non-resonantly, followed by one-photon ionisation from all those states. The population in any of the intermediate states is given by the strength of the excitation (which is sixth order in the intensity) and the detuning of the state, which appears in the denominator. The resulting superposition of intermediate states (sometimes known as the virtual intermediate state) is then ionised by the seventh photon.

From this description it is immediately clear that the seven-photon rate can not exceed the one-photon rate of the most rapidly ionising intermediate state. The point where LOPT predicts the seven-photon rate to grow higher than that is exactly the point where LOPT breaks down: For large-enough intensity the excitation step gets saturated, and the population of the states participating in the virtual intermediate approaches one and can not grow any further.

In order for a ground state to ionise faster than an excited state to which it is resonantly coupled, it is therefore necessary that this excited state is not on the dominant pathway of ionisation from the ground state. The ionisation through off-resonant intermediate states could then outperform that through the resonant states, and possibly the one-photon ionisation from the excited state itself. Of course off-resonant states are much at a disadvantage, because their contribution is inversely weighted with their detuning, but in theory it is possible that the matrix elements for excitation and ionisation more than compensate for this.

The situation sketched above is exactly the one that occurs in the case of seven-photon ionisation of xenon with circularly polarised light. The first experiments that used circular polarisation at 620 nm showed no sign of any resonant structure, but rather a broad nonresonant bump in the electron spectrum[13]. This could be easily explained, since the $\Delta m = +1$ selection rule allows only excited states with m=5, 6 or 7 to contribute to the virtual intermediate state when starting from a p-state. These states have fairly high energies, and by the time the light intensity is high enough to drive a six-photon transition to them with measurable efficiency, the ac-Stark effect has shifted them to a position far above the resonance energy. Thus for lack of resonance candidates the ionisation has to occur nonresonantly.

This explanation is specific to the wavelength used, because by tuning the laser one can always cause a resonance to occur at a suitable intensity. Surprisingly, though, if one tunes for a resonance with the 6h state, no obvious enhancement of the ionisation signal occurs. The electron spectrum just keeps its characteristic nonresonant shape, seen in figure 2a. Thus apparently the ionisation pathway through 6h is insignificant with respect tho those through off-resonant states, despite the more unfavourable detuning.

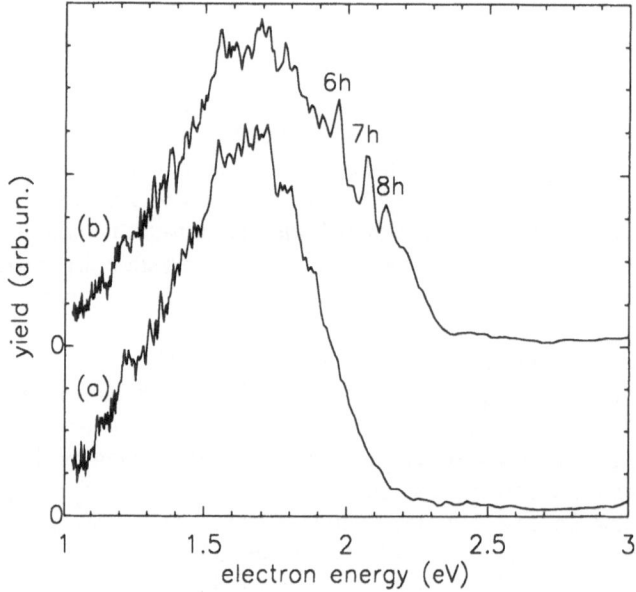

Figure 2. Electron spectra for multiphoton ionisation of xenon with circularly polarised 597 nm light in 100 fs pulses. No structure due to resonances can be seen in trace a. Nevertheless an additional probe beam (532 nm, 5 ns) shows in trace b that population does remain in the resonant Rydberg states.

It is interesting to ponder about whether it is the excitation or the ionisation step that makes the contribution of the 6h state so insignificant. This question can be addressed experimentally by probing the excited-state population after the original ionisation event by means of a second laser[14]. This reveals (figure 2b) that the 6h (as well as 7h and 8h) states do acquire some population, so excitation is definitely possible. Thus the main factor suppressing the importance of these states for REMPI must be the ionisation step from the states.

The physical explanation for the small ionisation rate of the excited states is based on the high angular momentum of the states. In the intermediate states the correspondingly strong centrifugal barrier pushes the wavefunctions out from the nucleus. At such a large distance from the nucleus the electron is nearly free, and ionisation is practically impossible. The same argument, however, applies to all available intermediate states, since the selection rules force the same angular momentum on each of them. Yet there must be off-resonant intermediate states that ionise much more rapidly than the 6h state in order to account for the non-resonant ionisation.

Since all Rydberg h-states suffer from the same problem as the 6h state, only more so, the only remaining candidates are continuum states. Provided their energy is large enough, electrons in these states are capable of penetrating the centrifugal barrier much deeper, which enhances the possibility for further photon absorption. Thus the strongly off-resonant continuum states act as the intermediate states for the seven-photon ionisation from the ground state.

To lend further credence to this hypothesis, we performed an LOPT calculation on a model system, namely six-photon ionisation of hydrogen through a five-photon resonance with the 6h state. At the resonance the cross-section diverges, because the energy denominator containing the detuning of the resonant state vanishes. This divergence, however, only occurs in the limit of infinitely slow switching of the light, which gives the resonance an infinitely long time to build up. In a femtosecond pulse the available time is a long way from infinity, and we take this into account by adding an imaginary part to the photon frequency, corresponding to an exponential rise time of 100 fs. This imaginary part prevents the denominators from vanishing, but at the resonance the contribution of the state now becomes finite and purely imaginary.

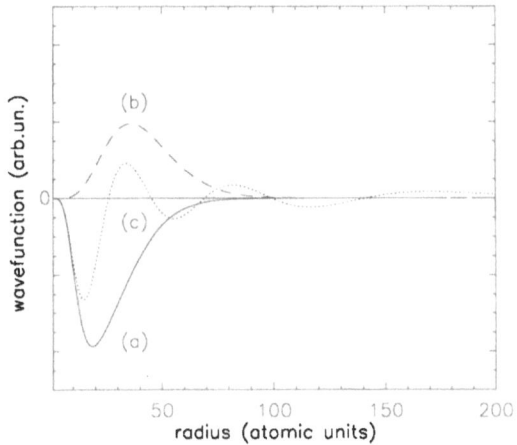

Figure 3. Calculated radial wavefunctions for a hydrogenic model. The solid line shows the virtual intermediate state at the five-photon level. In this state the electron approaches the nucleus much more than for a typical Rydberg state like 6h, which is given by the dashed line. The dotted line shows the contribution of continuum states to trace a, which is approximately 50%. Near the nucleus almost all contribution comes from the continuum states.

Since we were only interested in the contribution of each state, not on the relative phases, we replaced the imaginary energy denominator by an equally large real one. Furthermore we neglected the imaginary part in all the off-resonant energy denominators. The wavefunction of the virtual intermediate state (the fifth-order perturbed wavefunction) then is real, and can be plotted.(figure 3) As can be seen, the maximum of this wavefunction occurs much closer to the origin than that the first maximum of a typical Rydberg h-state. The contribution made by continuum states (obtained by projecting out the Rydberg states) actually represents about 50% of the virtual intermediate state, and this 50% approaches the origin much closer, so that it completely dominates the ionisation.

In conclusion we can say that we have shown that MPI with short pulses can lead to situations were the role of an exactly resonant intermediate state is reduced and becomes completely unimportant with respect to that of strongly off-resonant continuum states. Such a situation allows the non-resonant MPI rate from the ground state to exceed the single-photon ionisation rate from an excited state. Under the right conditions this excited state could then trap population.

ACKNOWLEDGEMENTS

This work is part of the research programme of the Stichting voor Fundamenteel Onderzoek der Materie (Foundation for the Fundamental Research on Matter) and was made possible by the financial support of the Nederlandse Organisatie voor Wetenschappelijk Onderzoek (Netherlands Organisation for the Advancement of Research) and the European Community through Grant No. SCI-0103C.

REFERENCES

[1] R.R. Freeman, P.H. Bucksbaum H. Milchberg, S. Darack, D. Schumacher and M.E. Geusic, Phys. Rev. Lett. **59**, 1092 (1987).

[2] H.G. Muller and H.B. van Linden van den Heuvell, Laser Phys., submitted for publication.

[3] F.H.M. Faisal, this volume.

[4] M. Pont and M. Gavrila, Phys. Rev. Lett. **65**, 2362 (1990).

[5] G. Alber, H. Ritsch and P. Zoller, Phys. Rev. A **34**, 1058 (1986).

[6] R.B. Vrijen, J.H. Hoogenraad, H.G. Muller and L.D. Noordam, Phys. Rev. Lett., submitted for publication.

[7] J.G. Story, this volume; J.G. Story, D.I. Duncan and T.F. Gallagher, Phys. Rev. Lett., submitted for publication.

[8] M.P. de Boer and H.G. Muller, Phys. Rev. Lett. **68**, 2747 (1992).

[9] G.N. Gibson, R.R. Freeman and T.J. McIlrath, Phys. Rev. Lett. **69**, 1904 (1992).

[10] M. Dörr, R.M. Potvliege and R. Shakeshaft, Phys. Rev. A **41**, 558 (1990).

[11] M. Dörr, D. Feldmann, R.M. Potvliege, H. Rottke, R. Shakeshaft, K.H. Welge and B. Wolff-Rottke, J. Phys. B **25**, L275 (1992).

[12] M.V. Fedorov and A.M. Movsesian, J. Opt. Soc. Am. B 6.

[13] P. Agostini, A. Antonetti, P. Breger, M. Crance, A. Migus, H.G. Muller and G. Petite, J. Phys. B 22, 1971 (1989).

[14] M.P. de Boer, L.D. Noordam and H.G. Muller, Phes. Rev. A **47**, 45 (1993).

ATOMIC PHYSICS WITH INTENSE, SUBPICOSECOND

HALF-CYCLE PULSES

R. R. Jones, D. You, and P.H. Bucksbaum

Physics Department
Randall Laboratory
University of Michigan
Ann Arbor, MI   48109-1120

INTRODUCTION

An intense laser system is a necessary prerequisite for experimental investigation of super-intense laser/atom physics. The highest laser intensities are realized by producing a large amount of laser energy in a short pulse. The amount of energy which can be produced by a laser system is limited by its physical size. However, the length of the laser pulse has only one physical constraint: the duration of the pulse cannot be significantly shorter than a single "optical-cycle". This limit has yet to be realized for visible radiation, where the shortest pulses used in intense laser systems are at least tens of cycles long.

We have created, half-cycle infrared pulses using 100 fs laser pulses incident on a large-aperture photoconductive switch.[1,2] These freely propagating electromagnetic pulses have central frequencies around 0.5 THz ($\sim$16cm^{-1}). We have succeeded in producing pulse energies of $1\mu J$ and peak fields in excess of 100 kV/cm in a nearly unipolar 500 fs electric field pulse.[1] The interactions of these pulses with atomic systems differ dramatically from ordinary laser-atom interactions due to the enormous coherent bandwidth in the pulse.

HALF-CYCLE PULSE (HCP) GENERATION TECHNIQUE

The ultra-short laser system used to generate the HCPs is quite involved, and has been described elsewhere.[3] Briefly, the 100 fsec, 770 nm laser pulses originate in a self-mode-locked Ti: Al$_2$O$_3$ oscillator and are amplified in a Chirped Pulse Amplifier (CPA). Pulse energies of 10 mJ are available; however, a laser fluence of 40 μJ/cm^2 incident on the switch is sufficient to saturate the photoconduction process.[1]

The photoconducting switch[2] which is the source of the HCP is a thin (0.5 mm) GaAs semiconductor wafer with a surface area of $\sim$3cm^2. An electric field (F < 10 kV/cm) is applied parallel to the surface of the GaAs. The electric field is then shorted across the semiconductor surface by illuminating one side of the wafer with a 100 fs, 770 nm laser pulse, which drives the GaAs into conduction. A substantial fraction of the radiated energy from the rapidly accelerating electrons in the photoconductor is transmitted through the wafer in the form of a spatially coherent electromagnetic pulse.[4] At least 90% of the energy in the coherent radiation resides in a $\tau_{HCP}\sim500$ fs, single-polarity pulse. This HCP is polarized in the direction of the bias field in the wafer and has a bandwidth of ~1 THz. The peak electric field in the HCP is proportional to the bias field. An electric field autocorrelator[5] allows us to determine the bandwidth of the HCP, and cross correlation techniques are used to infer temporal pulse shapes.[1] A typical interferogram, frequency spectrum, and fitted electric field shape are shown in Fig. 1.[1]

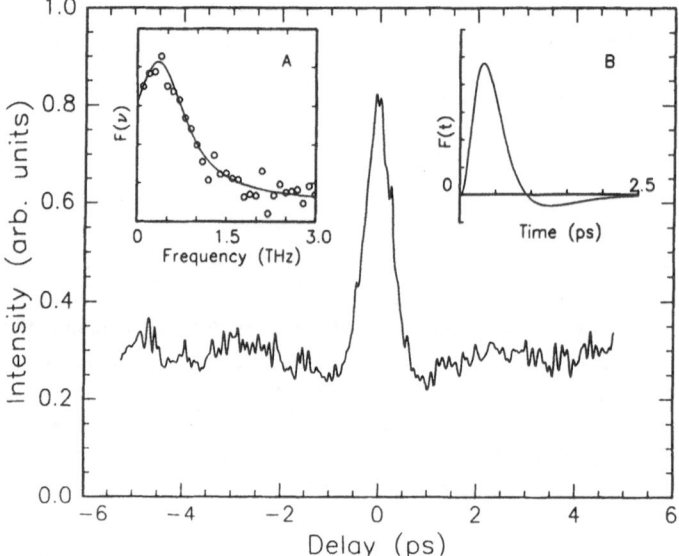

Figure 1. Electric field autocorrelation as a function of delay in one leg of the Michelson interferometer. Inset A: Frequency spectrum of the HCP. Inset B: Inferred temporal pulse shape of the HCP (Reprinted from Ref. 1).

Examination of the inferred electric field shape in Fig. 1 shows that the electric field is nearly unipolar, but does have a long negative tail. Although 90% of the energy in the pulse resides in the large half-cycle component of the pulse, we are confident that the pulse has no DC frequency components because of aperturing of the beam by mirrors and sheilding. Therefore, the time integral of the electric field in the pulse must be zero. However, over the time scales during which the atoms in our system interact with the pulse, the electric field is essentially unidirectional. Nonetheless, we will have to consider the effects of the negative tail of the electric field on the results of any experiment.

RYDBERG ATOM EXPERIMENTS

Although the peak electric fields in the Half-Cycle Pulses (HCPs) are too small to ionize the ground states of atoms, they are large enough to ionize highly excited Rydberg states. Furthermore, the interaction of Rydberg atoms with a HCP can easily demonstrate the effects of broadband coherence in the pulse. Consider the interaction of an initially populated 12s Rydberg state in Na with a HCP as shown in Fig. 2. The nearest dipole coupled level to 12s is 12p which is ~85cm^{-1} away. Inspection of Fig. 1 shows that there is very little energy in the HCP at this frequency. Hence, any transitions or ionization will

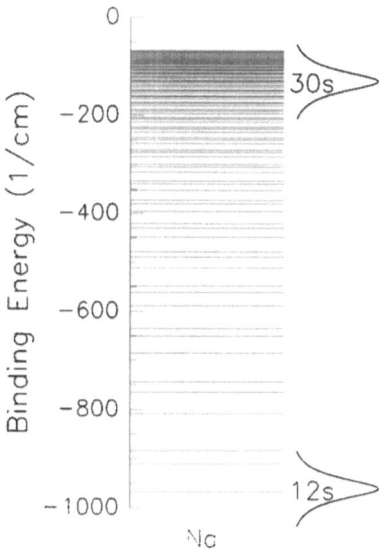

Figure 2. Diagram of part of the Rydberg series in Na showing ns, np, and nd levels for 10 < n < 40. Note that for an initial 12s state that the bandwidth of the HCP significantly overlaps only the 11d state. For an initial 30s state, hundreds of dipole allowed transitions lie within the frequency spectrum of the HCP.

probably occur through higher order multi-photon processes. If we use 30s as the initial state instead, there are many dipole allowed transitions to states with 22 < n < 40, and n < l. Here we expect that population can be coherently redistributed via many resonant single photon transitions. Alternatively, one can choose to analyze the effects that the HCP has on different states in the time domain. The 12s state has a classical motion with a Kepler period (184 fs) which is much shorter than the 500 fs HCP. The Kepler period of the 30s state (3.6 ps) is much longer than the HCP duration. The ratio of laser pulse duration to Kepler period has been found to be important for both excitation to and photoionization of Rydberg states.[6]

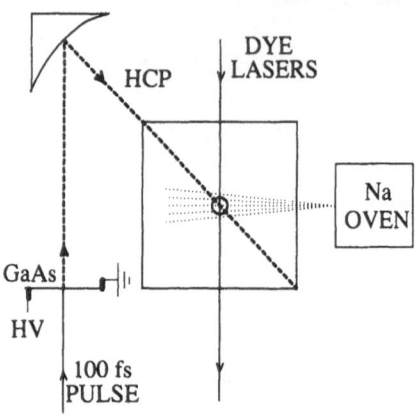

Figure 3. Schematic diagram of the interaction region shown from above. The HCP is produced by applying a 100 fs laser pulse to the GaAs wafer which shorts the high voltage (HV) across the photoconducting surface. The transmitted radiation is weakly focused with a parabolic mirror. The two pulsed dye lasers and the HCP interact with the Na beam between two parallel conducting plates. (Reprinted from Ref. 7)

In the remainder of this report, we will discuss three different experiments involving Na Rydberg states and HCPs. The first is the observation of ionization of Rydberg states under the influence of a HCP. Next we discuss the effect of altering the phase of the HCP on the ionization probability. We conclude with a brief discussion of coherent population transfer among Rydberg states by a HCP.

The experimental setup for all of the experiments is nearly identical.[7] Na atoms are produced in Rydberg states via two photon excitation through the $3p_{1/2}$ resonance line. The frequency of the second photon is tuned to excite a single Rydberg state between n = 12 and 50. The lasers interact with a thermal Na beam between a set of parallel capacitor plates in a vacuum chamber with a base pressure of 5×10^{-7} Torr. The HCP is generated in the vacuum chamber and the orientation of the various beams and components is shown in Fig. 3. The HCP enters the interaction region ~50 ns after the Rydberg states have been excited. The pulse is linearly polarized in the same direction as the dye lasers and is weakly focused by a gold parabolic mirror to a 6 mm waist at the laser/atom interaction region.

Ionization Experiment

In order to study the ionization of Rydberg states due to the HCP we apply a small (~50 V/cm) electric field between the capacitor plates, forcing any ions formed by the HCP through a 2 mm diameter hole in the upper plate towards a micro-channel plate (MCP) detector. The small extraction hole allows us to ignore any spatial variation in the HCP. We then monitor the ion current as a function of peak field in the HCP. Several typical electric field scans are shown in Fig. 4. Note that the ionization thresholds are quite broad - an effect which is inherent in the ionization process as discussed below.

Fig. 5 shows the measured electric fields needed to ionize 10% and 50% of the Rydberg state population. The field needed to ionize 10% of the s or d states follows a clear n^{*-2} scaling, where the effective quantum number n * is defined in terms of the Rydberg state

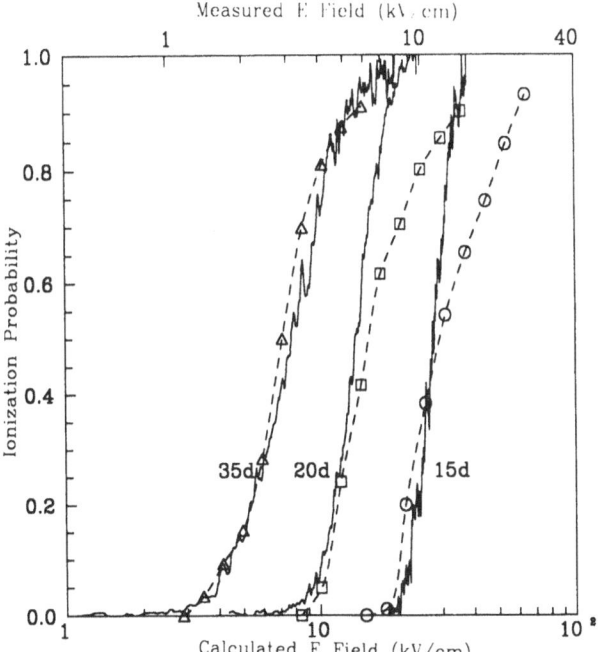

Figure 4. Measured (solid curves) and calculated (dashed curves) ionization probabilities for the 35d, 20d, and 15d states as a function of the peak electric field in the HCP. The electric field scales for the numerical and experimental data differ by a factor of 2.5. Note the broad ionization thresholds which are intrinsic to the ionization process. (Reprinted from Ref. 7)

energy $E = -1/2n^{*2}$ a.u. The field needed to ionize 50% of the atoms scales as $n^{*-3/2}$. Note that ionization does not occur at the classical diabatic field ionization threshold, $F = 1/9n^4$, which is shown as the dashed line in Fig. 4.[8,9] Apparently, ionization is occurring through a process which is distinct from ionization in a slowly varying DC field or tunneling ionization in the presence of an oscillating field.

Discussion of Ionization Experiment

A simple classical model explains the observed energy scaling of the ionization threshold.[7] The energy gained by a bound electron which is exposed to a time varying electric field pulse is,

$$\Delta E = -\int \mathbf{F}(t) \cdot \mathbf{v}(t) dt \qquad (1)$$

where $\mathbf{F}(t)$ the electric field in the pulse, and $\mathbf{v}(t)$ is the velocity of the electron (atomic units are used unless otherwise noted). The minimum field required for ionization is obtained by setting the energy gain equal to the binding energy of the atom and maximizing the integral in Eq. (1). Obviously, the integral is maximized if the velocity of the electron is anti-parallel to the direction of the electric field throughout the HCP. Hence, the threshold field required to ionize an electron in a one-dimensional orbit sets a lower limit on the field required to ionize an electron in three dimensions.

At small distances from the nucleus, the velocity of a Rydberg electron is a maximum and is virtually independent of its total energy ($v = \sqrt{2/x}\,[1 - O(x/n^{*2})]$, $x > 0$). Therefore, the velocity is also independent of energy shifts comparable to the binding energy due to the HCP and Eq. (1) simplifies to

$$\frac{1}{2n^{*2}} = \int F(t)\sqrt{2/x(t)}\,dt \qquad (2)$$

Clearly, the smallest value of the peak field F_0 which satisfies Eq. 2 is obtained if the

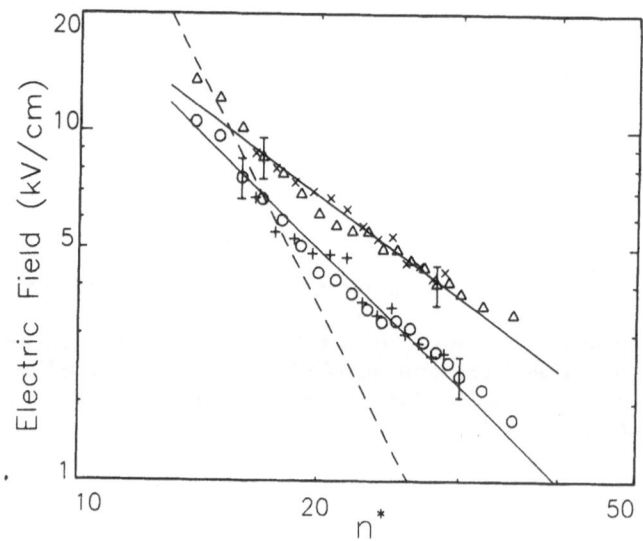

Figure 5. Experimental ionization thresholds as a function of peak field in the HCP. (o -d states, 10% ionization; + -s states, 10% ionization; Δ -d states, 50% ionization; × -s states, 50% ionization). The data clearly show the broadening of the thresholds at higher values of n *. The solid lines drawn through the data have slopes of -2 and -3/2 for 10% and 50% ionization, respectively. The dashed line shows the "long pulse" diabatic field ionization limit $1/9n^{*4}$. (Reprinted from Ref. 7)

electron is located at the nucleus at the beginning or end of the HCP. Using this condition and modeling the HCP as an inverted parabola, Eq. 2 gives

$$F_0 \approx \frac{2}{5n^{*2}\,\tau_{HCP}^{2/3}}. \qquad (3)$$

For a 500 fsec HCP, $F_0 = 2.7 \times 10^6 n^{*-2}$ V/cm which is in good agreement with the experimental 10% ionization result of $F_0 = 2.0 \times 10^6 n^{*-2}$ V/cm.

We can also perform a numerical integration of the full three dimensional classical equations of motion for an electron in a Rydberg orbit which is subjected to a HCP. First, an electron is placed at some position in a Rydberg orbit of given energy and angular momentum. The electron is then exposed to a HCP and the position and velocity of the electron are recorded after the HCP has turned off. The procedure is repeated for a different initial position in an allowed classical orbit. The final state distributions are normalized according to the classical probability for finding the electron at the position which is the starting point of the integration. This normalization accurately reproduces the quantum mechanical expectation values of $< r^2 >$, $< r >$, $< r^{-1} >$, and $< r^{-2} >$. In this way we can evaluate the probability for ionizing a quantum electron using the classical equations of motion. The good agreement between the measured and calculated ionization probabilities is evident in Fig. 4. Note that the experimental and theoretical electric field scales in Fig. 4 differ by a factor of 2.5 so that the agreement is only qualitative. We note that the results of the classical model predict a $\sim 1/9n^4$ ionization threshold for states with $\tau_{Kepler} < \tau_{HCP}$ and therefore, agree with previous long pulse field ionization experiments.

Only those electrons which are near the nucleus during the HCP can ionize at the classical threshold. Hence, the n^{*-2} scaling is only observed for small ionization fractions due to the small probability for finding the electron near the ion core during the peak of the electric field pulse. However, unlike a short laser pulse, the HCP changes the energy of a *free* electron. Therefore, the atomic electron can increase (or decrease) its energy at any distance from the nucleus. The extra energy needed to ionize the electron can be acquired over the entire classically allowed region and not just at small radii. In general, a higher peak field is needed to ionize electron probability which is at a large distance from the nucleus giving an inherent width to the ionization probability. The thresholds become increasingly broader with increasing values of n due to the decreasing probability for finding the electron near the nucleus ($P \sim n^{-3}$). This effect causes the different scaling of the 10% and 50% ionization data shown in Fig. 5.

The theoretical results shown in Fig. 4 are based on a model HCP which has a Gaussian pulse shape. If we choose a pulse shape which is the derivative of a Gaussian (i.e. a full cycle pulse), the ionization probability curves are dramatically different. Fig. 6 shows ionization probabilities for the same three states shown in Fig. 4, using a Gaussian derivative form for the HCP. The most dramatic qualitative difference between Fig. 4 and Fig. 6 appears in the 35d curve. With a Gaussian pulse complete ionization is possible, while only partial ionization occurs for a Gaussian derivative pulse. As discussed above, a true HCP may ionize the electron at any part in its orbit, but a full cycle pulse can only impart energy to the electron when it is near the nucleus. Therefore, since the Kepler period for the 35d state is much longer than the HCP duration, it is very improbable for the electron to be near the nucleus during the HCP. Hence, the ionization probability saturates at a low level.

Phase Dependent Ionization Experiment

The direction in which an electron leaves an atom due to ionization by a long ramped electric field is clearly opposite to the direction of the electric field. However, since the HCP does not ionize Rydberg states in the same way as a static field (at least for states with $\tau_{Kepler} > \tau_{HCP}$), electrons might leave the atom in any direction. An interesting experiment would be to measure the angular distribution of HCP ionized electrons. Unfortunately, the angular distribution of electrons which are ionized by the large half-cycle component of the electric field will be altered by a small negative tail in the electric field profile. Because we expect the ejected electrons to have extremely small energies ($<<$ 1eV), the effects of the negative tail in the HCP could completely dominate the measured electron energy and angular distribution.

Alternatively, we can measure the probability for ionizing a Rydberg atom as a function of which side of the atom the electron resides during the pulse. Of course for an angular momentum eigenstate, the electron has no perferred position about the xy plane. However, in an electric field angular momentum is no longer conserved and the eigenstates of the

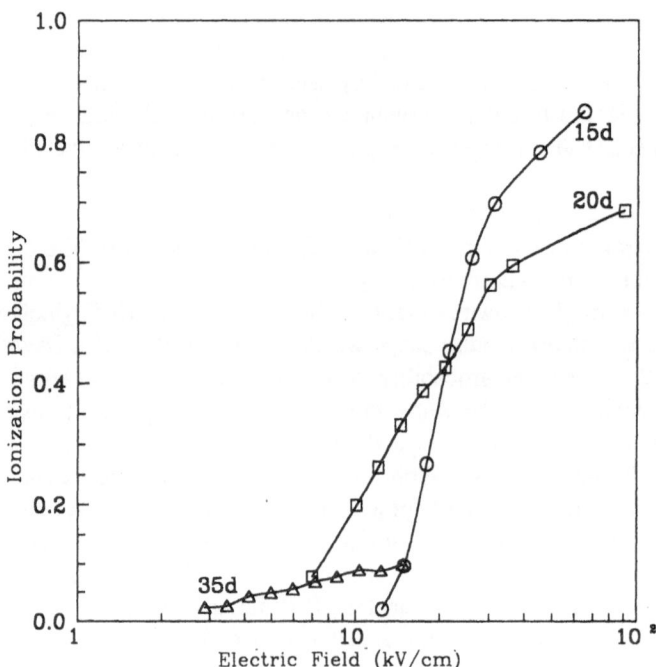

Figure 6. Calculated ionization probabilities for the 35d, 20d, and 15d states as a function of the peak electric field in the HCP. These curves were calculated using a Gaussian derivative form (i.e. full cycle) for the HCP. Note the dramatic difference in the shape of the threshold curves from those shown for the Gaussian pulse shape in Fig. 4.

system are parabolic states which are polarized with respect to the xy plane. Consider an atom in a static electric field in the positive z direction. States whose energies increase with increasing field (blue states) have sizable probability densities only for positive values of z, and the states whose energies decrease (red states) have most of their wave function in regions of space characterized by negative z values. By applying a HCP to "red" or "blue" states, we may be able to determine whether electrons "above" or "below" the nucleus are easier to ionize.

Fig. 7A shows the cross section for excitation of the n = 18 Stark manifold in a 500 V/cm DC electric field (in the negative z direction) as a function of photon energy. The "red" and "blue" sides of the manifold are labelled. Figs. 7B and 7C display the HCP ionization signal for HCP fields in the negative and positive z directions, respectively, with a peak amplitude of 3.8 kV/cm. The assymetry between the two figures is obvious if one normalizes the ionization signals by the excitation cross section in 7A. Clearly, blue states ionize more readily if the HCP field is in the direction of the static field and red states ionize easier for opposing field directions. The observed assymetry is less dramatic if the static field or the HCP field amplitudes are increased.

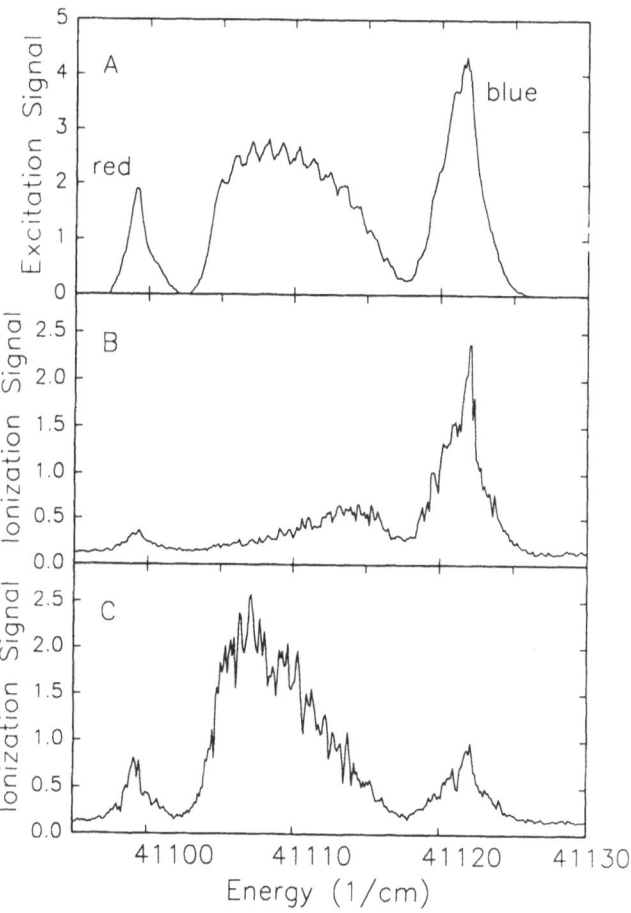

Figure 7. Ionization of the n = 18 Stark states in a 500 V/cm static field for for different HCP field amplitudes and directions. (A) No HCP (Signal proportional to the excitation cross section for the Stark states). (B). 3.8 kV/cm HCP field in the direction of the static field. (C). 3.8 kV/cm HCP field in a direction opposite to the static field.

A modification of our classical analysis of ionization in zero static field reproduces some of our observations in a non-zero DC field. We assume that the electric field changes the Rydberg system in two ways. First, the static field breaks the wave-function symmetry about the xy plane. Second, the effective ionization limit is lowered in the presence of a DC field by an amount $\Delta E = 2\sqrt{F_{DC}}$ due to static field ionization. We can then use the classical simulation to give the probability for ionization as a function of the orientation (i.e. angle between the periatom of the Kepler ellipse and the z axis) of the classical orbit. We assume the electron is ionized if it has an energy greater than the effective ionization limit at the end of the HCP. This simple model produces an assymetry in the ionization probabilities for red and blue state orientations. The calculation does reproduce the experimental observation that the assymetry becomes less dramatic for increasing HCP or static fields. Unfortunately, the magnitude and direction of the assymetry is highly dependent on the temporal HCP shape. Therefore, we are not completely confident that the classical model captures the essential physics.

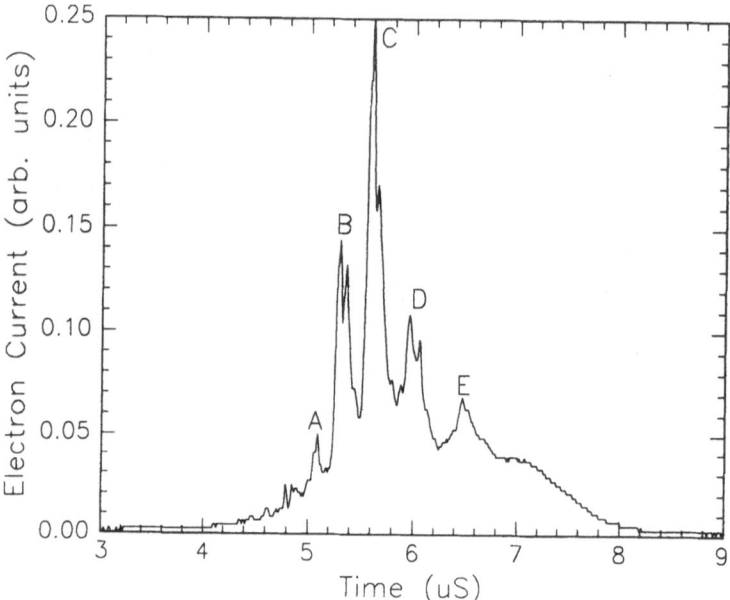

Figure 8. Oscilloscope trace showing the TOF final state distribution after the interaction of a HCP with an initial 20s state. Primarily low angular momentum states are excited at this field amplitude. Specific feature labels: (A) 22s, 21d, and 21p; (B) 21s, 20d, and 20p; (C) 20s and 19p; (D) 19s, 18d, and 18p; (E) 18s, 17d, 17p. Although the s and d states are not resolved, partial resolution between the ns and np states is visible.

Coherent Population Transfer

Up to this point, we have only discussed ionization of atoms using HCPs. Of course, one expects that for HCP fields below the ionization threshold, that population might be redistributed from the initial state to one or more final states. To study this effect, long pulse electric field ionization has been used to analyze the final state population after the interaction of a particular initial state with a HCP. Using this technique each angular momentum state follows an adiabatic (or nearly adiabatic) path to ionization during the 5 μs electric field ramp. Hence, electrons which are ionized out of different final states reach the MCP detector at different times. When a single Rydberg state is excited we observe a single peak in the time-of-flight (TOF) electron spectrum. However, if the HCP transfers population to several different Rydberg states we observe multiple peaks in the electron spectrum.

Fig. 8 shows a typical TOF spectrum which is observed when a 20s initial state is exposed to a HCP. Clearly, a substantial amount of population is transferred from the 20s state (labelled C) to many neighboring levels. The majority of the states populated in Fig. 8 have relatively low angular momentum, l < 3, but for higher fields the population becomes more evenly distributed over all the angular momentum states. Note that the redistribution

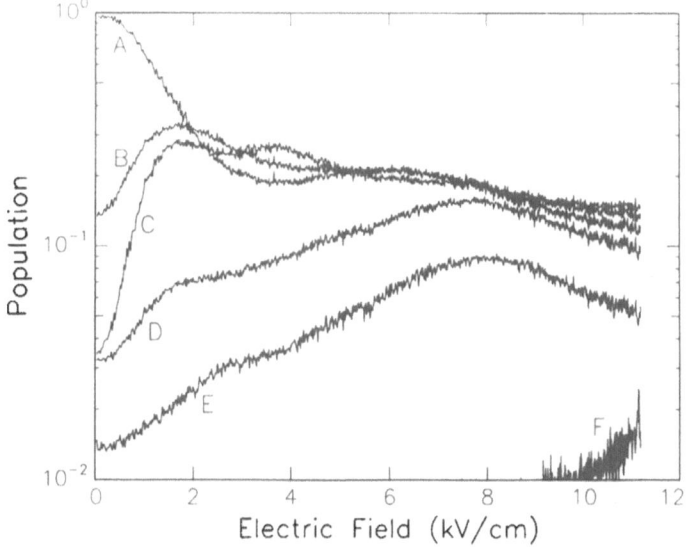

Fig 9. Final state probabilities as a function of HCP field for an initial 16d state. Note that just as ionization begins to occur all states are nearly uniformly populated. Specific feature labels: (A) 16d; (B) 16l, l > 2; (C) 17p; (D) 18s, 18p, 17d (E) 19s and 18d (F) Ionization Signal.

can occur nearly symmetrically to higher and lower energy states. The long pulse field ionization technique is very powerful and can be used to detect population in individual angular momentum states for a given principle quantum number. Fig. 10 is a relatively low resolution spectrum, but the np to ns splitting can still be observed in the large sharp features.

Interestingly, for some initial states resonances are observed for certain HCP field amplitudes. In this case, the initial state is almost completely depleted by the HCP and the majority of population is transferred to a single nearby state. The resonances are generally a very steep function of the electric field. At the onset of ionization, population is generally distributed from very high n states to levels with energies well below that of the initial state.

Fig. 9 shows the probability for finding population in a given state as a function of the HCP field amplitude. In general the population in a given state may show several maxima and minima as the peak electric field is increased. It is clear that just before the onset of ionization, that population is spread nearly uniformly over all available states!

A full quantum analysis is necessary to unravel the complex population redistribution results. Unfortunately, an integration of the full time-dependent Hamiltonian is difficult because the number of states resonantly coupled to the initial state by the HCP is divergent. Single photon transitions are allowed between almost any pair of states with opposite parity. A Floquet description of the atom dressed by the time varying HCP is completely invalid since the bandwidth of the pulse is greater than its central frequency.

CONCLUSION

In conclusion, we have created nearly unipolar, 500 fs electromagnetic pulses with energies of nearly $1\,\mu J$. We have begun to use these pulses to perform several atomic physic experiments on Rydberg atoms. We have observed a new n^{*-2} threshold field scaling in the ionization of Rydberg atoms by the HCPs. The scaling which applies to states whose classical periods are greater than the pulse duration, can be explained in terms of a classical picture. We have also studied the dependence of the HCP phase on ionization of Stark states and have found dramatic assymetries. The classical model also predicts these effects. We have measured the coherent transfer of population from one Rydberg state to another using these pulses. A detailed quantum calculation is formidable, but is a pre-requisite for a complete understanding of our experimental observations.

ACKNOWLEDGEMENTS

This work has been supported by the National Science Foundation.

REFERENCES

1.   D. You, R.R. Jones, D.R. Dykaar, and P.H. Bucksbaum, Opt. Lett. (in press).

2.   G. Morou, C.V. Stancampiano, and D. Blumenthal, Appl. Phys. Lett. 38, 470 (1981). D.H. Auston, K.P. Cheung, and P.R. Smith, Appl. Phys. Lett. 45, 284 (1984). A.P. DeFonzo, M. Jarwala, and C. Lutz, Appl. Phys. Lett. 50, 1155 (1987). Ch. Fattinger and D. Grischkowsky, Appl. Phys. Lett. 53, 1480 (1988). P.R. Smith, D.H. Auston, and M.C. Nuss, IEEE J. Quan. Elec. 24, 255 (1988). B.B. Hu, J.T. Darrow, X.-C. Zhang, and D.H. Auston, Appl. Phys. Lett. 56, 886 (1990). B.I. Greene, D.R. Dykaar, and J.D. Wynn, IEEE J. Quan. Elec., (in press).

3.   J. Squier, F. Salin, G. Mourou, and D. Harter, Opt. Lett., 16, 324 (1991).

4.   A smaller amount of THz radiation is also formed by parametric optical rectification of the incident laser pulse. D.You, R.R. Jones, D.R. Dykaar, and P.H. Bucksbaum (to be published).

5.   B.I. Greene, J.F. Federici, D.R. Dykaar, R.R. Jones, and P.H. Bucksbaum, Appl. Phys. Lett. 59, 893 (1991).

6. H. Gratl, G. Alber, and P. Zoller, J. Phys. B: At. Mol. Opt. Phys. 22, L547 (1989). J. Parker and C.R. Stroud, Jr., Phys. Rev. A 41, 1602 (1990). K. Burnett, P.L. Knight, B.R.M. Piraux, and V.C. Reed, Phys. Rev. Lett. 66, 301 (1991). R.R. Jones and P.H. Bucksbaum, Phys. Rev. Lett. 67, 3215 (1991). L.D. Noordam, H. Stapelfeldt, D.I. Duncan, and T.F. Gallagher, Phys. Rev. Lett. 68, 1496 (1992).

7. R.R. Jones, D. You, and P.H. Bucksbaum, Phys. Rev. Lett. (in press).

8. J.A. Bayfield and P.M. Koch, Phys. Rev. Lett., 33, 258 (1974).

9. R.F. Stebbings, C.J. Latimer, W.P. West, F.B. Dunning, and T.B. Cook, Phys. Rev. A, 12, 1453 (1975). T.F. Gallagher, L.M. Humphrey, R.M. Hill, and S.A. Edelstein, Phys. Rev. Lett., 37, 1465 (1976). M.G. Littman, M.M. Kash, and D. Kleppner, Phys. Rev. Lett., 41, 103 (1978). T.H. Jeys, G.W. Foltz, K.A. Smith, E.J. Beiting, F.G. Kellert, F.B. Dunning, and R.F. Stebbings, Phys. Rev. Lett., 44, 390 (1980).

R-MATRIX-FLOQUET THEORY OF
MULTIPHOTON PROCESSES

C.J. Joachain and M. Dörr

Physique Théorique
Université Libre de Bruxelles, C.P. 227
B-1050 Bruxelles
Belgium

1. INTRODUCTION

The study of atomic systems interacting with intense laser fields has attracted considerable interest in recent years[1]. In particular, the availability of increasingly powerful lasers has made possible the observation of multiphoton phenomena such as multiphoton ionization, harmonic generation and laser-assisted electron-atom collisions. Most of the early theoretical treatments of multiphoton processes relied on the use of perturbation theory. However, for higher laser intensities, non perturbative approaches are required.

In this article we describe a new method - the R-matrix-Floquet theory - which has been proposed[2,3] and developed[4] recently to analyze multiphoton processes in strong monochromatic, spatially homogeneous laser fields. This fully non perturbative theory treats multiphoton ionization, harmonic generation and laser-assisted electron-atom collisions in an unified way. It is completely ab initio and is applicable to an arbitrary atom. It also takes advantage of the natural R-matrix division of configuration space into an internal region and an external region[5,6], which implies that in each region the most appropriate form of the laser-atom interaction can be selected by performing unitary transformations on the wave function. In addition, since in the internal region the spectrum is entirely discrete, the standard Hermitian Floquet theory can be used. Moreover, use can be made of existing R-matrix computer codes[7] which have been developed to analyze field-free electron-atom collisions, one-photon ionization and one photon free-free transitions.

The organization of this paper is as follows. Section 2 is devoted to a general discussion of the R-matrix-Floquet theory. After introducing the basic equations and subdividing configuration space into an internal and an external region, we discuss the solutions of the time-dependent Schrödinger equation in both regions, using the Floquet method to reduce the problem to a time-independent one. By matching the internal and external solutions on the boundary, and using appropriate boundary conditions at the origin and at infinity, the theoretical quantities (ionization rates, branching ratios, cross sections, ...) can be calculated and compared with the experimental data. Finally, in Section 3 we illustrate the method by considering multiphoton ionization of atomic hydrogen and laser-assisted electron-proton collisions. Atomic units will be used unless otherwise stated.

2. THEORY

2.1 Basic Concepts

Let us consider an atomic system composed of a nucleus of atomic number Z and $N + 1$ electrons, in a laser field. We shall deal with the following three processes : multiphoton single ionization of atoms and ions, harmonic generation in laser-atom (ion) interactions and laser-assisted electron-atom (ion) elastic and inelastic collisions.

We assume that the laser field is treated classically as a spatially homogeneous, linearly polarized, monochromatic and single mode electric field

$$\mathcal{E}(t) = \hat{\epsilon} \mathcal{E}_0 \cos \omega t \tag{1}$$

where $\hat{\epsilon}$ is a unit vector along the polarization direction, $\mathcal{E}_0$ is the electric field strength and ω is the angular frequency; the corresponding vector potential is $\mathbf{A}(t) = \hat{\epsilon} A_0 \sin \omega t$, with $A_0 = -c\mathcal{E}_0/\omega$.

In the non-relativistic limit, the atomic system in the presence of this laser field is described by the time-dependent Schrödinger equation

$$i\frac{\partial}{\partial t}\Psi(\mathbf{X}, t) = \left[H_{at} + \frac{1}{c}\mathbf{A}(t) \cdot \mathbf{P} + \frac{N+1}{2c^2}A^2(t) \right] \Psi(\mathbf{X}, t) \tag{2}$$

where H_{at} is the non-relativistic Hamiltonian of the $(N + 1)$-electron atomic system in the absence of the laser field, $\mathbf{X} \equiv (\mathbf{x}_1, \mathbf{x}_2, \ldots \mathbf{x}_{N+1})$, where $\mathbf{x}_i \equiv (\mathbf{r}_i, \sigma_i)$ are the space and spin coordinates of the ith electron, and $\mathbf{P} = \sum_{i=1}^{N+1} \mathbf{p}_i$ denotes the total momentum operator.

According to the R-matrix method[5,6], we subdivide configuration space into two regions (see Fig. 1). The internal region is defined by the condition that the radial coordinates r_i of all $N + 1$ electrons are such that $r_i \leq a$ $(i = 1, 2 \ldots N + 1)$ where the sphere of radius a envelops the charge distribution of the target atom states retained

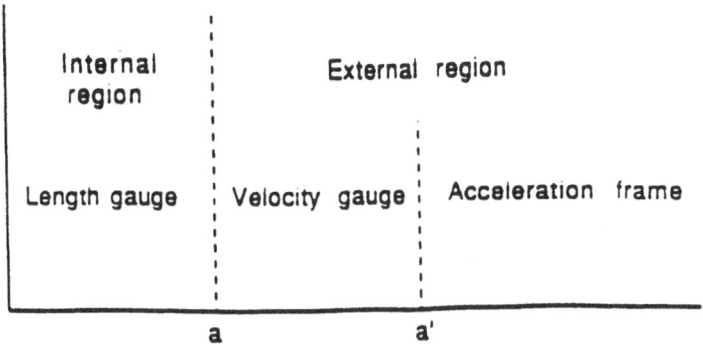

Figure 1 . Partitioning of configuration space in the R-matrix-Floquet theory.

in the calculation. Hence in this region exchange effects involving all $N + 1$ electrons are important. The external region is defined so that one of the $N + 1$ electrons lies on or outside the sphere of radius a while the remaining N "target" electrons are confined within this sphere. Thus in this region exchange effects between this one electron and the remaining N "target" electrons are negligible. The Schrödinger equation (2) is solved in these two regions separately, and the solutions are then connected on the boundary at $r = a$. In the internal region it is convenient to use the length gauge since in this gauge the laser-atom coupling tends to zero at the origin. In the external region it is advantageous to use the velocity gauge out to a radius a' which may extend to infinity. In certain cases it is useful to transform at $r = a'$ to the acceleration frame[8,9] which enables simple asymptotic boundary conditions to be defined.

2.2 The Internal Region Solution

In the internal region, the Schrödinger equation (2) is transformed to the length (L) gauge defined by the unitary transformation

$$\Psi(\mathbf{X}, t) = \exp\left[-\frac{i}{c}\mathbf{A}(t) \cdot \mathbf{R}\right] \Psi_L(\mathbf{X}, t) , \qquad (3)$$

where $\mathbf{R} = \sum_{i=1}^{N+1} \mathbf{r}_i$. The wave function $\Psi_L(\mathbf{X}, t)$ satisfies the Schrödinger equation

$$i\frac{\partial}{\partial t}\Psi_L(\mathbf{X}, t) = [H_{at} + \mathcal{E}(t) \cdot \mathbf{R}]\Psi_L(\mathbf{X}, t) . \qquad (4)$$

To solve this equation, we introduce the Floquet-Fourier expansion

$$\Psi_L(\mathbf{X}, t) = e^{-iEt} \sum_{n=-\infty}^{+\infty} e^{-in\omega t}\psi_n^L(\mathbf{X}) \qquad (5)$$

303

where E is the pseudo-energy. The harmonic components ψ_n^L satisfy the infinite set of coupled, time-independent equations

$$(H_{at} - E - n\omega)\psi_n^L + D(\psi_{n-1}^L + \psi_{n+1}^L) = 0 \tag{6}$$

where $D = \mathcal{E}_0 \hat{\epsilon}.\mathbf{R}/2$ is the dipole coupling operator. The functions ψ_n^L can be regarded as the components of a vector $|\psi^L\rangle$ in photon space. We then obtain the eigenvalue equation

$$(\mathbf{H}_F - E)\,|\psi^L\rangle = 0 \tag{7}$$

where the Floquet Hamiltonian is a block-tridiagonal infinite matrix of operators given by

$$\mathbf{H}_F = \begin{pmatrix} \ddots & & & & \\ D & H_{at} - (n-1)\omega & D & & \\ & D & H_{at} - n\omega & D & \\ & & D & H_{at} - (n+1)\omega & D \\ & & & & \ddots \end{pmatrix}. \tag{8}$$

In order to solve eq. (7) in the internal region we first remark that the Hamiltonian $\mathbf{H}_F$ is not Hermitian in this region due to surface terms at $r = a$ arising from the kinetic energy operator in H_{at}. These surface terms can be eliminated by introducing a Bloch operator[10] $\mathbf{L}_b$ so that $\mathbf{H}_F + \mathbf{L}_b$ is Hermitian in the internal region. We then rewrite eq. (7) in the formally equivalent way

$$(\mathbf{H}_F + \mathbf{L}_b - E)\,|\psi^L\rangle = \mathbf{L}_b\,|\psi^L\rangle \ . \tag{9}$$

The Bloch operator $\mathbf{L}_b$ is diagonal in photon space and has the components

$$L_b = \frac{1}{2}\sum_{\Gamma_{ij}} |\bar{\phi}_i^\Gamma(r_j^{-1})\rangle \delta(r_j - a)\left(\frac{d}{dr_j} - \frac{b-1}{r_j}\right)\langle \bar{\phi}_i^\Gamma(r_j^{-1})| \ , \tag{10}$$

where b is an arbitrary constant and the channel functions

$$\bar{\phi}_i^\Gamma(r_j^{-1}) \equiv \bar{\phi}_i^\Gamma(\mathbf{x}_1,\ldots,\mathbf{x}_{j-1},\mathbf{x}_{j+1},\ldots,\mathbf{x}_{N+1},\hat{\mathbf{r}}_j,\sigma_j) \tag{11}$$

are formed by coupling the atomic target states $\phi_i(\mathbf{x}_1,\ldots,\mathbf{x}_{j-1},\mathbf{x}_{j+1},\ldots,\mathbf{x}_{N+1})$ and possibly pseudostates $\varphi_i(\mathbf{x}_1,\ldots,\mathbf{x}_{j-1},\mathbf{x}_{j+1},\ldots,\mathbf{x}_{N+1})$ retained in the calculation with the spin-angle functions of the scattered or ejected electron (j) to give a state whose quantum numbers are collectively denoted by Γ. The equation (9) can then be solved as follows. In analogy with the R-matrix method in the absence of an external field[6], we introduce the functions

$$\psi_{kn}^L(\mathbf{X}) = \mathcal{A}\sum_{\Gamma i\ell} \bar{\phi}_i^\Gamma(r_j^{-1})r_j^{-1}u_\ell^\Gamma(r_j)a_{i\ell kn}^\Gamma + \sum_{\Gamma i}\chi_i^\Gamma(\mathbf{X})b_{ikn}^\Gamma \ , \tag{12}$$

where $\mathcal{A}$ is the antisymmetrization operator, the u_ℓ^Γ are radial basis functions which are non-vanishing on the boundary of the internal region, the χ_i^Γ are quadratically integrable antisymmetric functions which vanish by the boundary of the internal region, and $a_{i\ell kn}^\Gamma$ as well as b_{ikn}^Γ are coefficients obtained by diagonalizing $\mathbf{H}_F + \mathbf{L}_b$ in the internal region.

Let $|\psi_k^L\rangle$ be the vector in photon space whose components are ψ_{kn}^L. We have

$$\langle \psi_{k'}^L | \mathbf{H}_F + \mathbf{L}_b | \psi_k^L \rangle = E_k \delta_{k'k} \tag{13}$$

so that, using the spectral representation of the operator $(\mathbf{H}_F + \mathbf{L}_b - E)$, we can rewrite eq. (9) in the form

$$|\psi^L\rangle = \sum_k |\psi_k^L\rangle \frac{1}{E_k - E} \langle \psi_k^L | \mathbf{L}_b | \psi^L \rangle \ . \tag{14}$$

Projecting this equation onto the channel functions $\bar\phi_i^\Gamma(r_{N+1}^{-1})$ and onto the nth component in photon space and evaluating on the boundary $r_{N+1} = a$ yields

$$F_{in}^\Gamma(a) = \sum_{\Gamma' i' n'} {}^L R_{ini'n'}^{\Gamma\Gamma'}(E) \left(r \frac{dF_{i'n'}^{\Gamma'}}{dr} - b F_{i'n'}^{\Gamma'} \right)_{r=a} , \tag{15}$$

where we have introduced the R-matrix calculated in the length gauge

$$^L R_{ini'n'}^{\Gamma\Gamma'}(E) = \frac{1}{2a} \sum_k \frac{{}^L w_{ink}^\Gamma \, {}^L w_{i'n'k}^{\Gamma'}}{E_k - E} , \tag{16}$$

the reduced radial functions

$$F_{in}^\Gamma(r_{N+1}) = r_{N+1} \langle \bar\phi_i^\Gamma(r_{N+1}^{-1}) | \psi_n^L \rangle \tag{17}$$

and the surface amplitudes

$$^L w_{ikn}^\Gamma = a \langle \bar\phi_i^\Gamma(r_{N+1}^{-1}) | \psi_{kn}^L \rangle |_{r_{N+1}=a} \tag{18}$$

Equations (15) and (16) are the basic equations describing the solution of eq. (2) in the internal region. Having obtained the R-matrix from eq. (16), we can use eq. (15) to calculate the logarithmic derivative of the reduced radial functions on the boundary $r = a$, which provide the initial conditions for solving eq. (2) in the external region.

2.3 The External Region Solution

We now consider the solution of eq. (2) in the external region. In this region, we have only one electron ($r_{N+1} \geq a$) which we describe by using the velocity gauge, while the remaining electrons ($r_i \leq a$, $i = 1, 2, \ldots N$) are still treated by using the length gauge. Thus, starting from eq. (2), we perform the unitary transformation

$$\Psi(\mathbf{X},t) = \exp\left[-\frac{i}{c}\mathbf{A}(t)\cdot\mathbf{R}' - \frac{i}{2c^2}\int^t A^2(t')dt'\right]\Psi_V(\mathbf{X},t) \qquad (19)$$

where $\mathbf{R}' = \sum_{i=1}^{N}\mathbf{r}_i$. The corresponding Schrödinger equation for $\Psi_V(\mathbf{X},t)$ is

$$i\frac{\partial}{\partial t}\Psi_V(\mathbf{X},t) = [H_V + \boldsymbol{\mathcal{E}}(t)\cdot\mathbf{R}']\Psi_V(\mathbf{X},t) \qquad (20)$$

where

$$H_V = H_{at} + \frac{1}{c}\mathbf{A}(t)\cdot\mathbf{p}_{N+1} \qquad (21)$$

It is sometimes convenient to work beyond some large radius $a' \gg a$ in the Kramers (acceleration) frame[8,9]. We then perform a second unitary transformation

$$\Psi_V(\mathbf{X},t) = \exp\left[-i\boldsymbol{\alpha}(t).\mathbf{p}_{N+1}\right]\Psi_A(\mathbf{X},t) \qquad (22)$$

where

$$\boldsymbol{\alpha}(t) = \frac{1}{c}\int^t \mathbf{A}(t')dt' = \hat{\boldsymbol{\epsilon}}\alpha_0\cos\omega t \qquad (23)$$

and $\alpha_0 = \mathcal{E}_0/\omega^2$. We thus obtain for $\Psi_A(\mathbf{X},t)$ the Schrödinger equation

$$i\frac{\partial}{\partial t}\Psi_A(\mathbf{X},t) = [H_A + \boldsymbol{\mathcal{E}}(t).\mathbf{R}']\Psi_A(\mathbf{X},t) . \qquad (24)$$

where

$$H_A = H'_{at} - \frac{1}{2}\nabla^2_{N+1} - \frac{Z}{|\mathbf{r}_{N+1}+\boldsymbol{\alpha}|} + \sum_{i=1}^{N}\frac{1}{|\mathbf{r}_{N+1}+\boldsymbol{\alpha}-\mathbf{r}_i|} , \qquad (25)$$

H'_{at} being the field-free Hamiltonian describing the nucleus and the N inner electrons.

In order to solve the Schrödinger equation in the velocity gauge (V) or in the Kramers acceleration frame (A), we make a Floquet-Fourier expansion of the wave function :

$$\Psi_R(\mathbf{X},t) = e^{-iE_Rt}\sum_{n=-\infty}^{+\infty}e^{-in\omega t}\psi_n^R(\mathbf{X}), \ R \equiv V \text{ or } A . \qquad (26)$$

The harmonic components ψ_n^R satisfy the infinite set of coupled, time-independent equations

$$(H_R - E_R - n\omega)\psi_n^R + D'(\psi_{n-1}^R + \psi_{n+1}^R) = 0 \tag{27}$$

where $D' = \mathcal{E}_0\hat{\epsilon}.\mathbf{R}'/2$. The functions ψ_n^R can be considered as the components of a vector $|\psi^R\rangle$ in photon space, and eqs. (27) can be recast in the form

$$(\mathbf{H}_F^R - E_R)|\psi^R\rangle = 0 , \quad R \equiv V \text{ or } A . \tag{28}$$

To solve eq. (28), we introduce the close coupling expansion for the harmonic components ψ_n^R :

$$\psi_n^R(\mathbf{X}) = \sum_{\Gamma i} \bar{\phi}_i^\Gamma(r_{N+1}^{-1}) r_{N+1}^{-1} {}^R G_{in}^\Gamma(r_{N+1}) .$$

$$r_{N+1} \geq a \text{ if } R = V , \quad r_{N+1} \geq a' \text{ if } R = A. \tag{29}$$

It is worth noting that expansion (29), unlike expansion (12), is not antisymmetrized. This is due to the fact that the $(N + 1)$th electron now lies outside the sphere of radius a, so that exchange effects between this electron and the remaining N electrons (confined within the sphere) are negligible. Substituting the expansion (29) into eq. (28), projecting onto the channel functions $\bar{\phi}_i^\Gamma(r_{N+1}^{-1})$ and considering the nth component in photon space yields for the reduced radial functions ${}^R G_{in}^\Gamma$ a system of coupled differential equations describing the motion of the $(N + 1)$th electron in the external region:

$$\left(\frac{d^2}{dr^2} - \frac{\ell_i(\ell_i + 1)}{r^2} + \frac{2z}{r} + k_{in}^2\right) {}^R G_{in}^\Gamma(r) = 2 \sum_{\Gamma' i' n'} {}^R W_{ini'n'}^{\Gamma\Gamma'}(r) {}^R G_{i'n'}^{\Gamma'}(r) \tag{30}$$

with $r \geq a$ when $R = V$, $r \geq a'$ when $R = A$ and we have written $r \equiv r_{N+1}$ to simplify the notation. In eq. (30) ℓ_i is the orbital angular momentum of the $(N + 1)$th electron in the ith channel, $z = Z - N$, $k_{in}^2 = 2(E_R - w_i + n\omega)$, $\langle\phi_i|H'_{at}|\phi_j\rangle = w_i\delta_{ij}$ and ${}^R W_{ini'n'}^{\Gamma\Gamma'}(r)$ is the long range potential coupling the channels.

The coupled equations (30) must be solved subject to boundary conditions at $r = a$, $r = a'$ and $r \to \infty$. At $r = a$, the external reduced radial functions ${}^V G_{in}^\Gamma$ must be matched to the internal reduced radial functions F_{in}^Γ, while at $r = a'$ the matching of the external reduced radial functions ${}^V G_{in}^\Gamma$ and ${}^A G_{in}^\Gamma$, obtained respectively in the velocity gauge and in the acceleration frame, must be performed. Finally, for $r \to \infty$, the external solution must obey asymptotic Siegert boundary conditions for the case of multiphoton ionization and harmonic generation or asymptotic S-matrix (or T- or K-matrix) boundary conditions for the case of laser-assisted electron-atom collisions.

The solution of the coupled equations (30) in the external region has been obtained recently by Dörr et al.[4] using two new computational methods. The first one generalizes the Light and Walker[11] R-matrix propagator approach to include the first derivative terms which arise in the velocity gauge. In this gauge, the set of coupled equations (30) can be written in matrix form as

$$\left[\frac{d^2}{dr^2} + \underline{\underline{P}}\frac{d}{dr} - \underline{\underline{V}}(r) + 2E_V\right] {}^V\underline{\underline{G}}(r) = 0 \tag{31}$$

where $\underline{\underline{P}}$ and $\underline{\underline{V}}$ are square matrices and ${}^V\underline{\underline{G}}$ is a matrix whose columns correspond to independent solution vectors. Now, $\underline{\underline{P}}$ is independent of r and anti-Hermitian, with purely imaginary eigenvalues. We can therefore write

$$\underline{\underline{P}} = 2i\ \underline{\underline{T}}\ \underline{\underline{D}}\ \underline{\underline{T}}^\dagger \tag{32}$$

where $\underline{\underline{D}}$ is a real diagonal matrix and $\underline{\underline{T}}$ is a unitary transformation matrix. Using the operator identity

$$\exp(\underline{\underline{P}}\ r/2)\left[\frac{d^2}{dr^2} + \underline{\underline{P}}\frac{d}{dr}\right] = \left[\frac{d^2}{dr^2} - \frac{1}{4}\underline{\underline{P}}^2\right]\exp(\underline{\underline{P}}\ r/2) \tag{33}$$

where the matrix

$$\exp(\underline{\underline{P}}\ r/2) = \underline{\underline{T}}\exp(i\ \underline{\underline{D}}\ r)\underline{\underline{T}}^\dagger \tag{34}$$

is real and orthogonal, we can remove the d/dr term in equations (31) and rewrite these equations in the standard form

$$\left[\frac{d^2}{dr^2} - \underline{\underline{V}}_{eff}(r) + 2E_V\right]\exp(\underline{\underline{P}}\ r/2){}^V\underline{\underline{G}}(r) = 0 \tag{35}$$

where the effective potential

$$\underline{\underline{V}}_{eff}(r) = \frac{1}{4}\underline{\underline{P}}^2 + \exp(\underline{\underline{P}}\ r/2)\underline{\underline{V}}(r)\exp(-\underline{\underline{P}}\ r/2) \tag{36}$$

is Hermitian. The original approach of Light and Walker[11] can now be applied to eq. (35). For propagation from $r = a$ to $r = \check{a}$ the interval $[a, \check{a}]$ is subdivided into several sub-intervals in which $\underline{\underline{V}}_{eff}$ is nearly constant, so that the solution is readily found by using the Green's function approach. We remark that the R-matrix corresponding to the solution $\tilde{\underline{\underline{G}}}(r) = \exp(\underline{\underline{P}}\ r/2){}^V\underline{\underline{G}}(r)$ of eq. (35) is given at $r = a$ by

$$\underline{\underline{R}}(a) = \underline{\underline{\tilde{G}}}(a)\left[a\frac{d\underline{\underline{\tilde{G}}}}{dr}(r=a)\right]^{-1} \tag{37}$$

where we have chosen the constant $b = 0$. We can readily express $\underline{\underline{R}}(a)$ in terms of $^{V}\underline{\underline{G}}(r)$ and its derivative. Using the fact that

$$\frac{d\underline{\underline{\tilde{G}}}}{dr} = \exp(\underline{\underline{P}}\,r/2)\left[\frac{1}{2}\underline{\underline{P}}\,{}^{V}\underline{\underline{G}} + \frac{d^{V}\underline{\underline{G}}}{dr}\right] \tag{38}$$

we find that

$$\underline{\underline{R}}(a) = \exp(\underline{\underline{P}}\,a/2)\left[a\frac{d^{V}\underline{\underline{G}}}{dr}(r=a)\,{}^{V}\underline{\underline{G}}^{-1}(a) + \frac{a}{2}\underline{\underline{P}}\right]^{-1}\exp(-\underline{\underline{P}}\,a/2) \tag{39}$$

and we note that this modified R-matrix will stay symmetric under propagation.

The second method extends the Burke and Schey[12] asymptotic expansion to calculate the solution at $r = \tilde{a}$ satisfying given boundary conditions for $r \to \infty$. These boundary conditions are conveniently expressed in the acceleration frame where the equations are asymptotically uncoupled. In particular, using the asymptotic form of the frame transformation, the asymptotic solutions in the velocity gauge are such that

$$^{V}\underline{\underline{G}}(r) \underset{r\to\infty}{\longrightarrow} \underline{\underline{A}}^{(asy)}\underline{\underline{\Phi}}(r) \tag{40}$$

where $\underline{\underline{\Phi}}(r)$ is the diagonal matrix of asymptotic solutions in the acceleration frame, and $\underline{\underline{A}}^{(asy)}$ is a square unitary matrix. We therefore write for $^{V}\underline{\underline{G}}(r)$ an asymptotic expansion of the form

$$^{V}\underline{\underline{G}}(r) = \sum_{n=0}^{\infty} r^{-n}\underline{\underline{A}}^{(n)}\underline{\underline{\Phi}}(r) \tag{41}$$

where, for multiphoton ionization

$$\underline{\underline{\Phi}}(r) = (\underline{\underline{q}})^{-1/2}\exp[i\underline{\underline{q}}r + i\underline{\underline{\zeta}}\,\underline{\underline{q}}^{-1}\ln(2\underline{\underline{q}}r) - i\underline{\underline{\ell}}\pi/2 + i\underline{\underline{\sigma}}]\,. \tag{42}$$

The diagonal matrices $\underline{\underline{\ell}}$ and $\underline{\underline{\sigma}}$ correspond to the channel orbital angular momenta and Coulomb phases of the outer electron. The diagonal matrices $\underline{\underline{q}}$ and $\underline{\underline{\zeta}}$ and the coefficients $\underline{\underline{A}}^{(n)}$ are determined by substituting the asymptotic expansion (41) into the coupled equations (31), using the multipole expansion of $\underline{\underline{V}}(r)$ and equating the coefficients of terms with the same power of r. We remark that in an exact treatment, where an infinite Floquet basis would be retained, the elements q_j of the matrix $\underline{\underline{q}}$ would correspond to the channel momenta and the elements ζ_j of the matrix $\underline{\underline{\zeta}}$ would reduce to the residual charge $z = Z - N$.

The two methods described above for solving the coupled equations (30) in the external region, combined with the solution in the internal region, enable the complete solution of the R-matrix-Floquet equations to be obtained.

3. RESULTS

3.1 Multiphoton ionization of H(1s) and H(2s)

We first consider the multiphoton ionization of H(1s) in the linearly polarized monochromatic laser field of eq. (1). In Table 1 we display the total ionization rates Γ and dynamic Stark shifts Δ for two angular frequencies ($\omega = 0.184$ a.u., corresponding to a Kr F laser, and $\omega = 0.65$ a.u.) and several values of the intensity, as obtained by Dörr et al.[4].

In Fig. 2 we show the total ionization rate, the dynamic Stark shift and the branching ratio into the lowest photon channel ($N = 3$), for multiphoton ionization of H(1s), at the angular frequency $\omega = 0.184$ a.u. ($\lambda = 248$ nm). The $N = 3$ branching ratio is defined as the probability for the electron to emerge with three photons absorbed

Table 1 Total ionization rates and dynamic Stark shifts for multiphoton ionization of H(1s) in a linearly polarized, monochromatic laser field. The threshold energy E_t has been kept fixed at the value $E_t = 0$.

Ang.freq. ω (a.u.)	Intensity $I(\text{W cm}^{-2})$	Ioniz. rate Γ (a.u.)	Stark Shift Δ (a.u.)
0.184	1.0×10^{13}	8.8×10^{-6}	-2.54×10^{-3}
	1.0×10^{14}	1.4×10^{-3}	-2.57×10^{-2}
	7.2×10^{14}	1.7×10^{-2}	-1.92×10^{-1}
0.65	1.0×10^{14}	2.6×10^{-3}	$+3.60 \times 10^{-4}$
	2.0×10^{16}	1.4×10^{-1}	$+1.95 \times 10^{-1}$

(the minimum number of photons) divided by the total ionization probability. It is seen that the R-matrix-Floquet results for the total ionization rate Γ differ significantly from the values obtained from lowest order perturbation theory. On the other hand, the R-matrix-Floquet calculations agree very well with the Sturmian-Floquet results of Potvliege and Shakeshaft[13]. In Fig. 3, we display the total ionization rate and the dynamic Stark shift for ionization of H(2s) at the same angular frequency $\omega = 0.184$ a.u.

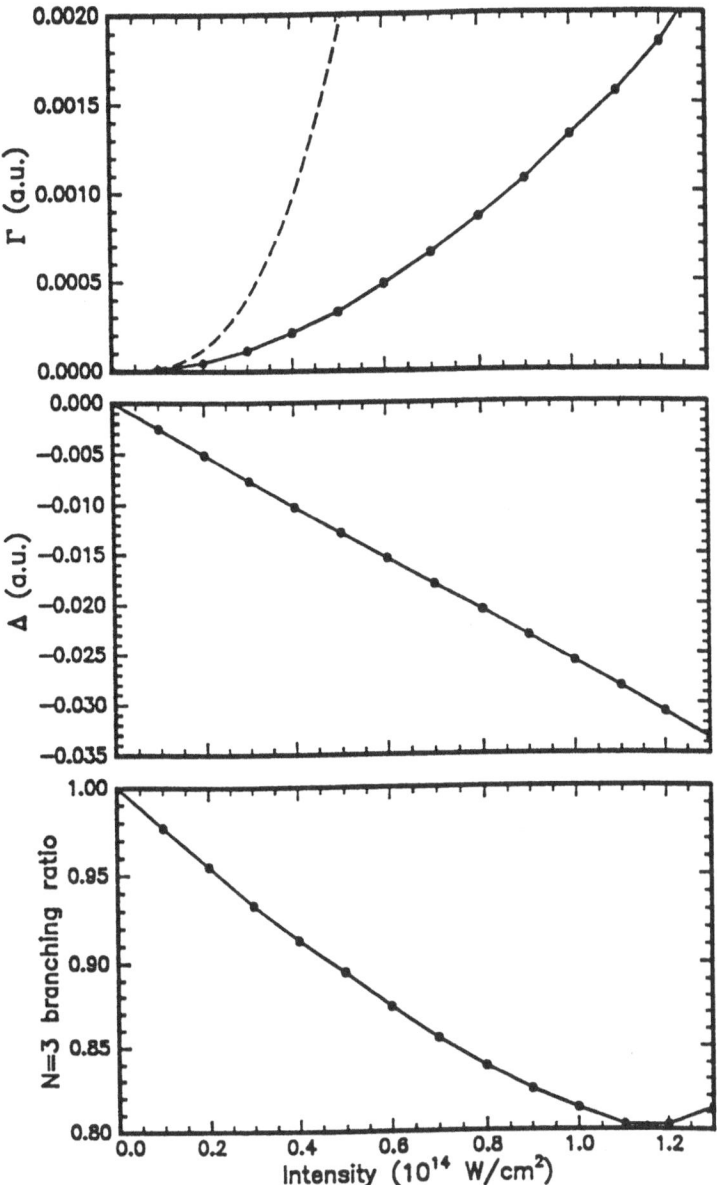

Figure 2 . The total ionization rate Γ, dynamic Stark shift Δ and branching ratio into the $N = 3$ photon channel, for H(1s) in a linearly polarized laser field of angular frequency $\omega = 0.184$ a.u. as a function of the intensity. The solid lines refers to the R-matrix-Floquet results. The dashed line corresponds to values obtained by using lowest order perturbation theory.

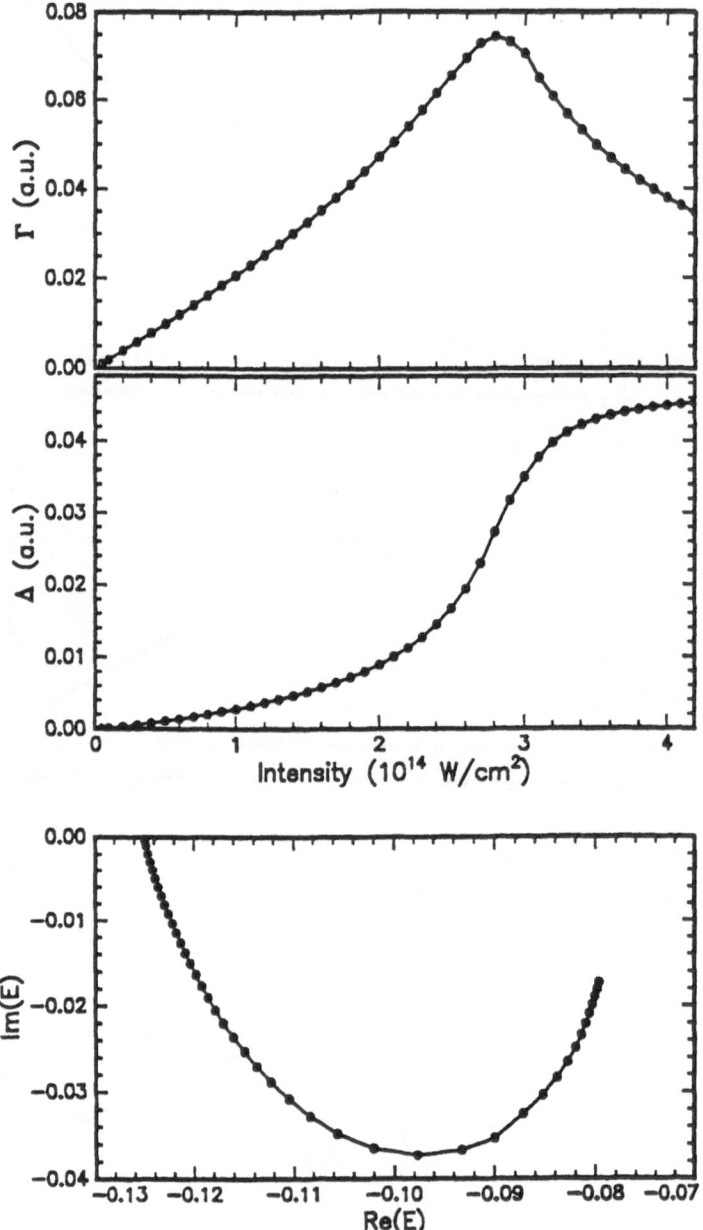

Figure 3. The R-matrix-Floquet results for the total ionization rate Γ and dynamic Stark shift Δ versus intensity for H(2s) in a linearly polarized laser field of angular frequency $\omega = 0.184$ a.u. The lower part shows the trajectory of the quasi-energy pole in the complex energy plane (for the same intensity points as on the upper figures).

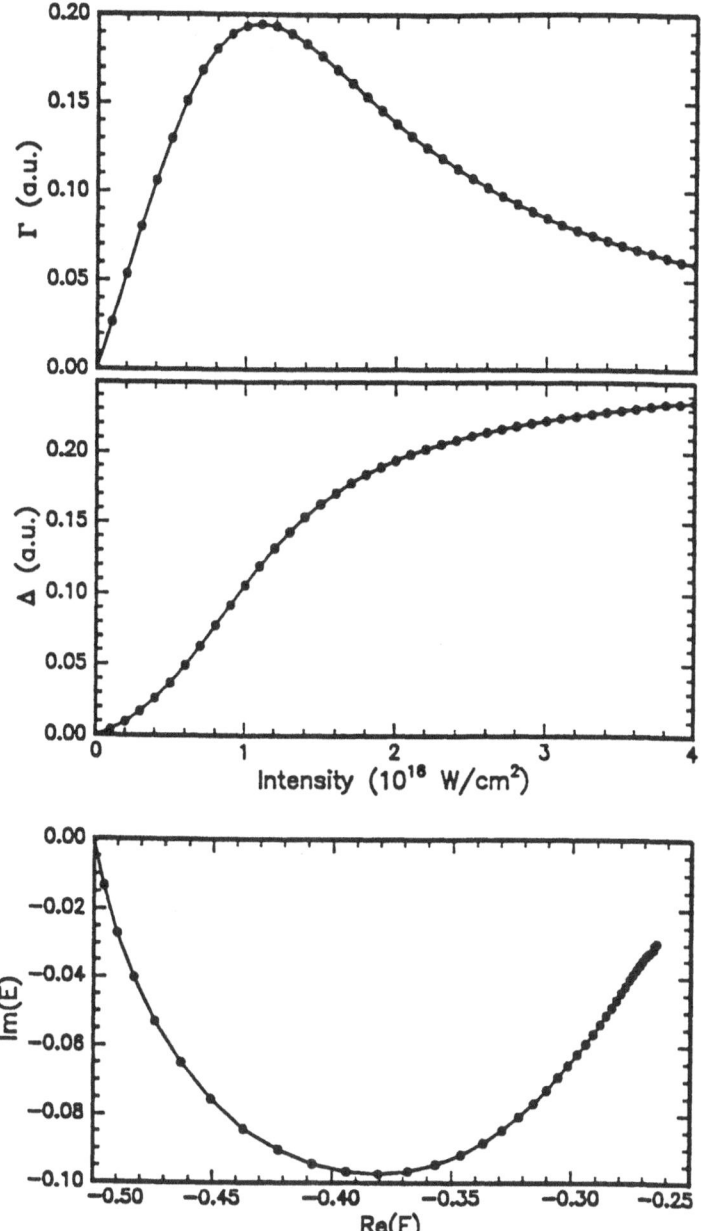

Figure 4 . Same as Fig. 3, but for H(1s) in a linearly polarized laser field of angular frequency $\omega = 0.65$ a.u.

At the bottom of Fig. 3 we have also plotted Im(E) vs. Re(E) to show the trajectory of the quasi-energy in the complex plane. The R-matrix-Floquet results for the total ionization rate are seen to exhibit a "stabilization" behaviour, i.e. a maximum followed by a decrease at higher intensities; this is to be expected since the value $\omega = 0.184$ a.u. belongs to the high-frequency regime for the $2s$ state. A similar stabilization behaviour

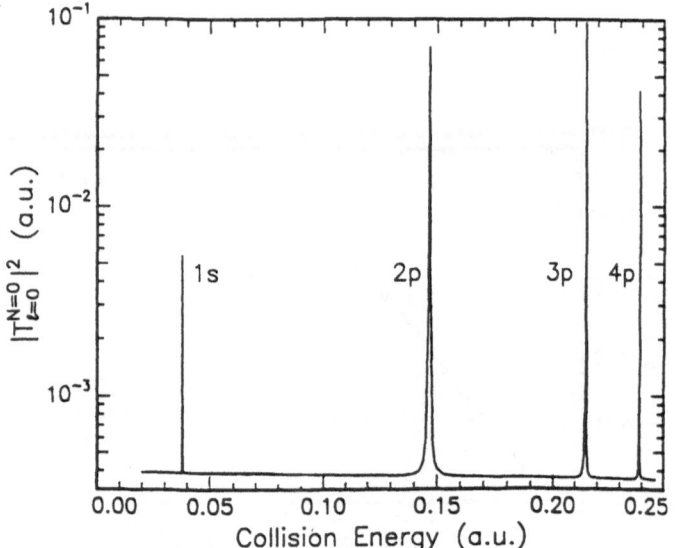

Figure 5. The R-matrix-Floquet results for quantity $|T_{\ell=0}^{N=0}|^2$, where T_{ℓ}^N is the ℓth partial wave T-matrix element for laser-assisted electron-proton scattering with the net exchange of N photons, as a function of the collision energy (in a.u.). The laser field is linearly polarized; its angular frequency is $\omega = 0.27$ a.u. and its intensity is $I = 1.5 \times 10^{13}$ W cm^{-2}.

is shown in Fig. 4 for the total ionization rate of H($1s$) at the high frequency $\omega = 0.65$ a.u. ($\lambda = 70$ nm).

3.2 Laser-assisted electron-proton scattering

We now turn to the application of the R-matrix-Floquet method to laser-assisted collisions. As an illustration, we consider electron-proton scattering in the presence of the linearly polarized monochromatic laser field of eq. (1), with $\omega = 0.27$ a.u., at an intensity $I = 1.5 \times 10^{13}$ W cm^{-2}. We display in Figs. 5 and 6 the quantities $|T_{\ell}^{N=0}|^2$, for $\ell = 0$ and 2, where T_{ℓ}^{N} is the ℓth partial wave T-matrix element for laser-assisted electron-proton scattering with the net exchange of N photons. The prominent feature in Figs. 5 and 6 is the presence of "capture-escape resonances"[14,15], in the present case a two photon 1s resonance, and one-photon np resonances, with $n = 2, 3, 4$. We note that the peak heights in Figs 5 and 6 are different, due to different branching ratios into the various orbital angular momentum channels. It is also worth noting that the position and width of the two-photon 1s resonance found in the present calculation are in excellent agreement with those obtained from the multiphoton ionization (Siegert) calculations, performed either by using the Sturmian-Floquet[13] or the R-matrix-Floquet[4] methods.

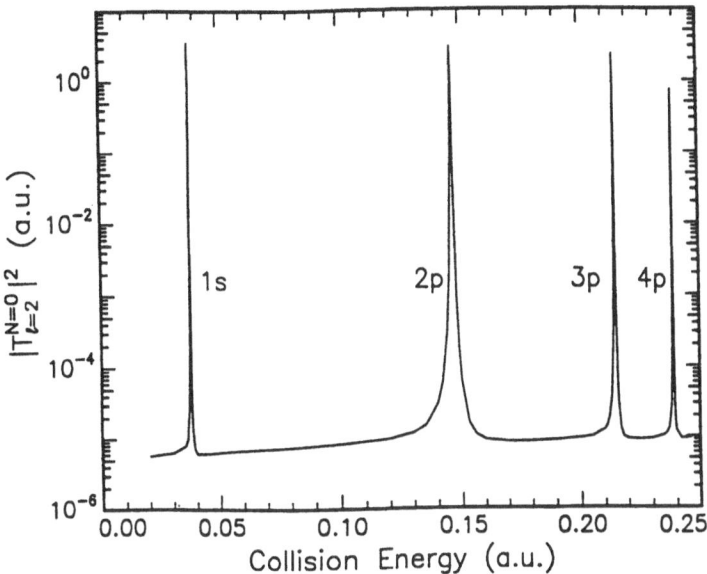

Figure 6 . Same as figure 5, but for the quantity $|T_{\ell=2}^{N=0}|^2$.

Acknowledgments

The work described in this article has been carried out in collaboration with P.G. Burke, P. Francken, C.J. Noble, J. Purvis and M. Terao-Dunseath. Support from the EC grants Nr SC1-0103-C and SC1-0168-C and from the British Council is gratefully acknowledged.

REFERENCES

1. C.J. Joachain, "Theory of Laser-Atom Interactions" in "Laser Interaction with Atoms, Solids and Plasmas", ed. by R.M. More (Plenum, New York, 1993).

2. P.G. Burke, P. Francken and C.J. Joachain, Europhys. Lett. 13, 617 (1990).

3. P.G. Burke, P. Francken and C.J. Joachain, J. Phys. B24, 761 (1991).

4. M. Dörr, M. Terao-Dunseath, J. Purvis, C.J. Noble, P.G. Burke and C.J. Joachain, J. Phys. B25, 2809 (1992).

5. E.P. Wigner, Phys. Rev. 70, 15 (1946).

6. P.G. Burke and W.D. Robb, Adv. At. Mol. Phys. 11, 143 (1975).

7. K.A. Berrington, P.G. Burke, M. Le Dourneuf, W. Robb, K.T. Taylor and Vo Ky Lan, Comput. Phys. Commun. 14, 367 (1978).

8. H.A. Kramers, "Collected Scientific Papers" (Amsterdam, North Holland, 1956) p. 272.

9. W.C. Henneberger, Phys. Rev. Lett. 21, 838 (1968).

10. C. Bloch, Nucl. Phys. 4, 503 (1957).

11. J.C. Light and R.B. Walker, J. Chem. Phys. 65, 4272 (1976).

12. P.G. Burke and H.M. Schey, Phys. Rev. 126, 147 (1962).

13. R.M. Potvliege and R. Shakeshaft, Phys. Rev. A 40, 3061 (1989).

14. L. Dimou and F.H.M. Faisal, Phys. Rev. Lett. 59, 872 (1987).

15. L.A. Collins and G. Csanak, Phys. Rev. A 44, R5343 (1991).

IONIZATION OF HYDROGEN ATOMS
BY CIRCULARLY POLARIZED E-M FIELD

D. Delande[1], R. Gębarowski[2], M. Kuklińska[3],
B. Piraux[4], K. Rzążewski[3] and J. Zakrzewski[2]

[1] Laboratoire de Spectroscopie Hertzienne de l'E.N.S., Université Pierre et Marie Curie, Tour 12, 4 place Jussieu, 75252 Paris Cedex 05, France.
[2] Instytut Fizyki, Uniwersytet Jagielloński, ulica Reymonta 4, 30-059 Kraków, Poland.
[3] Centrum Fizyki Teoretycznej PAN Aleja Lotników 32/46, 02668 Warsaw, Poland.
[4] Institute de Physique Corpusculaire, Université Catholique de Louvain, 1348 Louvain-la-Neuve, Belgium

INTRODUCTION

The ionization of a hydrogen atom which is initially prepared in a highly excited state by a linearly polarized microwave field (linearly polarized microwave ionization - LPMI) has been intensively studied for the last twenty years, since the early experiment of Bayfield and Koch[1]. This system has served as one of the two most prominent examples of "quantum chaos" in atomic physics (the other being the hydrogen atom in a strong static magnetic field[2]). The progress in the understanding of the physics of LPMI has been frequently reviewed[3-5].

By contrast, relatively little attention has been paid to the problem of ionization by a circularly polarized microwave field (CPMI) (see, however, early work on this subject[6]). Recently, experiments performed on Na atoms[7,8] have shown that in the low frequency limit the ionization threshold for circularly polarized field for low angular momentum initial states lies substantially higher than that known from LPMI and little differs from that obtained in the static limit. The latter conclusion has been challenged by Nauenberg[9] who predicted a relatively sharp drop of the ionization threshold with frequency for hydrogenic atoms. The experimental observation[7] that the departure from circular polarization towards elliptic polarization quickly decreases the threshold field has been verified in numerical studies[10]. Also the limit of large frequencies has been investigated classically[11] showing similarities with the LPMI case.

It is easy to understand the difficulty in studying the CPMI. While in LPMI the projection of the atomic angular momentum on the light polarization axis is conserved (and the problem becomes effectively two-dimensional), CPMI requires, in principle a fully 3-dimensional treatment. While in LPMI a simpler one-dimensional atomic model has frequently[3-5] been used making the quantum calculations much easier, in CPMI most of the classical studies have been restricted to a simplified two-dimensional model of hydrogen. The motion is then restricted to the plane defined by the rotation of the vector of polarization. Even in this case quite contradictory predictions, concerning ionization threshold, especially for low frequencies, have been reached[7-9]. One of our aims has been to resolve this problem.

In our classical[12] and quantum studies[13] we have presented preliminary results concerning the CPMI of hydrogen prepared initially in circular states. Such initial states, having large quantum numbers enable a direct comparison between classical and quantum results. In this report, the results obtained previously[12-13] in the simplified, 2-dimensional atom, are supported by additional classical simulations both for elliptic initial states as well as for truly 3-dimensional case (i.e., for arbitrary orientation between the Kepler orbit plane and the plane of light polarization). It is worth mentioning at this point that circular states may be obtained experimentally using consecutive switching of electric and magnetic fields[14]. In this method the resulting states are well oriented in space. Keeping that in mind we *do not* perform the classical average (see below) over the angle between the state orientation and the plane of polarization.

CLASSICAL AND QUANTUM RESULTS

Consider first the 2 dimensional model of hydrogen without any external perturbation. Its Hamiltonian (in atomic units) is obtained from the standard 3-dimensional case by suppressing the z-dependence:

$$H_0 = \frac{p_x^2 + p_y^2}{2} - \frac{1}{\sqrt{x^2 + y^2}}. \tag{1}$$

The quantum energies are given by the Rydberg formula $E_n = -1/2n_*^2$ where the effective quantum number $n_* = n - 1/2$, $n = 1, 2, ...$ rather than n appears. The states may be characterized by two quantum numbers: (n, m), where $m = -(n-1), .., (n-1)$ is the eigenvalue of the angular momentum operator $L_z = xp_y - yp_x$ and are degenerate with respect to m.

In the presence of a circularly polarized electromagnetic wave the Hamiltonian of the system, in the length gauge is:

$$H = H_0 - Ff(t)\big(x\cos(\omega t) + y\sin(\omega t)\big) \tag{2}$$

where $f(t)$ denotes the envelope of the light (microwave) pulse with unit maximum and F measures the amplitude of the pulse. In all the calculations reported here we have chosen a smoothly varying envelope of the form $f(t) = \sin^2(\pi t/T)$ where T denotes the pulse duration. Fast oscillations in eq.(2) may be removed by passing into the rotating frame[15,16]

$$\tilde{x} = x\cos(\omega t) + y\sin(\omega t)$$

$$\tilde{y} = -x\sin(\omega t) + y\cos(\omega t). \tag{3}$$

The corresponding unitary transformation in the quantum case is given by $U = \exp(i\omega L_z t)$. In the new frame,

$$\tilde{H} = \tilde{H}_0 - \omega\tilde{L}_z = \frac{\tilde{p}_x^2 + \tilde{p}_y^2}{2} - \frac{1}{\sqrt{\tilde{x}^2 + \tilde{y}^2}} - Ff(t)\tilde{x} - \omega\tilde{L}_z \tag{4}$$

and the degeneracy of the field-free Hamiltonian is removed. The energy of (n,m) eigenstate of $\tilde{H}_0$ becomes $E_{n,m} = -1/2n_*^2 - m\omega$. In the following we shall drop the tilde sign and work in the rotating frame only.

Consider first the classical mechanics approach. The ionization yield is dependent on four parameters: initial energy, E_i, frequency of the pulse, ω, its amplitude, F and the pulse duration, T. The well known[3-5] classical scaling property of the hydrogen atom allows one to fix one of these parameters, say initial energy at $E_0 = -1/2$ and scale accordingly the other parameters putting $\omega_0 = n_*^3\omega$, $F_0 = n_*^4 F$ and $T_0 = n_*^{-3}T$, where n_* corresponds to the initial state with energy (in the static frame) $E_i = -1/2n_*^2$. Once the ionization threshold dependence on F_0, ω_0 and T_0 is found, the classical predictions for arbitrary initial state (n,m) may be found using the above scaling relations. When the simulations are performed for fully 3-dimensional hydrogen, one should use n rather than n_* for scaling. Let us note at this point that the classical scaling does not hold in the quantum mechanical world as application of the scaling modifies commutation relations.

Ionization thresholds are obtained classically in the straightforward way. Initial points are selected randomly on the initial Kepler orbit, a circle for a circular state or an ellipse for states with nonvanishing eccentricity $e = (1-l^2)^{1/2}$ where l is the angular momentum ($l = 1$ corresponds to the circular orbit at $E_0 = -1/2$). Then, in three-dimensional case the orbit may be rotated with respect to the polarization plane by an angle, Θ. Equations of motion resulting from Hamiltonian, eq.(4) are integrated in time till the end of the pulse. A fraction of initial points leading to the positive final energy (after the pulse decayed to zero) to the total number of selected initial points yields an approximate classical ionization probability. The accuracy of

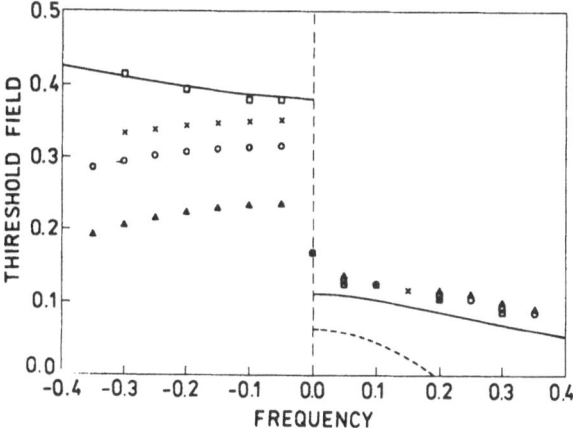

Figure 1. Scaled threshold amplitude, F_0, of the microwave pulse for ionization of the circular Rydberg orbit as a function of scaled frequency, ω_0. Squares correspond to the Rydberg orbit lying in the polarization plane, crosses, circles and triangles correspond to the angle between the orbit plane and the polarization plane being $\Theta = \pi/20, \pi/10, \pi/4$, respectively. Dashed line represents saddle point prediction of Nauenberg[9], solid line ($\omega_0 > 0$) represents corrected saddle point prediction while the solid line for negative ω_0 is a prediction based on the break-up of the stability of the periodic orbit lying opposite to the saddle.

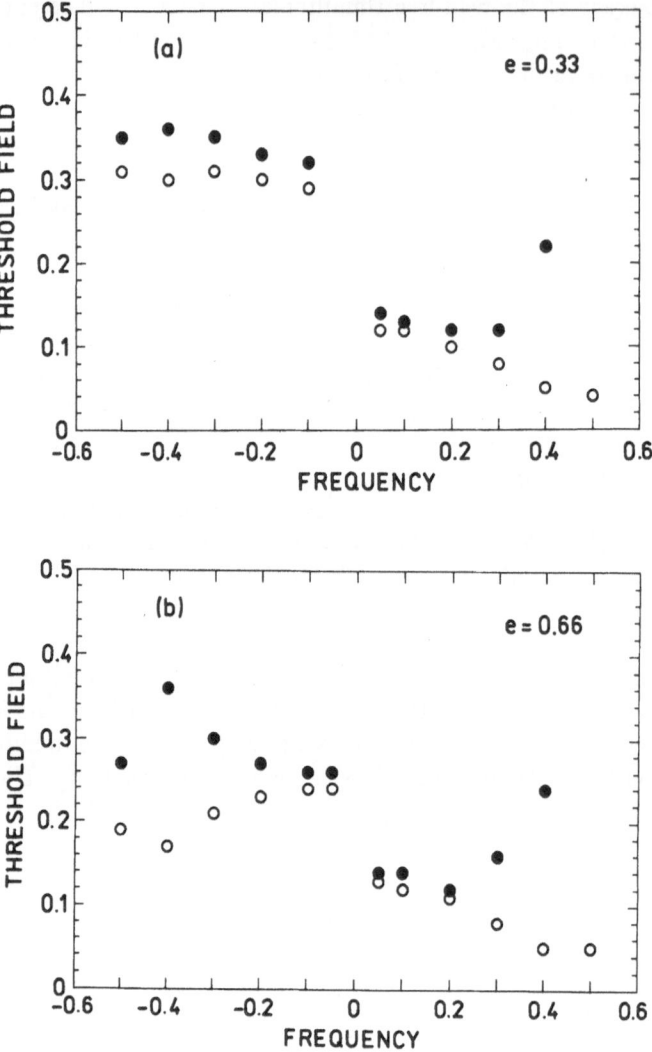

Figure 2. Same as Figure 1 but for elliptic Rydberg orbits with eccentricities indicated in the Figure. Open circles indicate minimal threshold for ionization while full circles give 100% ionization (all trajectories ionized).

integration has been checked by comparing trajectories integrated in the rotating frame with those obtained in the static frame. We have also applied the integration in semi-parabolic coordinates in which the Coulomb singularity is removed[2] to test the accuracy of numerical calculations. Typically around 200 initial points has been chosen, the pulse duration was assumed to be 100 Kepler periods. For frequencies close to zero the pulse duration has been taken even larger to assure the adiabatic behavior of the system.

The ionization threshold dependence on the frequency for initial circular states is presented in Figure 1. While positive frequencies correspond to right-hand polarization (σ^+), the negative frequencies correspond to σ^- polarization. Note the pronounced asymmetry between thresholds for positive and negative frequencies. In the region of frequencies presented in the picture practically all trajectories lead either to ionization (for sufficiently large pulse amplitude F) or return to the same orbit as the initial one (with accuracy in energy better than 1 part in 10^5). The situation is quite different for a static field ($\omega_0 = 0$). In a range of field amplitudes $F_0 \in (0.111, 0.384)$ only part of initial conditions leads to ionization, the threshold is then defined as a 50% ionization. Solid lines present classical estimates for the threshold based on the structure of the corresponding phase space (see below).

Figure 1 shows also the ionization thresholds obtained for several orientations of initial Kepler circular orbit with respect to the polarization plane. Although the asymmetry is the largest for the orbit located in the polarization plane, the significant asymmetry exists even for $\Theta = 45°$. The asymmetry disappears, of course, for $\Theta = 90°$ (not shown). Clearly, the effect is quite robust and not restricted to specific orientation of the initial orbit.

Before explaining the origin of this asymmetry let us consider initial elliptic states. Results of numerical simulations are presented in Figure 2. Open circles denote the field amplitude value at which at least one of trajectories is ionized while at values indicated by solid circles all initial points lead to ionization. The asymmetry becomes smalled for large eccentricity (it should disappear completely for $e = 1$), also the threshold amplitude becomes less sharply defined for elliptic orbits.

The asymmetry appears also on the quantum level as shown by us recently[14]. To find the quantum ionization probability we have used the single state approximation[17] (SSA) and diagonalizated the Hamiltonian, eq.(4) for $f(t) = 1$ and different values of field amplitude F using complex rotation method. This approach yields the ionization rates at fixed values of F and given frequency as twice the imaginary part of the complex energy of a state, adiabatically evolving from the initial state. The total ionization probability, P, in the SSA, is found to be

$$P = 1 - \exp\left[-\int_0^T \Gamma\big(Ff(t)\big)dt\right]. \qquad (5)$$

For direct comparison with classical simulations, we have assumed that initially the atom is in the circular state $(13, 12)$. Figure 3 presents the ionization thresholds, defined as 10% ionization (triangles) and 90% ionization (squares) for the pulse duration of 100ps (which for $n_* = 13$ corresponds roughly to 100 Kepler periods). The frequency and the maximum pulse amplitude have been rescaled using the scaling introduced above to facilitate comparison with classical calculations. To guide the eye, the solid lines giving classical prediction (same as in Fugure 1) are included for comparison. $n = 13$ was chosen to compromise between the attempted study of the ionization of Rydberg states and the sizes of matrices leading to converged energies. Let us note parenthetically that for $n = 13$ and scaled frequency $\omega_0 = 0.2$ more than 30 photons are required to ionize the atom. This provides an idea on a size of matrices diagonalized to obtain the data in Figure 3.

Comparison of Figure 3 with results of classical simulations shows that the asymmetry, although a bit smaller, appears also in quantum case. Although we diagonalized the simplified 2-dimensional problem we are convinced that the similar behaviour would be observed had we treated the full 3-dimensional case. This conjecture is based on the understanding of the origin of asymmetry (see below) and is supported by classical simulations (Figure 1 and Figure 2) which show little sensitivity to both the orbit orientation in space and its precise shape.

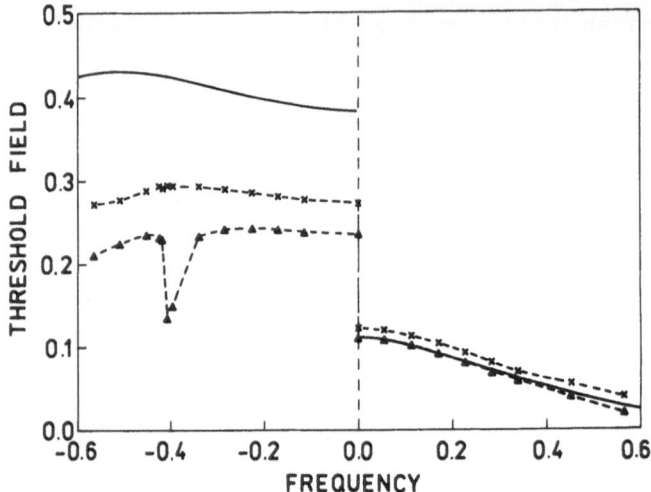

Figure 3. Same as Figure 1 but obtained from fully quantum-mechanical calculation within the SSA. Crosses denote 50% ionization while triangles the 10% ionization thresholds. Solid lines as in Figure 1 give classical estimate for the threshold.

Let us note that the dip observed for 10% ionization at $\omega_0 = -0.4$ is due to the two-photon resonance between (13,12) initial state and (14,10) state. The traces of the corresponding 2:1 classical resonance are visible also in Figure 2, the quantum effect is more robust. The shape of the resonance and its effect on the ionization threshold depends on the definition of the threshold (note that it disappears for 50% ionization curve) and is strongly dependent on the pulse duration.

ORIGIN OF THE ASYMMETRY

The origin of the observed, striking at the first glance, asymmetry between left-hand and right-hand polarization observed in all the results presented is easy to understand realizing that the Hamiltonian, eq.(4) represents in the static limit ($\omega \to 0$) the well known Lo Surdo-Stark effect. The eigenstates of eq.(4) are then not (n,m) but (n,k), and their energies are given in the perturbation limit by $E_{n,k} = -1/2n_*^2 + 3/2kn_*F$ with $k = -(n-1),..,(n+1)$. The states with maximal negative k are the lowest states of a given n_* manifold and lie close to the saddle of the potential. Their ionization threshold is given by $F_{neg} = 1/9n_*^4$ (see Ref.7). On the contrary, highest-energy states of the manifold, with maximal positive k, are located far from the saddle on the other side of the nucleus (in the region where the potential is maximal at negative x). They do not feel the influence of the saddle at all. As the Lo Surdo-Stark problem is separable, these states lie on well defined tori and they are not sensitive to whether their energy is lower or higher than the classical escape energy, given by the saddle energy. For $\omega \neq 0$, the first order perturbative result yields[18]

$$E_{n,k} = -1/2n_*^2 + k\sqrt{(3/2n_*F)^2 + \omega^2}. \qquad (6)$$

Clearly for both ω and F small and positive and for maximal positive m (negative k), as is the case in Figure 3, the initial state evolves into the state located close to the saddle (and its ionization threshold should be close to the static value of $F_0 = 1/9$) while for low negative ω it evolves into the state with maximal positive k and its ionization threshold should be much higher (compare Figs.1,3). Due to the symmetry of eq.(4) a change of $m \to -m$ is equivalent to the substitution $\omega \to -\omega$.

At the same time one should be aware of the fact that the SSA adopted here breaks down in the immediate vicinity of $\omega = 0$ when the pulse duration becomes comparable with the period of the electromagnetic wave. Thus the region $|\omega T| < 1$ is not well described by the present method. This is the region where the mixing of different m states forming the k state occurs just at the beginning of the pulse. For a $T = 100$ps pulse and $n = 13$ it corresponds to the small interval $|\omega_0| < 0.005$, invisible on the scale of Figure 3 but contributing to the apparent discontinuity at $\omega = 0$. One may use the SSA in the immediate vicinity of $\omega = 0$, but then as the initial state one should not use the m states but rather the k states. In this way *two* values for the threshold at $\omega = 0$ have been obtained. One corresponds to $k = -(n-1)$ and represents the limit for positive ω, while the other one, for $k = (n-1)$, is the limit for negative ω. To obtain the threshold for the circular initial state quantum mechanically, a fully time-dependent approach would be necessary.

It is clear from the above that classical predictions based on the saddle point criterium may give reasonable estimates for $\omega > 0$ only [with initial state $(n,m) = (n, n-1)$]. Such a prediction has been obtained in Ref.9. Unfortunately, the author compared the energy of the saddle with the unperturbed energy $E = -1/2n_*^2$ which yields (see dotted line in Figure 1) a wrong result. On the other hand one cannot use the method[9] to estimate the threshold for $m = 0$ states as they are located further from the saddle and ionize for even higher field amplitudes (as verified by our quantum mechanical calculations not shown here). However, using the same approach but utilizing eq.(6) as the proper energy of the state, the solid line in Figure 1 and Figure 3 ($\omega_0 > 0$ only) is obtained. It agrees resonably well with classical numerical simulations keeping in mind that it is obtained on the basis of first order perturbation theory approximation for the energy, eq.(6). a much better agreement with quantum numerical results is to some extend accidental as quantum threshold depends on the pulse duration (in the long pulse limit tunneling may lower quantum numerical results).

The quantum results agree quite well with classical simulations for $\omega > 0$, while for $\omega < 0$, classical thresholds are significantly higher (compare Figure 1 and Figure 3). This is to be expected following similar arguments to those already given. If, in the classical case, the field were switched slowly enough one would follow the periodic orbit corresponding to the initial circular orbit. For long pulses, the classical threshold then corresponds to the field at which this orbit looses its stability - represented by the solid line for $\omega < 0$ in Figure 1 and Figure 3. Note how well the break-up of periodic orbit stability reproduces the classical ionization threshold (Figure 1). It is quite interesting to notice that the periodic orbit looses its stability at higher values of F_0 for small negative ω_0 (when the classical problem is, strictly speaking, nonintegrable) than for the integrable static field case!

The quantum threshold is determined not by the break-up of the periodic orbit stability, but rather by the break-up of the torus corresponding to this state, (which surrounds the periodic orbit and its corresponding transverse action is given by $\hbar = 1/2$, instead of the zero value for the orbit itself). It indicates that the quantum threshold for $\omega_0 < 0$ does not scale classically for moderate n; the smaller $\hbar$ (the higher n), the "closer" the quantizable torus lies to the periodic orbit, and the higher will be the ionization threshold. This has been verified by studying ionization of different n states[13]. For positive ω_0, when the threshold is determined by the saddle point criterion, the quantum-classical correspondence is established[13] for n as low as 10.

CONCLUSIONS

We have presented the detailed study of low frequency ionization of an initially highly-excited hydrogen atom by strong circularly polarized light. The quantum results[13] obtained in a simplified two-dimensional model of hydrogen atom constitute, as far as we know, the first accurate quantum mechanical calculations for multiphoton ionization by circularly polarized radiation. The obtained quantum threshold are in good agreement with classical simulations, when possible. The origin of the remaining discrepancies has been explained. The results, although obtained originally[12,13] for a simplified 2-dimensional atomic model has been, at least classically, verified in a fully three-dimensional calculation. The asymmetry between positive and negative frequencies (or rather different circular polarizations) has been shown to be a robust phenomenon, appearing for various orientations of the Rydberg orbit with respect to the polarization plane as well as for elliptical and not only circular initial orbits. The origin of this asymmetry has been explained and traced back to the different spacial localization of the corresponding time-dependent wavefunctions.

It is worth stressing that the results presented here are restricted mainly to the hydrogen atom or hydrogen-like ions. For other atoms the residual interaction with the core may significantly affect the ionization process[7,8]. For such atoms the Lo Surdo-Stark problem, due to the core non-Coulombic potential, is no longer separable and the mixing of different k states (appearing via large avoided crossings) destroys localization properties of quantum wavefunctions for sufficiently strong field amplitude.

Acknowledgements

Presented quantum calculations were obtained when one of the authors (JZ) was on a sabbatical leave in Paris. He thanks his colleagues at the Laboratoire de Spectroscopie Hertzienne for hospitality and the Ministère de la Recherche et de la Technologie for their financial support. Support of the Polish Committee for Scientific Research under grant No 200799101 (J.Z.) and 204539101 (K.R.) is acknowledged. Le Laboratoire de Spectroscopie Hertzienne de l'Ecole Normale Supérieure et de l'Université Pierre et Marie Curie est un Unité Associée 18 du Centre National de la Recherche Scientifique.

REFERENCES

1. J. E. Bayfield and P. M. Koch, Phys. Rev. Lett. **33**, 258 (1974).
2. see e.g. D. Delande in *Les Houches Session LII, Chaos and Quantum Physics 1989*, eds. M.-J. Giannoni, A. Voros and J. Zinn-Justin, (North-Holland, Amsterdam) 1991, p.665.

3. G. Casati, I. Guarnieri, and D. L. Shepelyansky, IEEE J. Quant. Electron. **24**, 1420 (1988).

4. B. V. Chirikov in *Les Houches Session LII, Chaos and Quantum Physics 1989*, eds. M.-J. Giannoni, A. Voros and J. Zinn-Justin, (North-Holland, Amsterdam) 1991, p.443.

5. R. V. Jensen, S. M. Susskind, and M. M. Sanders, Phys. Rep. **201**, 1 (1991).

6. J. Mostowski and J. J. Sanchez-Mondragon, Opt. Commun. **29**, 293 (1979).

7. P. Fu, T. J. Scholz, J. M. Hettema, and T. F. Gallagher, Phys. Rev. Lett. **64**, 511 (1990).

8. T. F. Gallagher, Mod. Phys. Lett. B **5**, 259 (1991).

9. M. Nauenberg, Phys. Rev. Lett. **64**, 2731 (1990).

10. J. A. Griffiths and D. Farrelly, Phys. Rev. A **45**, R2678 (1992).

11. J. E. Howard, Phys. Rev. A **46**, 364 (1992).

12. K. Rzążewski and B. Piraux, Phys. Rev. A **47** Rapid Commun. *in press.*

13. J. Zakrzewski, D. Delande, J.C. Gay and K. Rzążewski, Phys. Rev. A **47** Rapid Commun. *in press.*

14. D. Delande and J. C. Gay, Europhys. Lett. **5**, 303 (1988).

15. F. V. Bunkin and A. Prokhorov, Sov. Phys. JETP **19**, 739 (1964).

16. A. Franz, H. Klar, J. T. Broad, and J. S. Briggs, J. Opt. Soc. Am. B **7**, 545 (1990).

17. M. Dörr, R. M. Potvliege, and R. Shakeshaft, Phys. Rev. A **41**, 1609 (1990).

18. D. Wintgen, Z. Phys. D **18**, 125 (1991).

PART III

DYNAMICS OF PHOTOIONIZATION AND PHOTODISSOCIATION OF MOLECULES

IONIZATION OF H_2^+ IN INTENSE LASER FIELDS

F. H. Mies,[1]   A. Giusti-Suzor,[2,3]   K. C. Kulander[4] and K. J. Schafer[4]

[1]N.I.S.T., Gaithersburg MD 20899, USA
[2]Lab. de Chimie-Physique, 11 rue P. et M. Curie, 75321 Paris
[3]Lab. de Photophysique Moleculaire,Universite Paris-Sud, France
[4]Lawrence Livermore National Lab., Livermore, CA 94550

INTRODUCTION

Recent calculations[1] of ionization rates for H_2^+ by an intense λ=228nm laser show a strong dependence of the rate constant on the internuclear distance R. We have performed comparable time-dependent calculations for H_2^+ in fields I≤1x10^{15}W/cm^2 for a wide range of fixed R (2-8au) and wavelengths (228-1064 nm). Indeed we obtain excellent agreement with the results of Chelkowski etal[1] at λ=228nm. At large R we find the ionization rate for all wavelengths rapidly approaches that of H(1s). However, at R≈2au and especially for the longer wavelength lasers, such as the λ=769nm Ti:sapphire laser, the large ionization potential (29eV) of the molecule allows it to easily survive intensities up to I=5x10^{14}W/cm^2 for typical short pulses of, say, 150 fsec.

Interestly, the ionization rates obtained from the exact numerical solution of the time-dependent Schroedinger equation conform beautifully with a very simple representation of H_2^+ which only involves its two lowest electronic states $1s\sigma_g$ and $2p\sigma_u$ and an optical potential to represent the ionization. Moreover, the two-state model only conforms to the exact results using a length gauge coupling, and yields extremely poor results if the asymptotically pleasing velocity gauge coupling used in reference 2 is employed. These fully dynamic results confirm the conclusion inferred from quasi-energy variational calculations[3] in a periodic field that indeed a two-state length-gauge model is good in H_2^+ at least out to R≈6au.

The motivation for the present ionization calculations is to test the reliability of a recent study of H_2^+ photodissociation[4] which employed such a two electronic state model and neglected any competition with ionization. The photodissociation calculations indicate that in intense short pulsed laser fields appreciable populations of stable vibrational states can survive the pulse. This survival effect can be attributed to the trapping of portions of the initial vibrational wavepacket in transient laser-induced potential wells at intermediate R≈3-4au distances. Since the calculated ionization rates exhibit a marked decrease at short R, they already lend some credence to the vibrational trapping effect. Having accurate R-dependent rates enables us to estimate the competitive influence of the ionization on the stabilized population, and may ultimately allow us to predict the contribution of the Coulomb 'explosion' channel to observed proton kinetic energy distributions.

In this paper we will demonstrate the effectiveness of the two-state length gauge model in <u>interpreting</u> the ionization rates that we extract from the numerically <u>exact</u> solutions of the time-dependent Schroedinger Equation. A more elaborate presentation of the theory and the results for the full range of distances and wavelengths will be presented elsewhere[7].

THEORY

We assume that H_2^+ has a fixed bond-length R with the bond oriented along the $\hat{z}$ axis, which coincides with the polarization vector of the linearly polarized electric field of the laser

$$\mathbf{E}(t) = E_{max}f(t)\sin(\omega t+\delta)\ \hat{z} = E(t)\ \hat{z}, \tag{1}$$

where f(t) defines the envelope of the laser pulse. This orientation preserves the cylindrical symmetry of the molecule[1,6] and allows us to perform calculations using finite difference methods in a two-dimensional cylindrical grid.[6] Of course we use the dipole approximation.

The numerically exact calculations are actually performed in the Coulomb (≈ velocity) gauge,

$$[\tfrac{1}{2}(\mathbf{p} + \tfrac{e}{c}\mathbf{A})^2 + V_{coul}](e^{-i\Lambda}\Psi) = i\hbar\frac{\partial}{\partial t}(e^{-i\Lambda}\Psi). \tag{2}$$

Using the following gauge transformation

$$\Lambda(t) = \frac{e}{c\hbar}\ \mathbf{A}(t)\cdot\mathbf{r} \qquad\qquad \hbar\frac{\partial\Lambda}{\partial t} = \mathbf{E}(t)\cdot(e\mathbf{r}) \tag{3}$$

where $\mathbf{r}$ is the electronic coordinate conjugate to $\mathbf{p}$, we obtain the usual time-dependent Schroedinger eq. in the length gauge,

$$[H_o + \mathbf{E}(t)\cdot(e\mathbf{r})] \ \Psi = i\hbar\,\frac{\partial}{\partial t}\Psi. \tag{4}$$

Obviously the field-free electronic Hamiltonian $H_o \equiv \frac{1}{2}(\mathbf{p}^2 + V_{coul})$ for H_2^+ is implicitly dependent on R. We calculate the ionization rates for the ground $H_2^+(1s\sigma_g)$ and the first excited $H_2^+(2p\sigma_u)$ states which are represented by the electronic wavefunctions φ_g and φ_u with R-dependent eigenvalues (potentials) $V_g(R)$ and $V_u(R)$ respectively. Identical numerical results are obtained from integrating either Eq. (2) or Eq. (4) for a given initial $\Psi(0)$, but only if one is very careful to include the proper gauge trans- formation factor $e^{-i\Lambda(t)}$. The same consideration applies when we discuss various projected probabilities in the Results Section. Actually φ_g and φ_u both dissociate to $H^+ + H(1s)$ and are asymptotically degenerate. At large R they correlate respectively with a $\pm$ combination $[\varphi_{1s}(\mathbf{r}-\mathbf{R}/2)\pm\varphi_{1s}(\mathbf{r}+\mathbf{R}/2)]/\sqrt{2}$ of atomic ground state wavefunctions φ_{1s} centered on either of the two protons. As expected, we find that their ionization rates at large R are identical and coincide with the rate for the isolated hydrogen atom.

Our concern in this paper is to understand the effect of radiative coupling between these two states which is induced by the transition dipole $\mu(R)_{gu} = \langle\varphi_g|(ez)|\varphi_u\rangle$. In the length gauge the radiative coupling is

$$\Omega_{gu}(R,t) = \langle\varphi_g|\mathbf{E}\cdot(e\mathbf{r})|\varphi_u\rangle = E(t)\mu_{gu}(R). \tag{5}$$

Because of the asymptotic degeneracy $\mu_{gu} \to eR/2$ and Ω_{gu} actually diverges as $R\to\infty$. In contrast, the asymptotic radiative coupling in the velocity gauge,

$$\Omega_{gu}^{vel}(R,t) = \langle\varphi_g|\mathbf{p}\cdot\frac{e}{c}\mathbf{A} + (\frac{e}{c}\mathbf{A})^2|\varphi_u\rangle = \frac{(V_g - V_u)}{\hbar\omega}\Omega_{gu}(R,t), \tag{6}$$

vanishes since $(V_g - V_u)\to 0$ as $R\to\infty$. Thus, this gauge is often used in Floquet- type[2] calculations in order to avoid troublesome boundary conditions. Of course if a complete electronic basis is used then the solutions to (2) and (4) must give identical results. This is apparent in the interesting variational calculations of Muller[3] as well as in the present results.

If we chose to use only the two field-free electronic states φ_g and φ_u to diagonalize the time-dependent Hamiltonian in Eq. (4) we obtain a pair of adiabatic "field-following" dressed states, using length gauge coupling,

$$\varphi_s(\text{Length},t) = \cos\phi_L(t)\ \varphi_g - \sin\phi_L(t)\ \varphi_u \tag{7a}$$

$$\varphi_f(\text{Length},t) = \sin\phi_L(t)\ \varphi_g + \cos\phi_L(t)\ \varphi_u. \tag{7b}$$

The phase $\phi_L(t)$ is defined by the 2x2 orthogonal matrix obtained from the diagonalization of eq. (4), such that,

$$<\varphi_s(t)|H_o + \mathbf{E} \cdot (e\mathbf{r})|\varphi_s> = V_s(t) \underset{\mathbf{E} \to 0}{\tilde{}} V_g \tag{8a}$$

$$<\varphi_f(t)|H_o + \mathbf{E} \cdot (e\mathbf{r})|\varphi_f> = V_f(t) \underset{\mathbf{E} \to 0}{\tilde{}} V_u \tag{8b}$$

$$<\varphi_s(t)|H_o + \mathbf{E} \cdot (e\mathbf{r})|\varphi_f> \equiv 0. \tag{8c}$$

These dressed states and their diagonalized potentials V_s and V_f are now implicit functions of t and adiabatically follow the time-dependent field $\mathbf{E}(t)$. In particular $\phi_L(t)$ is periodic over an optical cycle, and has been defined to insure that $\varphi_s \to \varphi_g$ and $\varphi_f \to \varphi_u$ whenever $\mathbf{E}(t) \to 0$. For any finite R the potential V_f always exceeds V_s and hence the ionization potential of φ_s always exceeds that of φ_f. The f,s notation is chosen to remind us that we can expect a 'faster' ionization rate for the φ_f state and a generally 'slower' rate for the φ_s state.

If we choose to use the same two "bare" electronic states φ_g and φ_u to diagonalize the velocity gauge Hamiltonian in Eq. (2), then we will obtain a <u>different</u> pair of field dressed states $\varphi_s(\text{Velocity}, t)$ and $\varphi_f(\text{Velocity}, t)$. Although these adiabatic states have the same form as Eq. (7), the phase $\phi_L(t)$ is now replaced by a distinctly difference phase $\phi_V(t)$ obtained from the orthogonal matrix which diagonalizes Eq.(2).

In either the length or velocity gauge representation the adiabatic states φ_f and φ_s are coupled by non-adiabatic terms involving $\partial\phi/\partial t$. In addition we can simulate the net effect of ionization by introducing an <u>adhoc</u> optical potential for each state in the pair of coupled equations. We have solved these coupled equations[7] using two alternative forms for the optical potentials, one defined by the projections of the diabatic or bare states φ_g and φ_u

$$V_{op} = -i\Gamma_g/2|\varphi_g><\varphi_g| - i\Gamma_u/2|\varphi_u><\varphi_u| \tag{9}$$

and a second involving the projections of the time-dependent adiabatic dressed states φ_s and φ_f

$$V_{op}(t) = -i\Gamma_s/2 |\varphi_s><\varphi_s| - i\Gamma_f/2|\varphi_f><\varphi_f|. \tag{10}$$

If we further assume that the factors Γ are independent of time, then, for a given laser wavelength and intensity, the particular electronic state associated with Γ will decay as $e^{-\Gamma t/2\hbar}$, and molecules in this pure state will have an ionization rate constant equal to $\Gamma/\hbar$. The question is, do any of these simple two-state models, dressed either in the length gauge o the velocity gauge, and ionizing as prescribed by Eq. (9) or (10) come even close to simulating the ionization rates obtained for the exact solutions of Eq. (2)? In fact we have found[5] that simply using the two state solutions in the <u>length</u> gauge, and employing the optical potential in Eq. (10) yields a surprisingly good approximation to the exact results obtained from Eq. (2), and gives us confidence in extracting and interpreting state specific rate constants from the exact time-dependent wavepackets.

RESULTS

The time-dependent Schroedinger equation is solved in a box[6] of dimension z=±100au in the field direction, and with transverse cylindrical coordinate $\sqrt{(x^2+y^2)} \leq 28au$. Beyond these distances a gradually absorbing boundary is installed which does not allow reflection of the wavefunction,[6] and hence simulates the loss of electron density due to ionization. The deviation of the norm within this box

$$P_{norm}(t) = \int_{box}|\Psi(t)|^2 d\mathbf{r} \tag{11}$$

from its initial unit value at t=0 gives a measure of the total probability for ionization, $P_{ion}(t) = 1-P_{norm}(t)$. In addition to P_{norm} we evaluate the instantaneous contribution of various states φ_j to the exact wavefunction within the box

$$P_j(t) = |<e^{-i\Lambda}\varphi_j|e^{-i\Lambda}\Psi(t)>|^2 \equiv |<\varphi_j|\Psi(t)>|^2. \tag{12}$$

Specifically the probabilities for the bare states j=g,u and the field-dressed states j=s,f in both the length and velocity gauge are obtained. Although we use particular states $\Psi(0)=\varphi_j$ to <u>initialize</u> the wavepacket at t=0, <u>all the projections we present are made on the exact solutions</u>, and we are in <u>no way imposing or using the two-state models discussed above</u>. What we want to demonstrate is that the exact solutions do in fact conform nicely to one particular two-state model for times extending over many optical cycles. .

As a first example we present the probabilities for H_2^+ fixed at its equilibrium distance R=2au and exposed to a λ=769nm laser with intensity $I=2x10^{14}W/cm^2$. In each case we start the solution of Eq.(2) with some prescribed initial state. In Fig. 1 we have the results for $\Psi(0)=\varphi_g$, and also for an initial field dressed state $\Psi(0)=\varphi_s(0)$ defined by Eq. (7a), using the length gauge coupling in Eq. (5). In all the examples we present the field is turned on as a step function at t=0, with f(t)=0 for t<0 and

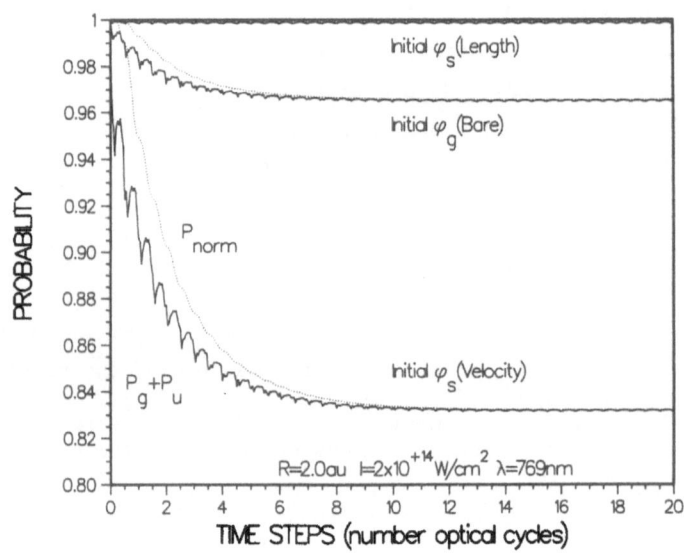

Figure 1. Decay of initially prepared φ_g, φ_s(Length) and φ_s(Velocity) states for $\lambda=769$nm, $I=2\times10^{14}$W/cm^2, at R=2.0au. Solid curves give the total probability (P_g+P_u) of the wavepacket being in either the φ_g or φ_u state. This sum is identical to (P_s+P_f). The dotted curves give P_{norm}. An initial φ_s(Length) state decays slowly and fairly exponentially, while an initial φ_g state always exhibits a rapidly decaying φ_f(Length) component. An initially prepared φ_s(Velocity) state contains an even larger φ_f component.

f(t)=1 for t≥0, in Eq. (1). Furthermore, the initial field is taken at its maximum with δ=π/2 in Eq. (1), which produces the maximum dressing of the field-dressed states $\varphi_s(0)$ and $\varphi_f(0)$ in Eq.(7).

We see that preparing the initial wavepacket in the φ_s(Length) state leads to a very slow, and very exponential decay of the packet over 20 optical cycles of time. In every example that follows wavepackets initially prepared in either the slow φ_s(Length) or the fast φ_f(Length) field-dressed state exhibits a nice, almost perfectly exponential decay. Thus they appear

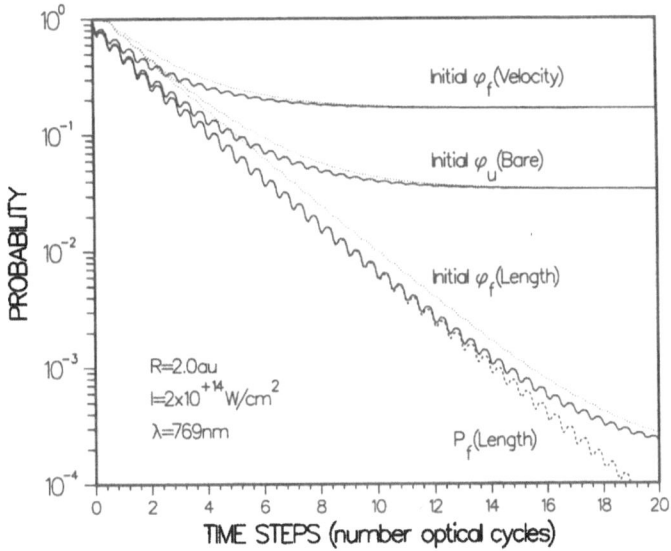

Figure 2. Same conditions as Fig. 1, but for initially prepared ω_u, φ_f(Length) and φ_f(Velocity) states. Again, an exponential decay for the field dressed ω_f state in the length gauge is observed, while φ_u and φ_f(Velocity) introduce various mixtures of fast and slow components. In contrast, note how P_f(Length) associated with the initial φ_f(Length) state remains exponential out to 20 optical cycles.

as almost pure "stationary states" of Eq. (4), which are uncoupled from each other, and separately experience direct ionization into the continuum. In contrast wavepackets initiating in the field-free or bare states, φ_g and φ_u invariably exhibit a fast and a slow decaying component which is well represented as a linear combination of decaying φ_s(Length) and φ_f(Length) states. It should be noted that if we adiabatically "turn on" the field, with f(t) rising gradually over a number of optical cycles in Eq. (1), the φ_g states predominantly forms the field-dressed φ_s(Length,t) state.

A striking feature of the figures we show for $\lambda=769$nm is that the sum $(P_g+P_u)\equiv(P_s+P_f)$ closely follows P_{norm}, and the wavepacket within the 'box' is predominantly in the two lowest H_2^+ electronic states, φ_g and φ_u, or equivalently, φ_s and φ_f. There is a lag between (P_g+P_u) and P_{norm} due to the 'free' electron component of $\Psi(t)$ which has been ionized, but has not yet reached the absorbing boundary and is still included in P_{norm}. For shorter wavelengths, such as $\lambda=228$nm studied by Chelkowski etal[1] we indeed find a more pronounced discrepancy between (P_g+P_u) and P_{norm} which suggests that excited electronic states of H_2^+ might be contributing to $\Psi(t)$ as well.

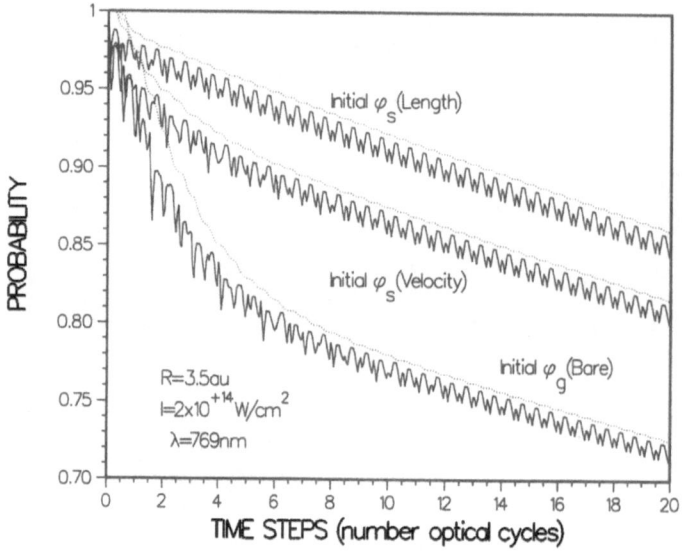

Figure 3. Identical conditions and notation as Fig. 1, but for a fixed internuclear distance of R=3.5au.

In Fig. 1 we also include the decay of a wavepacket initially prepared in a velocity dressed φ_s(Velocity) state. Here we see that a very large φ_f(Velocity) component contributes to the decay, strongly confirming the validity of using the two-state length gauge model. This is exactly the same conclusion that we have extracted from the calculations of Muller[1] for $\lambda=396$nm. Even more dramatic confirmation is seen in Fig. 2 where the initial φ_f(Length) state remains beautifully exponential over three decades, while the initial φ_f(Velocity) state is strongly coupled to the slowly decaying φ_s(Velocity) state.

This same behavior is still prevalent at R=3.5au as seen in Fig. 3 and Fig. 4 again for λ=769nm and I=2x10^{14}W/cm^2. Once more, in the length gauge φ_s and φ_f are predominantly uncoupled, and decay exponentially, while φ_g and φ_u and now to a lesser extent the φ_s(Velocity) and φ_f(Velocity) states are still coupled to each other and represent poor "stationary states" in the laser field.

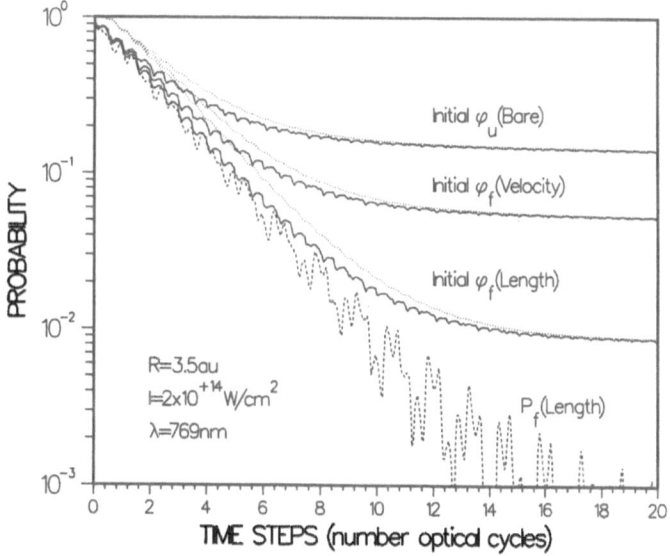

Figure 4. Identical conditions and notation as Fig. 2, but for a fixed internuclear distance of R=3.5au.

Table I gives a summary of some ionization rate constants that have been extracted from the observed exponential decay of the <u>exact</u> solution initially prepared in the φ_s(Length) state. If this rate constant is equated to $\Gamma_s/\hbar$ in Eq. (9), and if the comparable rate for $\Gamma_f/\hbar$ extracted from the exact solutions is also used, then indeed the solution of the simple two-state model does an excellent job[7] of simulated the exact results. Hopefully an adiabatic turn-on of the laser field predominantly converts $\varphi_g \rightarrow \varphi_s$(Length) and these tabulated rates are a good measure of what we can expect in a real pulsed laser experiment. The question is what do these results tell us about the possibility of producing vibrationally stabilized populations of H_2^+ molecules[4] without having them destroyed by ionization?

Table 1. Calculated ionization rate constants ($fsec^{-1}$) for H_2^+ at fixed R. Rates are given for the slow φ_s(length) field-following dressed state which adiabatically correlates with the field-free φ_g state for $H_2^+(1s\sigma_g)$.

R(a.u)	Intensity (W/cm²) for linearly polarized λ=769nm laser					
	5×10^{13}	1×10^{14}	2×10^{14}	3×10^{14}	5×10^{14}	1×10^{15}
2.0	-	-	2×10^{-7}	2×10^{-5}	0.0009	0.004
2.5	-	-	0.0002	0.0008		0.14
3.0		0.0001	0.0006	0.003	0.04	
3.5	0.0003	0.0010	0.003	0.016	0.36	
4.0		0.009	0.017			
4.5[a]		0.013	0.04			
5.0		0.012	0.06			
6.0[b]		0.014	0.10			
7.0[b]		0.022	0.10			
8.0[b]		0.023	0.14			

R(a.u)	Intensity (W/cm²) for linearly polarized λ=228nm laser[c]					
	5×10^{13}	1×10^{14}	2×10^{14}	3×10^{14}	$3.9\text{x}0^{14}$	5×10^{14}
2.0	2×10^{-5}	0.0001	0.0004	0.004	0.018	0.028
3.0[a]	0.014	0.022	0.11	0.12	0.12	0.16

[a] At these distances $V_u - V_g \approx \hbar\omega$ and length and velocity couplings are equivalent.
[b] At large distances the φ_s(length) and φ_f(length) states give identical rates and approach the H(1s) rate defined by the dissociation limit H(1s) +H$^+$.
[c] Wavelength and distances chosen for comparison with comparable rates obtained by Chelkowski etal[1]. The two sets of calculations are in excellent.

The situation for λ=228 in Table 1 is as presented in reference 1. At R=2au the ionization rate is negligible for I$\leq 2\times10^{14}$W/cm², but rapidly increases as R→3au. Indeed, at this wavelength vibrationally stabilized molecules are expected to be predominantly trapped in the vicinity of R$\approx$3au[4], and this enhanced ionization rate puts a severe upper bound on how long such trapped molecules can survive without being ionized. In this case, as already emphasized by Chelkowski etal[1], either very short pulses, or peak intensities appreciable below 5×10^{13}W/cm² are required if stabilized populations of H_2^+ are to survive without being ionized and experiencing the effect of Coulomb explosion.

The results for $\lambda=769$nm are much more encouraging. The population of vibrationally trapped H_2^+ molecules[3] for $\lambda=769$nm is predominantly located at $R\approx3.5$-4.0au. The results presented in Table 1 suggest that at $I=2\times10^{14}$W/cm^2 and certainly for $I=1\times10^{14}$W/cm^2 the stabilized molecules might survive a 150fsec pulse without appreciable ionization. For higher intensities the ionization of the trapped molecules will again lead to Coulomb explosion. Such processes are suggested by recent experiments with $\lambda=769$ radiation[7]. It is our hope that the ionization rates provided by these calculations can be used to predict the proton kinetic energy distribution associated with the explosion.

ACKNOWLEDGMENTS

We thank H.G. Muller for his generousity in sharing his results and code with us. FHM and AGS acknowledge travel support by a NATO grant for International Collaborative Research.

REFERENCES

1. S. Chelkowski, T. Zuo, and A.D. Bandrauk, Phys. Rev. A **46**, R5342 (1992)

2. A. Giusti-Suzor, X. He, O. Atabek, and F.H. Mies, Phys. Rev. Lett. **64**, 515 (1990).

3. H.G. Muller, in "Coherence Phenomena in Atoms and Molecules in Laser Fields" Ed. A.D. Bandrauk and S.C. Wallace, Plenum Press, NY,(1992) pages 89-98.

4. A. Giusti-Suzor, and F.H. Mies, Phys. Rev. Lett. **68**, 3869 (1992)

5. J.L. Krause, K.J. Schafer, and K.C. Kulander, Chem. Phys. Lett. **178**, 573 (1991).

6. A. Zavriyev, P.H. Bucksbaum, J. Squier,and F. Saline, Phys. Rev. Lett. **70**, 1077 (1993).

7. F.H. Mies, A. Giusti-Suzor, K.C. Kulander, and K.J. Schaffer, submitted to Phys. Rev. A.

DYNAMICS OF DISSOCIATION VERSUS IONIZATION
IN STRONG LASER FIELDS

L. F. DiMauro and Baorui Yang

Chemistry Department
Brookhaven National Laboratory
Upton, NY 11973

INTRODUCTION

The behavior of isolated atoms in strong radiation fields has been a topic of intensive theoretical and experimental interest over the last decade and has resulted in a cumulative literature rich in new phenomena, unresolved issues, and exciting new predictions for future studies.[1] Recently, some investigations have begun to examine the general behavior of molecules in intense fields, with special relevance placed on strong-field photodissociation dynamics. The fundamental motivation for such studies derives its interest from the ability to control chemical dynamics by external variation of a laser field. The traditional chemical physics approach relied upon the laser frequency as the only external field parameter for achieving state-selectivity but has met with little success. However, current efforts[2] have focused on examining all aspects of the laser field for achieving control. Although premature in realization, an understanding of the influence of intensity, coherence, phase, and pulse duration and shaping in concert with frequency may ultimately lead to our ability or inability to control chemical dynamics.

The special challenge encountered in studying molecules in strong fields results from the difficulty of sorting out the general behavior of field-induced effects from the details or specifics of the molecular structure. Investigations on atoms[3,4] clearly demonstrate the importance of the atomic structure in describing the ionization dynamics in the multiphoton regime. Consequently, it is not surprising to expect enhanced structural effects in molecules due to the many internal degrees of freedom as compared to atoms. However, the anticipated complications posed by molecules are in practice quite tractable. In fact, the field-molecule interaction can lead to some surprising insights into the molecular structure.[5,6] Furthermore, molecular strong-field studies raise some interesting and unique questions concerning the role and interplay of ionization and dissociation, and can result in studies of atoms in unusual circumstances.[7]

In this paper, experimental results are presented which clearly demonstrate the effectiveness that an external field has in altering the dissociation dynamics. The experiment examines the strong-field dissociation dynamics of molecular hydrogen ions and its deuterated isotopes. These studies involve multiphoton excitation in the intensity regime of $10^{11\text{-}14}$ W/cm^2 with the fundamental and second harmonic of a Nd:YAG or Nd:YLF laser system. Measurements include energy resolved electron and mass spectroscopy which provide useful probes in elucidating the interaction dynamics predicted by existing models. The example discussed in this paper, examines the strong-field dissociation of H_2^+, HD^+, and D_2^+ at green (0.5 μm) and infrared (1 μm) frequencies. The diatomic

Super-Intense Laser-Atom Physics, Edited by
B. Piraux *et al.*, Plenum Press, New York, 1993

ions are formed via multiphoton ionization of the neutral precursor which is physically separable from the dissociation process. This study provides the first observation of the dynamics associated with the above threshold dissociation (ATD) process and analogies will be made with the more familiar above threshold ionization (ATI) phenomenon.

EXPERIMENTAL

The photon sources used are well characterized Nd:YAG (1.06 μm, 10 nsec, 10 Hz) and Nd:YLF (1.05 μm, 50 psec, 1 kHz) lasers. Details of these laser systems have been described previously.[8] The second harmonic is generated using standard nonlinear techniques either in BBO or KD*P crystals. The intensity ranges for 0.532 and 0.527 μm radiations are 8×10^{11} to 9.7×10^{12} W/cm^2 and 5×10^{12} to 4×10^{13} W/cm^2, respectively. The apparatus consists of an ultrahigh vacuum chamber equipped with time of flight electron and mass spectrometers. The field-free photoelectron spectrometer has an energy resolution of about 50 meV for 1 eV electrons and an acceptance angle of 1×10^{-3} sr. It is calibrated using the well known ATI spectrum of xenon atoms. The mass spectrometer consists of a 42 cm long drift tube with a series of electric field acceleration plates. Fragment kinetic energy analysis is achieved by applying an uniform extraction field across the laser focus with the laser polarization parallel to the flight tube axis. This produces two nearly symmetric peaks in the time of flight spectrum corresponding to the two velocity components of the same fragment initially directed towards and away from the detector. The kinetic energy of the fragment is simply determined by measuring the arrival time difference between the peaks. The mass spectrometer energy calibration was performed by recording the weak-field photodissociation of chlorine[9] molecules. The resolution was estimated to be about 100 meV for 1 eV protons. The background pressure in the ultrahigh vacuum system was 1×10^{-9} torr. The data collection system is always operated in the pulse counting regime with count rates much less than one event per laser shot.

RESULTS AND DISCUSSION

Diatomic hydrogen ion represents the *"model"* molecular system to study strong-field behavior because of its simple and well-known one-electron structure and as a result should provide the foundations for future molecule-field studies. A physically distinct motivation for studying the strong-field behavior of molecules instead of atoms is the presence of dissociative channels. Questions addressing the validity of extending our knowledge on intense field atomic ionization to describe molecular dissociation seem warranted. For example, does high field dissociation proceed in an analogous manner to ATI, that is, via absorption of additional photons beyond the minimum needed to break a molecular bond and are the dynamics similar. We will address these issues through out this paper and contrast the differences between dissociation and ionization.

Dressed-State Model

Photodissociation of a molecule in an intense field can lead to a phenomenon which is analogous to above threshold ionization (ATI) observed in the photoionization of atoms and molecules. The general effect results in the absorption of additional photons beyond the minimum necessary to break the chemical bond and has been dubbed above threshold dissociation (ATD). The traditional weak-field representation of the photodissociation process is shown in Fig. 1(a) for the first two potential curves of the H_2^+ molecule. The vertical arrows show the internuclear separation for the Franck-Condon (FC) favored 1- and 3-photon absorption. Thus, the absorption of additional photons results in the production of a series of hydrogen atoms and protons with kinetic energies differing by $(1/2)\hbar\omega$ or $\hbar\omega$ total energy. Energetically this representation is equivalent to the more standard ATI picture. However, this picture offers no intuitive understanding of the dynamics of the dissociation process nor potential distortion or vibrational shifts in the strong field limit.

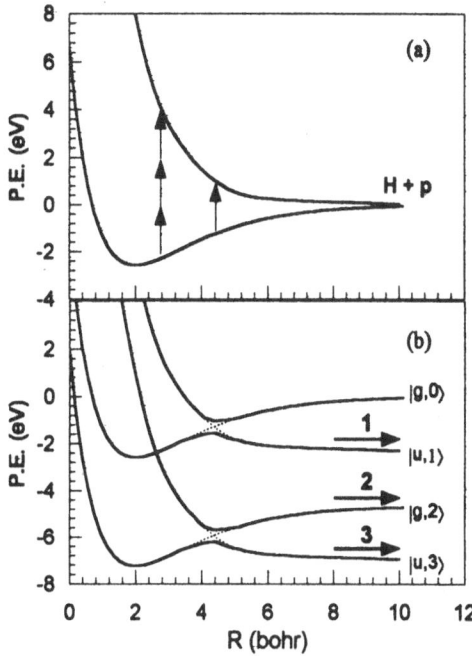

Figure 1. Potential diagrams for (a) traditional photodissociation and (b) dressed-state model for 0.5 μm excitation.

In order to obtain a more intuitive feel for this physics it is advantageous to reformulate the problem in a dressed state basis. In this model, the laser field is easily incorporated as a strong perturber to the molecular potentials. A number of papers[10,11,12] have been published concerning this approach and the readers are referred to them for more details. A finite number of molecule-field states $|d\rangle$ are constructed as the product of the molecular state $|\phi\rangle$ and the photon-number state $|N-n\rangle$, where $\phi \equiv 1s\sigma_g$ or $2p\sigma_u$, $n = 0, \pm 1, \pm 2...$ and N is the total photon number. Therefore, the weak-field two state picture shown in Fig. 1(a) can be transformed into a infinite series of paired dressed states, as illustrated in Fig. 1(b) for excitation with 2.3 eV photons. The photodissociation is represented by passage along the potential curves to large internuclear separation at the limits corresponding to 1-photon ($|u,1\rangle$), 2-photon ($|g,2\rangle$), and 3-photon ($|u,3\rangle$) absorption (indicated by the horizontal arrows). The potential curve crossings occur at internuclear separations corresponding to multiples of the photon energy (dotted diabatic curves). Since the ground and repulsive ionic state have gerade and ungerade symmetry, only level crossings with an odd difference in photon number can interact because of parity conservation. Note, that the crossings in Fig. 1(b) occur at the odd photon absorption in Fig. 1(a) at the FC favored internuclear separation. Consequently, the molecular potential couple and distort (solid curves in Fig. 1(b)) via the strong dipole interaction. Any populated vibrational levels within the gap will become unstable and predissociate. The number of unstable vibrational states increases with larger field strength since the avoid-crossing gap continues to open. Bond-softening occurs when the field-induced dissociation is at the 1-photon crossing. Furthermore, ATD can occur due to level crossings at higher order couplings. For example, 3-photon dissociation can diabatically dissociate along curve $|u,3\rangle$ or adiabatic 2-photon dissociation along path $|g,2\rangle$. Consequently, a *net even number* of photon dissociation can also occur due to the absorption and stimulated emission of photons. Thus, we see that the dressed-state picture renders an intuitive feel for predicting the dissociation dynamics and we will examine this connection throughout the experimental results.

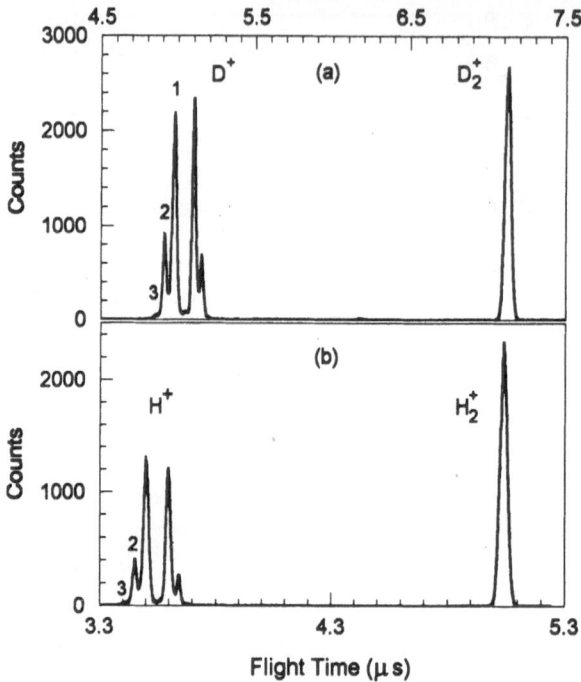

Figure 2: Time of flight mass spectrum resulting from 0.5 μm excitation of (a) D_2 and (b) H_2 at 9.7 x 10^{12} W/cm^2. Peak labels 1, 2, and 3 correspond to the number of absorbed photons leading to dissociation.

H_2^+ dissociation: 0.53 μm excitation

Figure 2 shows a typical time of flight mass spectrum resulting from the 0.532 μm nonresonant MPI of H_2 and D_2 molecules at an intensity of 9.7 x 10^{12} W/cm^2. The molecular ions formed by the MPI of the neutral molecule produces the peak at long arrival times in the spectra of Fig. 2. The peaks at the short arrival times are the atomic fragments (protons and deuterons) formed via dissociation of the molecular ion. The symmetric substructure results from the forward and backward directed fragments and the splitting is proportional to their initial kinetic energy, as discussed in Experimental Section. The three fragment peaks are labeled according to the number of photons being absorbed by molecular ions leading to dissociation. Figure 3 shows the proton spectrum converted to total kinetic energy. The tick marks indicate the proton's kinetic energy for different initial unperturbed vibrational energy. Accordingly, at low intensity (Fig. 3(b)) most H_2^+ ions are dissociated from the $v_{avg}^+ = 5$ vibrational level following the absorption of a single photon via bond-softening. The maximum of this peak is observed to shift from $v^+ = 5$ to $v^+ = 4$ at higher intensities, as seen in Fig. 3(a). At higher intensity another peak emerges corresponding to two photon dissociation (peak at $2h\nu$ in Fig. 3) via $v_{avg}^+ = 2$. Consequently, the spacing of the 1- and 2-photon peak is less than a photon total energy due to differing initial vibrational states and is intensity dependent. *Thus, an immediate difference is apparent between ATD and ATI which is related to the fact that different order odd photon processes have their origin in different initial states or in the dressed state language, at different internuclear separation.* At still higher intensities a third peak appears one photon total energy away from the second peak corresponding to a three photon dissociation. It is clear from the dressed state model that the peak separation for any pair of Floquet states is one photon total energy since dissociation occurred from the same initial vibrational state. The same general behavior is observed for D_2 and HD molecules[13]. Similar kinetic

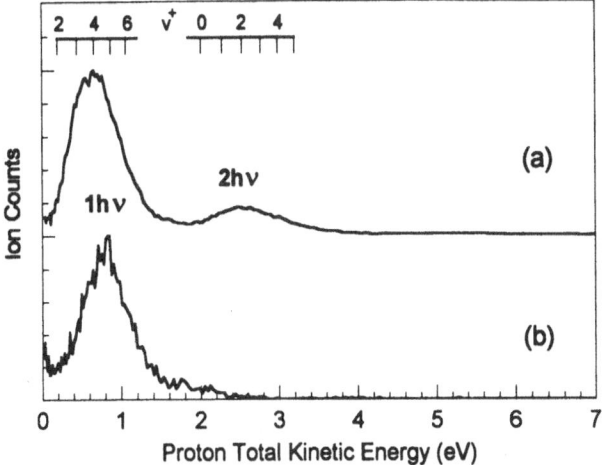

Figure 3: Total kinetic energy (twice the measured proton kinetic energy) spectrum of H$^+$ fragments resulting from 0.532 μm dissociation of H_2^+ at (a) 9.7 TW/cm^2 and (b) 2.7 TW/cm^2. The tick marks show the position of the unperturbed ionic ground state's vibrational levels in one (left) and two (right) photon dissociation processes.

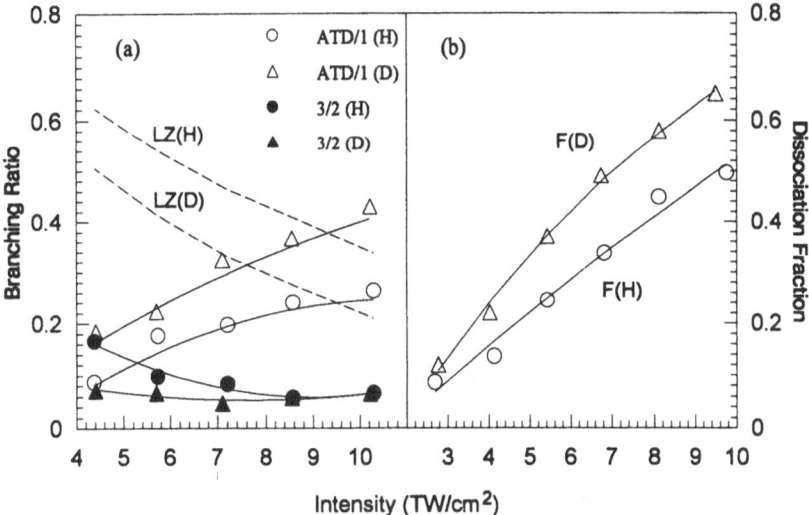

Figure 4: A plot of the (a) ATD branching ratios, R, and (b) dissociation fraction, F, for H_2^+ and D_2^+ as a function of 0.53 μm intensity.

345

energy shifts are observed in the 0.527 μm (psec) spectra over an intensity range of 5×10^{12} to 4×10^{13} W/cm^2. Shown in Fig. 4(b) is the dissociation fraction F^H and F^D as a function of laser intensity for H_2^+ and D_2^+, respectively and where $F^H = [H^+/(H^+ + H_2^+)]$ and $F^D = [D^+/(D^+ + D_2^+)]$. Note, that the dissociation fraction for each molecule increases with intensity and that F^D is larger than F^H at all intensities shown.

We have carefully examined the photodissociation of H_2^+ molecules by recording the proton kinetic energy spectra and the H_2 photoelectron spectra simultaneously as a function of laser intensity. Our results verify that the mechanism responsible for H^+ and D^+ formation is the photodissociation of the parent molecular ion. It appears that the amount of H^+ produced is proportional to that of photoelectrons due to (7+1) molecular ionization. In addition, energy conservation implies that seven second harmonic photons of YAG(YLF) can populate only up to $v^+ = 3(4)$ vibrational levels in the ionic ground state, whereas the lowest intensity data have proton energies peaked at $v^+ = 5$. Thus, we conclude that the H^+ peak labeled 1 in Fig. 2 (0.7 eV total energy) is due to the eight photon ionization of H_2 followed by one photon dissociation of H_2^+.

The dressed-state model predicts that as the field strength increases the fragments kinetic energy will decrease as lower vibrational levels become unstable in the field. Physically, the distort potential corresponds to a softened potential at the outer turning points which decreases the vibrational frequency with increasing field strength. In the current experiment, we observed apparent negative shifts in proton kinetic energies with increasing laser intensity with both 532 nm (ns), as shown in Fig. 3, and 527 nm (ps) pulses. The maximum population of the 1-photon proton kinetic energy peak shifts by ~ 200 meV over the entire intensity range. Likewise, the maximum of the ionic vibrational state distribution in the photoelectron spectroscopy (PES) changes from high to low vibrational levels in a similar intensity range, although the shift in the amplitude of the vibrational distributions is independent of the molecule studied. Thus, the bond--softening mechanism to first order is predominantly an *electronic coupling* and not vibrational. However, we could not observe any commensurate positive energy shifts in the PES peak positions with increasing intensity although a broadening is clearly evident. An asymmetric broadening in our PES could result from spatial or temporal averaging in our experiment masking a "pure" energy shift but the sign of the observed broadening is opposite to that predicted by a bond-softening model.[13] This inconsistency could imply a shortcoming in our numerical method or an indication that the PES probes the potential at a low intensity before significant potential distortion has occurred. Another contributing factor to the peak width could result via strong-field rotational pumping with increasing laser intensity, causing the photoelectron peaks to broaden towards lower energies.

Referring to Fig. 4(a), the ratio ($R_{2+3,1}$) of dissociation, labeled ATD/1, via the sum of the 2- and 3-photon (ATD) versus the 1-photon channels (peaks labeled 1, 2, and 3 in Fig. 2) for H_2^+ (D_2^+) changes from 7% to 25% (18% to 42%) as the intensity is increased to 1.3×10^{13} W/cm^2. Thus, the *total amount of ATD* increases with intensity in a way analogous to the intensity dependence of ATI. The difference observed in the dissociation fraction, F, for H_2^+ and D_2^+ at any given intensity is a vibrational effect which results in a better FC match in the unperturbed vibrational levels for D_2^+ at low intensity. Consequently, D_2^+ dissociates more readily than H_2^+ at this wavelength.

Further insight into the physically distinct dynamics between strong-field dissociation and ionization becomes evident by examining the ATD branching ratio R_{32} (labeled 3/2) for 3- versus 2-photon dissociation of H_2^+ in Fig. 4(a). This ratio decreases from 16.5 to 6.5% over the entire increasing intensity range. Thus, the branching ratio among the high energy fragment channels favors the production of low energy protons with increasing intensity, in sharp contrast to the behavior of ATI. Furthermore, the D_2^+ ratio R_{32} is smaller than H_2^+ and relatively constant (~ 6%) over the entire intensity range. These branching ratio effects can be all understood with the dressed-state model. Physically, the increasing laser intensity corresponds to a larger avoided-crossing gap resulting in a decrease in the 3-photon diabatic transition rate while favoring 2-photon adiabatic passage (3-photon absorption, 1-photon emission). Since H_2^+ and D_2^+ have a relatively large difference in vibrational frequencies due to their different masses, therefore at low intensities where the gap is not large enough to completely shut off the 3-photon diabatic path as compared to H_2^+ due to the fact that a smaller vibrational frequency implies more adiabatic motion. Specifically, $R_{32}(H) > R_{32}(D)$ at low intensity. As the gap continues to widen, one should also observe the ratio R_{32} to decrease for both H_2^+ and D_2^+ molecules. Such behavior is clearly demonstrated in our experiments.

Figure 4(a) also shows for comparison the results of the calculated fragment $R^{LZ},_{32}$ ratios (labeled LZ) predicted by a simple Landau-Zener (LZ) theory.[14] The theory predicts well the general behavior of the ratios as the light intensity changes. However, the calculated ratios are approximately three times larger than the experimental values which could imply that the degree of deformation of the calculated potential curves is beyond the limits of applicability of simple LZ theory or a short coming of our numerical Floquet analysis.

H_2^+ dissociation: 1.06 μm excitation

Figure 5 is a plot of a typical time of flight mass spectrum resulting from dissociation of H_2 with 1.06 μm radiation. The dissociation fraction F^H for H_2^+ molecules, where $F^H = [H^+/(H^+ + H_2^+)]$, changes from 0.25 to 0.43 for the entire dynamic range of laser intensities $2.3 - 4.6 \times 10^{12}$ W/cm². The peaks labeled a and b correspond to a total proton kinetic energy of 0.49 and 2.88 eV, respectively, and are assignable to the dissociation of H_2^+ following absorption of 2- and 4-photons. For both processes, the measured proton kinetic energies give an initial vibrational distribution with an average value of $v_{avg} = 3$. *No odd number photon dissociation processes are observed* in the present experiment at all intensities studied. Furthermore, the ATD ratio, $R_{42}(H)$, for the 4-photon and 2-photon dissociation channels are intensity dependent and increase from 0.11 to 0.14 with increasing intensity.

The photoelectron spectra of H_2 taken at two different laser intensities with 1.06 μm radiation is shown in Fig. 6. The tick marks indicate the number (order) of photons absorbed by the H_2 molecule. Since the IP of H_2 is 15.42 eV, a minimum of 14 photons are required to ionize. The low-intensity photoelectron spectrum plotted in Fig. 6(b) can be characterized as a broad distribution centered at 5 eV which exhibits suppression of low energy electrons up to 2.8 eV and a maximum kinetic energy at 15 eV. However, the spectrum still shows structure which is vibrationally resolvable and assignable to the first four vibrational levels of the H_2^+ ground state. As the laser intensity is increased in Fig. 6(a), the photoelectron distribution tends to become structureless and shifted towards higher kinetic energy (peaked at 7.5 eV). Likewise, the low-energy electron peaks associated with the 16 and 17-photon ionization processes show additional suppression. The MPI and dissociation of H_2 at 1.06 μm have been studied by Zavriyev *et al.*[13] using 100 ps pulses. Their photoelectron distributions show a similar shape to those shown in Fig. 6 but with no resolvable structure. The difference between the two experiments probably reflects the change in saturation intensity *via* the pulse width dynamics which unfortunately results in an inability to extract any

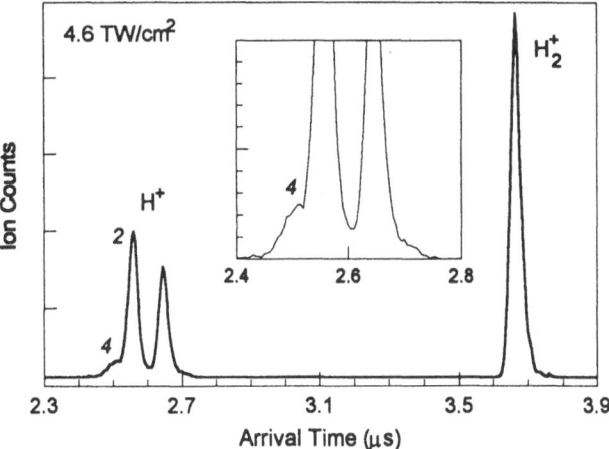

Figure 5: Time of flight mass spectrum of H_2 resulting from 1.06 μm excitation at an intensity of 4.6 TW/cm². The peak labelled 2 and 4 are the forward velocity components formed via the 2- and 4-photon dissociation of H_2^+. The insert is an expanded view of the proton distribution.

detailed information from the photoelectron spectrum. Consequently, long pulse studies can provide a useful look at the mapping of the molecule-field interaction. The mechanism resulting in the suppression of low energy electrons in Fig. 6 is not understood. The suppression is analogous in its behavior to *"channel closure"* observed in rare gas atoms[15] which is caused by the increasing binding energy of the atom in the field and proportional to the ponderomotive potential. However, the ponderomotive potential available at our highest intensity is only 0.5 eV, which is obviously not enough shift to account for our degree of electron suppression.

The proton kinetic energy spectrum of Fig. 5 shows no evidence that 1-photon dissociation is occurring at 1 μm. Thus, bond-softening is not playing any significant role in the dissociation dynamics at this wavelength, although it dominates at 0.53 μm. This result is not difficult to understand, since the 1-photon curve crossing associated with the dressed-state picture illustrated in Fig. 7, occurs at large enough internuclear separation (~ 5 bohr) which necessitates the need for significant population in the higher vibrational (v = 7) levels of the molecule. However, our photoelectron spectrum analysis shows that only the lower vibrational states are populated via the MPI process.

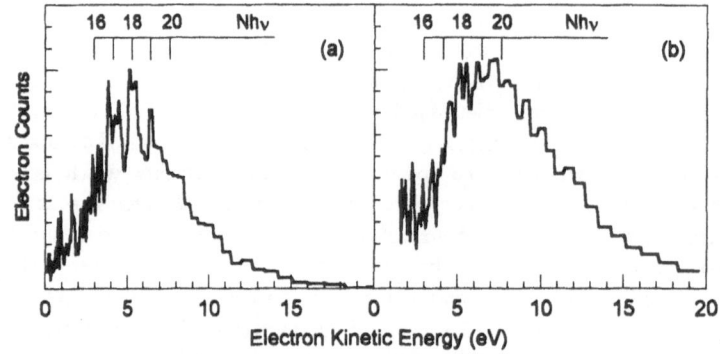

Figure 6: Photoelectron spectra of H_2 molecule taken at 1.06 μm intensities of (a) 2.3 TW/cm^2 and (b) 4.1 TW/cm^2. The tick marks indicate the number of absorbed photons above the H_2 ground state.

As discussed above for 0.53 μm excitation, the ATD ratio (R_{32}) decreases with increasing laser intensity, that is, dissociation occurs more readily through the lower-order channels as the laser intensity increases. This behavior is just the opposite at 1 μm, where the ATD ratio (R_{42}) increases with rising intensity. The 1 μm behavior is analogous to the more familiar intensity dependence observed in ATI but with a major difference being the formation of fragments with energy spacing $2\hbar\omega$. Although this strong wavelength dependence results in such different dynamics, the interpretation is all consistent with the dressed-state model discussed above. According to the dressed-molecular states model, illustrated in Fig. 7 for 1 μm radiation, the diabatic dressed-states $|u,3>$ and $|u,5>$ cross the ground state $|g,0>$ at 3.6 and 3.1 bohr, respectively. Furthermore, the intersections are electric-dipole allowed ($g \rightarrow u$) which result in an intensity-dependent avoid-crossing. The gap also occurs in the potential region which are vibrational populated by the MPI process. The same argument applies to the diabatic crossing (dashed line in Fig. 7) between the $|g,2>$ and $|u,3>$ curves. The 1-photon interaction results in an avoid-crossing which is more strongly laser dependent than the higher order processes. In addition, the bond-softening gap intensity dependence is greater at 1 μm than 0.5 μm because of the increasing electric dipole moment at larger internuclear separation. Consequently, the high intensity and longer wavelength causes the 1-photon gap to become so large (solid lines) that adiabatic passage dominates (3-photon absorption, 1-photon emission) resulting in only 2-photon dissociation. The same logic then follows for any dressed-state pairs, *i.e.* $|g,4>$ and $|u,5>$, which always result in even-photon dissociation. This is physically manifested in our proton kinetic energy spectrum as the series of peaks separated by the 2-photon (total) energy. Likewise, the increase in the ATD ratio (R_{42}) results from the fact that each proton peak in the series occurs via a different order photon process. Thus, the dynamics, as manifested by the increasing ATD ratio, results in more efficient production of higher-order protons with increasing intensity. This has similar dynamics compared with production of high energy ATI electrons.

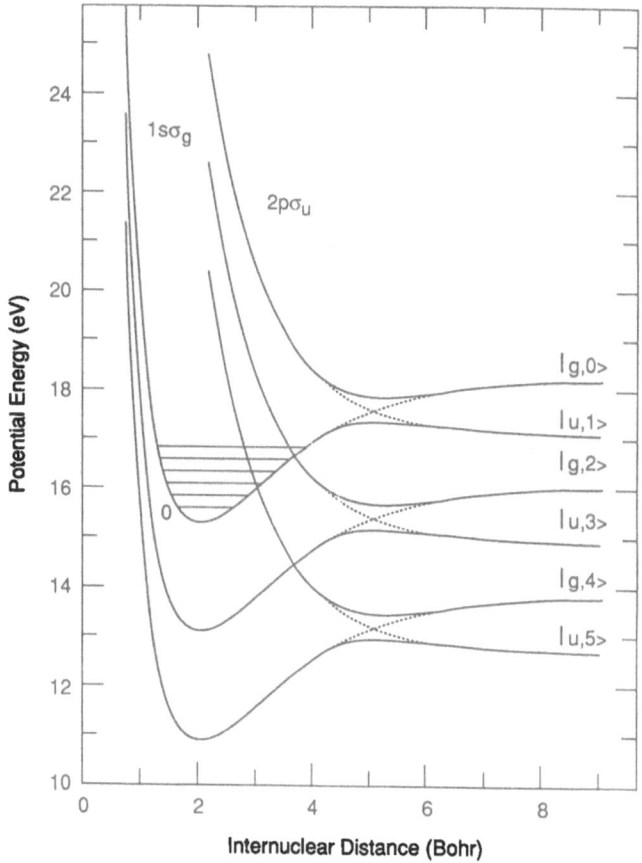

Figure 7: The dressed-state H_2^+ molecular potentials for a 1.06 µm laser field. The dotted lines shows the diabatic crossings which distort (full lines) due to the molecule-field interaction.

CONCLUSIONS

In conclusion, the experiments presented above clearly demonstrate the effectiveness that an external field has in altering the dissociation dynamics. Furthermore, the strong-field dissociation dynamics manifest very distinctive behavior compared to ionization. The dressed-state picture has proven to be a potent intuitive model for understanding the high-field dynamics and should provide a valuable predictive model for future studies.

ACKNOWLEDGMENTS

We would like to thank Dr. F. Mies and Prof. A. Giusti-Suzor for their useful discussions. This research was carried out at Brookhaven National Laboratory under contract No. DE-AC02-76CH00016 with the U.S. Department of Energy and supported by its Division of Chemical Sciences, Office of Basic Energy Sciences.

REFERENCES

1. See for example, "Atoms in Intense Fields", edited by M. Gavrila, Academic Press, Orlando, 1992.
2. See for example, "Coherence Phenomena in Atoms and Molecules in Laser Fields", A. D. Bandrauk and S. C. Wallace, eds., Plenum Press, New York, 1992.
3. R. R. Freeman, P. H. Bucksbaum, H. Milchberg, S. Darak, D. Schumacher, and M. E. Geusic, *Phys. Rev. Lett.* **59**, 1092 (1987).
4. Dalwoo Kim, S. Fournier, M. Saeed, and L. F. DiMauro, *Phys. Rev.* **A41**, 4966 (1990).
5. B. Walker, M. Saeed, T. Breeden, B. Yang, and L. F. DiMauro, *Phys. Rev. A* **44**, 4493 (1991).
6. L. F. DiMauro, B. Yang, and M. Saeed, *in*: "Coherence Phenomena in Atoms and Molecules in Laser Fields", A. D. Bandrauk and S. C. Wallace,eds., Plenum Press, New York (1992).
7. M. Saeed, B. Yang, X. Tang, and L. F. DiMauro, *Phys. Rev. Lett.* **68**, 3519 (1992).
8. M. Saeed, D. Kim, and L. F. DiMauro, *Appl. Opt.* **29**, 1752 (1990); L. F. DiMauro, D. Kim, M. W. Courtney, and M. Anselment, *Phys. Rev.* **A38**, 2338 (1988).
9. L. Li, R. Lippert, J. Lobue, W. Chupka, and S. Colson, *Chem. Phys. Lett.* **151**, 335 (1988).
10. M. V. Fedorov, O. V. Kudrevatova, V. P. Makarov, and A. A. Samokhin, *Opt. Comm.* **13**,2991975).
11. A. Bandrauk and M. Sink, *J. Chem. Phys.* **74**, 1110 (1981).
12. A. Guisti-Suzor, X. He, O. Atabek, and F. Mies, *Phys. Rev. Lett.* **64**, 515 (1990).
13. A. Zavriyev, P. Bucksbaum, H. Muller, and D. Schumacher, *Phys. Rev.* **A42**, 5500 (1990); B. Yang, M. Saeed, L. F. DiMauro, A. Zavriyev, and P. Bucksbaum, *Phys. Rev.* **A44**, R1458 (1991).
14. L. D. Landau, and E. M. Lifshitz, "Quantum Mechanics", Pergamon Press, New York, 1965.
15. R. R. Freeman, P. H. Bucksbaum, and T. J. McIlrath, *IEEE J. Quant. Elec.* **24**, 1461 (1988).

MOLECULES IN STRONG LASER FIELDS:

WHY ARE THE FRAGMENTS ALIGNED?

D. Normand and C. Cornaggia

Service de Photons des Atomes et des Molécules (SPAM)
Centre d'Etudes de Saclay
91191 Gif-Sur-Yvette Cedex
France

INTRODUCTION

When submitted to an intense laser field, a molecule is first multiply ionized without significant modification of its internuclear distance[1,2]. Except for the singly charged molecular ions, the multiply charged molecular ions are very instable and they cannot be detected. In fact, the only ions accessible to the experiment are the atomic fragments coming from the dissociation of the molecular ions.

In the ionization process, the molecular ion does not gain any kinetic energy since the outgoing electron takes away the excess photon energy, but it does get potential energy which converts into kinetic energy of the fragments during the dissociation process. By analyzing carefully the characteristics of the fragments, one can trace back the fragmentation channels and identify the molecular ion precursors. From the measurement of the fragment energies, one thus has access to the potential energy of these precursors and one can infer the processes which happened at the very beginning of the interaction. In particular, one should be able to determine if the molecular ions are formed, or not, on Coulombic states[1].

In this paper we will focus our attention on a very basic and fruitful property of the interactions of molecules with strong laser fields, namely that the fragments are essentially emitted along the laser polarization ε. The experiments are performed on CO molecules, alternatively with a 30 ps YAG laser and with a 1 ps Ti-Sapphire laser. In a first section we show that the fragments are indeed aligned and we give the angular distributions of the different decay paths. In the second section, we explore the origin of the fragment confinement. Using a double pulse experiment, we prove that the linearly polarized laser induces a torque that forces the molecules to align with the ε vector[3].

ANGULAR DISTRIBUTIONS OF FRAGMENTS

We first describe briefly the experimental setup and then we give the angular distributions of the fragments issued from the Multielectron Dissociative Ionization (MEDI) of CO for two laser pulse durations (1 and 30 ps), at roughly the same laser wavelength (1053 and 1064 nm respectively).

Experimental Set-up.

The experiment basically consists of an intense picosecond infrared laser (YAG or Ti-Sapphire) tightly focused in a vacuum chamber by a 60 mm lens[4]. The base pressure is 2 10^{-10} Torr, whereas the CO operating pressure is adjusted between 10^{-9} and 5 10^{-6} Torr to optimize the ion counts as a function of the laser intensity.

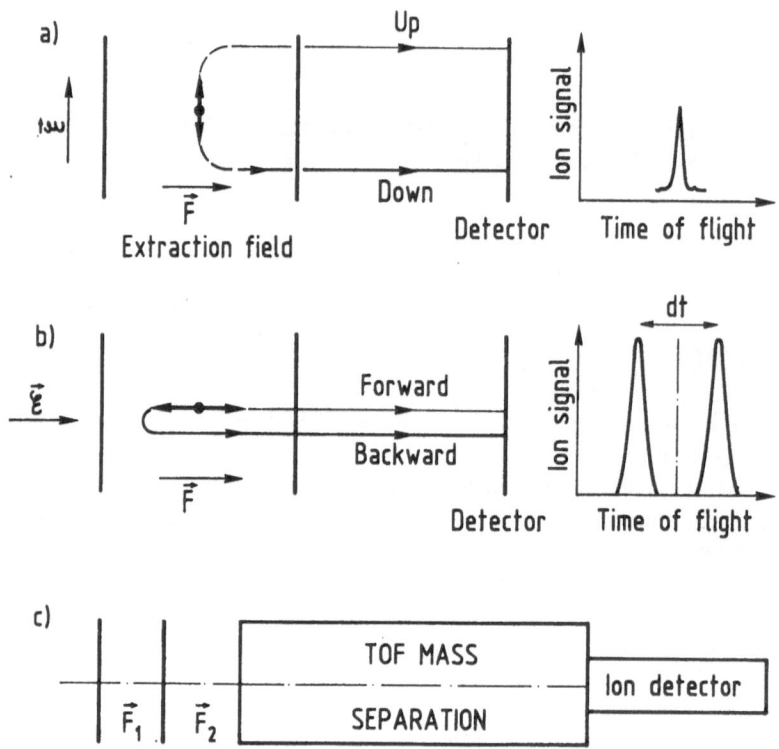

Figure 1. Schematic ion collection set up showing the ion trajectories and the corresponding TOF ion spectra when the laser polarization vector ε is: a) perpendicular b) parallel to the electric extraction field F. c) shows the ion spectrometer with the double acceleration chamber.

Fig.1 shows the principle of the ion measurements. The ions created at the laser focus are accelerated by a dc electric field F_1 and directed towards the detector, via a drift tube where they are separated according to their mass over charge (m/q) ratio. When the laser polarization vector ε is set perpendicular to the detection axis (horizontal axis), the fragments are sent up and down in the drift tube. Due to the presence of a 3 mm aperture in front of the detector, these ions cannot hit the detector (Fig.1a). Only the fragments emitted with very weak tranverse velocities will reach the detector. This configuration is used to identify the masses and charges of all the fragments, since the time-of-flight (TOF) is independent of the initial velocities. Fig.2a gives an illustration of a TOF spectrum obtained with ε perpendicular to F_1.

On the contrary, when the ε vector is set parallel to F_1, the ions stay confined on the detector axis and most can hit the detector (Fig.1b). Statistically, 50% of the fragments are emitted towards the detector: these forward ions arrive at the detector after a time of flight shorter than that of the zero energy ions with same m and q. The other 50% are emitted backwards, but they are reflected back by the dc field F_1 and arrive later at the detector.

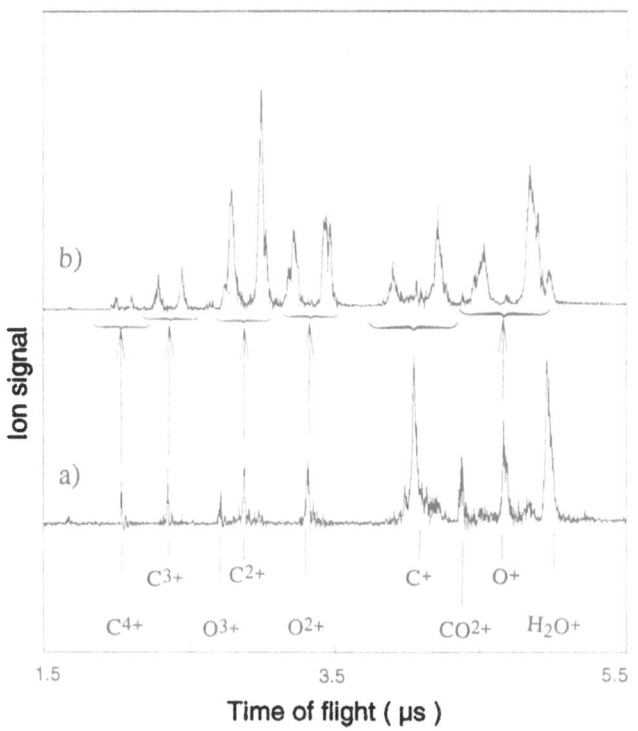

Figure 2. Two TOF ion spectra of CO, recorded at 1053 nm, for a laser intensity of 10^{16} W/cm² and a pulse duration of 1 ps. The laser polarization vector is a) perpendicular; b) parallel to the detection axis. In case a), the CO pressure is increased by a factor of 5; the ion signals are also magnified by a factor 5

Thus an ion class, with given charge, mass and energy, yields two ion peaks on the TOF spectrum, whose separation time δt allows to determine the initial energy of the ion with

$$E_0 \; = \; e \, q^2 \; F_1^2 \; \delta t^2 \, / \, 8 \; m \qquad\qquad (1)$$

where e is the electron charge. For example, the spectrum of Fig.2b shows that, with the ε vector parallel to F_1, all the ions coming from the dissociation of molecular ions appear as double peaks. Note that the ion peaks observed in fig.2a are also present in fig.2b, but barely visible in that case since the CO pressure is lower by a factor 5 and the energetic fragments much more abundant than the cold fragments. In order to ensure a good energy resolution, we use low F_1 values, whereas a stronger dc field F_2 drives the ions to the detector (Fig.1c).

The choice of the diameter of the aperture inserted at the end of the drift tube realizes a compromise between the collection efficiency and the angular resolution. The ion angular resolution is function of the initial kinetic energy. It is determined by computer simulation to be (Half Width at Half Maximum) 7° for 0.5 eV ions, 3° for 3 eV ions and 2° for 10 eV ions.

Ion Angular Distribution Measurements

In order to determine the angular distributions of the fragments, we record a series of TOF ion spectra of CO while rotating the angle α between the laser polarization vector ε and the detection axis (Fig.3). The laser intensity is kept constant at 6 10^{14} W/cm^2 during the experiment, the pulse duration being 30 ps at 1064 nm. For each α value, we control the CO$^+$ ion signal (not shown), which ensures that all parameters remain constant.

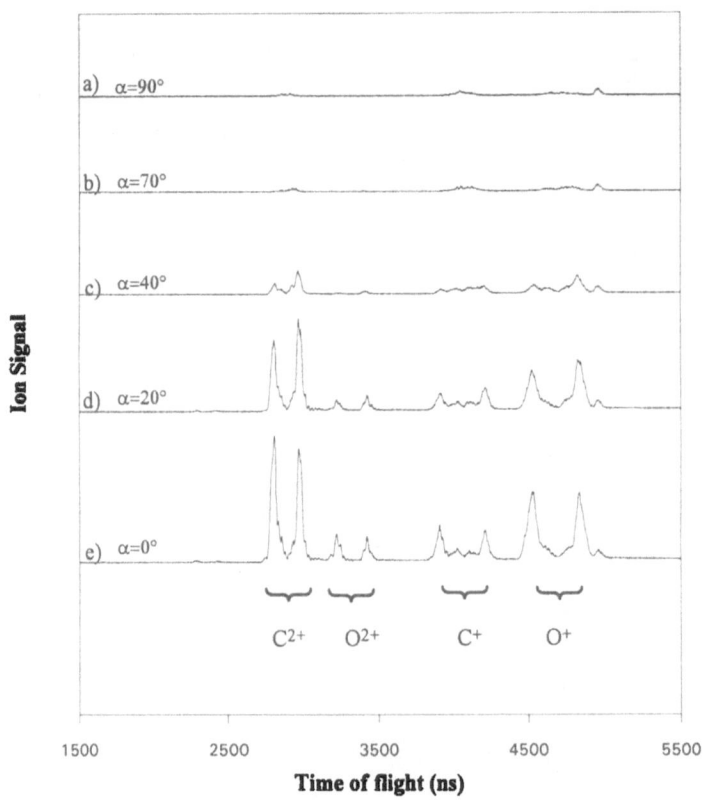

Figure 3. Series of TOF ion spectra of CO recorded for different values of the angle α between the laser polarization ε and the detection axis. The laser intensity is 6 10^{14} W/cm^2 at 1064 nm, with a 30 ps pulse duration.

By integrating the ion signals corresponding to a given fragmentation channel, we can determine the angular distributions for all channels. We correct the signals for the detector sensitivity by dividing the doubly charged ion signals by a factor 2. The results are presented in Table 1. For each decay path, we give the kinetic energy release, the branching ratio and the half-width $\alpha_{1/2}$ of its angular distribution. The channels are ordered according to increasing values of the threshold laser intensity. We observe that the fragment confinement along the ε vector increases when going from the $C^+ + O$ channel (that which appears at the lowest laser intensity) to the $C^+ + O^{++}$ channel. In short, the higher the intensity required to observe a channel, the tighter the confinement of the fragments.

To give a better visualization of the angular distributions, we show in fig.4 the number of fragments as a function of the angle α, in polar diagrams. For sake of clarity, the distributions for the carbon ions (Fig.4a) and the oxygen ions (Fig.4b) are presented separately. We observe that, once corrected for the detector sensitivity, the most abundant species are the "O^+ rapid", which makes sense since these ions are involved in the two main decay channels (channels 3 and 4 in table 1).

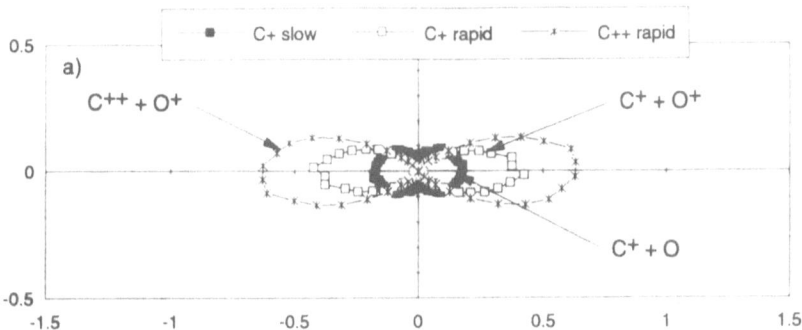

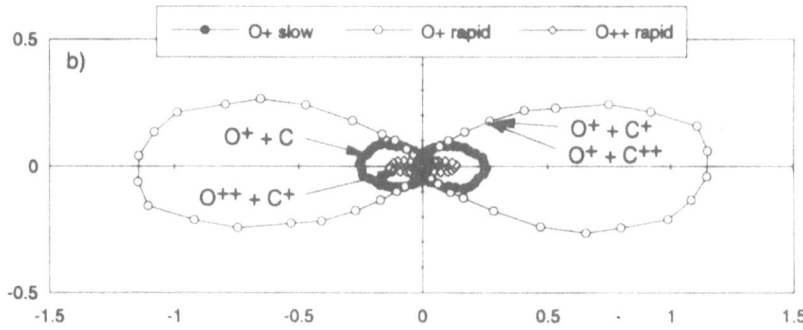

Figure 4. Polar diagrams showing the angular distributions of the main fragments observed in the MEDI of CO for: a) the carbon ions, b) the oxygen ions. The slow C^+ and O^+ ions are respectively associated with neutral O and C atoms. The rapid O^+ ions are associated with C^+ ions and with C^{++} ions. The laser intensity is $6\ 10^{14}$ W/cm^2 at 1064 nm with 30 ps pulses. Similar distributions have been obtained with 1 ps pulses at 1053 nm.

Table 1. Decay paths of the CO molecule irradiated by a 30 ps YAG laser at a peak intensity of $6 \ 10^{14}$ W/cm^2, arranged in ascending threshold laser intensities order. For each channel we give the kinetic energy released, the branching ratio, the fragment emission angle $\alpha_{1/2}$. The laser wavelength is 1064 nm.

	decay paths of CO molecules	energy released (eV)	branching ratio (%)	$\alpha_{1/2}$
(1)	$CO^+ \rightarrow C^+ + O$	1.2	11	55°
(2)	$CO^+ \rightarrow C + O^+$	2	15	39°
(3)	$CO^{2+} \rightarrow C^+ + O^+$	6	24	23°
(4)	$CO^{3+} \rightarrow C^{2+} + O^+$	10	39	23°
(5)	$CO^{3+} \rightarrow C^+ + O^{2+}$	11	8	18°
(6)	$CO^{4+} \rightarrow C^{2+} + O^{2+}$	20	3	-

ORIGIN OF THE FRAGMENT ALIGNMENT

The Two Possible Explanations

The fact that diatomic molecules irradiated by an intense picosecond laser yields fragments essentially aligned along the laser polarization vector ε can be understood in two different ways. In the first case (case A), it is assumed that the laser ionizes preferentially those molecules whose internuclear axes are more or less parallel to its ε vector. The fragment alignment thus results from a selective ionization of well oriented molecules, the effect being due to the elongated shape of the molecule. In the second hyptheses (case B), the laser field is supposed to rotate the molecules and force them to align with the ε vector. In that case all molecules present in the interaction volume interact with the laser.

To discriminate between these two hypotheses, we designed a pump-probe experiment[3] to test whether there remains neutral molecules in the interaction volume after the laser interaction. If there remains molecule, then case A is correct; but if the interaction volume is empty, then case B is correct.

The Double Pulse Experiment Design

For this purpose, we send successively two identical laser shots, with crossed polarizations, on the same point of the interaction chamber. The time delay (typically 100-800 ps) between the two pulses is chosen to be short enough so that the interaction volume cannot be refilled by new molecules (due to thermal agitation) between the two shots. Fig.5 shows the optical set-up used to split the incident (YAG or Ti-Sapphire) laser beam. The incident beam is polarized horizontally. It first travels through a half-wave plate (HW) that permits to balance the energy between the two arms of a Michelson type interferometer. The beam is then split by a dielectric polarizer (P) at Brewster incidence that reflects totally the vertical polarization but transmits the horizontal component. Each arm contains a quarter-wave plate (QW) and a totally reflecting mirror (M_1, M_2) at normal incidence. When travelling back to the polarizer, the two partial waves have now a direction of polarization turned by 90 degrees. Consequently, beam 1 is now transmitted by the polarizer whereas beam 2 is totally reflected.

Beyond the polarizer, we thus have two copropagating beams with crossed polarizations, whose delay can be chosen by translating mirror M_2. Beam 1 has an ε vector perpendicular to the detection axis, and will be noted as the ⊥ beam, whereas beam 2 will be noted as the ∥ beam. We use the leaks of light transmitted through mirror M_3 and visualize the focused spots of the two beams on a video camera via a x63 microscope objective (MO). We adjust the tilt angles of mirror M_2 to provide a perfect superposition of the spots, keeping a permanent control during the experiments. To reproduce the focalization conditions of the two laser beams, the focal spots measurements are performed with a lens (L), through a window (W), both identical to those used for the interaction chamber.

The Double Pulse Experiment Results

The experimental results are presented in the four following figures (Figs.6-9). Each of theses figures shows the TOF ion spectra obtained with the ⊥ beam alone (a), with the ∥

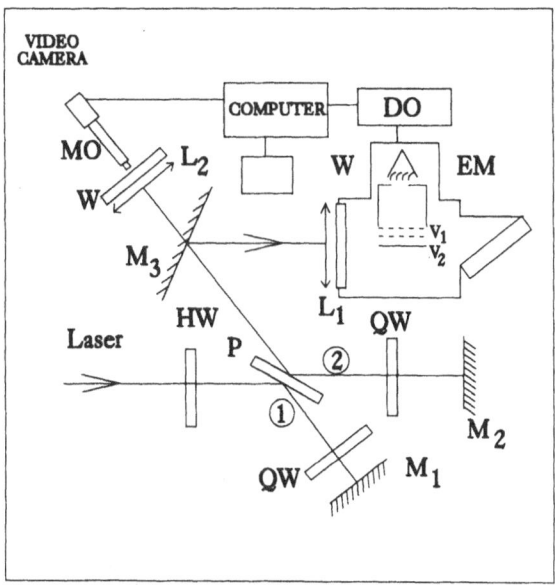

Figure 5. Experimental setup: HW is a half-wave plate, P is a dielectric polarizer, QW is a quarter-wave plate, M_i are totally reflecting mirrors, L is a 60 mm spheroparabolic lens, W is the chamber entrance window, MO is a x63 microscope objective and EM an electron multiplier.

beam alone (b), and with the two beams (c). The first three figures concern the interaction of CO molecules with the YAG laser (pulse duration 30 ps), whereas the last one is obtained with the Ti-Sapphire laser (pulse duration 1 ps).

The YAG laser experiment. Let us start by describing the results obtained with the YAG laser whose pulse duration is 30 ps. In figs.6-8, the intensity of the ∥ beam is kept roughly constant, but the intensity of the ⊥ beam is strongly varied.

In Fig.6, the intensities of the two beams are made equal to 3 10^{14} W/cm^2. As expected from the ions trajectories, most energetic ions are lost with the ⊥ beam (Fig.6a), whereas nice forward-backward splittings of the ion peaks are observed with the ∥ beam (Fig.6b). Note that, in both cases, the CO$^+$ signals are roughly the same. Fig.6c shows the

TOF spectrum of CO obtained when the two YAG laser shots are successively fired with a time delay of 100 ps. Comparison of the TOF ion spectra of Fig.6b and 6c permits to draw several conclusions:

1) The two beams are indeed focused at the same point of the interaction volume since the CO^+ signal measured with the two shots, is only 10% higher than that created by the first shot alone. The second laser interacts with an ionized gas and does not contribute significantly to the CO^+ production.

2) The net number of fragments collected by the detector is drastically reduced by the presence of the first shot (Fig.6c). In fact, all the ion populations are not affected identically: the slow O^+ and C^+ ions (channels 1 and 2 in Table 1) are almost unchanged, whereas the fast fragments (channels 3 to 6) are all the more decreased as they require a higher laser intensity to be created.

The conclusion is that the first laser shot interacts with all molecules in the interaction volume whatever their orientations are: the interaction volume is thus depleted of CO^+ ions by the first shot, which cuts down the ion yield of the second shot. Because of

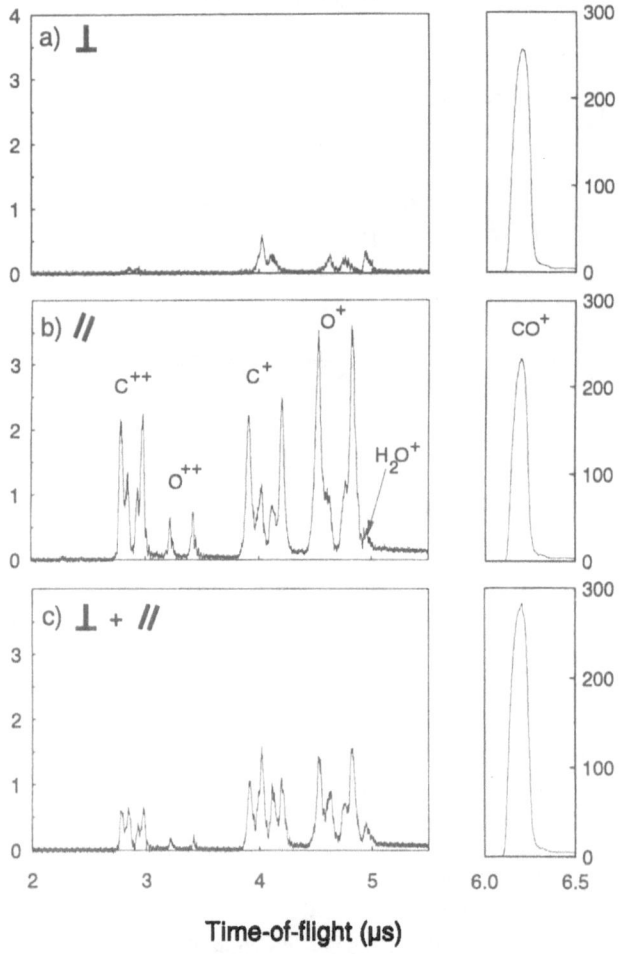

Figure 6. Ion TOF spectra of CO obtained at 1064 nm with 30 ps pulses: a) the ⊥ laser beam is fired alone, b) the ∥ laser beam is fired alone, c) the two laser beams are successively fired with a time delay of 100 ps. In each case the molecular CO^+ signals are displayed in a separate box, on the right of the figure. The two beams have the same laser intensity (3 10^{14} W/cm^2).

the spatial intensity distribution of the first laser shot, the CO^+ ion depletion is particularly severe in the centre of the interaction volume where channels 5 and 6 take place. On the contrary, in outer parts of the interaction volume where the local intensity is not so high, part of the CO^+ ions can survive to the first shot, which explains why the less energetic decay channels produced by the second shot are less affected by the presence of the first one. To support this interpretation, we have repeated the double shot experiment with a twice higher intensity of the $\perp$ beam (Fig.7a-c). The effect of the perturber beam is obviously reinforced since the fast O^{2+} and C^{2+} ions are completely suppressed, and that the fast O^+ and C^+ ions are more severely reduced (comparison of Fig.7b and 7c).

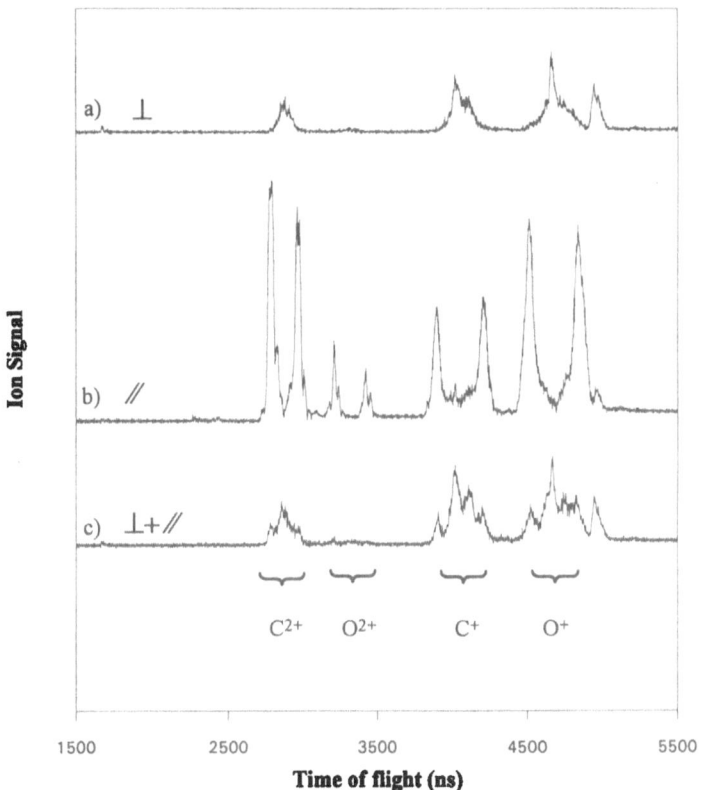

Figure 7. Ion TOF spectra of CO obtained at 1064 nm with 30 ps pulses: a) $\perp$ laser beam alone, b) $\diagup$ laser beam alone; c) the two laser beams are successively fired with a time delay of 100 ps. The $\perp$ beam intensity is twice that of the $\diagup$ beam , ($I_{\perp} = 6 \ 10^{14}$ W/cm^2 and $I_{\diagup} = 2.5 \ 10^{14}$ W/cm^2).

Role of the singly-charged molecular ions. It was unclear, at the beginning of these experiments, whether the fact that the $\perp$ beam may align the CO^+ ions could be, in part, responsible for the signal reduction observed in the double pulse configuration. Therefore we performed another pump-probe experiment under slightly different conditions to prepare a large volume of undissociated CO^+ ions with the first laser shot and check whether it affects the second beam ion yield. In Fig.8, the $\perp$ beam is slightly defocused so that it irradiates a large volume at a weak intensity level (a few 10^{13} W/cm²). Under these conditions, this first beam prepares a large target of CO^+ ions where at most 0.1% of the molecular ions are dissociated (Fig.8a).

Fig.8b shows the ion spectrum obtained with the ∕ beam alone, at an intensity level of $3 \cdot 10^{14}$ W/cm². When the two shots are fired successively with a time delay of 100 ps, we

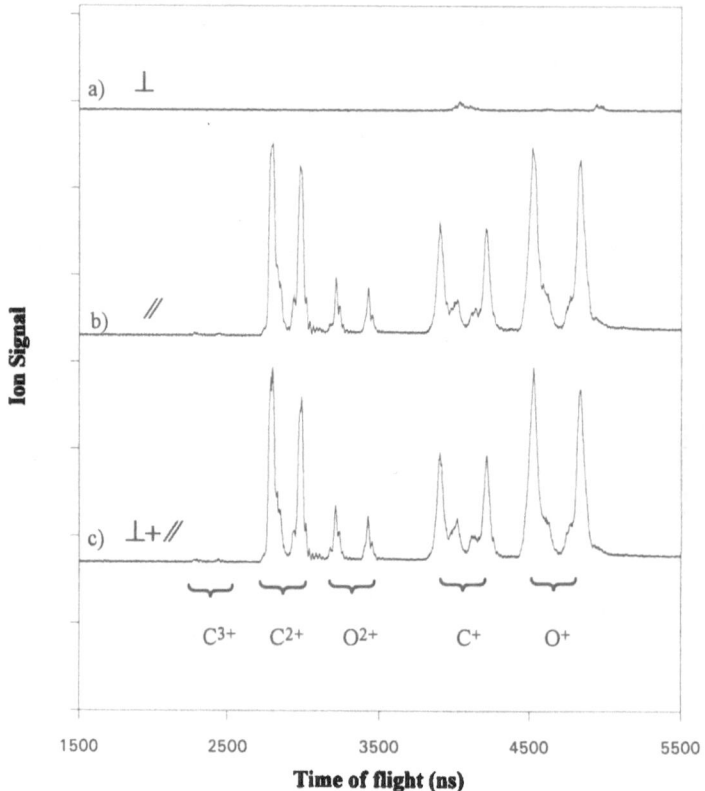

Figure 8. Ion TOF spectra of CO obtained at 1064 nm with 30 ps pulses: a) ⊥ laser beam alone and defocused, b) ∕ laser beam alone; c) the two laser beams are successively fired with a time delay of 100 ps. The ⊥ beam intensity is only a few 10^{13} W/cm², that of the ∕ beam is $3 \cdot 10^{14}$ W/cm².

observe that the ion yield of the second shot is absolutely unchanged (Fig.8c). In conclusion, when the laser field interacts with CO molecules in the ground electronic state or with CO^+ ions, the MEDI processes are the same. This observation is a direct evidence that the multiply charged ions are formed by sequential ionization.

The Ti-Sapphire laser experiment. Considering that the field-free rotational periods of the main J levels populated at room temperature are about 1ps, it could not be excluded that the CO molecules could get freely aligned with the laser ε vector during the 30 ps YAG laser pulse duration. Were it the case, the molecules could be preferentially ionized once they get parallel with ε without requiring any laser induced alignment.

To prove that the fragment alignment is indeed forced by the laser and not due to free rotations of the molecules during the pule rise, we have also studied the MEDI of CO with a 1 ps pulse (1053 nm Ti-Sapphire laser). The effective interaction time for the 12-photon ionization is now as short as 300 fs, that is definitely shorter than the radiationless rotational period of CO. The results are presented in Fig.9. The laser intensities are $3 \ 10^{15}$ W/cm^2 and $2 \ 10^{15}$ W/cm^2 for the $\perp$ and the $\parallel$ beams respectively. The TOF ion spectrum for the $\perp$ beam alone (Fig.9a) shows that the contribution of the slow fragments is strongly

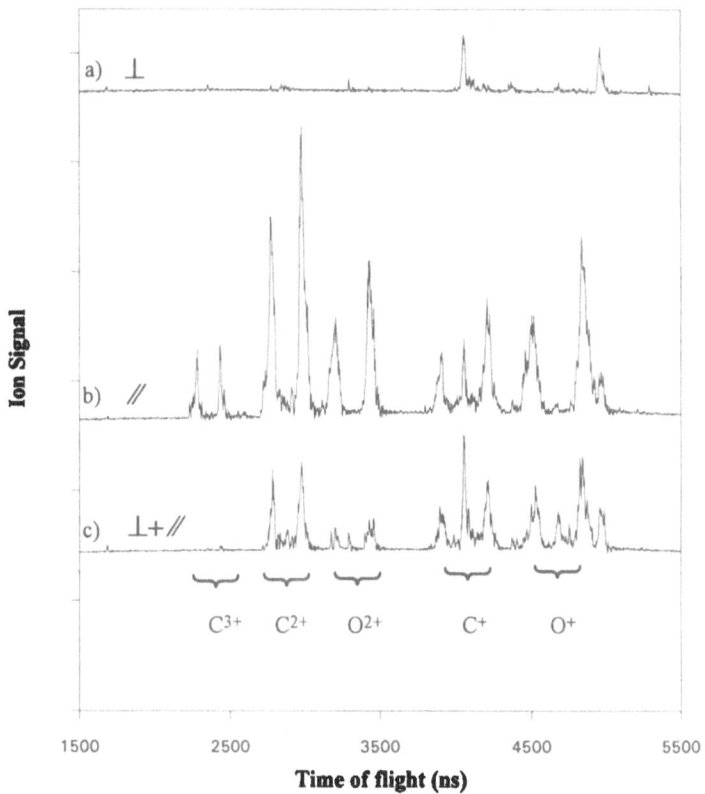

Figure 9. TOF ion spectra of CO obtained at 1053 nm with 1 ps pulses: a) $\perp$ laser beam alone, b) $\parallel$ laser beam alone; c) the two laser beams are successively fired with a time delay of 100 ps. The $\perp$ beam intensity is $3 \ 10^{15}$ W/cm^2, that of the $\parallel$ beam is $2 \ 10^{15}$ W/cm^2.

reduced with the 1 ps laser pulse duration (as compared to Fig.7a). The comparison of the TOF ion spectra of figs.9b and 9c confirms that $\parallel$ beam ion yield is very dramatically affected when the $\perp$ beam has been fired first. Indeed we observe in fig.9c that the triply-charged fragments have disappeared and that the doubly-charged fragment population are severely reduced. In conclusion, the 1 and 30 ps experiments yield identical results both for the fragment angular distributions and for the effects of the double beam configuration. The time scale of the laser forced molecular rotation is certainly shorter than 300 fs.

Interestingly, the ion spectrum of fig.9c exhibits a zero energy peak of O^+ ions which is absent in fig.9a and 9b. In addition fig.9a shows a zero energy peak of C^+ which we assign to the channel

$$CO^+ \quad \rightarrow \quad C^+ \quad + \quad O \quad\quad + 0\, eV \tag{2}$$

The zero energy O^+ ions must therefore proceed from the ionization of these zero energy neutral oxygen atoms, by the second beam. We thus prove directly that the dissociation of channel (2) is longer than the pulse duration, otherwise the neutral O atoms would have been ionized by the first laser itself.

CONCLUSION

The experiments that we have performed with the double pulse configuration prove that the linearly polarized laser field forces the molecules to align with its electric vector ε (case B). (This conclusion has also been obtained very recently by Dietrich et al., on iodine molecules[5]). The similarities of the results obtained with pulses of 30 ps and 1 ps indicate that the rotation time scale is at most a few hundreds fs long. In addition, we have shown that the fragments are all the more tightly confined along the laser polarization vector as their threshold laser intensities are higher. From a theoretical point of view, the laser induced molecular alignment can be understood in terms of optical pumping[6] or in terms of field ionization[7]. Consequently, although we could not get any informations on the CO^+ ion angular distribution (because they have no initial velocities), we assume that the CO^+ ions are also aligned by the laser, and probably slightly less than the slow C^+ fragments (channel 1 in Table 1). In the double shot experiment, the CO^+ ions must be first aligned ⊥ to the detection axis by the ⊥ beam and then turned at ninety degrees by the ∥ beam. Finally we have shown that the MEDI processes are identical when the laser field interacts with CO^+ ions or with neutral CO molecules. This is the first direct evidence that the multiple ionization is a sequential process.

REFERENCES

1. D. Normand, C. Cornaggia and J. Morellec, "Coherence Phenomena in Atoms and Molecules in Laser Physics", edited by Bandrauk A D and Wallace S C, Plenum Press, NY, 133 (1992).
2. W.T. Hill, J. Zhu, D.L. Hatten, Y. Cui, J. Goldhar and S. Yang, Phys. Rev. Lett. 69, 2646, (1992).
3. D. Normand, L.A. Lompré and C. Cornaggia, J. Phys. B Lett. 25, L497 (1992).
4. C. Cornaggia, J. Lavancier, D. Normand, J. Morellec and H.X. Liu, Phys. Rev.A42, 5464 (1990).
5. P. Dietrich P, D.T. Strickland and P.B. Corkum P B, to be published (1993).
6. A. Giusti-Suzor, X. He, O. Atabek and F.H. Mies, Phys. Rev. Lett 64, 575 (1990).
7. P.A. Hatherly, L.J. Frasinski, K. Codling, A.J. Langley, W. Shaikh, J.Phys.B 23, 291.(1990).

STABILIZATION AND COHERENCE IN THE PHOTODISSOCIATION OF DIATOMIC MOLECULES BY INTENSE LASERS

J F McCann [†], A D Bandrauk [‡] and J-M Gauthier [‡]

[†] Physics Department
University of Durham
Durham DH1 3LE
U. K.

[‡] Département de Chimie
Université de Sherbrooke
Sherbrooke, Québec
Canada J1K 2R1

INTRODUCTION

The study of the fragmentation of molecules exposed to very intense sources of light, has led to the discovery of some unusual effects. Some of these effects are common to any multielectron monoatomic target, others are unique features of polyatomic systems. For example, the photoelectron spectrum obtained from the multiphoton ionization of a molecule will exhibit a rich fine-structure, absent from that obtained from a single atom, due to the nuclear motion. While molecules are generally held to be more complicated systems than atoms, because of the extra degrees of freedom present; still, one can simplify the calculations to a great extent by decoupling, at least to first order, the nuclear and electronic motion within the Born-Oppenheimer approximation.

There is a simpler reason why molecules are different, and in some respects more interesting than atoms. As the laser field is amplified from weak to superstrong values, it is possible to scan across a hierarchy of different couplings. These couplings are characteristic of the various time-scales of movement within the molecule ranging from rotational, to vibrational and electronic motion. As the Rabi frequency increases from values associated with weak-fields to the strong-field limit, it tunes into resonance with each of these motions in succession. If the intensity increases beyond these levels until the Rabi frequency is comparable to the optical frequency, then coupling with the vacuum is a dominant feature of the interaction. Before this order of extreme nonlinearity is reached, we have to deal with the more mundane, yet still poorly understood régime in which the various mechanical excitation processes are in strong competition with one another. The analysis of the excitation and relaxation mechanism is complicated by the presence of radiative scattering and many continuum channels : multiple-electron multiphoton ionization and molecular dissociation. This area of photophysics has lately attracted much attention [1,2,3] . This paper discusses some of the effects which have been

Super-Intense Laser-Atom Physics, Edited by
B. Piraux *et al.*, Plenum Press, New York, 1993

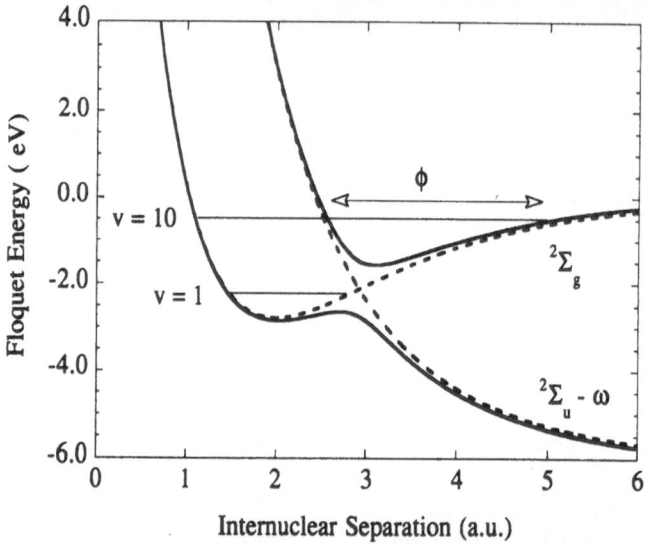

Figure 1. The dressed-state energy curves of the hydrogen molecular ion and their laser-induced avoided crossing ($\lambda = 205.3$nm). The field-molecule diabatic states are indicated by the dashed lines.

observed in experiment, and other nonperturbative effects which occur at rather modest intensities. For molecular targets, intensities of the order of 10^{12}-10^{13} W cm^{-2} are more than adequate for the observation of nonlinearities and multiphoton processes. At the same time, the troublesome and interesting ionization channels are still comparatively weakly-coupled at such intensities. The recent proliferation of research work in molecular science using short pulses of intense lasers has been stimulated by the feasibility of studying physical events which occur on femtosecond timescales, such as molecular collisions and vibrational relaxations. We discuss the behaviour of several simple diatomic systems in strong fields, and the effects arising from the strong coupling of the rovibrational manifolds of bound and continuum states.

In this paper, we concentrate upon evidence for strong-coupling effects in photodissociation. In particular, results are presented for coherent suppression of photodissociation through *bond-hardening* (adiabatic stabilization) and *coherence* (phase-controlled two-colour interference).

STABILIZATION THROUGH BOND-HARDENING

In the Floquet approximation, the bound and continuum field-molecule states are degenerate, in analogy to autoionization of atoms. The photodissociation process can be viewed as a laser-induced predissociation [4], in which the wavepacket hops from a bonding surface to an antibonding surface. The evolution of molecular wavepackets [5] is governed by the topography of these potential surfaces. In the field-molecule adiabatic representation, the time dependence of the pulse envelope means that these surfaces are dynamic as well as the wavepacket.

At extremely high laser intensities, the dressed-molecule potential surfaces become strongly avoiding and the associated surface deformation can lead to the supression of the potential barrier that binds the nuclei together. This mechanism of photodissociation is analgous to the *barrier-suppression ionization* of atoms. Recently experimental evidence has emerged for this *bond softening* effect in the photodissociation of the hydrogen molecular ion [3,6]. The dressed-molecule energy curves for the H_2^+ ion are shown in fig. 1. The lower dressed state possesses a small barrier which can support Feshbach resonance states [7]. As for those vibrational states which lie above the avoided crossing; even when the crossing is strongly avoided, the process of photodissociation is hampered by the presence of the well of the upper dressed-state. The coupling between the field-molecule (diabatic) state and the upper dressed (adiabatic) state is strongly enhanced when the bare molecule and dressed-molecule energies are almost equal. Under such conditions the molecule, instead of breaking up, resonates between the two wells and becomes trapped. This effect is familiar from the study of predissociation [8], and can be viewed as a diabatic-adiabatic resonance. The effect is reminiscent of the interference present in full-collisions between atoms, when the incoming nuclear wave has the choice of propagating on the diabatic and adiabatic surface. It chooses both paths; and the resulting interference appears as the famous Stueckelberg oscillations [9]. So, one can also take the view that this trapping is an interference effect. We term this suppression of dissociation by the upper dressed state as *bond hardening*. Such effects were recently observed [10,11] in a related curve-crossing problem.

In the dressed-state representation the perturbations, or nonadiabatic couplings, come from two sources: the kinetic energy of the nuclei, and the time-dependence of the pulse envelope. The coupling is proportional to the space and time derivatives of the adiabatic mixing-angle $\theta(r,t)$, defined by $\tan 2\theta = (E_1(r) - E_2(r) + \omega)^{-1}\Omega(r,t)$ where $\Omega(r,t)$ is the Rabi frequency, $(E_1(r) - E_2(r) + \omega)$ is the detuning, and ω is the frequency of the laser.

The value of the quasienergy E can be predicted by the semiclassical method [8,9,12] and in the two-state approximation is a simple refinement of Landau's formula :

$$E = E_n + \Delta_n - i\Gamma_n/2 \tag{1}$$

where the f.w.h.m. and a.c. shift are given by the expressions;

$$\Gamma_n = \frac{2\omega_n}{\pi} P \cos^2\phi \qquad \Delta_n = \frac{\omega_n}{\pi} P \sin\phi \cos\phi \tag{2}$$

and both the crossing probability, $P(E)$, and the action, $\phi(E)$, are evaluated at the energy E_n. In the weak-field limit, E_n corresponds to the unperturbed diabatic rovibrational energy of the potential well, P is given to a good approximation by the single-passage Landau-Zener formula, and the phase ϕ is equal to the action within the adiabatic well plus a small dynamic phase correction. The local vibrational frequency within the diabatic well is represented by ω_n. Thus trapping will occur ($\Gamma_n \to 0$) whenever ϕ is quantized according to the Bohr-Sommerfeld rule. In spite of the simplicity of the Landau-Zener model, the qualitative and quantitative agreement with the exact results is very good [10,13]. In fig. 2 we present our results for a calculation of the photodissociation of H_2^+ by intense pulses. The pulse profile is trapezoidal with a plateau lasting 20 fs, and rise and fall times equal to 1 fs each. The calculations have been performed within the two-state rotating-wave approximation using the split-operator FFT method. A sample of our results at a wavelength of 205.3 nm is given in fig. 2. The graph plots the variation of total dissociation yield as a function of laser intensity for four initial conditions. The set of initial vibrational states ($v = 0, 1, 4$ and 10) include levels below and above the crossing energy (fig. 1). For the state $v = 0$, we see that the dissociation is rather small due to the poor overlap with the continuum. At higher intensities (10^{12} W cm^{-2}). the yield increases almost linearly with intensity until bond-softening (or barrier suppression) comes into effect at around 20 TW cm^{-2}. The $v = 1$ level has a larger overlap with the continuum than $v = 0$, and thus a much higher yield at low laser intensity. The single-photon dissociation is strong, and the bond-breaking is not impeded by the very thin potential barrier. Our results for the

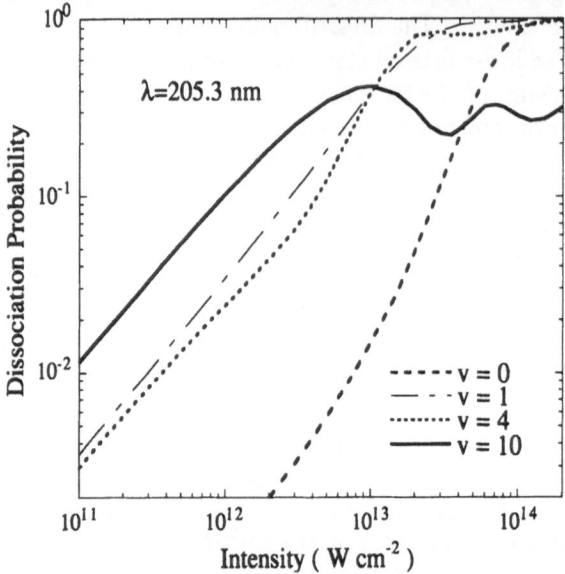

Figure 2. The photodissociation probability of H_2^+ ($^2\Sigma_g \rightarrow^2 \Sigma_u$) by an intense pulse lasting 20 fs of wavelength 205.3 nm. Results for initial vibrational states $v = 0, 1, 4$ and 10.

steady, smooth increase in the dissociation yield for $v = 1$, confirm the findings of Garraway and Stenholm [10], in that there is a smooth transition from under-barrier to over-barrier dissociation. The results for $v = 4$ indicate two features, a slight inflection of the curve near 2 TW cm^{-2} and a plateau near 20 TW cm^{-2}. The inflection is caused by *bond-hardening* due to resonance with $v_{ad} = 0$ (the ground-state vibrational level of the upper adiabatic curve associated with the upper dressed state [13]). The plateau is due to some survival of the probability in excited states of the $^2\Sigma_g$ well. We find this residual population tends to be created through pulse-rise transients, and to a lesser extent is a result of the wavepacket climbing the vibrational ladder of the ground state via Raman-type transitions to the excited state [14].

In contrast, the curve for $(v = 10)$ shows some structure where the yield drops sharply at certain critical high intensities. The double dip corresponds to two bond-hardening resonances. The first occuring at I=3.5×10^{13} W cm^{-2} corresponds to the quantization of ϕ due the degeneracy of the level $(v = 10)$ with the dressed state $v_{ad} = 3$. As the intensity increases, the adiabatic states are displaced upwards until a second stabilization minimum occurs at around 1.2×10^{14} W cm^{-2} which corresponds to resonance with $v_{ad} = 2$. This effect was also observed by Aubanel et al [13]. However, as was shown by Muller [6], the two-state model and rotating-wave approximation begin to break down at 10^{14} W cm^{-2} for this system; the coupling with other Floquet blocks and with ionization channels begins to take effect [15,16].

The semiclassical formulation also works extremely well when the nonadiabaticity is driven by ultrashort ultraintense pulse with rapidly-varying envelopes. In this case the coupling induces Rabi oscillations between the continuum and bound states. Suominen [17] has shown that whenever the crossing time and dispersion time are relatively small in comparision with the Rabi period, then the area-theorem applies and for a $2\pi n$-pulse, one gets zero dissociation yield, even at the highest intensities.

STABILIZATION THROUGH COHERENCE

We shall consider a second form of stabilization or trapping, namely that produced by interfering pathways to the same continuum [18]. This mechanism of stabilization is close to that discussed by Fedorov [19] and Noordam [20] at this meeting. It is similar to the stabilization and d.c. Stark effect in Rydberg states of atoms described by Chardonnet et al [21], and corresponds to the creation of *coherent states* by the redistribution of

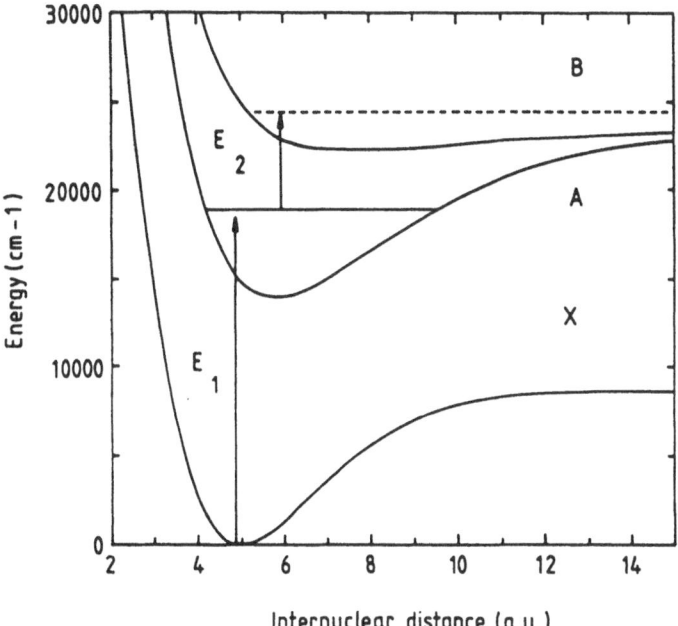

Figure 3. Two-colour selective photodissociation of Li_2:
$1^1\Sigma_g \rightarrow 1^1\Pi_u \rightarrow 1^1\Pi_g$

population among the rotational sublevels of the same vibronic manifold. Second-order transitions in a Raman Λ-system occur whenever the photodissociation linewidths, $\Gamma(v, J)$, become comparable to the rotational or vibrational separation, $E(v, J) - E(v, J')$. We investigated the two-step ($|X\rangle \rightarrow |A\rangle \rightarrow |B\rangle$) two-colour selective photodissociation of the lithium dimer by long intense pulses (fig. 3). Within the time-independent three-dimensional Floquet method, one can express the Hamiltonian for the system in the form :

$$H = H_m + H_{na} + H_f + H_{mf} \qquad (3)$$

of which the four components represent respectively the Born-Oppenheimer Hamiltonian, the nonradiative nonadiabatic coupling, the quantized fields and the field-molecule interaction. The term H_{na} contains the radial, rotational and spin-orbit perturbations.

Since the separation kinetic energies are very low, we can neglect electron-translation corrections *a priori*. The fields can be represented by the term :

$$H_f = \sum_k \omega_k a_k^\dagger a_k \tag{4}$$

and the coupling with the molecule by :

$$H_{mf} = \sum_j \left(\frac{2\pi\omega_j}{V}\right)^{\frac{1}{2}} \mathbf{e}^{(j)} \cdot \mathbf{d}^{(j)} (a_j^\dagger + a_j) \tag{5}$$

in which $\mathbf{e}^{(j)}$ represents the polarization of the laser, $\mathbf{d}^{(j)}$ represents the dipole moment of the jth transition, and V denotes the volume of the optical cavity. In the following discussion, we only consider linearly polarized light. The wave equation is expanded on an infinite set of Floquet (field-molecule) states which diagonalize the uncoupled terms $H_m + H_f$:

$$|j; \Lambda_j J_j M_j\rangle = r^{-1} F_j(J_j; r) \psi_j(q, r) |\Lambda_j J_j M_j\rangle |n_j\rangle \tag{6}$$

In this notation, ψ_j corresponds to the electronic eigenfunction with energy $E_j(r)$, q represents the ensemble of electron coordinates and spins, and r is the internuclear separation. The rotational motion of the system is represented here by a symmetric-top function; $|\Lambda_j J_j M_j\rangle$. The state of the field is defined according to the occupation numbers of the different modes of the lasers and the vacuum $|n_j\rangle = |n_{j1}, ..., n_{jk}, ..., n_{jN}\rangle$ such that:

$$H_f |n_j\rangle = \left(\sum_k \omega_k n_{jk}\right) |n_j\rangle \tag{7}$$

If the exact wavefunction is expanded in terms of the basis set (6) and the coupled equations are solved subject to the usual Siegert boundary conditions [22,23], the photodissociation rate as a function of angle can be obtained. In terms of the effective Hamiltonian of the intermediate bound states, the quasienergies can be defined as solutions of the implicit equations;

$$|\mathbf{H}_{eff} - E\mathbf{1}| = 0 \tag{8}$$

$$\mathbf{H}_{eff} = H_m + H_f + \Delta(E) - i\Gamma(E)/2 \tag{9}$$

$$\Delta_{ij}(E) = \sum_l P \int dE_l \, (E - E_l)^{-1} \langle A_i | H_{mf} | B_l(E_l) \rangle \langle B_l(E_l) | H_{mf} | A_j \rangle \tag{10}$$

$$\Gamma_{ij}(E) = \sum_l 2\pi \langle A_i | H_{mf} | B_l(E) \rangle \langle B_l(E) | H_{mf} | A_j \rangle \tag{11}$$

The intermediate states $|A_j\rangle$ have unperturbed energies E_j. If the coupling is weak, the excited states will dissociate directly to the continuum $|B_j\rangle$, or in terms of equation (11) $\Gamma_{ij}(E) = \Gamma_{jj}(E_j)\delta_{ij}$ and the decay rate and a.c. Stark shift are linearly proportional to the laser intensity $E = E_j + \Delta_{jj}(E_j) - i\frac{1}{2}\Gamma_{jj}(E_j)$. For stronger couplings, the off-diagonal elements of $\mathbf{H}_{eff}$ representing Raman transitions via the continuum, will perturb these values. The states associated with weak-field resonances then recouple to form a set of new linear combinations of the old states, some of which may have zero width. This is sometimes referred to as *coherent trapping*, and has been studied in detail by Knight [18] in the context of laser-induced continuum structure.

In figure 4, we show a simple example of trapping, for excitation from the $v = 0, J = 0$ level of the $1^1\Sigma_g$ state of Li_2, via the $1^1\Pi_u$ state, to the $1^1\Pi_g$ continuum (fig. 3). The dip indicates the critical intensity at which the coherent trapped state is formed. Its effect is more dramatic when the angular distribution of photofragments is analyzed [24]. The trapping is rather sensitive to the preparation of the $1^1\Pi_u$ state

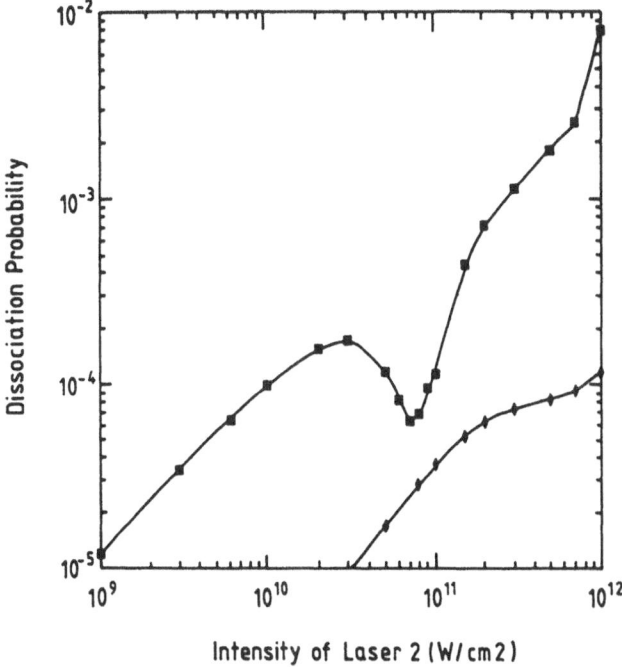

Figure 4. Stabilization minimum : dissociation yields for two isotopes of Li$_2$ as a function of intensity I_2 of laser 2. Excitation from the ground state $v = 0$, $J = 0$: □ ^{7}Li$_2$ ◇ ^{6}Li$_2$

through the primary laser excitation , and is not present for the same laser parameters in the species ^{6}Li$_2$ because of the detuning effect of the new isotope. On the other hand, upon inspection of fig. 4, this trapping is readily destabilized by increasing the laser intensity.

SUPRESSION THROUGH TWO-COLOUR COHERENCE

In photochemistry the laser is a unique tool for preparing excited states. The form of the wavepacket on the excited state potential surface can, in principle at least, be controlled by the judicious choice of laser pulse length, frequency, and intensity. The evolution of this packet is critically determined by the fact that chemical reaction pathways are sensitive to the phases of the wavefunction components. This idea has been the inspiration behind recent proposals towards reaction steering through coherence control [25]. One such proposal [26] is based upon two-colour coherence. Two lasers of commensurable frequencies are employed to excite the molecule to a common final state via two separate paths. The interference of the two transition amplitudes associated with these two paths leads to variation in the phase and probability amplitude of the final state. If the relative phase of the lasers can be controlled, then a truly coherent superposition of substates of the final wavepacket can be excited. We show a simple example of how this can be achieved in practice. The example chosen is a 2+4 photon process leading to predissociation of the chlorine dimer (Cl$_2$). We investigate how this yield can be either suppressed or enhanced through variations in the intensity and relative phase of the lasers [27]. The transition in question is a jump from the ground state (X Σ_g) to an excited vibrational level of the 2Π_g state achieved by a four-photon nonresonant excitation of frequency ω_4. The same 2Π_g level can be populated by the second harmonic of this laser, frequency ω_2, through a resonantly-enhanced absorption. This excited state then couples, via the usual nonradiative nonadiabatic coupling, to the neighbouring repulsive energy curve (1$^1\Pi_g$). Suppose we wish to close the predissocia-

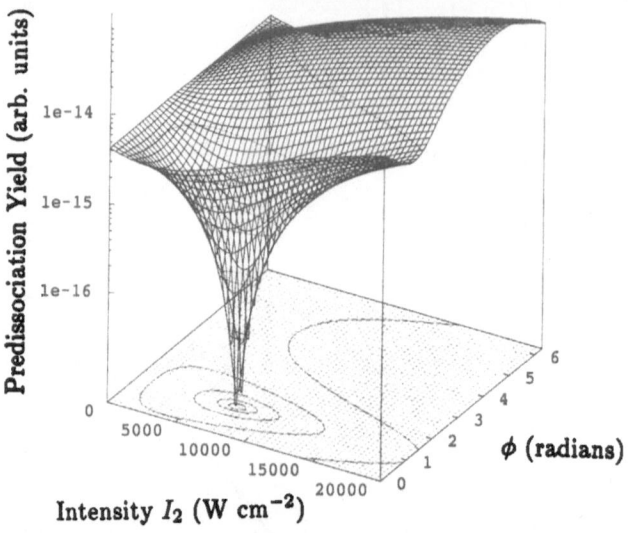

Figure 5. Suppression of predissociation through two-colour interference: the predissociation yield as a function of phase difference ,ϕ, and I_2 the intensity of the second-harmonic. The intensity I_4 is kept fixed at a value of 5×10^8 Wcm^{-2}. The critical value for trapping is given by $I_2 = 0.7359 \times 10^4$ Wcm^{-2} with $\phi = 0.37\pi$.

tion channel in order to favour another relaxation channel; for example, an anti-Stokes Raman scattering process. This can be achieved by using the lasers to create a destructive interference for predissociation. In figure 5, we show the result of a close-coupling calculation for excitation from the ground state X (Σ_g, $v = 0$, $J = 0$). The predissociation yield is plotted as a function of two variables, the second-harmonic intensity and its phase with respect to the laser operating at ω_4. The classical electric fields of the lasers can be described by $\mathbf{E}_4(t) = \mathbf{E}_{40} \cos(\omega_4 t)$ and $\mathbf{E}_2(t) = \mathbf{E}_{20} \cos(\omega_2 t + \frac{1}{2}\phi)$, respectively. Of course the interference effect is most dramatic whenever the two probability amplitudes are almost the same. In figure 5, the intensity of the laser of frequency ω_4 is fixed at a value of $I_4 = 5 \times 10^8$ W cm^{-2} and has an equivalent photon energy $E_p = \hbar\omega_4 = 16605.3$ cm^{-1} . Since the intensities involved are very low, the results agree well with perturbation theory. However, two-colour interference and coherence trapping can be effected at very high intensities [28,29]. The critical value for trapping, in the experiment described above, is given by $I_2 = 0.7359 \times 10^4$ W cm^{-2} at a phase difference $\phi = 0.37\pi$. Clearly, the predissociation yield is a sensitive function of both the relative intensities and the phase differences of the lasers.

ACKNOWLEDGMENTS

Thanks are due to Robert Potvliege for many useful discussions. JFMcC is also grateful to Teijo Åberg and Kalle-Antti Suominen for their input. This work was given financial support by the Conseil de Recherches en Sciences Naturelles et en Génie du Canada , UKSERC, and an RIC grant from the University of Durham.

REFERENCES

1. J.W.J. Verschuur, L.D. Noordam and H.B. Van Linden van den Heuvell, Anomalies in above-threshold ionization observed in H$_2$ and its excited fragments, *Phys. Rev. A* 40:4383 (1989).

2. S.W. Allendorf and A. Szöke, High-intensity multiphoton ionization of H_2, *Phys. Rev. A* 44:518 (1991).

3. P.H. Bucksbaum P H and A. Zariyev, Above-threshold photodissociation and bond-softening at high intensities, *in*: "Coherence phenomena in Atoms and Molecules in Laser fields", A.D. Bandrauk and S.C. Wallace eds. , Plenum, NY (1992).

4. A.D. Bandrauk and O. Atabek, Coupled-equation method for multiphoton transitions in diatomic molecules: bridging the weak- and intense-field limits, *Adv. Chem. Phys.* 73:823 (1989).

5. R.D. Coalson, Time-dependent wavepacket approach to optical spectroscopy involving nonadiabatically coupled potential surfaces, *Adv. Chem. Phys.* 73:605 (1989).

6. H.G. Muller, Calculation of potential curves for H_2^+ molecules in a strong laser field, *in*: "Coherence phenomena in Atoms and Molecules in Laser fields", Bandrauk A D and Wallace S C eds. , Plenum, NY (1992).

7. X. He, O. Atabek and A. Giusti-Suzor, Laser-induced resonances in molecular dissociation in intense fields, *Phys. Rev. A* 38:5586 (1988).

8. M.S. Child, *Molecular Collision Theory*, Academic, London (1974).

9. D.S.F. Crothers, Zwaan-Stueckelberg phase integral methods, *Adv. Phys.* 20:405 (1971).

10. B.M. Garraway and S. Stenholm, Wave-packet description of laser-induced crossings, *Opt. Comm.* 83:349 (1991).

11. K.-A. Suominen, B.M. Garraway and S. Stenholm, Wave-packet model for excitation by ultrashort pulses, *Phys. Rev. A* 45:3060 (1992).

12. A.D. Bandrauk A D and J.F. McCann, Semiclassical description of molecular dressed states in intense laser fields, *Comm. At. Mol. Phys.* 22:325 (1987).

13. E. Aubanel, A.D. Bandrauk and P. Rancourt, Pulse-shape effects and laser-induced avoided crossings in photodissociation, *Chem. Phys. Lett.* 197:419 (1992).

14. A. Giusti-Suzor and F.H. Mies, Vibrational trapping and suppression of dissociation in intense laser fields, *Phys. Rev. A* 68:3869 (1992).

15. S. Chelkowski, T. Zuo and A.D. Bandrauk, Ionization rates of H_2^+ in an intense laser field by numerical integration of the time-dependent Schrödinger equation, *Phys. Rev. A* 46:5342 (1992).

16. F.H. Mies, A. Giusti-Suzor, K.C. Kulander and K.J. Schaffer, Stabilization and ionization of H_2^+ in intense laser fields, *this meeting* , (1993).

17. K.-A. Suominen, Time dependent two-state models and wave packet dynamics, *Report Series in Theoretical Physics, University of Helsinki, HU-TFT-IR-92-1* (1992).

18. P.L. Knight, Laser-induced continuum structure, *Comm. At. Mol. Phys.* 15:193 (1984).

19. M.V. Fedorov, Intense field interference stabilization of atoms and molecules, *this meeting* , (1993).

20. L.D. Noordam, Transient stabilization, *this meeting* , (1993).

21. C. Chardonnet, D. Delande and J.C. Gay, Interference and stabilization in the quasibound Stark spectrum, *Phys. Rev. A* 39:1066 (1989).

22. R.M. Potvliege and R. Shakeshaft, Nonperturbative treatment of multiphoton ionization within the Floquet framework, *in Atoms in Intense fields*, Gavrila M ed. , Academic Press, NY (1992).

23. S.-I. Chu, Generalized Floquet theoretical approaches to intense-field multiphoton and nonlinear optical processes, *Adv. Chem. Phys.* 73:739 (1989).

24. J.F. McCann and A.D. Bandrauk, Laser induced stabilization and alignment in multiphoton dissociation of diatomic molecules, *J. Chem. Phys.* 96:903 (1992).

25. S.-P. Lu, S.-M. Park, Y. Xie and R.J. Gordon, Coherent phase control of the resonance enhanced multiphoton ionization of HCl and CO, *in*: "Coherence phenomena in Atoms and Molecules in Laser fields", Bandrauk A D and Wallace S C eds. , Plenum, NY (1992).

26. M. Shapiro, J.W. Hepburn and P. Brumer, Simplified laser control of unimolecukar reactions : simultaneous (ω_1, ω_3) excitation, *Chem. Phys. Lett.* 149:451 (1988).

27. A.D. Bandrauk, J.-M. Gauthier and J.F. McCann, Laser control of predissociation by (2+4) photon interference, *Chem. Phys. Lett.* 200:399 (1992).

28. R.M. Potvliege and P.H.G. Smith, Two-colour multiphoton ionization of hydrogen by an intense laser field and one of its harmonics, *J. Phys. B: At. Mol. Opt. Phys.* 25:2501 (1992).

29. J.F. McCann, Two-colour coherences in intense field dissociation, *to be published* (1993).

PART IV

INTERACTION OF TWO-ELECTRON SYSTEM
WITH A STRONG FIELD

SINGLE AND DOUBLE PHOTOIONIZATION OF TWO-ELECTRON SYSTEMS

Daniel Proulx, Zhong-jian Teng and Robin Shakeshaft

Physics Department,
University of Southern California,
Los Angeles, CA 90089-0484

1 INTRODUCTION

In this paper we describe a method for calculating rates for multiphoton detachment or ionization of a two-electron system by a perturbative field. we present results of an application to both H^- and He, and we also report on a study of double ionization of He by one photon.

We have calculated rates for 2- and 3-photon detachment/ionization of H^- and He by a linearly polarized field. In particular, we have explored the region of the $^1S^e$ and $^1D^e$ Feshbach resonances below the first and second excitation thresholds of H and He^+. These resonances are reached by two photons and lie in the "excess-photon detachment/ionization" region where one photon is already sufficient to remove the electron. In our calculations we employ a two-electron basis composed of products of one-electron complex radial Sturmian functions and spherical harmonics, and our results are, for most frequencies, fairly well converged with respect to increasing basis size.

The process of double ionization by a single photon cannot occur without the two electrons interacting with each other, in contrast to single ionization. At high photon energies the dynamics of double ionization continues to be of interest, particularly in view of a recent experiment on He.[1] We present below results of calculations of the angular distribution for double ionization of He by one 2.8 KeV photon. We include electron-electron correlation in both initial *and* final states, and we identify from the angular distribution three different mechanisms for double ionization of He. One well-known mechanism for double ionization is shakeoff — one of the electrons absorbs the photon and the other electron is shaken loose by the sudden change in the effective nuclear charge resulting from the swift departure of the first electron.[2] In collision terms, the shakeoff of the second electron is the consequence of a *soft* collision with the first one. Shakeoff is effective in producing slow electrons. However, two fast electrons can emerge *via* the "knockout" mechanism — as in shakeoff, one of the electrons absorbs the photon, but as this fast electron exits the atom it undergoes a *hard binary* collision with the other electron.[3]-[8] In both of these mechanisms the large net momentum carried away by the electrons must originate from a hard collision with the nucleus since the photon can impart energy but not momentum to the electrons (we neglect retardation throughout). However, there is a third mechanism whereby almost no momentum need be exchanged with the nucleus; the two electrons, by simultaneously sharing the photon, can leave with nearly equal but opposite momenta, and thereby carry away almost no net momentum. (In double ionization at low photon energies, near threshold, the two electrons leave with momenta that are almost equal and opposite; however, this is due to the fact that the stability of the process is maximized, rather than that the momentum exchanged with the nucleus is minimized — see Ref. [9]). Remarkably, this mechanism is inhibited (but not excluded) for the following reasons: Two identical charged particles which are otherwise free cannot absorb radiation since their electric dipole depends only on their center of mass coordinate; if there is no external force the center of mass, and therefore the dipole, does not accelerate, and hence cannot absorb radiation. Consequently, while an isolated electron-positron pair can absorb radiation, two electrons can absorb radiation only in the presence of a third body, in our case the nucleus. Furthermore, due to inversion symmetry, as shown in sect. IV, the two electrons cannot emerge with *exactly* equal and opposite momenta.

Many of the results presented here were recently reported elsewhere.[10, 11, 12] Unless specified otherwise, we use atomic units.

2 SINGLE IONIZATION: METHOD

We begin by describing the method we used to calculate the lowest (N-th) order amplitude, $A_{fi}^{(N)}$, for the atom to absorb N photons and undergo a transition from the initial unperturbed state i, in which both electrons are bound, to the final unperturbed state f in which one electron is bound and the other is free. If $V(t) = V_+ e^{-i\omega t} + V_- e^{i\omega t}$ is the interaction of the atom with a monochromatic classical field of frequency ω, within the dipole approximation, and if $|\Psi_f^-\rangle$ represents the final state f, we have

$$A_{fi}^{(N)} = \langle \Psi_f^- | V_+ | \mathcal{F}_{N-1}^{(N-1)} \rangle. \tag{1}$$

where the N-th order harmonic components $|\mathcal{F}_N^{(N)}\rangle$ satisfy the coupled (Dalgarno-Lewis) equations

$$(E_i + N\omega - H_a)|\mathcal{F}_N^{(N)}\rangle = V_+ |\mathcal{F}_{N-1}^{(N-1)}\rangle, \quad N \geq 1, \tag{2}$$

$$|\mathcal{F}_0^{(0)}\rangle = |\Psi_i\rangle, \tag{3}$$

where E_i and $|\Psi_i\rangle$ are the energy and state-vector of the atom in state i, and where H_a is the Hamiltonian of the atom. We now rearrange this expression for $A_{fi}^{(N)}$ by first expressing H_a as $H_a \equiv H_0 + W$, where H_0 is the *independent particle* Hamiltonian that describes complete screening of the "outer" electron (the one which becomes free) by the "inner" electron (the one which remains bound). Thus if we label the outer and inner electrons by 1 and 2, respectively, W is the "short"-range potential $W \equiv (1/r_{12} - 1/r_1)$, where r_{12} is the inter-electron separation and where r_1 is the distance of the outer electron from the nucleus. We neglect spin-orbit coupling so that we can factor the spin out of the problem. The final channel f is specified by the parity, by the total orbital angular momentum and magnetic quantum numbers L and M of the two-electron system, by the individual orbital angular momentum quantum numbers, l_1 and l_2, of electrons 1 and 2, and by the (positive) energy ϵ with which electron 1 emerges. Speaking loosely, we refer to W as the "final-state correlation" (FSC). If we were to neglect FSC, the final state would be represented by the direct product $|\psi^-_{l_1,\epsilon}\rangle \otimes |\phi_{l_2}\rangle$, appropriately symmetrized and summed over individual magnetic quantum numbers, where $|\psi^-_{l_1,\epsilon}\rangle$ represents electron 1 moving with energy ϵ in the Coulomb potential $-(Z-1)/r_1$ (with Z the atomic number of the atom, i.e. $Z = 2$ for He) and where $|\phi_{l_2}\rangle$ represents electron 2 bound in the isolated residual He$^+$ ion. Introducing the resolvent $G^\pm_a(E) = (E \pm i\eta - H_a)^{-1}$, where η is positive but infinitesimal, and defining $E_f \equiv E_i + N\hbar\omega$, the exact final-state vector is given by the Lippmann-Schwinger equation:

$$|\Psi^-_f\rangle = \mathcal{P}\left(1 + G^-_a(E_f)W\right)\left(|\psi^-_{l_1,\epsilon}\rangle \otimes |\phi_{l_2}\rangle\right), \tag{4}$$

where $\mathcal{P}$ is the symmetrization operator. Substituting the right-hand side of Eq. (4) into the right-hand side of Eq. (1), noting that $[G^-_a(E)W]^\dagger = WG^+_a(E)$ and that Eq. (2) implies that $G^+_a(E^{(0)}_f)V_+|\mathcal{F}^{(N-1)}_{N-1}\rangle = |\mathcal{F}^{(N)}_N\rangle$, yields $A^{(N)}_{fi} = B^{(N)}_{fi} + C^{(N)}_{fi}$, where

$$B^{(N)}_{fi} = \sqrt{2}\left(\langle\psi^-_{l_1,\epsilon}| \otimes \langle\phi_{l_2}|\right)V_+|\mathcal{F}^{(N-1)}_{N-1}\rangle, \tag{5}$$

$$C^{(N)}_{fi} = \sqrt{2}\left(\langle\psi^-_{l_1,\epsilon}| \otimes \langle\phi_{l_2}|W\right)|\mathcal{F}^{(N)}_N\rangle. \tag{6}$$

The quantity $B^{(N)}_{fi}$ is just the amplitude obtained when FSC is omitted, and $C^{(N)}_{fi}$ is the correction accounting for FSC. Writing $H_a \equiv H_0 + W$, and noting that $|\psi^-_{l_1,\epsilon}\rangle \otimes |\phi_{l_2}\rangle$ is an eigenvector of H_0 with eigenvalue $E^{(0)}_f$, we can replace W by $H_a - E_f + H^\dagger_0 - H_0$ on the right-hand side of Eq. (6); using Eq. (2) we see that the part of $C^{(N)}_{fi}$ involving $H_a - E^{(0)}_f$ cancels with $B^{(N)}_{fi}$, and hence we can express $A^{(N)}_{fi}$ as

$$A^{(N)}_{fi} = \sqrt{2}\left(\langle\psi^-_{l_1,\epsilon}| \otimes \langle\phi_{l_2}|\right)(H^\dagger_0 - H_0)|\mathcal{F}^{(N)}_N\rangle \tag{7}$$

$$= \sqrt{2}\left(\langle\psi^-_{l_1,\epsilon}| \otimes \langle\phi_{l_2}|\right)(E^{(0)}_f - H_0)|\mathcal{F}^{(N)}_N\rangle. \tag{8}$$

The last form, i.e Eq. (8), is particularly suitable for computation since H_0 does not contain the electron-electron interaction (and therefore matrix elements of H_0 can be calculated easily and rapidly, with minimum roundoff error) but, of course, the harmonic component $|\mathcal{F}^{(N)}_N\rangle$ does contain the electron-electron interaction. Note, further, that Eqs. (8) is exact and yet the final state is represented by the (symmetrized) direct product $|\psi^-_{l_1,\epsilon}\rangle \otimes |\phi_{l_2}\rangle$, which has a simple closed-form expression in position space; this is a substantial simplification, for which we pay only a modest price, namely, rather than calculate $|\mathcal{F}^{(N-1)}_{N-1}\rangle$, as required by expression (1) for $A^{(N)}_{fi}$, we must calculate $|\mathcal{F}^{(N)}_N\rangle$.

We solved Eq. (2) for the harmonic components on a two-electron basis consisting of terms $S^\kappa_{nl}(r_1)S^\kappa_{n'l'}(r_2)Y^{LM}_{ll'}(\hat{\mathbf{r}}_1, \hat{\mathbf{r}}_2)$, where $Y^{LM}_{ll'}(\hat{\mathbf{r}}_1, \hat{\mathbf{r}}_2)$ couples spherical harmonics and where $S^\kappa_{nl}(r)$ is a radial Sturmian function which is a polynomial of degree $n_r \equiv n - l - 1$ multiplied by $e^{i\kappa r}$. We chose the "wavenumber" κ to lie in the upper right quadrant of the complex κ-plane so as to simulate both outgoing-wave open channels and exponentially decaying closed channels.[13, 14] Our basis consisted of terms up to, at most, l, $l' \leq 3$ and n_r, $n'_r \leq 20$, and (depending on κ) gave binding energies of between 0.5274 and 0.5276 a.u. for the ground state of H$^-$ (compared to the Pekeris value of 0.52775102 a.u.) and between 2.90283 and 2.90316 a.u. for the ground state of He (compared to the Pekeris value 2.903724374 a.u.). The results presented below were obtained using Eq. (8), within the velocity gauge; for multiphoton transitions the convergence of the N-photon amplitude $A_{fi}^{(N)}$ with respect to increasing basis size is more rapid in the velocity gauge than in the length gauge due to the nature of the asymptotic boundary conditions of the harmonic components.[13, 14] We note that any

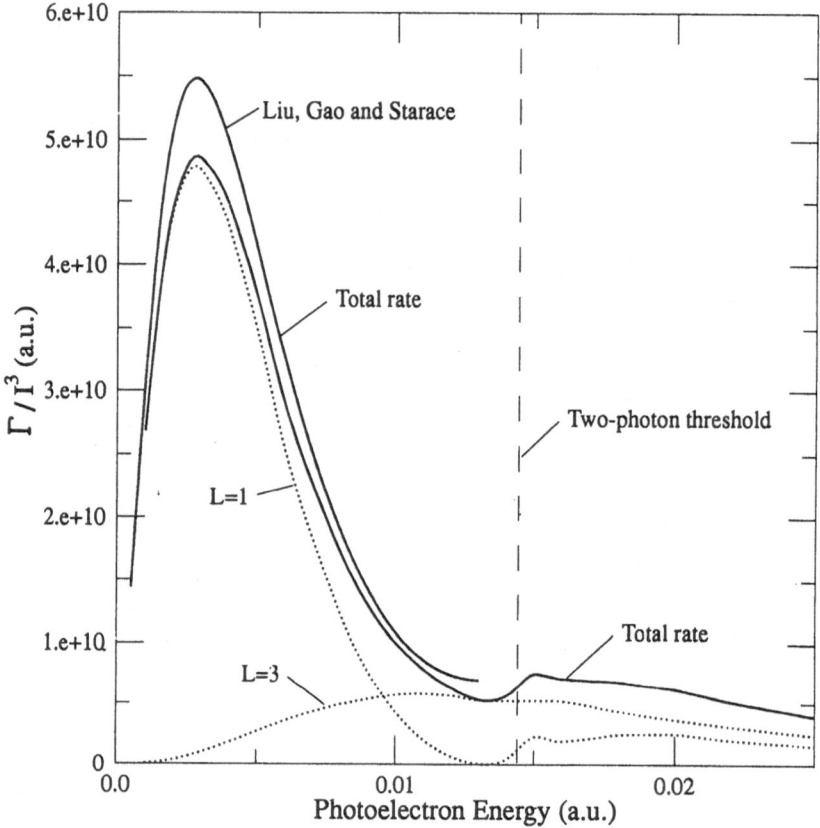

Figure 1. *Rate Γ, divided by the square of the intensity I, for 2-photon detachment of H^-, with the H atom left in the ground state. Broken line: results of Liu et al [16]; solid line: present results. Note Γ/I^2 is independent of I.*

estimate of $A_{fi}^{(N)}$ obtained by expansion on a discrete basis formally diverges, in some cases, as the number of radial basis functions increases since the final wavefunction of the ejected electron is a standing wave, composed of both ingoing and outgoing waves, while a discrete basis can only simulate ingoing- *or* outgoing-wave boundary conditions (not both); this matter has been discussed by Potvliege and Shakeshaft[13] and Dörr *et al* [14], and as in those works we used Padé summation[15] to analytically continue the divergent series.

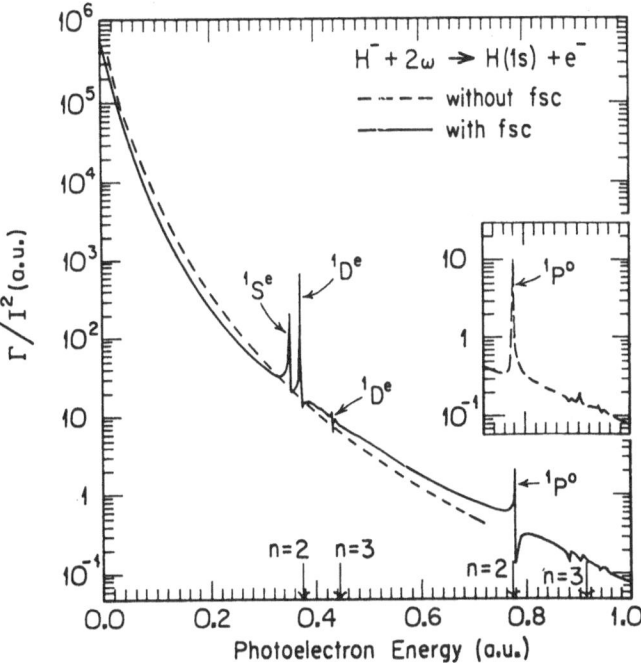

Figure 2. *Same as Fig. 1, but for 3-photon detachment, with* Γ *divided by* I^3.

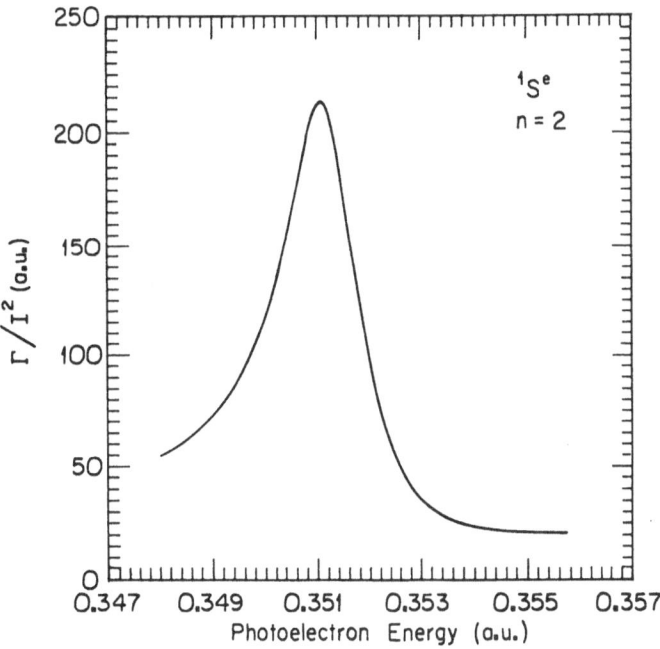

Figure 3. *Two-photon detachment of* H^- *in the excess-photon detachment region. The broken line shows results obtained when final-state correlation (FSC) is omitted.*

3 SINGLE IONIZATION: RESULTS

A. Negative Hydrogen Ion

In Figs. 1 and 2 we show the total rates (integrated over angles) for 2- and 3-photon detachment of H^- over the range of frequencies below the threshold for excess photon detachment. We compare our results with the semiempirical adiabatic hyperspherical results of Liu, Gao, and Starace[16]; the relative difference is 10 % or less, while the discrepancies with other results[17]-[19] are more serious. Our results are not completely reliable for photoelectron energies below, and in the vicinity of, the maximum of the rate. Note that as the photoelectron energy increases, the rate at first rises due to the increase in available phase space, but the rate quickly reaches a maximum long before the photoelectron energy reaches the electron affinity. There are two partial waves contributing to the total rate, the $L = 0$ and $L = 2$ waves for the 2-photon process and the $L = 1$ and $L = 3$ waves for the 3-photon process. The rapid rise in the total rate just above the threshold is due mainly to the rapid rise in the contribution from the partial wave with the smallest value of L, as expected from the Wigner threshold law — the angular momentum barrier precludes the emission of photoelectrons with linear momentum much less than $(L + \frac{1}{2})$ a.u. However, at photoelectron energies not far above the threshold the photoelectron is more likely to absorb the maximum angular momentum; this propensity rule was discussed by Fano,[20] and may be roughly understood by noting that as long as the photoelectron absorbs photons at a distance from the nucleus that is comaparable to the characteristic binding radius, the angular momentum of the photoelectron increases with its linear momentum. Thus, the contribution from the partial wave with the smallest L declines rapidly at photoelectron energies not far above the threshold, and this results in a decline in the total rate long before the photoelectron energy exceeds the electron affinity. Once the photoelectron energy does exceed the electron affinity, the linear momentum of the photoelectron exceeds the characteristic atomic orbital momentum of the electron in its initial bound state, and, since photons cannot impart momentum to the electron (within the dipole approximation), the photoelectron can aquire the necessary linear momentum only by absorbing photons while it is close to the nucleus: the improbability of this event results in a decrease in the contribution from all partial waves.

In Fig. 3 we show the 2-photon detachment rate over a wide range of frequencies above the threshold for excess photon detachment. The rate falls rapidly as the photoelectron energy increases, but rises sharply at the $^1S^e$ and $^1D^e$ Feshbach resonances below the $n = 2$ and $n = 3$ excitation thresholds of H. These resonances are in the final state, and are not reproduced when FSC is neglected. At still higher photoelectron energies the $^1P^o$ shape resonance above the $n = 2$ excitation threshold is seen. This shape resonance occurs in an intermediate state, reached by absorption of the first photon; it appears even when FSC is neglected (since correlation is included in $|\mathcal{F}_1^{(1)}\rangle$), although the line shape is incorrect when FSC is omitted. We did not locate the $^1P^o$ Feshbach resonances below the $n = 2$ threshold, but we did reproduce these resonances in the 1-photon detachment rate, in good agreement with earlier results.[21, 22]

In Figs 4-6 we show in more detail the three prominent Feshbach resonances seen in Fig. 3. Since the $^1S^e$ and $^1D^e$ resonances below the $n = 2$ threshold are approximately symmetric, we can graphically deduce the widths of the profiles, and they are in good agreement with earlier calculated values.[23] The $^1S^e$ and $^1D^e$ resonances below the $n = 3$ threshold combine to give a highly asymmetric profile, which we previously[10] misidentified as a $^1D^e$ resonance whose asymmetry was incorrectly attributed to strong interference with the 2-photon detachment background.

Table 1: Γ_a and Γ_b are, respectively, the rates for 2-photon ionization (below the threshold for 1-photon ionization) of ground-state He calculated from Eq.(8) and from the induced width of the ground-state level for various values of the photoelectron energy ϵ and a basis $l. l' \leq 3$, $n_r, n'_r \leq 20$, $|\kappa| = 1.0$ and $\arg(\kappa) = 80°$. All quantities are in a.u.

ϵ	Γ_a/I^2	Γ_b/I^2
0.15	0.23170	0.23199
0.30	0.14357	0.14373
0.45	0.09246	0.09254
0.60	0.09828	0.09839
0.75	0.02685	0.02684

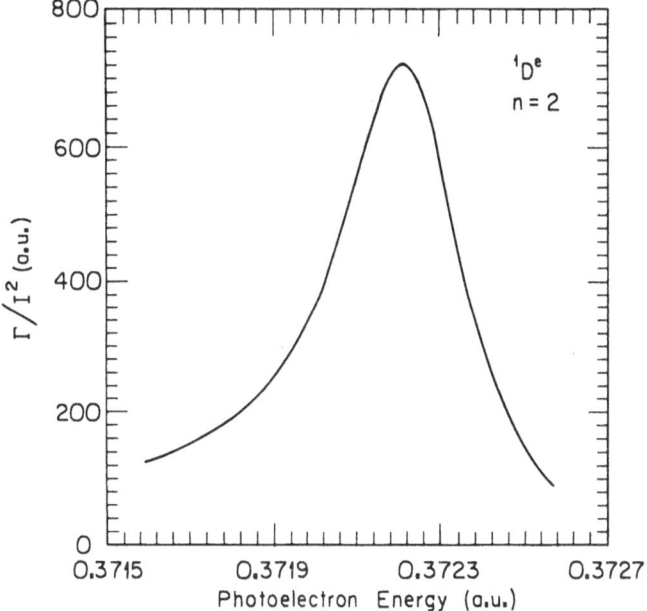

Figure 4. $^1S^e$ resonance profile, below the $n = 2$ excitation threshold of H.

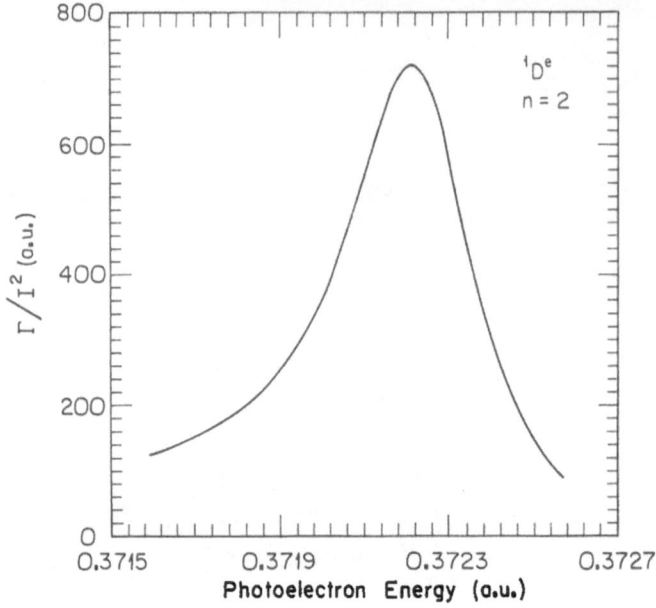

Figure 5. $^1D^e$ resonance profile, below the $n = 2$ excitation threshold of H.

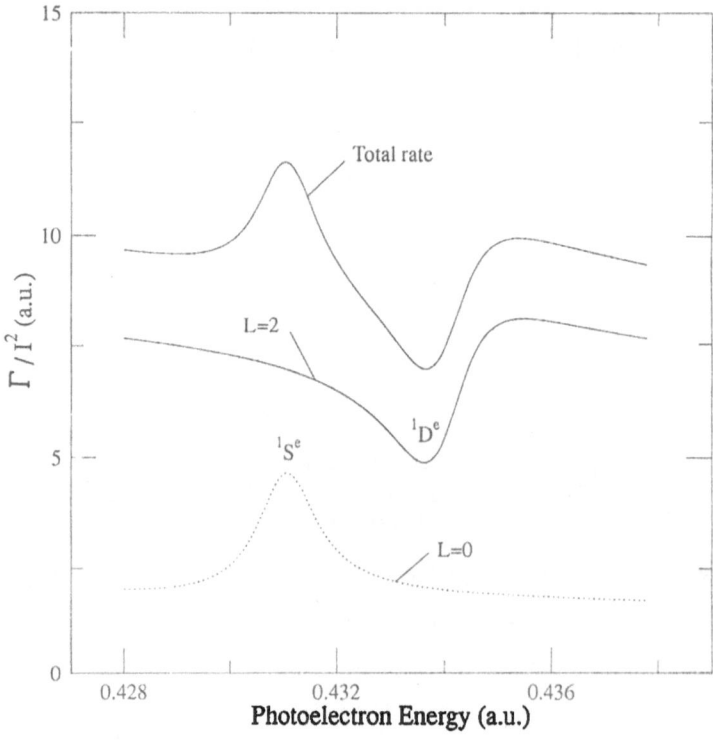

Figure 6. $^1S^e$ and $^1D^e$ resonance profiles, below the $n = 3$ excitation threshold of H.

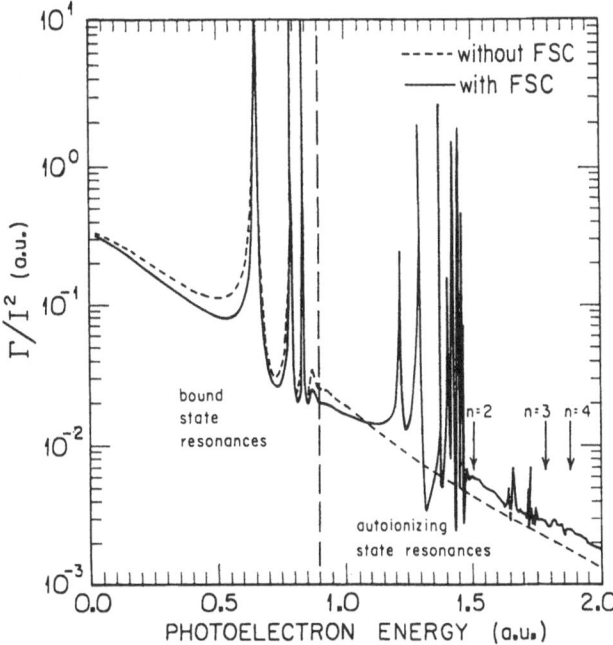

Figure 7. *Rate Γ, divided by the square of the intensity I, for 2-photon ionization from the ground-state of He, with the He^+ ion left in the ground state.*

Multiphoton detachment of H^- has been observed,[24] but not in the resonance region. (Moreover, rates were not measured, and the experiment was performed at pulse energies such that perturbation theory is probably unreliable.) Taking an "average" value of $100I^2$ a.u. for the 2-photon rate in the resonance region below the $n = 2$ threshold, and noting that the 1-photon rate in the same frequency range (i.e. $\omega \approx 5.4$ eV) is about $2I$ a.u., we obtain 1- and 2-photon rates of about 10^{12}/sec and 10^9/sec, respectively, at an intensity I of 10^{12} W/cm^2. Although the 2-photon rate is several orders of magnitude below the one-photon rate, the 1- and 2-photon signals are well separated in energy (by about 5 eV) so that the 2-photon resonances should be observable.

B. Helium

In Fig. 7 we show total rates (integrated over angles) for 2-photon ionization of the ground state of He by linearly polarized light over a range of frequencies extending from below to above the threshold for excess photon ionization. There is a lot of structure, arising from resonances due to 1-photon transitions to singly-excited bound states and 2-photon transitions to doubly-excited autoionizing states below the $n = 2$ and $n = 3$ excitation thresholds of He^+. There is only a limited set of published theoretical results with which we can compare. For frequencies below the threshold for 1-photon ionization of ground-state He (i.e. $\omega < 24.6$ eV), results have been obtained by Victor (1967), Ritchie (1977), L'Huillier *et al*,[27] Kulander,[28] and Bachau *et al.*[29] The agreement between all of these previous results, where comparison can be made, is fairly good, as is the agreement with our results. Bachau *et al*[29] also reported results for excess-photon ionization (frequencies > 24.6 eV) and we present a quantitative comparison in

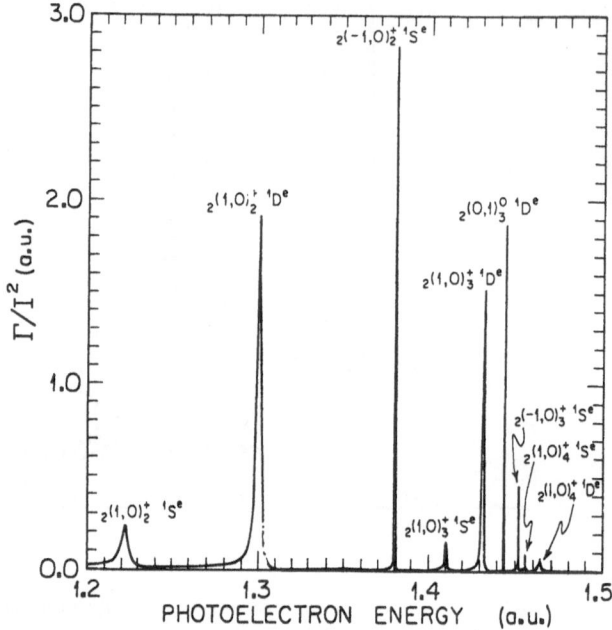

Figure 8. *Resonances below the* $n = 2$ *threshold of* He^+, *reached from the ground state of He.*

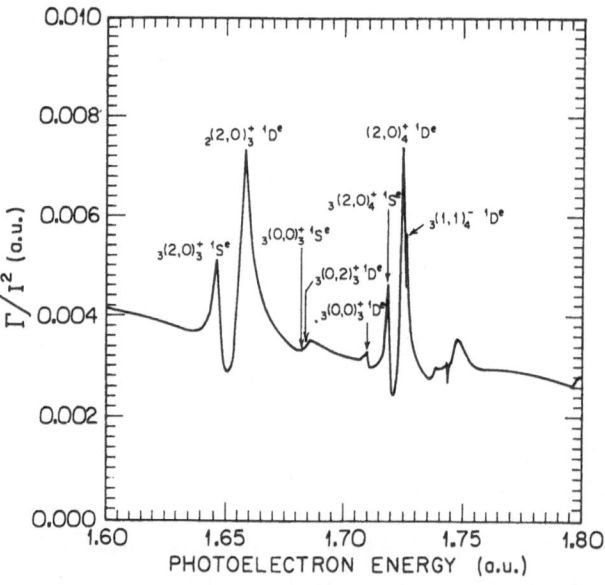

Figure 9. *Resonances below the* $n = 3$ *threshold of* He^+, *reached from the ground state of He.*

Table 1. The agreement is at best fair, and at one frequency, 1.1 a.u., the disagreement is rather striking, apparently due to the presence of a $^1D^e$ resonance (below the $n = 2$ excitation threshold of He^+) which is not accounted for in the calculations of Bachau *et al* who neglect strong correlation in the final state. We have checked that our results are rather well converged with respect to increasing basis size and changing κ.

In Figs. 8 and 9 we have expanded the scale to show the autoionizing resonance line-profiles more clearly, and we have labelled these resonances using the K and T quantum numbers of Herrick and Sinanoglu (1975). The positions and widths of these resonances have been calculated by e.g. Lipsky *et al*,[31] Ho,[32] and Ho and Callaway.[33] These previous results for the positions are in good agreement with ours. For those reso-nances that are approximately symmetric, we can graphically deduce the widths of the profiles, and they are also in good agreement with earlier calculated values. Some of the resonances are highly asymmetric, indicating strong interference with the 2-photon detachment background.

To give some idea of the competition between 1- and 2-photon ionization we note that for ionization from the ground state, at frequencies around 30 eV, where the resonances below the $n = 2$ threshold are reached by absorption of 2 photons, the 1-photon ionization rate is about 10^{12}/sec at an intensity of 10^{13} W/cm^2, while (at the same intensity) the resonant 2-photon rate is about 10^9/sec if we take an "average" resonant 2-photon rate (see Fig. 8) of $1 \times I^2$ a.u.

4 DOUBLE IONIZATION: METHOD

We now discuss the method we used to calculate cross sections for double ionization, and in the following section we report results for double photoionization of He at high photon energies. We assume that the light is linearly polarized (along the z-axis) and that the atomic states are spin-singlet (we factor out the spin). The differential cross section for double ionization, with one electron emerging into the solid angle $d\Omega_1$ with momentum $\mathbf{k}_1$ and energy $E_1 \equiv k_1^2/2$, and the other electron emerging into the solid angle $d\Omega_2$ with momentum $\mathbf{k}_2$ and energy $E_2 \equiv k_2^2/2$, is

$$\frac{d\sigma}{dE_1 d\Omega_1 d\Omega_2} = \frac{4\pi^2}{\omega c} k_1 k_2 |f(\mathbf{k}_1, \mathbf{k}_2)|^2, \tag{9}$$

where, with $|\Psi_i\rangle$ and $|\Psi_{\mathbf{k}_1,\mathbf{k}_2}^{(-)}\rangle$ the initial- and final-state vectors of the atomic system,

$$f(\mathbf{k}_1, \mathbf{k}_2) = \langle \Psi_{\mathbf{k}_1,\mathbf{k}_2}^{(-)}| \left(\frac{d}{dz_1} + \frac{d}{dz_2}\right) |\Psi_i\rangle. \tag{10}$$

The indistinguishability of the electrons implies $f(\mathbf{k}_1, \mathbf{k}_2) = f(\mathbf{k}_2, \mathbf{k}_1)$, and inversion symmetry implies $f(\mathbf{k}_1, \mathbf{k}_2) = -(-1)^P f(-\mathbf{k}_1, -\mathbf{k}_2)$, where P is the parity of the initial state. Putting $\mathbf{k}_1 = \mathbf{k} = -\mathbf{k}_2$ gives $f(\mathbf{k}, -\mathbf{k}) = -(-1)^P f(\mathbf{k}, -\mathbf{k})$, and since P is even $f(\mathbf{k}, -\mathbf{k}) = 0$, as noted above. [However, if the atom were to absorb an *even* number of photons, $f(\mathbf{k}, -\mathbf{k})$ would not vanish.] Integrating over all directions of $\mathbf{k}_2$, gives [34]

$$\frac{d\sigma}{dE_1 d\Omega_1} = \frac{1}{4\pi} \frac{d\sigma}{dE_1}[1 + \beta(E_1) P_2(\cos\theta_1)], \tag{11}$$

where θ_1 is the angle between $\mathbf{k}_1$ and the z-axis. The total final energy is $E_f \equiv E_1 + E_2 = E_i + \omega$. The energy distribution $d\sigma/dE_1$ is symmetric about the midpoint $E_f/2$; if one electron emerges with energy E_1, the other emerges with energy $E_f - E_1$. However, the angular asymmetry parameter, $\beta(E_1)$, is not symmetric about $E_f/2$ since in arriving at Eq. (11) we integrated over the angles of $\mathbf{k}_2$ but not $\mathbf{k}_1$.

We took the initial-state wavefunction, $\Psi_i(\mathbf{r}_1, \mathbf{r}_2)$, to have the form

$$\Psi_i(\mathbf{r}_1, \mathbf{r}_2) \approx C e^{-\alpha(r_1+r_2)-\gamma r_3} \sum_{i+j+k \leq N} c_{ijk} r_1^i r_2^j r_3^k, \tag{12}$$

where $\mathbf{r}_1$ and $\mathbf{r}_2$ are the electron coordinates and $\mathbf{r}_3 = \mathbf{r}_2 - \mathbf{r}_1$, with $r_1 = |\mathbf{r}_1|$, $r_2 = |\mathbf{r}_2|$, and $r_3 = |\mathbf{r}_3|$. The integer N was chosen to be 0, 1, 2, or 3, and the two nonlinear parameters α and γ and the linear parameters c_{ijk} were chosen to minimize E_i. The 2-parameter ($N = 0$), 4-parameter ($N = 1$), 8-parameter ($N = 2$), and 14-parameter ($N = 3$) wavefunctions yield the E_i-estimates -2.8896, -2.8915, -2.90347, and -2.903641 a.u, respectively, compared to the Pekeris energy -2.903724374 a.u. However, as others have emphasized,[35, 36] the two Kato cusp conditions [37] are a more relevant measure of the accuracy of $\Psi_i(\mathbf{r}_1, \mathbf{r}_2)$. The first condition pertains to the confluence of one electron and the nucleus, and the second condition pertains to the confluence of the two electrons. For the 8-parameter wavefunction, the relative errors in the first and second cusp conditions are, respectively, less than 5% and 35%. The second condition, particularly in the region where the two electrons are near the nucleus, is significant in the description of the third (photon sharing) mechanism mentioned above.

We took the final-state wavefunction, $\Psi_{\mathbf{k}_1,\mathbf{k}_2}^{(-)}(\mathbf{r}_1, \mathbf{r}_2)$, to have the form

$$\Psi_{\mathbf{k}_1,\mathbf{k}_2}^{(-)}(\mathbf{r}_1, \mathbf{r}_2) \approx \chi_{\mathbf{k}_1,\mathbf{k}_2}^{(-)}(\mathbf{r}_1, \mathbf{r}_2) + a e^{ib(\mathbf{k}_1 \cdot \mathbf{r}_1 + \mathbf{k}_2 \cdot \mathbf{r}_2)} \Psi_i(\mathbf{r}_1, \mathbf{r}_2), \tag{13}$$

where $\chi_{\mathbf{k}_1,\mathbf{k}_2}^{(-)}(\mathbf{r}_1, \mathbf{r}_2)$ has the correct asymptotic form [38, 39]:

$$\chi_{\mathbf{k}_1,\mathbf{k}_2}^{(-)}(\mathbf{r}_1, \mathbf{r}_2) = (2\pi)^{-3} e^{i\mathbf{k}_1 \cdot \mathbf{r}_1 + i\mathbf{k}_2 \cdot \mathbf{r}_2} \prod_{j=1-3} e^{-\pi\gamma_j/2} \Gamma(1 - i\gamma_j) \, {}_1F_1(i\gamma_j, 1, -ik_j r_j - i\mathbf{k}_j \cdot \mathbf{r}_j),$$

$$\tag{14}$$

where $\mathbf{k}_3 = \frac{1}{2}(\mathbf{k}_2 - \mathbf{k}_1)$, $\gamma_1 = -Z/k_1$, $\gamma_2 = -Z/k_2$, and $\gamma_3 = 1/(2k_3)$, with $Z = 2$. The second term on the right-hand side of Eq. (13) is a correction in the inner region. Its justification will be given elsewhere; but we note here that the correction term is applicable only when $E_f \gg |E_i|$, that the coefficient a was determined from the orthogonality of $\Psi_i(\mathbf{r}_1, \mathbf{r}_2)$ and $\Psi_{\mathbf{k}_1,\mathbf{k}_2}^{(-)}(\mathbf{r}_1, \mathbf{r}_2)$, and that $b = \sqrt{\omega/E_f}$ (≈ 1 if $E_f \gg |E_i|$).

5 DOUBLE IONIZATION: RESULTS

We now show results based on the 8-parameter ground-state wavefunction. We first look at the angular distribution $d\sigma/dE_1 d\Omega_1 d\Omega_2$, for a photon energy of 2.8 KeV. We fix the momentum $\mathbf{k}_1$ of one electron — electron 1, say — to be at an angle $\theta_1 = \cos^{-1}(1/\sqrt{3}) = 54.73°$ with the z-axis, and we allow the momentum $\mathbf{k}_2$ of the other electron — electron 2, say — to vary in the plane of $\mathbf{k}_1$ and the electric field (z-) axis. Our choice of θ_1 implies $P_2(\cos\theta_1) = 0$, and hence $\int d\Omega_2 \, (d\sigma/dE_1 d\Omega_1 d\Omega_2)$ is, from Eq. (11), the energy distribution $d\sigma/dE_1$ (reduced by $1/4\pi$). We denote the angle between $\mathbf{k}_1$ and $\mathbf{k}_2$ as θ_{12} (see Fig. 10, lowest box). In Fig. 10 we show the angular distribution (or rather a slice of it in one plane) for 4 different partitionings of the energy E_f (= 2.721 KeV). We first focus on the case where one electron carries away almost all the energy, i.e. either E_1 or E_2 equal to 10^{-6} eV. When $E_2 = 10^{-6}$ eV we see a peak centered where $\theta_{12} = \pi$. This peak corresponds to shakeoff: Electron 1 absorbs the photon and as it soars out of the atom it tickles electron 2. Since electron 2 does not experience the photon, the only direction relevant to this slow electron is the direction of emission of electron 1. Therefore, by symmetry, electron 2 must emerge parallel or

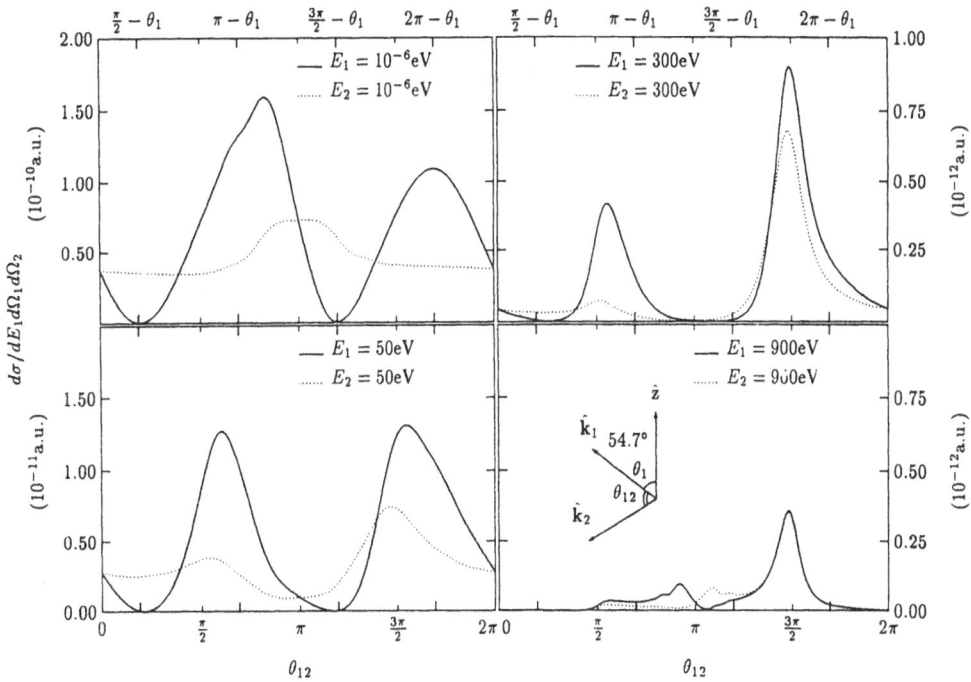

Figure 10. *Double ionization of He; angular distribution in plane of electric field (z-axis) and momentum $\mathbf{k}_1$ of one electron, with $\mathbf{k}_1$ fixed at angle 54.73° with z-axis. Photon energy is 2.8 KeV. Thicker tick marks on θ_{12}-axis correspond to values indicated on upper θ_{12}-axis.*

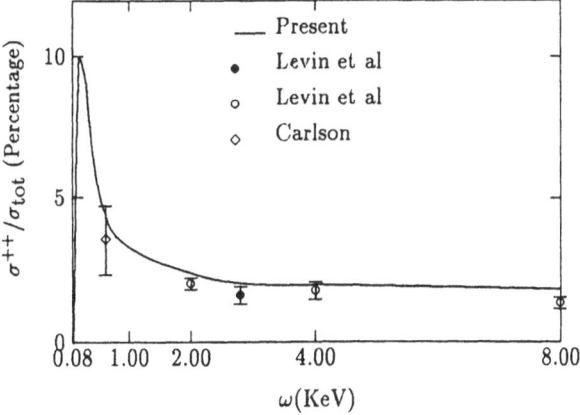

Figure 11. *Ratio of cross sections for double ionization to total ionization of He vs photon energy ω, compared to data of Refs. [1,2]*

antiparallel to electron 1, and since the electrons repel each other the preferred orientation is antiparallel, i.e. $\theta_{12} = \pi$. When $E_1 = 10^{-6}$ eV we see two minima and two peaks. The minima, which are almost zeroes, occur at $\theta_{12} = \frac{1}{2}\pi - \theta_1$ and $\theta_{12} = \frac{3}{2}\pi - \theta_1$, corresponding to the unlikely emission of the *fast* electron -- now electron 2 -- perpendicular to the electric field axis. The fast electron would prefer to emerge along the electric field axis, i.e. where $\theta_{12} = \pi - \theta_1$ or $\theta_{12} = 2\pi - \theta_1$, but in fact the peaks are shifted somewhat closer to π since we have fixed the angle θ_1 of the slow electron, and shakeoff is maximum when $\theta_{12} = \pi$. Now consider the case E_1 or E_2 equal to 50 eV. When $E_2 = 50$ eV we see two peaks, one centered close to $\theta_{12} = \pi/2$, the other close to $\theta_{12} = 3\pi/2$. While the energy, 50 eV, of the slower electron is not large, it is an appreciable fraction of the initial total binding energy, and the slower electron cannot easily acquire 50 eV via the shakeoff mechanism. Rather, it acquires this energy via the knockout mechanis to the fact that when two particles of equal mass undergo a binary collision they emerge with a relative angle of $\pi/2$. When $E_1 = 50$ eV the minima once again occur close to where the fast electron emerges perpendicular to the electric field, but now the left and right peaks are shifted somewhat closer to $\pi/2$ and $3\pi/2$, respectively, due to knockout being the more efficient mechansim for ejection of the slower electron (whose angle θ_1 is fixed). Looking now at the case E_1 or E_2 equal to 300 eV, where even the slower electron is moving fast, knockout dominates, and all peaks are centered not far from where $\theta_{12} = \pi/2$ or $3\pi/2$. Note that the peaks near $3\pi/2$ are more pronounced than those near $\pi/2$, a feature which may be understood as follows: The electron which absorbs the photon prefers to be ejected along the electric field axis, i.e. the positive or negative z-axis. Furthermore, this fast electron is deflected through a relatively small angle in the knockout collision. Now, the second electron cannot be knocked out at a large angle (i.e. $> 90°$) relative to the direction of incidence of the first electron. Hence, after the knockout collision the two electrons are most likely to emerge within a $90°$ cone that includes either the negative or, as in our case (since $\theta_1 = 54.73°$), the positive z-axis; therefore $\theta_{12} = 3\pi/2$ is preferred. Finally we consider the case E_1 or E_2 equal to 900 eV, where both electrons are moving very fast and with not very different speeds (note $E_f/2 = 1360$ eV). Knockout at $\theta_{12} = \pi/2$ is now almost insignificant, while the knockout peaks at $\theta_{12} = 3\pi/2$ are almost independent of which electron is fastest. More interestingly, we see a new peak not far from $\theta_{12} = \pi$. This new peak arises from the photon sharing mechanism, i.e. the two electrons simultaneously share the photon and leave in nearly opposite directions, carrrying away almost no net momentum. The minimum close to $\theta_{12} = \pi$ is a vestige of the inversion-symmetry zero of $f(\mathbf{k}_1, \mathbf{k}_2)$ which, when $E_1 = E_2$, occurs at $\theta_{12} = \pi$. Finally, in Fig. 11, we show our estimate of the ratio, R, of cross sections for double to total ionization (our values for the total-ionization cross section at low photoelectron energies were taken from experimental data compiled by J. A. R. Samson, priv. comm.) and we compare with various experimental data. [1, 2] Within the nonrelativistic framework R approaches a constant as ω increases [40, 3, 35], and different theoretical estimates of this constant have been given, e.g.: 1.6 % [8], 1.7% [3, 35, 36], and 2.3 % [4]. We find $R = 1.8$ % at 8 KeV while the measured value [1] is 1.5 ± 0.2 %. We believe our estimate would be reduced, but only slightly, by using an improved ground-state wavefunction.

Acknowledgements

The work on single ionization was supported by the National Science Foundation under grant number PHY-9017079; the work on double ionization was supported by the Office of Basic Energy Sciences of the Department of Energy under contract number DE-FG03-92ER14266. One of us (D.P.) gratefully acknowledges a scholarship from the Natural Sciences and Engineering Research Council of Canada.

References

[1] J. C. Levin, D. W. Lindle, N. Keller, R. D. Miller, Y. Azuma, N. Berrah Mansour, H. G. Berry, and I. A. Sellin, Phys. Rev. Lett. **67**, 968 (1991); J. C. Levin, I. A. Sellin, B. M. Johnson, D. W. Lindle, R. D. Miller, N. Berrah Mansour, Y. Azuma, H. G. Berry, D. H. Lee, Phys. Rev. A in press (1993).

[2] T. A. Carlson, Phys. Rev. **156**, 142 (1967).

[3] F. W. Byron and C. J. Joachain, Phys. Rev. **164**, 1 (1967).

[4] M. Ya. Amusia, E. G. Drukarev, V. G. Gorshkov, and M. Kazachkov, J. Phys. B **8**, 1248 (1975).

[5] T. N. Chang and R. T. Poe, Phys. Rev. A **12**, 1432 (1975).

[6] S. L. Carter and H. P. Kelly, Phys. Rev. A **24**, 170 (1981).

[7] J. A. R. Samson, Phys. Rev. Lett. **65**, 2861 (1990).

[8] T. Ishihara, K. Heno, and J. H. McGuire, Phys. Rev. A **44**, R6980 (1991).

[9] G. H. Wannier, Phys. Rev. **90**, 817 (1953); **100**, 1180 (1955).

[10] D. Proulx and R. Shakeshaft, Phys. Rev. A **46**, R2221 (1992).

[11] D. Proulx and R. Shakeshaft, J. Phys. B **26**, L7 (1993).

[12] Z.-j. Teng and R. Shakeshaft, Phys. Rev. A (in press).

[13] R. M. Potvliege and R. Shakeshaft Phys. Rev. A **39**, 1545 (1989).

[14] M. Dörr, R. M. Potvliege, and R. Shakeshaft, JOSA B **7**, 433 (1990).

[15] S. Klarsfeld and A. Maquet Phys. Lett. **78A**, 40 (1980).

[16] C-R. Liu, B. Gao, and A. F. Starace Phys. Rev. A **46**, 5985 (1992).

[17] M. G. Fink and P. Zoller, J. Phys. B **18**, L373 (1985).

[18] T. Mercouris and C. A. Nicolaides, J. Phys. B **21**, L285 (1988).

[19] M. Crance, J. Phys. B **23**, L285 (1990).

[20] U. Fano, Phys. Rev. A **32**, 617 (1985).

[21] J. T. Broad and W. P. Reinhardt, Phys. Rev. A **14**, 2159 (1976).

[22] H. R. Sadeghpour, C. H. Greene, and M. Cavagnero, Phys. Rev. A **45**, 1587 (1992).

[23] T. Scholz, P. Scott, and P. G. Burke, J. Phys. B **21**, L139 (1988).

[24] C. Y. Tang, P. G. Harris, A. H. Mohaghegi, H. C. Bryant, C. R. Quick, J. B. Donahue, R. A. Reeder, S. Cohen, W. W. Smith, and J. E. Stewart, Phys. Rev. A **39**, 6068 (1989).

[25] G. A. Victor, Proc. Phys. Soc. **91**, 825 (1967).

[26] B. Ritchie, Phys. Rev. A **16**, 2080 (1977).

[27] A. L'Huillier, L. Jönsson, and G. Wendin, Phys. Rev. A **33**, 3938 (1986).

[28] K. C. Kulander, Phys. Rev. A **36**, 2726 (1987).

[29] H. Bachau, P. Lambropoulos, and X. Tang, Phys. Rev. A **42**, 5801 (1990).

[30] D. R. Herrick and O. Sinanoglu, Phys. Rev. A **11** 97 (1975).

[31] L. Lipsky, R. Anania, and M. J. Conneely, Atomic Data and Nuclear Data Tables **20**, 127 (1977).

[32] Y. K. Ho, Physics Reports **99**, 1 (1983).

[33] Y. K. Ho and J. Callaway, J. Phys. B **18**, 3481 (1985).

[34] C. N. Yang, Phys. Rev. **74**, 764 (1948).

[35] T. Åberg, Phys. Rev. A **2**, 1726 (1970).

[36] A. Dalgarno and H. R. Sadeghpour, Phys. Rev. A **46**, R3591 (1992).

[37] T. Kato, Comm. Pure. Appl. Math. **10**, 151 (1957).

[38] M. Brauner, J. S. Briggs, and H. Klar, J. Phys. B **22**, 2265 (1989).

[39] F. Maulbetsch and J. S. Briggs, Phys. Rev. Lett. **68**, 2004 (1992).

[40] A. Dalgarno and A. L. Stewart, Proc. Phys. Soc **76**, 49 (1960).

SEQUENTIAL VS. SIMULTANEOUS IONIZATION OF
TWO-ELECTRON ATOMS BY INTENSE LASER RADIATION

Sayoko Blodgett-Ford,[$] Jonathan Parker,[†]
and Charles W. Clark

Electron and Optical Physics Division
National Institute of Standards and Technology
Gaithersburg, MD 20899 USA
[$] also School of Law, Yale University,
New Haven, CT 06520 USA
[†] also Institute for Physical Science and Technology
University of Maryland
College Park, MD 20742

THE BASIC ISSUE

The first observations[1] of multiple ionization of atoms by strong laser fields (intensity $I > 10^{13}$ W cm^{-2}) gave rise to a variety of speculations on the sequence of ionization events. Enthusiasm for collective or statistical models,[2] in which an electron shell responded as an entity to the driving field, was dampened by the seminal paper of Lambropoulos.[3] He drew attention to the fact that as the electric field amplitude in the laser pulse grows from zero to its peak value, individual electrons will be stripped off sequentially in a manner more or less following the predictions of perturbation theory. This view was validated by experiments on laser ionization of molecules by Frasinski et al.,[4] who found no evidence for the high energy fragments expected from the Coulomb explosion of a multiply-charged molecular ion created by simultaneous ejection of several electrons. However, this picture is predicated upon the rise time of the pulse being long compared to typical multiphoton ionization rates. Although this was certainly so in the first generation of experiments, the progressive development of more intense, sub-picosecond laser sources will require the issue to be revisited.

THEORETICAL TREATMENT OF MANY-ELECTRON ATOMS

Significant progress has been made in the theory of one-electron atoms in intense radiation fields, but the treatment of many-electron atoms remains rudimentary. We are aware of only one calculation to date[5] in which more than one electron (or one electronic orbital) of a three-dimensional atom is allowed to evolve under the influence of the radiation field. This paper reports results from numerical integration of the time-dependent Schrödinger equation for two-electron atoms in a model radiation field, in which the spatial coordinates of both electrons are represented explicitly. Its main conclusion is that there is a relatively sharp threshold for simultaneous ionization of both electrons: if the field amplitude is less than a critical value, only sequential ionization is seen; if it exceeds this value, and the pulse is ramped on sufficiently rapidly, then simultaneous ionization proceeds from the beginning of the pulse. Furthermore, significant effects of electron correlation are observed at this threshold, so that it is inappropriate to view the double ionization process simply in terms of the rapid stripping of two independent electrons. For He irradiated at the frequency corresponding to KrF and ArF lasers (wavelengths of 248 and 196 nm), the threshold in this model is found to correspond to a laser intensity of about 10^{16} W cm^{-2}. This value is roughly consistent with experimental observations.[6]

THE HELIUM ATOM IN A MODEL RADIATION FIELD

We proceed by solving the initial value problem for the Schrödinger equation of motion for a two-electron atom,

$$i\frac{\partial}{\partial t}\Psi(\vec{r}_1, \vec{r}_2, t) = \left[H_0(\vec{r}_1, \vec{r}_2) + V(\vec{r}_1, \vec{r}_2)\, cos(\omega t)\, f(t)\right]\Psi(\vec{r}_1, \vec{r}_2, t) \; ,$$

(1)

where H_0 is the nonrelativistic atomic Hamiltonian, V is the spatial interaction of the two electrons with the model radiation field, $f(t)$ describes the pulse envelope ($0 \le f(t) \le 1$), and $\Psi(\vec{r}_1, \vec{r}_2, t=0)$ is the ground state of the system. For linearly-polarized radiation, treated in the electric dipole approximation, the true electron-field interaction takes the form

$$V_{\text{true}}(\vec{r}_1, \vec{r}_2) = \vec{F} \cdot \left[\vec{r}_1 + \vec{r}_2\right] \; ,$$

where $\vec{F} = F\hat{z}$ is the electric field vector. In this paper, we employ instead a pure scalar model of this potential

$$V(\vec{r_1}, \vec{r_2}) = F\,[r_1 + r_2]\,. \tag{2}$$

This provides a "breathing mode" of excitation, in which a force of constant strength (throughout all space) is directed radially with respect to the atomic nucleus, rather than along a fixed direction as for V_{true}. This driving field cannot impart angular momentum to the electrons, which is both a potentially serious deficiency in its relevance to actual radiative excitation, and a significant advantage in facilitating computation. The latter aspect provides the primary motivation for our adoption of the V of eq. (2); the following comments are not intended to extenuate its deviations from V_{true}, which we recognize as being significant. At large distances r, V and V_{true} are identical in their effects upon a localized wavepacket, to the extent that the radius of curvature in the force represented by V can be ignored; in atomic hydrogen, V gives photoelectron spectra that have above-threshold ionization peaks similar to those produced by V_{true}; the static ($\omega = 0$) "polarizability" of the ground state of hydrogen under the influence of V, as determined from the second-order level shift, is 3 a.u., vs. 4.5 a.u. for V_{true}. Thus there are some respects in which atomic excitation by either V or V_{true} leads to roughly similar results. We also wish to emphasize that since V is isotropic (and since $\Psi(\vec{r_1}, \vec{r_2}, t = 0)$ has ${}^1S$ symmetry), all angular dependence of the two-electron wavefunction derives from electron correlations.

From eqs. (1) and (2) it is obvious that the total angular momentum L is a constant of the motion, so the wavefunction can be expanded in partial waves as

$$\Psi(\vec{r_1}, \vec{r_2}, t) = \sum_{l_1, l_2} \Psi_{l_1 l_2}(r_1, r_2, t)\, |\, l_1\, l_2\, L\, M >\,. \tag{3}$$

In the cases treated here, $L = M = 0$, so $l_1 = l_2$. Up to four partial waves have been retained in the calculations reported here; this appears give satisfactory convergence in the cases we have treated. Eqs. (1), (2), and (3) then yield a set of coupled partial differential equations for the $\Psi_{ll}(r_1, r_2, t)$. We retain only the monopole and dipole terms in the expansion of the electron-electron interaction, i.e.

$$r_{12}^{-1} \rightarrow r_{>}^{-1}\,(1 + (cos\,\theta_{12})\,r_{<}/r_{>})\,. \tag{4}$$

This approximation is not of central importance, but is convenient in our computational approach. We discretize the electronic radial coordinates on a rectangular grid of uniform spacing δ, $(r_i, r_j, l) \rightarrow (i\delta, j\delta, l)$, and approximate the partial differential equations by finite-difference equations, accurate to $O(\delta^2)$, for $\Psi_{ijl}(t) \equiv \Psi_{ll}(i\delta, j\delta, t)$. Eqs. 1-3 then reduce to a system of linear equations

$$i\frac{\partial}{\partial t}\Psi_{ijl} = \sum_{i'j'l'} H(t)^{i'j'l'}_{ijl} \, \Psi_{i'j'l'} .$$

(5)

In the representation we adopt here, the two-electron configuration

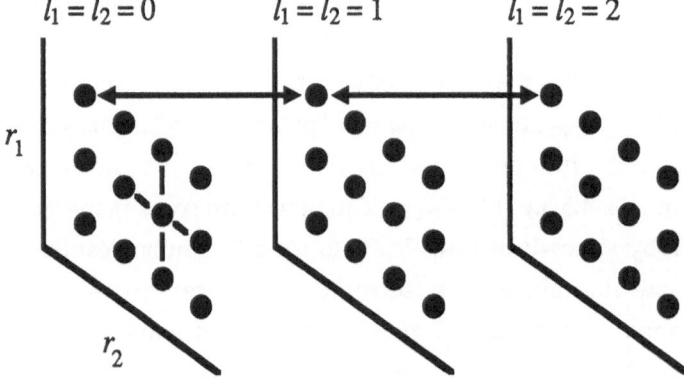

Fig. 1 Schematic of spatial representation of the wavefunction. Each point is coupled to its nearest neighbors in its own plane by the kinetic energy operator (simple lines), and to its neighbors in adjacent planes by the electron-electron interaction. Our "radiative" interaction is *local*, in that it does not couple different points. Treatment of the true radiative interaction requires that this representation be extended to include states of non-zero L; in this case, the radiative interaction does couple neighboring planes.

space is to be visualized as a three-dimensional lattice of parallel planes, each plane corresponding to a specific value of l, and containing the square lattice of points (i,j) that represent the two-electron radial coordinates for that partial wave. The operator H in eq. (5) then couples the wavefunction at the point (i, j, l) to its nearest neighbors on the lattice, $(i\pm1, j\pm1, l\pm1)$: the coupling in i,j derives from the two-electron kinetic energy operator; the coupling in l from the dipole term in the expansion of r_{12}^{-1}. This representation is depicted in Fig. 1.

SOLUTION OF THE EQUATIONS OF MOTION ON A MASSIVELY PARALLEL PROCESSOR

We have carried out the time integration of eq. (5) on a Thinking Machines, Inc. CM-2 Connection Machine, a computer which provides a facility for assigning a dedicated "virtual" central processing unit (with its own local memory) to each point (i, j, l), with efficient communication between neighboring processors. The work of the virtual processors is handled by a large number of independent physical processors (up to 2048 floating-point units). This computational model contrasts to the familiar approach to solving vector equations, in which the configuration space is mapped onto a memory array that is handled by a single processor. In the solution of eq. (5), all virtual processors execute the *same* global instruction set, but each applies the instructions to the contents of its own local memory and to those of its nearest neighbors. This scheme is extensible to less restrictive approximations to this system, including the treatment of V_{true} in a partial wave expansion, though the size of the calculation increases significantly when L is permitted to vary in a general manner.

Eq. (5) was integrated using a split-operator method described by Richardson.[7] It provides an explicit propagator, accurate to $O(\tau^2)$ in the time step τ, which is unitary and unconditionally stable, and which preserves nearest-neighbor coupling on the lattice. The calculations reported here involve time integration over four optical periods, with the pulse ramped linearly on and off over a single period. We have used a spatial grid size δ of 0.5 a.u., and time steps τ between 3×10^{-3} and 2×10^{-2} a.u. The initial state of the system was obtained by an arbitrary initialization of the wavefunction on the grid, and then integrating the Schrödinger equation in imaginary time,

$$\partial \psi / \partial t = - H_0 \, \psi \; .$$

At sufficiently large times, the solution of this equation converges to the ground state of H_0. We note that is important to employ identical integration methods to solve this equation and the time-dependent equation (5), so that no spurious excitation occurs when $f(t) = 0$. For $\delta = 0.1$ a.u. and with inclusion of the first four partial waves ($l = 0 - 3$), we find the ground-state energy of He to be -2.725 a.u., vs. the experimental value of -2.904 a.u. Most of the difference between calculated and experimental values derives from the use of a finite spatial grid, rather than from the multipole approximation of the potential (4) or truncation of the partial wave expansion. For example, the ground state energy of He$^+$ 1s on a finite uniform grid is known in closed form[8], and in this case is equal to -1.8885 a.u., deviating by 6% from the exact value of -2 a.u. Thus our calculated total energy corresponds within this approximation to a He ionization potential of 0.837 a.u., vs. an exact value of 0.904 a.u.; the relative error is thus consistent with that expected to be induced by spatial discretization only. The two largest partial wave populations in the ground state are similar to those obtained in multiconfiguration Hartree-Fock calculations;[9] we find populations of 0.9942, 0.0058, 9.21 x 10^{-6}, and 2.15 x 10^{-7} for the s, p, d, and f waves respectively, compared to MCHF values of 0.9958, 0.00398, 1.8 x 10^{-4}, 1.7 x 10^{-5}. We believe the large deviations in the d and f wave populations derive from the truncation expressed by eq. (4), since these partial waves would presumably be populated primarily by higher multipole interactions with the dominant s wave.

The probabilities for single and double ionization are determined by computing the probability that one or both electrons are in the region $r > 10$ a.u. respectively; we therefore do not distinguish between final states in which the system is ionized or excited into a Rydberg state. These probabilities will be the only quantitative data presented in this paper. We have also observed the time evolution of the two-electron wavefunction by taking "snapshots" of the probability $|\Psi(r_1, r_2, t)|^2$ at discrete time intervals, converting them to false-color, two-dimensional maps in TIFF format, and animating the sequence. Although we shall not present those images here, they are available in electronic form on request from the authors.[10] A qualitative summary of our observations is as follows. We view electron wavepackets in the two-dimensional quadrant r_1, r_2. At low field strengths ($F < 0.25$ a.u.), we see wavepackets emerge from the atom and travel outward, remaining close to the axes $r_1 = 0, r_2 = 0$; almost no probability is observed in the "Wannier region" $r_1 \sim r_2$. This corresponds to single ionization. As the field strength increases ($F < 0.75$ a.u.), we begin to see probability flow into the Wannier region, but it tends to follow rectilinear trajectories. For example, a wavepacket will be ejected outwards close to the axis $r_2 = 0$, and at a relatively large value of r_1 it will split, some of it continuing along the original trajectory and the rest moving out close to the line $r_1 = $ constant. This corresponds to sequential ionization, in which the second electron leaves the atom after the first has moved to large distances. In cases of both weak and intermediate field strengths, the wavepackets exhibit well-developed, regular nodal structure, the nodal lines being parallel to the coordinate axes. For field strengths roughly in the region $0.75 < F < 1$ a.u., we see dramatically different behavior. At the onset of the laser pulse, the wavepacket moves directly into the Wannier region, and its nodal structure becomes highly irregular. This is a manifestation of simultaneous two-electron ionization.

Figures 2 and 3 show results for a helium atom driven at a frequency corresponding to the 248 nm wavelength of the KrF laser, i.e. $\omega = 0.184$ a.u. The field is ramped on linearly over a single optical cycle, after which its amplitude remains constant. Figure 2 shows the time-dependent populations of the ground state and of singly- and doubly-ionized He (determined in the approximate manner described above). One can see that for $F = 0.25$ and 0.55 a.u., double ionization occurs as a sequential process, whereas at $F = 1$ a.u. it starts at about the same time as single ionization.

We have not computed the photoelectron energy distributions, since this requires a projection of the wavepacket upon the zero-field eigenstates of the two-electron Hamiltonian, which themselves are quite difficult to compute. The autocorrelation method described by Millack at the present meeting might provide a very efficient way of getting information on the distribution of total energy, but it would seem to fall short of being able to describe how it is partitioned between the two electrons. However, we can make some observations about the correlations between the ejected electrons by looking at the angular distributions. This is done in a rudimentary manner by simply fixing the mutual angle θ_{12} and integrating the probability distribution over the double-ionization region. The results, shown in Fig. 3, indicate a strong tendency for the electrons to emerge on opposite sides of the nucleus in the case of sequential ionization. When the field is sufficient to cause simultaneous ionization, one still sees an anisotropic distribution of the photoelectrons, with a preference for $\theta_{12} = 180°$. However, the anisotropy is washed out at higher field strengths.

CONCLUSIONS

Our results agree with what one would expect from relatively simple considerations. Clearly, in the weak-field limit we should observe only single ionization; and, in the case of an ultra-strong field (which attains its peak value over a few optical cycles), both electrons ought to be stripped rapidly from the atom. However, we draw attention to two aspects that are perhaps less obvious. First, the transition from sequential to simultaneous ionization occurs over a range of field strengths sufficiently narrow for us to identify an intensity *threshold* for simultaneous ionization. Secondly, at this threshold, the field is not so strong that one is seeing rapid independent stripping of both electrons from the atom; there are pronounced effects of electron correlation, which appear in the angular distribution of the electrons.

We emphasize that results obtained in this model cannot be compared directly with experiments, since it does not utilize the true interaction of the electrons with the radiation field, and has also been investigated in cases where

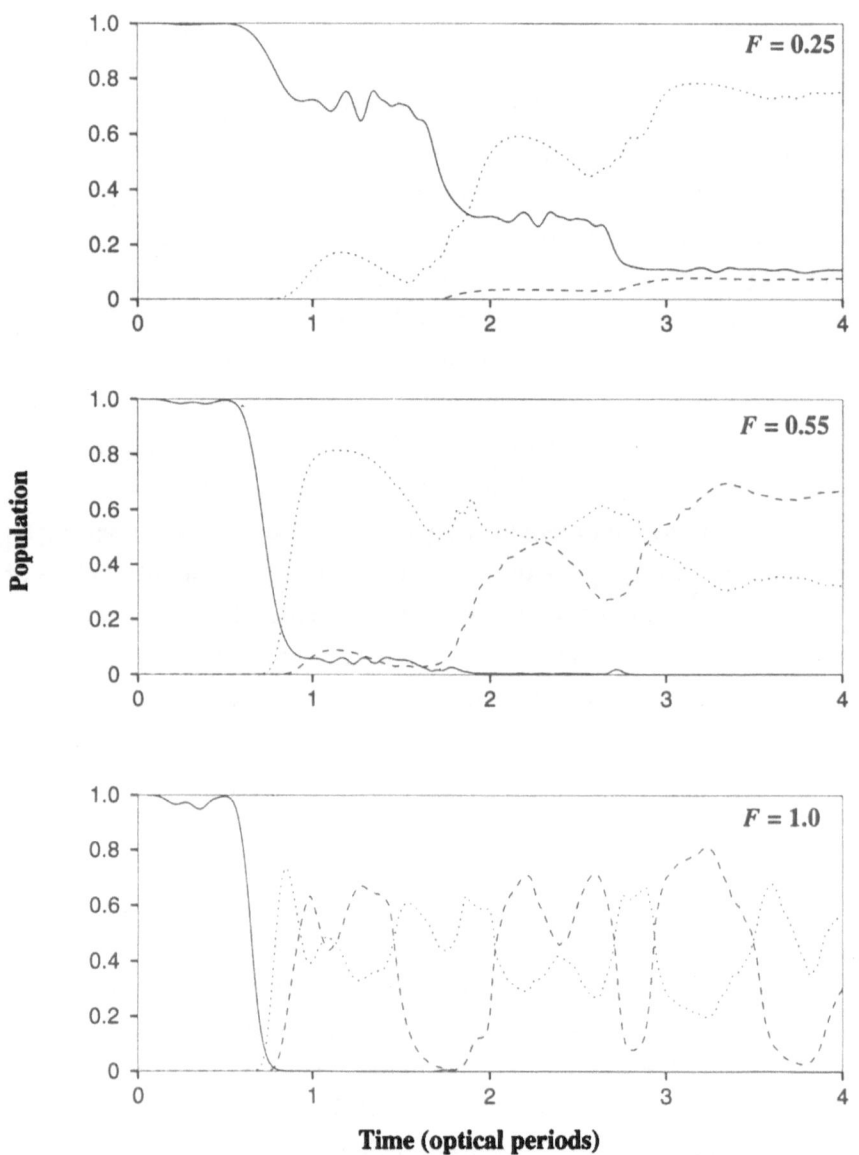

Fig. 2   Populations of ground state (solid line), singly-ionized (dotted line) and doubly-ionized He as a function of time, for peak field amplitudes as indicated.

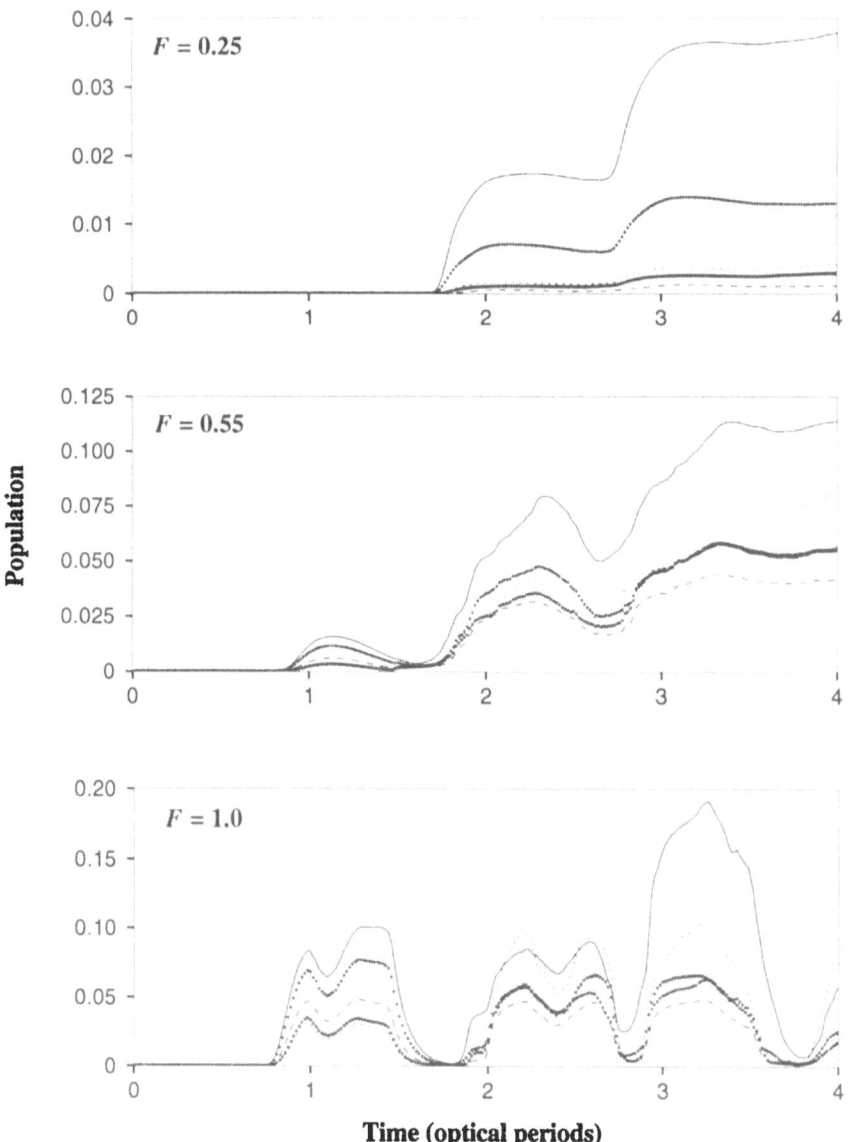

Fig. 3 Angular distribution of the photoelectrons in the double-ionization channel at three field strengths. Curves show the probability densities at angles θ_{12} of 180° (solid line), 135° (plus), 90° (dotted line), 45° (crosses) and 0° (dashes).

the radiation pulse is much shorter than those used in practice. However, it is interesting to note that the threshold intensity we find for double ionization is similar to that observed in the experiments reported in ref. 6.

Extension of this approach to treat the true radiative interaction is in progress. It requires a considerable expansion of the framework indicated in Fig. 1, and its feasibility as a practical computational approach remains to be determined.

ACKNOWLEDGEMENTS

We are grateful for the provision of computer time by the Northeast Parallel Architectures Center at Syracuse University, and the University of Maryland Institute for Advanced Computer Studies, and for technical assistance by the staff of those institutions. Part of this research has been supported by the U. S. Air Force Office of Scientific Research.

REFERENCES

1. A. L'Huillier, L. A. Lompré, G. Mainfray, and C. Manus, Phys. Rev. Lett. **48**, 1814 (1982); T. S. Luk, H. Pummer, K. Boyer, M. Shakidi, H. Egger, and C. K. Rhodes, Phys. Rev. Lett. **51**, 110 (1983)
2. A review is given by M. Crance, Phys. Reports **144**, 117 (1987)
3. P. Lambropoulos, Phys. Rev. Lett. **55**, 2141 (1985)
4. L. J. Frasinski, K. Codling, P. Hatherly, J. Barr, I. N. Ross, and W. T. Toner, Phys. Rev. Lett. **58**, 2424 (1987)
5. X. Tang, H. Rudolph, and P. Lambropoulos, Phys. Rev. A **44**, R6994 (1991)
6. A. McPherson, G. Gibson, H. Jara, U. Johann, T. S. Luk, I. A. McIntyre, K. Boyer, and C. K. Rhodes, J. Opt. Soc. Am. B **4**, 595 (1987)
7. J. Richardson, "Numerical solutions to the time-dependent Schrödinger equation," preprint, Thinking Machines, Inc. (1990)
8. K. T. Taylor and C. W. Clark, submitted to J. Phys. B: At. Mol. Opt. Phys.

9. C. Froese Fischer, *The Hartree-Fock Method for Atoms: A Numerical Approach* (John Wiley and Sons, New York 1977)

10. Requests may be made by e-mail to clark@bruce.nist.gov

TWO-ELECTRON ATOMS IN ULTRA SHORT, INTENSE PULSES

P. Lambropoulos [a,b, c] and X. Tang [c]

(a) Max-Planck-Institut für Quantenoptik, D-8046 Garching, Germany
(b) FORTH-IESL and University of Crete, Heraklion 71110, Crete, Greece
(c) University of Southern California, Los Angeles, CA 90089-0484

I. INTRODUCTION

Although most of multiphoton strong field experiments are performed on multi-electron atoms, it appears that the behavior can be understood in terms of Single Active Electron (SAE) models. Multiple ionization (ejection of as many as ten electrons) is of course a well-established phenomenon[1], above a certain range of intensity. But since (with the exception of one or two cases which are still unclear) the electrons are emitted sequentially, the SAE picture can be said to be valid as well in that case. On the other hand, it is known that 2-electron excitations have played a significant role in the multiphoton excitation and ionization of alkaline earth atoms[2-4]. This has happened in experiments with intensities in the range of 10^9 - 10^{13} W/cm^2 where transition probabilities calculated through perturbation theory (including resonance effects) has been capable of providing a good interpretation of the underlying behavior[4]. Thus it can be safely said that multiphoton 2-electron excitation has played an important role in certain atoms, although 2-electron direct ejection remains to be established.

A brief digression on some general aspects of 2-electron transitions may be useful at this point. First of all, recall that in traditional single-photon physics, a 2-electron transition through the absorption of a single photon can take place only if there is correlation[5]. Although the technical meaning of correlation may vary, in order to make the point here, we will define as absence of correlation the independent particle model. Thus, to state it once more, a single photon cannot cause the transition of two independent electrons from one state to another, continuum or not. By contrast, correlation in the final or intermediate states is not necessary at all for N-photon q-electron ionization, as long as N>q. Again, to stress the point, a two or more photon transition can eject two even independent electrons. The reason is recognized easily upon examination of the simple case of 2-photon 2-electron ionization of ground state He ($1s^2$). The relevant transition amplitude is

$$M\left(\varepsilon\varepsilon';1s^2\right) = \sum_n \frac{<\varepsilon p\varepsilon' p|\vec{r}_1 + \vec{r}_2|1snp><1snp|\vec{r}_1 + \vec{r}_2|1s^2>}{E_{np} - E_{1s^2} - \hbar\omega} \tag{1}$$

Super-Intense Laser-Atom Physics, Edited by
B. Piraux *et al.*, Plenum Press, New York, 1993

where the summation (integration) extends over single-electron excited states of the discrete as well as the continuum part of the spectrum. We leave two-electron intermediate excitations out of this summation because they require correlation and it is the correlation-free part of the amplitude that is of interest here. The subscripts 1 and 2 in r label the two electrons and fully antisymmetrized wavefunctions are assumed. Inspection of Eq. (1) shows that for $np = \varepsilon p$ or $\varepsilon' p$ and completely uncorrelated wavefunctions, the numerator reduces to the form

$$<\varepsilon p / r / 1s> \; <\varepsilon' p / \varepsilon' p> \; <\varepsilon' p / r / 1s> \; <1s / 1s>$$

where the overlap integrals are equal to unity and the matrix elements are single-electron matrix elements. Although the above derivation has been rather sketchy, it does show the essence of the argument. Another part of the amplitude due to correlation will also contribute, but the matrix elements will tend to be much smaller. In the event that $E_{1s^2} + \hbar\omega$ is near a doubly excited state, the corresponding contribution may be more significant. In any case, however, correlation will provide an additional contribution and not the essential part of multiphoton ionization.

The qualitative extention of the above argument to q-electron ionization is now self-evident. It also follows that the magnitude of an N-photon ionization generalized cross section for q-electron ejection is expected to be similar to that of an N-photon single-electron cross section[6].

The case of a two-photon 2-electron transition from one bound state to another is a bit more complicated[5,7]. Consider, for example, a two-photon transition from $1s^2$ to $2p^2$. Clearly, it can take place without correlation; as an argument along lines analogous to those of Eq. (1) easily reveals. A two-photon transition from $1s^2$ to $2s^2$, on the other hand, cannot take place without correlation. If we have, however, a four-photon transition from $1s^2$ to $2s^2$, it can certainly take place without correlation.

In reality, states like the $1s^2$, $2s^2 \left({}^1S_o \right)$ or $2p^2 \left({}^1S_o \right)$ of He involve significant correlation; in other words the configurations $2s^2$ and $2p^2$ are significantly present in both of the above doubly excited states. But if we examine a singly excited state of the form $1snl$ in He, we will find that it is pretty much a single configuration; its quantum defect is unperturbed, as is the case in a single-valence-electron atom. For alkaline earth atoms like Mg, Ca, Sr (for which extensive multiphoton studies exist) the situation is entirely different. Even singly excited states[8] (of the type $3snl$, for example, in Mg) contain significant configuration mixing (correlation). The main reason is that, unlike He, doubly excited states are very close to the first ionization threshold (some of them are even below in Ca and Sr). In He the first doubly excited state is as far above the first ionization threshold as the ground state is below.

The discussion above has been based on arguments which assume a perturbation theory approach. What should we expect if the atom is exposed to an intensity well above the perturbative limit? On physical grounds, one would think that correlation (which is the result of electron-electron interaction) would loose its importance, since the interaction with the field would be stronger than that with the other electron(s). The initial (ground) state does of course have full correlation, but it would seem that the transition out of that correlated state should occur as if the electrons were independent and non-interacting. This creates an unusual situation with no obvious way of approaching its solution. Clearly, two essential elements are necessary: First, a good description of the structure of the bare atom, and second a method for handling the temporal development under the strong field without recourse to perturbation theory.

In traditional (weak-field) single-photon absorption, much insight has been gained through a number of sophisticated techniques such as R-matrix[9], L²-bases[10] and RPA[11] in the rare gases. They provide good description of the atomic structure and of the transition probability within perturbation theory but not all of them are equally adaptable to the time dependent (TD) non-perturbative formalism. In this respect, we have shown earlier that an L²-basis constructed in terms of B-splines usefully combines both desired features, and by using hydrogen as a test case[12], we have also compared our results favorably with other non-perturbative calculations[13]. We feel thus confident in tackling a system such as Mg which involves strong correlation in its excited state spectrum and for which there are no other non-perturbative calculations.

As far as the atomic structure is concerned, the method has been tested in calculations[14] of single-photon transitions and ionization, two- and three-photon ionization and autoionization, as well as in the comparison with experimental angular distributions including one corresponding to ATI, all within the framework of perturbation theory. The above tests have included comparisons between length and velocity calculations (with excellent agreement), and comparisons with autoionization widths and positions of resonances obtained in experiments were available. The structure of this two-electron system and its dipole transitions are therefore represented with this method as well as the state of the art allows.

II. NON-PERTURBATIVE TD THEORY

For the implementation of the TD calculation[12,15], we recall that the wavefunctions $\varphi_{nLM}\left(r_1, r_2\right)$ of the two active electrons satisfy the Schrödinger equation $H^\circ \varphi_{nLM} = E_{nL}\varphi_{nLM}$ with H° being the atomic Hamiltonian, and the boundary conditions $\varphi_{nLM}(0,0) = \varphi_{nLM}\left(r_1^{max}, r_2\right) = \varphi_{nLM}\left(r_1, r_2^{max}\right) = 0$. We use here the abbreviation n for the pair $n_1 n_2$ of the principal quantum numbers for the single-particle wavefunctions $\varphi_{n_1 l_1 m_1}\left(r_1\right)$ and $\varphi_{n_2 l_2 m_2}\left(r_2\right)$ which enter in the antisymmetrized, linear combinations of products in terms of which the two-particle wavefunctions are constructed. The single-particle orbital functions are constructed in terms of B-splines on a frozen Mg^{++} Hartree-Fock core. A Configuration Interaction (CI) procedure is employed for the construction of all two-particle wavefunctions which constitute an L²-basis spanning the energy region below as well as above the ionization threshold. A major difference between this problem and our recent work[15] on He is the challenge in obtaining reliably a significant part of doubly excited states whose proximity to the first ionization threshold cause, as we shall see, qualitatively different behavior even at high intensity.

The TD wavefunction is now expressed as $\Psi\left(r_1, r_2, t\right) = \Sigma_{nLM} b_{nLM}(t)\varphi_{nLM}\left(r_1, r_2\right)$ and is substituted into the TD Schrödinger equation $\frac{\partial}{\partial t}\Psi\left(r_1, r_2, t\right) = -iH(t)\Psi\left(r_1, r_2, t\right)$, subject to the initial condition $\Psi\left(r_1, r_2, 0\right) = g\left(r_1, r_2\right)$ i.e. the ground state $3s^2\left({}^1S_o\right)$ of the atom. The total TD Hamiltonian H(t) is given by $H^\circ + V(t)$ where $V(t) = -e\vec{e}\cdot\left(r_1 + r_2\right)E(t)\sin(\omega t)$ is the interaction between the field of frequency ω and the two electrons in the dipole approximation, with $\vec{e}$ being the unit polarization vector of the field taken here to be linearly polarized. The time dependence of the amplitude E(t) of the pulse is expressed as $Ef\left(\dfrac{t}{\tau}\right)$ where $f\left(\dfrac{t}{\tau}\right) = 1/\cosh\left(\dfrac{t}{\tau}\right) or\ e^{-\frac{1}{2}\left(\frac{t}{\tau}\right)^2}$ depending on which is more appropriate for a given experiment. We refer to τ as the pulse duration or width, but in all cases the integration of the differential equations is carried out from -T to +T where |T| is sufficiently larger than τ for the result of the calculation to be insensitive to the further increase of |T|.

The TD Schrödinger equation is then reduced to the following set of coupled linear differential equations

$$i\frac{d}{dt}b_{nLM}(t) = \sum_{n'L'M'}\left(E_{nLM}\delta_{nn'}\delta_{LL'}\delta_{MM'} - <\varphi_{nLM}\left|V(t)\right|\varphi_{n'L'M'}>\right)b_{n'L'M'}(t) \quad (2)$$

for the expansion coefficients b_{nLM}. The dipole form of V(t) leads to matrix elements only

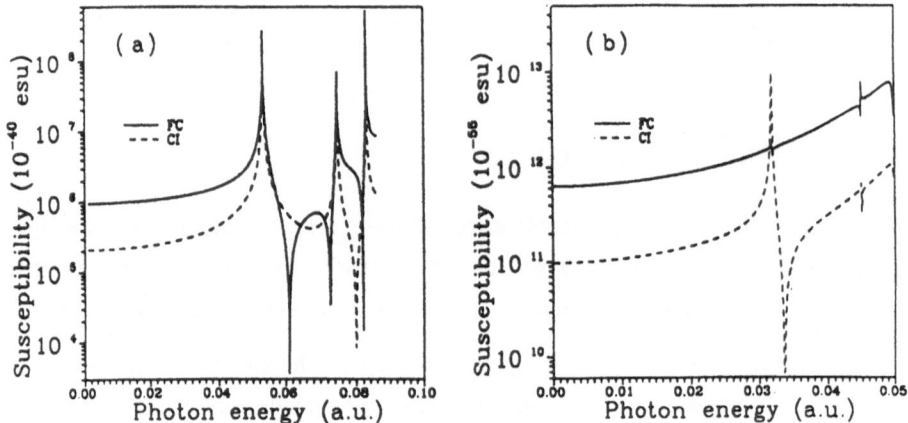

Figure 1. Nonlinear susceptibilities $\chi^{(3)}$ and $\chi^{(5)}$ [(a) and (b), respectively] of Mg as a function of photon energy as given by FC and CI calculations. Since FC and CI give somewhat different energies for bound states, we have shifted the FC results by that difference so as to provide a visually direct comparison.

between the single-particle functions φ_{nlm}, while the assumption of linear polarization reduces the values of M' to only zero since for the initial (ground) state M = O and the selection rule requires M' = M.

The probability of ionization after time t is obtained through $P(t) = \Sigma_{nL(E\geq0)}\left|b_{nL}(t)\right|^2$ where the summation extends over all states of positive energy of a given basis set. A result is considered converged when it becomes insensitive to the further increase of the size of the box within which the functions of the L^2-basis are non-zero (i.e. the values of $r_1^{max} and\ r_2^{max}$), the number of basis functions per angular momentum, and the number of angular momenta. We do not need to present in the short space available here results for the perturbative calculation of multiphoton ionization and autoionization to document the level of quality of the atomic model, as it has been documented in work already published. We will therefore focus upon the TD non-perturbative behavior, with only a brief parenthesis on perturbative results, namely the calculation of the non-linear susceptibilities $\chi^{(3)}$ and

$\chi^{(5)}$ corresponding to 3rd and 5th harmonic generation for a range of frequencies such that 3ω and 5ω remain below the first ionization threshold. These results are shown in Fig. 1 for two different models: A complete CI and a Frozen Core (FC). FC here is the same as SAE and means that a separate calculation is performed which includes only configurations of the form 3snl properly antisymmetrized. As expected, the resulting ground state energy is somewhat different from that obtained through CI. In Fig. 1, we have shifted the FC results

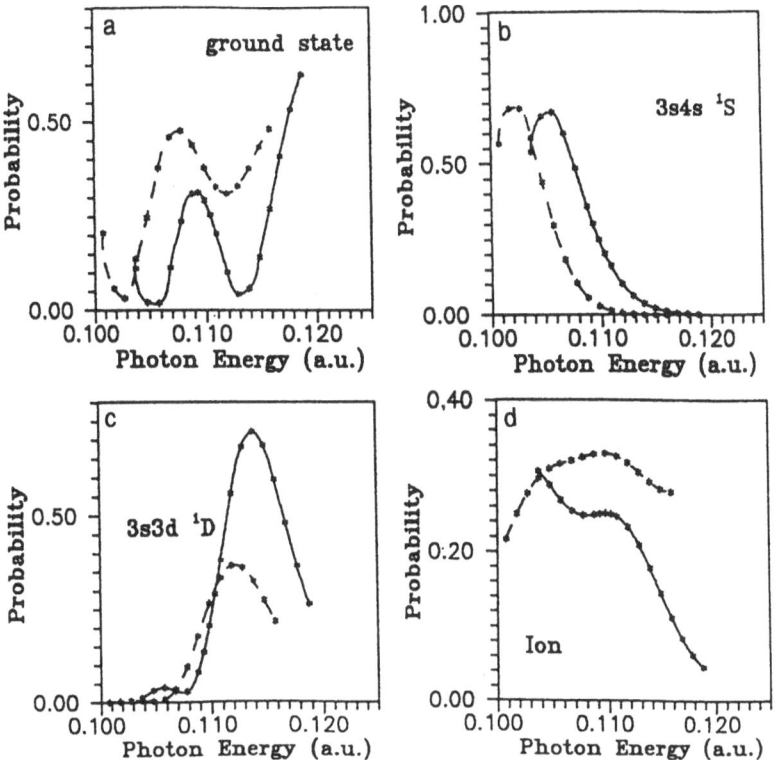

Figure 2. Ionization probability and populations in ground state, 3s4s^{1}S, and 3s3d ^{1}D as a function of photon energy as given by FC (solid line) and CI (dashed line) at the end of a pulse with = 5 optical cycles at the intensity I=10^{13} W/cm^2.

by that difference so as to provide direct visual comparison of the susceptibilities. The effect of electron correlation is quite dramatic in both $\chi^{(3)}$ and $\chi^{(5)}$ amounting to one order of magnitude. Thus although harmonic generation is an elastic photon scattering process, below the ionization threshold for the frequencies of Fig. 1, the atom responds with both valence electrons. By contract, in our calculation[15] for He, we had found $\chi^{(3)}$ and $\chi^{(5)}$ obtained through CI or FC calculations to be practically indistinguishable.

We proceed now with results on the main theme of this paper, the importance of correlation in the strong field regime. We present results on the amount of ionization for a frequency range corresponding to 3-photon ionization, and for relatively short pulses as indicated on the figures. It should be kept in mind that in order to investigate non-perturbative behavior at high intensity the pulse has to be short, otherwise the atom ionizes during the rise of the pulse. In addition to ionization, the AC Stark shifts of some key states, as well as the population left in such states provide important indicators of correlation and of non-perturbative behavior.

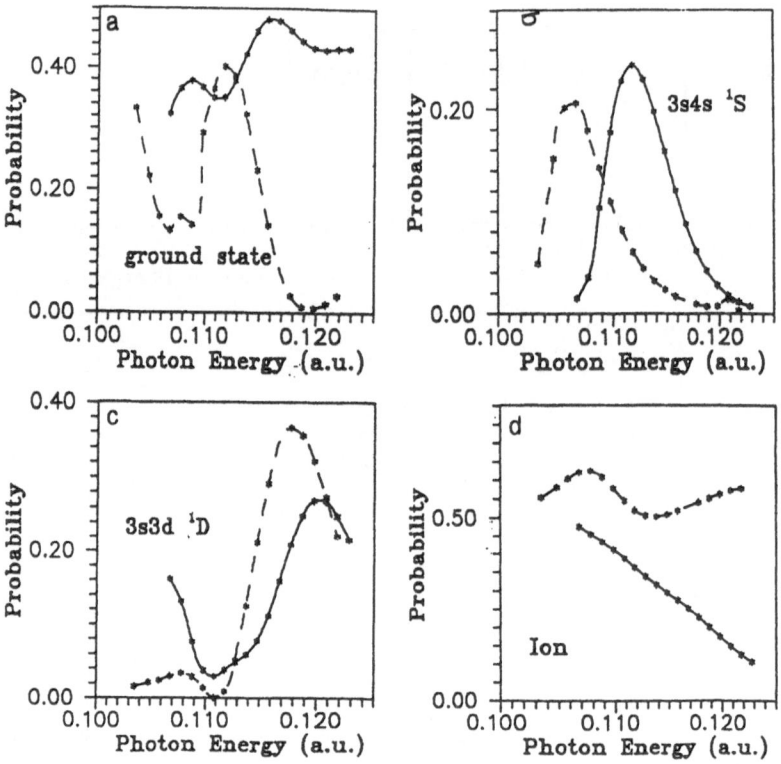

Figure. 3. The same as in Fig. 2 calculated at the intensity I=2x10^{13} W/cm².

A sample of results that encapsulate the importance of correlation under high intensity is shown in Figs. 2 and 3 which represent two critical quantities in any strong field interaction: Populations left in the ground and representative excited states and of course the amount of ionization, all at the end of the pulse. In both figures those quantities have been calculated through CI as well as in the FC approximation as defined earlier. The range of frequencies chosen for the calculation would correspond to 3-photon ionization, in the perturbative regime. For some of those frequencies, the energy of two photons is such that the states 3s3d and 3s4s may shift through resonance during the pulse. The amount of population they retain as well as their AC Stark shifts (on which those populations depend) represent significant benchmarks for the assessment of correlation. The comparison of the quantities plotted in Fig. 2 and 3 in the FC approximation (with those for which fully correlated wavefunctions have been employed) is quite revealing. Not only the absolute magnitude of each quantity plotted, but also their dependence on photon frequency show rather dramatic dependence on correlation. For example, in Figs. 2d and 3d the amount of ionization as given by CI is larger than that given by the FC approximation by almost one order of magnitude for photon energies around 0.118 a.u. $\cong 3.2$ eV. The corresponding populations left in the ground state, on the other hand, exhibit an intuitively expected behavior in Fig. 3a (smaller for CI than for FC) but a counterintuitive behavior in Fig. 2a (larger for CI than for FC). Thus an intricate interplay between intensity and correlation, at $10^{13}/cm^2$ (Fig.2) causes the atom to produce more ionization and at the same time retain more population in the initial state when correlation is included. This reflects the effect of correlation on the role of intermediate states from which the population found in the ground state and the ion must be drawn.

A measure of the departure from lowest order perturbation theory (LOPT) is the magnitude of the AC Stark shift, which for the TD theory must be inferred from the apparent position of the resonances with intermediate states. These must then be compared with their respective values obtained through LOPT. Such a comparison has been summarized in Table I.

Table I. AC Stark shift for two excited states of Mg at photon energies 2.88 and 3.07eV

States	Intensity W cm^{-2}	LOPT CI	LOPT FC	TD CI	TD FC
3s4s ^{1}S	10^{13}	0.288eV	0.440eV	0.146eV	0.364eV
	2×10^{13}	0.576eV	0.881eV	0.418eV	0.691eV
3s3d ^{1}D	10^{13}	0.473eV	0.706eV	0.330eV	0.494eV
	2×10^{13}	0.946eV	1.413eV	0.657eV	0.766eV

For each intensity, we note a significant difference between the values of the shifts for FC and CI as obtained in LOPT. This difference demonstrates the importance of correlation in the determination of the shifts at low intensity which simply reflects the properties of the bare atom. For the same intensities, the TD calculation yields shifts whose values are unequivocally much different from those obtained through LOPT, which provides an additional signature of non-perturbative behavior. At the same time, we note significant differences between the TD CI and TD FC values of the shifts, which reconfirm the effects noted in Figs. 2 and 3, namely the persistence of configuration interaction (correlation) in the non-perturbative regime.

III. CONCLUDING DISCUSSION

We have thus demonstrated that intensity sufficiently high to cause non-perturbative behavior does not obliterate intraatomic interactions. Why then had we reached different conclusions[15] in He? The reason, we believe, has to do with the photon energy in relation to the energy range where configuration interaction (as reflected in the perturbation of the quantum defects of excited states or equivalently the proximity of doubly excited states to the first ionization threshold) is manifested. In He, a photon energy which is one third or one fourth of the first ionization energy is much smaller than the energy of the first doubly excited state which is at about 56 eV above the ground state. In Mg and the other alkaline earths on the other hand, the first doubly exited state is below or at the first ionization threshold. It would then require a photon energy much smaller than the ionization energy to obtain ionization behavior which is insensitive to configuration interaction, probably one tenth or less of the ionization potential. It follows then that photons of energy 30 eV or more should produce similar effects in He as well. But sources of intensity sufficiently large to produce non-perturbative phenomena are not available at such short wavelengths, and it is difficult to forecast when they may become available.

The problem we addressed in this paper, on the other hand, is accessible to immediate experimental investigation. We have seen that clear correlation effects in non-perturbative behavior are manifested at 10^{13}W/cm². And sources at significantly higher intensity are readily available in the broad wavelength range that we have investigated. One would need to examine the behavior of ionization as a function of intensity and frequency, as well as, photoelectron angular distributions in comparison with calculations such as the one presented here. One additional condition of course is that the pulse be short (~100fs) for the atom to experience the high intensity. Needless to say that we certainly expect equally pronounced correlation effects in absorption above the threshold where Mg and the other alkaline earth have many doubly excited (autoionizing) states. Recent experiments[16] have established that short pulses (~100fs) make possible extensive absorption above the threshold (ATI) even in low order ionization. In view of the above results, the behavior of a two-electron atom under intense femtosecond pulses ought to pose rather interesting questions pertaining to ionization, ATI, harmonic generation, and stability (or lack of it) of atoms in strong fields.

The physical arguments we pursued in the introduction led to the idea that at sufficiently high intensity all signs of correlation should disappear. Yet as we have seen above, if correlation is significant in the excited bound states of the bare atom, it persists at intensities sufficiently high for non-perturbative behavior to have set in; at least for the range of frequencies examined here. It may well be that the highest intensity we studied is not sufficient to obliterate correlation in this case. But keeping in mind that we had more than 50 % ionization even for pulses as short as 5 optical periods, we would need to go to much shorter pulses if we were to raise the intensity much more. We may then reach the limit of complete ionization within a quarter of a cycle, which means tunneling. This would then bring us to the case recently discussed by Fittinghoff et al[17] where the two electrons tunnel out (almost) simultaneously but correlation plays no role. Of course in that case the frequency was low so that the SAE behavior was dominating anyway. The question then is: Can the behavior of the type reported in ref. 17 be seen in a situation of strong correlation (as described in Mg) or will the necessary pulse duration be too short to be realistic? That remains to be seen.

Acknowledgements

This work was supported by NSF under Grant N° PHY-9013434 and DOE under Grant N° DE-FG03-87ER60504. The authors also acknowledge use of the resources of the NERSC at Livermore and the State University of Florida Supercomputer Centers supported by the DOE.

References

1. For a collection of articles on multielectron excitation and ionization under strong fields, see JOSA B, 4 (1987) pp. 705-862.

2. D. Feldmann and K. H. Welge, J. Phys. B,15 (1982) 1651.

3. G. Petite and P. Agostini, J. Physique 47 (1986) 795 and references therein.

4. P. Lambropoulos, X. Tang, P. Agostini, G. Petite and A. L'Huillier, Phys. Rev. A38 6155 (1988).

5. H. Bachau and P. Lambropoulos, Phys. Rev. A, 44, R 9 (1991).

6. P. Lambropoulos, Comments on Atomic and Molecular Physics, 20, 199 (1987).

7. H. Bachau and P. Lambropoulos Z. f. Physik D 11, 37 (1989).

8. T. N. Chang and X. Tang, Phys. Rev. A, 46, R2209 (1992).

9. M. J. Seaton, J Phys. B, 18, 2111 (1985). C. H. Greene and L. Kim, Phys. Rev. A, 36 2706 (1987).

10. R. Moccia and P. Spizzo, J Phys. B,21, 1145 (1988). T.N. Chang and X. Tang, Phys. Rev. A, 38, 1258 (1988).

11. M. G. J. Fink and W. R. Johnson Phys. Rev. A,42 3801 (1990). C. D. Lin and W. Johnson Phys. Rev. A, 15 1046 (1977).

12.. X. Tang, H. Rudolph and P. Lambropoulos, Phys. Rev. Lett. 65, 3269 (1990).

13. S. I Chu and J. Cooper, Phys. Rev. A, 32, 2769 (1985).

14. X. Tang, T. N. Chang, P. Lambropoulos, S. Fournier and L. F. DiMauro, Phys. Rev. A, 41, 5265 (1990).

15. X. Tang, H. Rudolph and P. Lambropoulos Phys. Rev. A, 44 R6994 (1991).

16. W. Nicklich, H. Kumpfmüller, H. Walther, X. Tang, Huale Xu, and P. Lambropoulos. Phys. Rev. Lett.

17. D. N. Fittinghoff, P. R. Bolton, B. Chang and K. Kulander. Phys. Rev. Lett70, 2642 (1992).

QUASICLASSICAL APPROACH TO IONIZATION OF ATOMS BY STRONG LASER PULSES: COMPARISON WITH THE "SIMPLEMAN'S" MODELS AND NEWTONIAN DYNAMICS

P. B. Lerner, K. LaGattuta, and James S. Cohen

Los Alamos National Laboratory
Los Alamos, NM 87545

INTRODUCTION

The current paper is a continuation of our previous paper [1], in which we proposed to study electron-electron correlations in the quasiclassical model of Kirschbaum and Wilets (also called Fermi Molecular Dynamics (FMD)) [2] with respect to multiphoton ionization. This model has been already implemented in the paper of Wasson and Koonin [3] for the study of multiphoton ionization rates for multielectron atoms. In the present paper we provide the comparison between quasiclassical FMD and purely classical models for multiphoton ionization. Here, we concentrate on the second simplest quantum system, the He atom.

The closest benchmark available for comparison is the set of so-called "simpleman's theories", or SM's (a name suggested by Gallagher [4]). These theories become applicable for extremely high laser irradiances, where over-the-barrier ionization (OBI) becomes dominant [5]. This regime is characterized by the property that, at least for a part of the period of the external field, the field becomes so strong that no bound states can exist in the combined potential of the atom and laser. As was demonstrated in Ref. [6], this regime can be connected with tunneling formulas, which are applicable for fields much weaker than atomic. A notable feature of the SM theories is that their predictions depend only on parameters of the laser pulse (intensity, frequency, and duration) and on the phase of the electron motion relative to the phase of the laser field. Another benchmark will be comparative simulation of the energy gain in the classical model of the He atom (Section 4).

The successful implementation of the classical and semiclassical methods in other, closely related fields of physics, such as ionization of Rydberg atoms [7], scattering problems [8], and interaction of ions with surfaces [9] prompts the use of quasiclassical computer calculations for this problem. In the pioneering work by Gaida, Grochmalicki,

Super-Intense Laser-Atom Physics, Edited by
B. Piraux *et al.*, Plenum Press, New York, 1993

Lewenstein and Rzazewski, the methods of classical mechanics were used for three-dimensional simulation of the hydrogen atoms in strong fields [10].

However, in this case the purely classical methods suffer a great deficiency. They cannot provide for the stability of the systems more complicated then the hydrogen atom and hydrogen-like atoms and ions in external fields. The known stable multielectron systems have a high degree of symmetry [11] which is destroyed by the external field. Unlike collision problems, the stability of atoms should be assured for the characteristic times of thousands and tens of thousands of atomic periods even for the shortest available pulses.

THE "SIMPLEMAN'S" THEORY OF MULTIPHOTON IONIZATION

The SM model of multiphoton ionization has its origin in ideas first expressed by van den Heuvell and Müller [12] and Corkum [6]. Both Corkum and Gallagher have performed actual experiments on real atoms which seem to have provided good evidence for the correctness of the model, within its domain of applicability.

The SM describes ionization in the regime of laser irradiances $I_0 > Z^6(2n)^{-8}$ a.u. for electrons bound in H-like ions with principal quantum number n and ionic core charge Z. This is the regime of so-called over-the-barrier ionization (OBI). This regime is characterized by the property that, at least for part of the period of the external field, the field becomes so strong that no bound states can exist in the combined potential of the atom and laser. For smaller irradiances, only stochastic ionization and tunneling are possible. In the relatively weak fields these two mechanisms are complementary: the ground state is ionized by tunneling (non-existent in the proposed model) and the highly excited states are ionized by stochastic mixing with the continuum [7].

An assumption of adiabaticity is also usually made, so that the laser frequency must be low, $\omega \ll Z^2/n^3$, which means that the electron will execute many orbits during one laser period. Finally, this essentially classical approach to ionization also requires that $n \gg 1$, if a comparison to real atomic systems is contemplated.

For linear polarization, the laser electric field takes the form

$$\vec{F}(t) = \hat{x} F_0 \sin(\omega t + \beta) \tag{1}$$

having assumed, for convenience of the ensuing analysis, an abrupt turn-on at t=0 with phase β. The electron is imagined to emerge into the continuum at a distinct time t'. Once having become unbound, it is considered to be unaffected by the binding potential, moving only under the influence of electric field. No possibility of recapture is admitted.

Therefore, for $t \geq t'$, the velocity of the ionized electron is just

$$\vec{v}(t) = \hat{x}\frac{F_0}{\omega}[\cos(\omega t' + \beta) - \cos(\omega t + \beta)] + \vec{v}(t') \tag{2}$$

where $\vec{v}(t')$ is the velocity at the moment of release ($|\vec{v}(t')| \leq Z/n$). The cycle-averaged kinetic energy of the unbound electron is then

$$U(t') = [v_x(t') + \frac{F_0}{\omega}\cos(\omega t' + \beta)]^2/2 + [v_y^2(t') + v_z^2(t')]/2 + U_p , \tag{3}$$

where U_p is the ponderomotive potential

$$U_p = \frac{F_0^2}{4\omega^2} . \tag{4}$$

414

For cases in which $F_0 \le (Z/n)^3$, it is "intuitive" to expect that ionization will occur primarily at a maximum of the driving field; i. e. when $\omega t' + \beta = \pi/2, 3\pi/2, ...$ The cycle-averaged kinetic energy of the unbound electron must then be $U(t') = v^2(t')/2 + U_p$. In the opposite extreme, when $F_0 \gg (Z/n)^3$, the ionization should occur at the instant of turn-on ($t' = 0$). Consequently, in this limit, the cycle-averaged kinetic energy is $U(t') = [v_x(0)^2 + (F_0/\omega)^2 \cos^2(\beta)/2] + U_p$, assuming no correlation between $v_x(0)$ and the driving field.

All of the preceding is in accordance with the strict SM theory. However, it is also clear that OBI can occur whenever it happens that $F(t') \ge Z^3/16n^4$. This slightly different approach to determining t' implies the kinetic energy

$$U(t') = \frac{v^2(t')}{2} + U_p + \frac{1}{2\omega^2}\left[F_0^2 - \frac{Z^6}{256n^8}\right]. \tag{5}$$

The difference between these two formulations can be great if $\omega \ll 1$ and $F_0 < (Z/n)^3$. Numerical calculations performed for classical systems seem to validate the SM, rather than the notion of simple OBI, as described here.

If, subsequent to ionization, the driving field is turned off "slowly", then electron kinetic energies will be modified to $U = v^2/2$, in the case that $F_0/\omega \le (Z/n)^3$, and $U = (F_0/\omega)^2 \cos^2(\beta)/2 + v^2/2$, if $(F_0/\omega) \gg (Z/n)^3$. In either case, an energy equal to U_p will be returned to the driving field. Our numerical calculations have all been performed in this mode.

The situation for circular polarization is even simpler. If the laser electric field has the form

$$\vec{F}(t) = \frac{F_0}{\sqrt{2}}[\hat{x}\sin(\omega t) + \hat{y}\cos(\omega t)] \tag{6}$$

for $t \ge 0$, then the velocity of the unbound electron must be

$$\vec{v}(t) = \frac{F_0}{\sqrt{2}}\{\hat{x}[\cos(\omega t' + \phi) - \cos(\omega t + \phi)] + \hat{y}[\sin(\omega t + \phi) - \sin(\omega t' + \phi)]\} + \hat{v}(t') \tag{7}$$

for $t \ge t'$. The cycle-averaged kinetic energy is then

$$\begin{aligned} U(t') &= \frac{v^2(t')}{2} + 2U_p + \frac{F_0}{\sqrt{2}}[v_x(t')\cos(\omega t' + \phi) - v_y(t')\sin(\omega t' + \phi)] \\ &\approx \frac{v^2(t')}{2} + 2U_p \end{aligned} \tag{8}$$

since no correlation is expected between $\hat{v}(t')$ and $\hat{F}(t')$. And, as previously, if the driving field is turned off slowly before the kinetic energy is sampled, then $U(t') \approx v^2(t')/2 + U_p$. For circular polarization the form of the energy spectrum of the emitted electron is invariant under changes in magnitude of the driving field F_0 or its initial phase. Finally, we note that the strict SM makes no predictions about the *rate* of photoionization, except in the very strong field domain ($F_0 \gg Z^3/16n^4$) where ionization is considered to be immediate.

However, a slight modification of the SM, suggested by Corkum, enables one to make a precise prediction of the emitted-electron kinetic-energy spectrum, for linear polarization, even if the turn-on is not abrupt. Recourse can be made the well-known formula for the rate of (static) electric field ionization, R_{FI}, from the ground state of atomic hydrogen,

$$R_{PI} = \frac{4}{|F(t')/F_{at}|}\exp\left(-\frac{2}{3|F(t')/F_{at}|}\right) \tag{9}$$

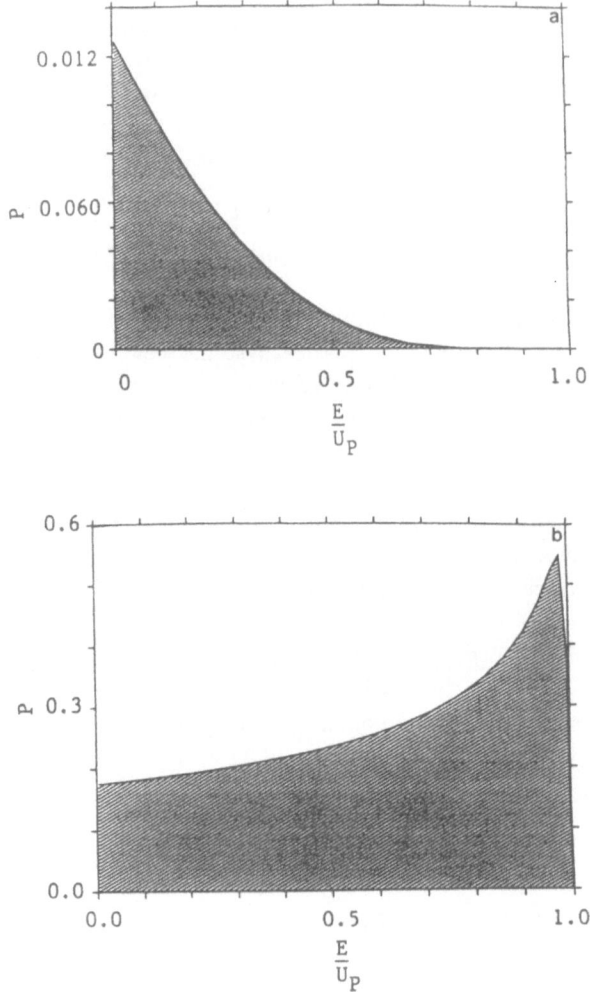

Figure 1. Spectrum of ionized electrons in a strong field of constant intensity and linear polarization according to the SM model of Brunel and Corkum. The probablity of ionization $P(E)$ is plotted as a function of the ratio of the kinetic energy and the quiver energy $x = E/U_p$. The tunneling formula for constant fields is applied both for low-field and high-field regimes: a) Low-field regime, $F_0 = 0.1$ $a.u.$ The spectrum is concentrated around zero energy; b) High-field regime, $F_0 = 5.0$ $a.u.$ The spectrum has a peak around the quiver energy value. The difference between scales of Fig. 1b and Fig. 2 is that in Fig. 2 the measurements of energy were provided in the aftermath of the pulse; during the pulse turn-off the electron gains another U_p of energy. Note that the spectrum is much wider in the low-energy region than in the quasiclassical spectrum of Fig. 2. This is a feature of the SM model that is close to purely classical simulations, but is not confirmed in our Fermi Molecular Dynamics approach.

generalized to a time-dependent field, where $F(t')$ may be given by Eqs. (1) and (6) (if desired, a pulse envelope function can be included). Then, the relative probability $N(U)$ of appearing in the continuum with energy $U(t')$, where $U(t')$ is given by Eq. (5), is determined parametrically as a function of $F(t')$ by

$$N(U) = \frac{F_{at}}{|F(t')|} \exp\left(-\frac{2F_{at}}{3|F(t')|}\right) \qquad (10)$$

for $0 \leq t' \leq \pi/2$. An example of the application of this semiclassical procedure appears in the following figure. Note that the formula for R_{PI} can sensibly be applied only if $F_0 < 1$. The energy spectra for Corkum's version of SM and a pulse of constant intensity are reproduced on Figs. 1a-c. They will serve as a benchmark for a comparison with results of quasiclassical and classical simulations.

MULTIPHOTON IONIZATION IN THE FIELD OF CONSTANT INTENSITY

For the FMD to have some validity it is necessary that its results should be compatible with the SM theory, which has been proven so successful in explanations of photoionization of atomic gases [5, 6]. The detailed comparison, however unphysical (because all real ultrashort pulses have smooth envelopes), is the best available benchmark for our quasiclassical program, because all of the one-electron phenomena can be easily interpreted in terms of SM observations.

The simulation was performed as follows. We ran statistically representative (usually several hundred trajectories) dynamical evolutions of the quasiclassical He in an external field with randomized initial conditions. Further, we selected some special orbit, like Langmuir or Wannier, as an initial state (both are stable in FMD approach). Then we randomize the phase of atomic oscillations (one angular parameter), and the direction of the field with respect to the atomic orbit (two angular parameters). Naturally, the choice of quasiergodic initial state [10, 14, 15] will be based on more solid physical grounds. However, our simulations provide evidence that, at least for moderate fields, the final bound state is close to quasiergodic.

We analyzed the data in two ways: (1) by histograms of the electronic energy, similar to the of real spectra of electronic energies in experiment and (2) by construction of "covariance maps" [16]. The values for energies, angles, and angular momenta were established in the aftermath of the pulse when all ionized electrons are in free motion, while all electrons with negative energies correspond to some bound zero-field states. The immediate rise of the pulse doesn't present any difficulties. To avoid computer difficulties in aftermath of the pulse, we add a gaussian tail to the back of the rectangular pulse: $\exp\left(-\frac{(t-t_{imp})^2}{\Delta t^2}\right)$ with $\Delta t \sim 100$ $a.u.$. This provides smooth fall of the intensity after the end of the essential part of the pulse $t < t_{imp}$. The energy spectrum for linear polarization and the field of the form $F(t) = F_0 \sin(\omega t)$ is given in the Fig. 2a. For the rectangular pulse the typical energy scale is provided by twice the value of the quiver energy (1250 $a.u.$ for our case), $E_i = 2U_p$ (the results of Section 2 for the electric field stronger then atomic; see also Fig. 1); yet, there is substantial dispersion of energies below this value (see [1] and Section 4). This suggests that ionization happens almost immediately after the pulse is turned on. Then, the motion of the electron is quasifree.

The case of circular polarization is even simpler. The energy spectrum is almost monoenergetic with $E_i = 2U_p$ (Fig. 2b). The difference between linear and circular

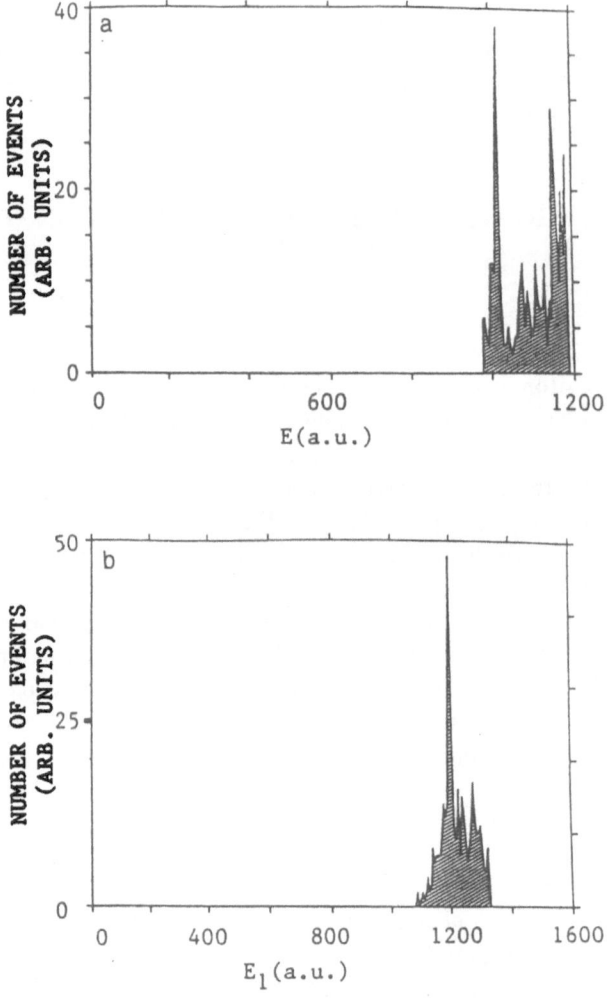

Figure 2. Energy spectra of a single ionized electron: a) case of linear polarization, conditions corresponding to Fig. 1c. The spectrum is a relatively narrow band of energies near $2U_p = 1250\ a.u.$; b) case of circular polarization. The peak is much narrower than for the case of linear polarization (Fig. 2a).

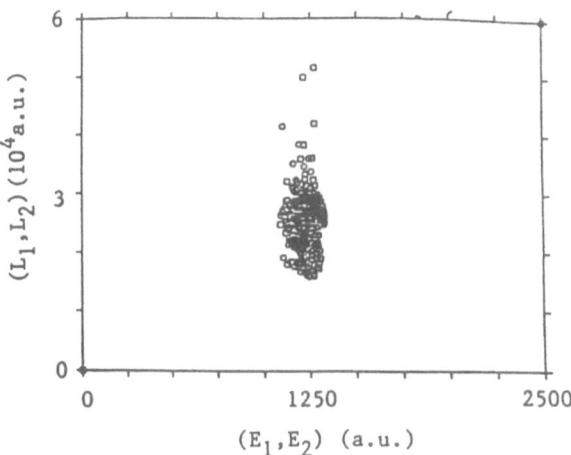

Figure 3. Energies and angular momenta for circular polarization of the field ((E_1, L_1) are squares, (E_2, L_2) are circles). The electrons are only statistically different from each other. $F_0 = 5.0\ a.u.$, $\omega = 0.1\ a.u.$, the energy is very close to $2U_p$. Each electron is described by 400 trajectories.

polarization can also be demonstrated through the relationship between energy and angular momentum. While for the linear polarization, angular momentum is more or less randomly distributed in the vicinity of relatively small values, the angular momentum for the case of circular polarization is essentially proportional to the energy (or intensity, for that matter) with a rather big proportionality coefficient of order $L_i = \frac{F_0}{2\omega} E_i \sim 25 E_i$ (see Fig. 3).

The previous results for linear polarization were produced for the field proportional to $\sin(\omega t)$. "Simpleman's" theories predict drastic dependence of the energy spectrum on the initial phase of the field. Because the electrons, under extreme diabatic conditions of excitation, are freed almost immediately, the initial phase of the electric field becomes exceedingly important. We display the results for the different initial phases for the laser field: $\sin(\omega t + \beta)$, $\beta = 0$, $\frac{\pi}{4}$, $\frac{\pi}{2}$; other choices of $\beta = \frac{n\pi}{4}$ are similar (Figs. 2 and 4). We see that, under strong-field conditions, for the sinusoidal field the energies are close to twice the value of the quiver energy; for the phase shift of 45° the electron absorbs $E_i \sim U_p$, and the energy spectrum for the cosinusoidal field is localized around zero.

The results of our computations, in general, corroborate the conclusions of "simpleman's" theories of multiphoton ionization. This is a bit surprising, because these theories are based on assumptions that are, rigoriosly speaking, valid only for fields weaker than the atomic field, in fact, the tunneling formulas are deduced for the tunneling exponent much larger than unity and, consequently, small probabilities of ionization.

THE COMPARISON BETWEEN THE WILETS-KOONIN MODEL AND PURELY NEWTONIAN SOLUTION FOR THE HELIUM ATOM

In the previous section, the Wilets-Koonin model for the multiphoton ionization was justified by its comparison with the widely used "simpleman's" model for multiphoton ionization.

To investigate advantages (and possible shortcomings) of the new approach, we compare the results of the Wilets-Koonin model with purely classical simulations. In a very strong field the atom and molecule are inherently unstable. The energy spectrum for a very strong field is mostly determined by the dynamics of the interaction between quasifree fragments of the particle, and this is precisely the type of behavior upon which the "simpleman's" theory is based. At least under extremely high field the results of the Wilets-Koonin and purely Newtonian dynamics should be similar.

The two problems, however close they may seem analytically, are completely different numerically. In the case of FMD close approaches to the nucleus with small relative angular momenta (billiard-ball type of collisions) are highly improbable because of emergence of large repulsive forces. The trajectory in the FMD case, though, is very irregular and one needs rather precise integration to be confident in the results. In the Newtonian case the situation is quite different. The orbits, outside the close vicinity of the nucleus are smooth and this allows for great tolerance of roundoff error. Vice versa, in the case of relatively small fields, the attraction of the nucleus dominates and unphysically close approaches of the electron to the nucleus become frequent. The situation, when the particles of the opposite charge can form states with indefinitely large binding energies is typical for classical mechanics, while it is prohibited by the quantization of the angular momenta in quantum mechanics (it is statistically very improbable in our quasiclassical approach). The ionization and ejection of the electrons from these very tightly bound states produces occasionally some electrons with high energies even for very low fields. In very high fields, because of the large amplitude of the quiver motion, these collisions are discriminated against for purely geometrical reasons and the validity of classical simulation is restored.

To prevent artificial stiffness of the Newtonian model under moderate-to-low intensity conditions we make a cutoff of the Coulomb potential at relatively large distances from the nucleus $r_{cutoff} = 0.141 \ a.u.$. This falls very well into the contemporary paradigm, prescribing to work with the Coulomb potential with a smoothed core [17]. While the Newtonian model has 100% ionization for moderate intensities ($\sim 2.0 \times 10^{14} \ W/cm^2$), the quasiclassical model shows no ionization at all. This was discussed in the Section 3, where the typical critical field strength for the ionization in the model with Wilets-Koonin potential was found to be $\sim 0.12 \ a., u.$ At $F_0 \sim 0.45 \ a.u.$ the ionization becomes much faster ([1], Section 3b). It happens when the strength of external field becomes comparable with the atomic field strength and the OBI from the ground state dominates. Vice versa, the purely classical helium is inherently unstable in the external field. This means that the electrons are essentially ionized from the beginning, and later undergo only stochastic acceleration [7] in the combination of laser and atom fields.

The energy spectrum for the Newtonian model is given by the Fig. 5, and the covariance map for the $E_1 - E_2$ correlation in Fig. 6. One observes that the single-electron spectrum for the classical model is, in fact, much wider and extends to the region of low energies, as the SM spectrum in Fig. 1c. A plausible explanation of this is related to the absence of the threshold in a Newtonian model. The dispersion of the ejection times in Eq. (5) is much greater for the quasiclassical case, resulting in a much wider spectrum of energies.

The comparison between the classical and quasiclassical simulations makes a very important point about purely classical models. Namely, the classical models accurately describe the quantum dynamics of the atom in moderate fields *only if* the Coulomb potential is regularized. In the classical simulations of [18], the static magnetic field was added to regularize the model.

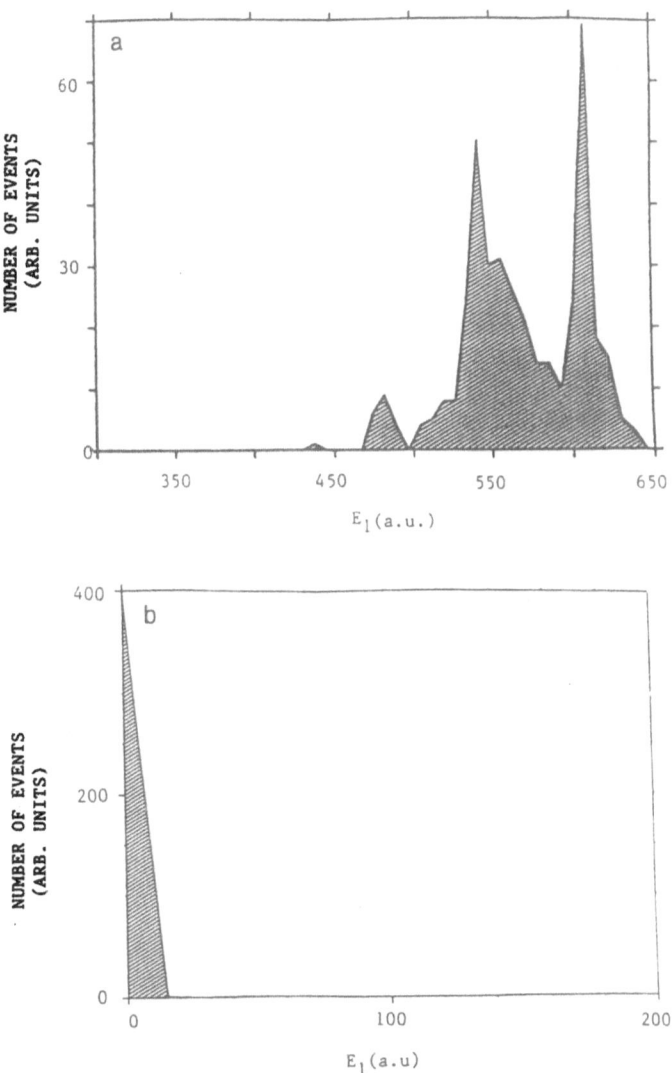

Figure 4. a) Same as in Fig. 2a except for an initial phase of the linearly polarized field equal to 45°, i.e. $F(t) = F_0 \sin(\omega t + \frac{\pi}{4})$. The energy has a peak around *one* quiver energy; b) same, for cosinusoidal field $F(t) = -F_0 \cos(\omega t)$. The peak is around *zero*.

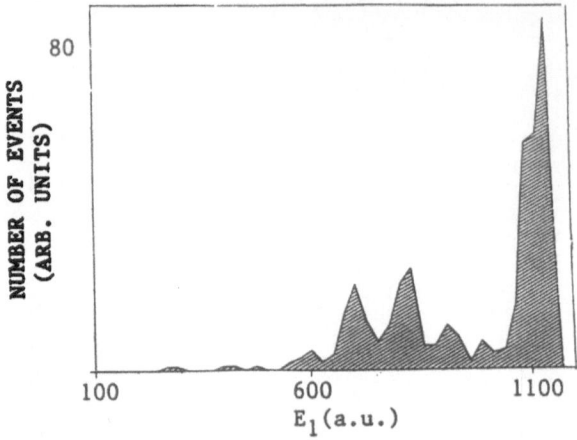

Figure 5. Energy gain for Newtonian dynamics of He. Conditions correspond to Fig. 1c.

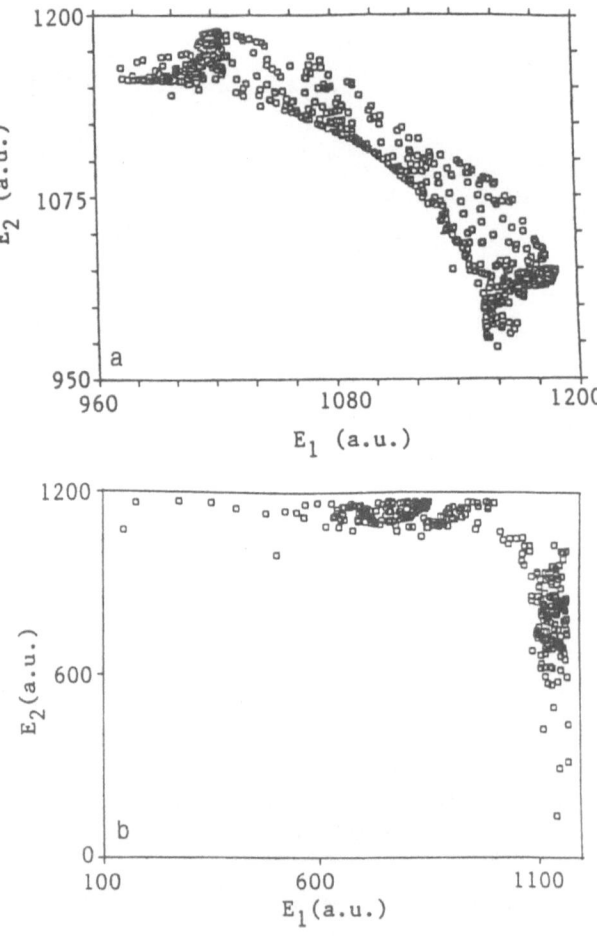

Figure 6. "Covariance maps" for the FMD and Newtonian He: a) $E_1 - E_2$ is plotted for FMD simulation; b) same except without stabilization of He by the Wilets' potentials. One observes substantially larger dispersion due to the absence of threshold intensity in the case b).

Now, we can point out differences between the purely classical and quantum situations. First, in classical mechanics there is a possibility of very close approach of the electron to the nucleus, with reasonably low relative angular momentum. These collisions can result in two possible effects: very sharp change in the direction of the electron with subsequent ionization, or capture of electron by the nucleus on a very low orbit with the binding energy much smaller than 1 $a.u.$ Classical mechanics makes such orbits perfectly permissible, while quantum mechanics prohibits them altogether. In the second case, the ionization out of a tightly bound state when the field rises enough produces electrons with very high energies. Another difference is applicable for multi-electron systems, while it doesn't exist for hydrogen. The multielectron atomic systems are generally unstable classically and this means that there is no threshold condition influencing the time of ejection. This results in much broader energy features (Figs. 5 and 6).

CONCLUSION

In the present paper we investigated the Fermi Molecular Dynamics (FMD) model for multiphoton ionization and compared it with the SM theory and purely classical simulations for the He atom. We observe that for the very high fields ($F \geq 5.0$ $a.u.$) these theories are qualitatively similar. This is very important, because our simulation deals with two-electron atoms and field intensities in which at least some of the underlying presuppositions of the SM theories are not applicable. The robust character of the OBI mechanism provides that one-electron phenomena will be satisfactorily described by any of these theories, even for multielectron atoms. The phenomena of electron correlation, however, manifest a significant difference related to the absence of the threshold in multiphoton ionization of a classical multielectron atom.

REFERENCES

1. P. B. Lerner, K. LaGattuta, J. S. Cohen, Laser Physics (to be published).

2. C. L. Kirshbaum, L. Wilets, Phys. Rev. A **21**, 834 (1980).

3. D. W. Wasson, S. E. Koonin, Phys. Rev. A **39**, 5676 (1989).

4. T. Gallagher, in *The Electron: New Theory and Experiment*, eds. D. Hestenes and A. Weigartshofer, vol. 45 (Kluwer Academic, Boston, 1991); see also D. A. Tate *et al.*, Phys. Rev. A **42**, 5703 (1990).

5. S. Augst *et al.*, Phys. Rev. Lett. **63**, 2212 (1989).

6. P. Corkum, N. Burnett, F. Brunel, Phys. Rev. Lett. **62**, 1259 (1989).

7. G. Casati, I. Guarneri, D. L. Shepelyansky, IEEE J. Quant. Electr. **QE-24**, 1420 (1988); R. Jensen, S. M. Susskind, M. M. Sanders, Phys. Rep. **201**, 1 (1991).

8. J. S. Cohen, to be published.

9. J. Burgdörfer, P. Lerner, F. Meyer, Phys. Rev. A **44** 5674 (1991).

10. M. Gaida, J. Grochmalicki, M. Lewenstein, K. Rzazewski, Phys. Rev. A **46**, 1638 (1992).

11. J. Müller Phys. Rev. A **45** 1471 (1992); R. Richter, D. Wintgen, J. Phys. B **23**, L197 (1990).

12. H. B. van Linden van den Heuvell, H. G. Müller, in *Multiphoton Processes*, eds. S. J. Smith, P. L. Knight (Cambridge Univ. Press, Cambridge, 1989), p.25.

13. Q. Su, J. H. Eberly, Phys. Rev. A **43**, 2474 (1991).

14. J. G. Leopold, I. C. Percival, J. Phys. B **12**, 709 (1979).

15. J. S. Cohen, J. Phys. B **18**, 1759 (1985).

16. L. J. Frasinski, K. Codling, P. A. Hatherly, Science **246**, 1029 (1986).

17. Q. Su, J. H. Eberly, Phys. Rev. A **43**, 2474 (1991).

18. B. Ritchie *et al.*, Phys. Rev. A **41**, 6114 (1990); C. M. Bowden *et al.*, Phys. Rev. A **46**, 592 (1992).

STABILIZATION OF MULTIELECTRON ATOMS: CLASSICAL APPROACH

Maciej Lewenstein[1,2], Kazimierz Rzążewski[2],
and Pascal Salières[1]

[1] Centre d'Etudes de Saclay
Service des Photons, Atomes et Molécules
Bat. 22, 91191 Gif sur Yvette Cedex, France
[2] Centrum Fizyki Teoretycznej, Polska Akademia Nauk
02-668 Warsaw, Poland

INTRODUCTION

Stabilization of atoms in ultrastrong fields, as proposed by Gersten and Mittleman[1] and Gavrila and Kamiński[2], became recently a major topic of multiphoton physics. Computer simulations of 1–dimensional[3] and 3–dimensional[4] atoms confirmed that the electron stays localized in the vicinity of the nucleus when the incident laser light has a sufficiently large frequency and intensity. Moreover, it is possible to introduce and keep the electron in the stabilized configuration even if the laser pulse has very rapid turn–on and –off times.

Stabilization suggests an amazing possibility of forming a new kind of stable medium consisting of atoms in the laser field. One can speculate about physics and chemistry of such a medium. Unfortunately, most of the studies of this subject[1-7] deal with single electron models, whereas in reality one has to deal in most cases with multielectron atoms and to consider effects of electron correlations. The studies of two-electron atoms and ions have just been initiated. Grobe and Eberly, who studied a one dimensional model with smoothed potential[8], reported an onset of the 1–electron stabilization in the negative ion for moderate field strengths[9]. The effect vanishes, however, for stronger fields. Mittleman, using a variational approach in the Kramers–Henneberger (K–H) frame, proposed fully stabilized two electron trial functions, with electrons localized in the two minima of the K–H potential[10]. Kulander initiated 3–dimensional calculations with the help of the time–dependent Hartree–Fock method[11]. It is hard to believe that fully 3–dimensional results for all interesting regimes of parameters become available soon. Therefore it is natural to seek for alternative approaches that can be realized with accessible computer power.

Classical phase space averaging methods became a very usefull tool in the studies

Super-Intense Laser-Atom Physics, Edited by
B. Piraux *et al.*, Plenum Press, New York, 1993

of stabilization[12-14] for 1–electron systems. In this paper we extend it to multielectron atoms and ions (see also Ref. 15). Our motivations are threefold: (a) there
is an obvious need for investigations of electron correlation effects in the context of
stabilization; (b) classical methods, although time consuming because of the need for
Monte Carlo (MC) sampling, present no difficulties in terms of stability of numerical
codes etc.; (c) classical methods provide us with physical intuitions which are worth
studying, even if they turn out to be invalid in quantum mechanics.

PHASE SPACE AVERAGING METHOD

The phase space averaging method is based on the assumption that the initial
state of a quantum mechanical system may be approximated by an appropriate probability distribution in a phase space. One then solves the Newton equations for the
system using MC sampling for the initial state distribution. Quantum mechanical averages of observables of interest are identified with classical averages over the initial
sample.

The status of classical methods in the context of 1–electron stabilization is as
follows. We consider here a regime of parameters when initial energy corresponds to
a ground state (i.e. $E = -1/2$ for the hydrogen), the laser frequency ω is of the order
of 1 a.u., and the electric field $\mathcal{E}$ is of the order of a few a.u., so that the free electron
excursion amplitude $\alpha = \mathcal{E}/\omega^2 \simeq 1 - -10$. One introduces often smooth potentials,
like

$$V(x) = -Z/\sqrt{r_{reg}^2 + x^2} \tag{1}$$

in 1-D, and

$$V(r) = -Z/(r_{reg} + r) \tag{2}$$

in 3–D, with Z being the charge of the nucleus. Regularization radius is taken to be
of the order of 1 a.u. The laser pulse duration τ_D is taken to be $20 - -100$ optical
cycles, and the turn–on and turn–off times, τ_{on} and τ_{off}, are of the order of a few
cycles.

Note that all those numbers have in classical mechanics a relative meaning only
due to a scaling invariance of the model. Namely, classical equations of motion are
invariant with respect to the scaling transformation $E \rightarrow a^{2/3}E$, $\omega \rightarrow a\omega$, $\mathcal{E} \rightarrow a^{4/3}\mathcal{E}$,
$r_{reg} \rightarrow a^{-2/3}r_{reg}$ and $\tau \rightarrow a^{-1}\tau$. This scaling relation remains valid for multielectron
systems as well.

Quantum mechanical results[3,4] in the described regime of parameters indicate
quite substantial stabilization in the sense of Mittleman and Gavrila. Moreover[4], there
are no essential differences between the results obtained for smoothed potentials and
the Coulomb potential. Classical results agree fairly well with quantum mechanical
ones for smoothed potentials both in 1–D (see Ref. 13), as well as in 3–D (see Refs.
12 and 14). Classically stabilized electrons bounce typically between the two minima
of the K–H potential. Their trajectories extend from -2α to 2α in the direction of
polarization, and have very small transverse size of the order of 1 a.u. One should
distinguish this kind of behavior from the situation when the electron absorbs initially
some energy, lands on a higher Rydberg orbit, and remains far from nucleus without
a chance to absorb more. As has been pointed out by us, the classical results differ
very strongly from the quantum ones in the case of Coulomb potential. The reason is
that phase space averaging method gives too much weight to the regions of the phase
space that are close to the nucleus. The electron is under the influence of very strong

Coulomb forces during the onset of the laser pulse. It absorbs thus a lot of energy and is immediately ionized. Classical stabilization in Coulomb potential is practically impossible in the discussed regime of parameters, except for rare cases of trapping in higher Rydberg orbits.

One may thus speculate that the main effect of QM consists in smoothing of the potential! There are several hand–waving arguments (based on Heisenberg principle or on hydrodynamic formulation of QM) that one can invoke to support such a point of view. Wilets[16] proposed a long time ago another method of effective potential smoothing, introducing velocity dependent forces to account for both Heisenberg and Pauli principles, and to restrict electrons' access to unphysical regions of the phase space (small values of $|\mathbf{p}|$ and $|\mathbf{r}|$). Note that within this approach the effective regularization radius grows with the increase of the electron's energy as $\simeq \hbar/\sqrt{|E|}$, which is in accordance with the scaling discussed above[17]! We shall therefore use in the following smoothed potentials keeping in mind that there are deep physical reasons that they mimic well the quantum Coulomb potential.

PROBLEMS TO BE SOLVED

Before we apply the classical methods, we must solve two important problems, concerning initial distributions and detection of final states. The problem of initial distribution is usually associated with an intrinsic instability of multielectron systems. Note, however, that this problem is absent when we deal with smooth potentials. Physics suggests here itself a proper solution. For instance, the classical two–electron potential

$$V(r_1, r_2) = -\frac{Z}{r_1 + r_{reg}} - \frac{Z}{r_2 + r_{reg}} + \frac{1}{r_{12} + r_{reg}} \qquad (3)$$

has well defined thresholds, just like its quantum mechanical counterpart. Namely, all states with energies $E_0 = -(2Z-1)/r_{reg} \leq E < E_1 = -Z/r_{reg}$ are two electron bound states, almost all of the states with the energies $E_1 \leq E < 0$ are 1–electron stable (i.e. correspond either to autoionizing states or to 1–electon ionized states). Finally, all states with $E > 0$ are one– or two–electron ionized. We propose thus two methods of constructing the initial phase space distribution:

a) *Periodic orbits.* Recent studies in quantum chaos suggest that many quantum mechanical states resemble classical (stable or unstable) periodic orbits ("quantum scars"). We may thus use such periodic orbits as initial phase space distributions. These could be for instance the orbits for which $\mathbf{r}_1 = -\mathbf{r}_2$, $\mathbf{p}_1 = -\mathbf{p}_2$, and the electrons move in the radial direction (Wannier $L = 0$ orbits) or on a circle (Langmuir $L = 2$ orbits). Such distributions have also been used in the Ref. 15. Note that such an approach may also be even adopted in the case of pure Coulomb potential.

b) *Dynamical sampling.* In principle, for $E_0 \leq E < E_1$ one can construct a microcanonical ensemble. This is, however, a rather difficult task. We propose a simple way to achieve it invoking arguments that stem from ergodic theory. Namely, we may toss two initial positions and momenta of the electrons independently, check the total energy of such a state, and let it evolve in time. The resulting trajectory for a uniform sequence of time instants provides a dynamical realization of the microcanonical sample.

Let us mention finally that the physics of the problem turns out to depend weakly on the details of the initial distribution, simply because we are in the regime of very strong laser fields. Only global properties of the initial state (such as its energy) are relevant for the dynamics in the field.

The second problem concerns detection of final states. In principle, smoothing of the potential solves as a byproduct this problem as well. We think, however, that it is important to distinguish final 1-electron ionized states created *during* the interaction with the laser, and final autoionizing states. Those two types of states correspond do different photoelectron spectra, characteristic decay times of autoionizing states can be much different from pulse duration time, etc. We propose thus the two methods of the final state detection

a) *Detection of global energy.* Here we identify final states with $E_0 \leq E_{fin} < E_1$ as two–electron stable, those with $E_1 \leq E_{fin} < 0$ as 1–electron ionized (or 1–electron stable), and those with $0 \leq E_{fin}$ as two electron ionized.

b) *Detection of individual energies.* Here we measure final energies of individual electrons, some time after the pulse is gone, $E(i) = -Z/(r_i + r_{reg}) + 1/(r_{12} + r_{reg})$ for $i = 1, 2$. We identify electrons as stabilized if $E(i) \leq 0$, and ionized otherwise.

In principle the physics suggests that the two methods should give similar results, and indeed it is so in 1–D. In 3–D the second method tends to overestimate the number of two–electron bounded states, whereas the first one underestimates the number of 1–electron stabilized states.

RESULTS IN 1–D

Let us now turn to discussion of the results in 1–D. In Fig. 1 we present an initial distribution obtained via dynamical sampling for the potential

$$V(x_1, x_2) = -\frac{Z}{\sqrt{x_1^2 + r_{reg}^2}} - \frac{Z}{\sqrt{x_2^2 + r_{reg}^2}} + \frac{1}{\sqrt{x_{12}^2 + r_{reg}^2}}. \tag{4}$$

Here $r_{reg} = 1$, $Z = 2$ and initial energy $E = -2.375$. The distribution is meant to mimic a ground state of the He atom.

As we see, the distribution in the momentum space has the rotational symmetry, whereas the one in the configuration space has not. The latter gives more weight to the states with $x_1 \simeq -x_2$, expressing the effects of electron correlations. Note that this sample which consists of 6000 points provides a very good representation of the microcanonical ensemble, due to ergodicity of the 1–D problem.

In Fig. 2 we present results of MC simulation for the He atom interacting with a laser pulse of a duration of 100 cycles, and $\tau_{on} = \tau_{off} = 5$ cycles. Laser frequency is here above the two–electron ionization threshold $\omega = 3$. We have detected the global final energy of the system and classified final states accordingly. We present probabilities of ionization as a function of α.

Note that one can easily distinguish three characteristic regimes:

A) This is a perturbative regime ($\alpha \ll 1$), when single electron ionization becomes dominant.

B) This is a regime of dominance of two–electron ionization ($\alpha \simeq 1$). Note that there are practically no two–electron bound states in this regime. 1–electron stable states start to resemble Mittleman–Gavrila states already.

C) Finally, for $\alpha > 1$ we enter the regime of stabilization. First, two–electron ionization tends to zero, so that most of the final states are 1– or 2—electron stable.

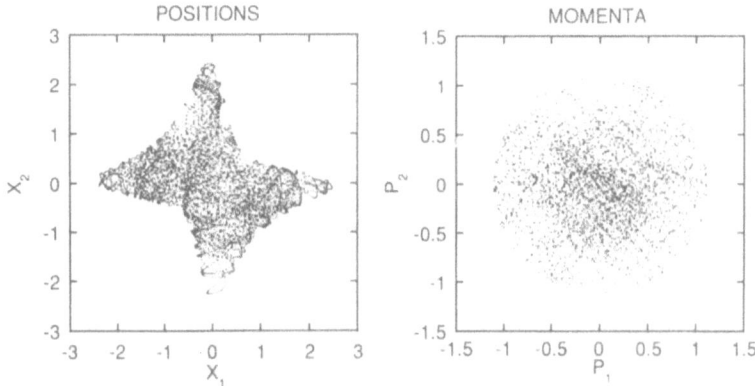

Figure 1. Initial distribution in the (a) configuration and (b) momentum space for the 1-D model; $r_{reg} = 1$, $Z = 2$ and $E = -2.375$.

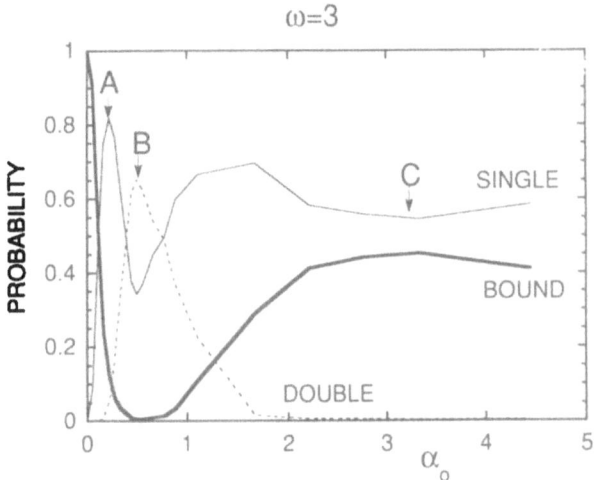

Figure 2. Probabilities of two-electron bound states, single ionized states and two-electron ionized states as a function of α. The laser pulse has $\omega = 3$, $\omega\tau_D/2\pi = 100$, $\omega\tau_{on}/2\pi = \omega\tau_{off}/2\pi = 5$.

In Fig. 3 we present final distributions of the total energies for the regimes A, B, and C. In the regime A the distribution is sharply peaked right above the threshold for single electron ionization. In the regime B, there appears a broad peak for positive energies. Note, that this peak is slightly shifted toward the negative energies due to the ponderomotive effects. Finally, in the stabilization regime C the energy distribution becomes concentrated at negative energies close to and below E_1. Fig. 4 shows the corresponding final spatial distributions of electrons. They exhibit characteristic features: a cross shape in the regime A, a "rotational symmetry" with the gap at $x_1 = x_2$ in the regime B, and a strong concentration close to the origin in the regime C.

For 1–electron stable states, the trapped electron bounces between the minima of the K–H potential at $-\alpha$ and α. The two–electron bound states are classical

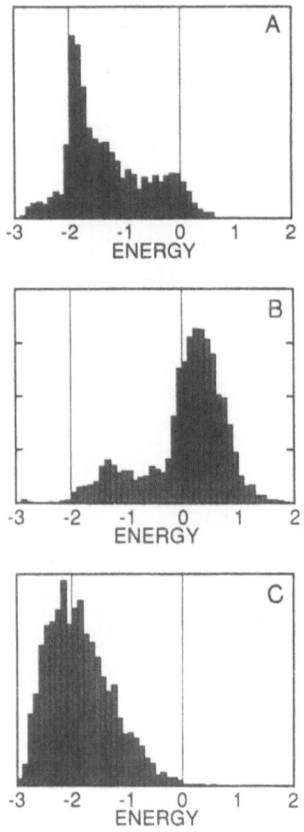

Figure 3. Distributions of final energies in the regimes A, B, and C.

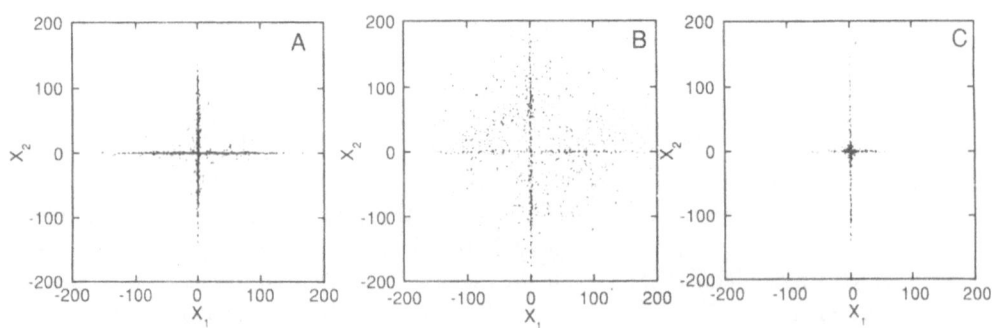

Figure 4. Spatial distributions of both electrons in the regimes A, B, and C.

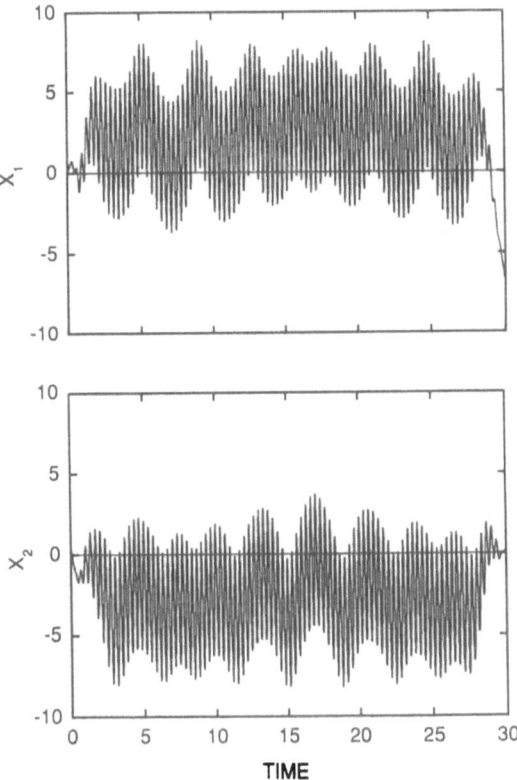

Figure 5. Trajectories of electrons trapped on the two opposite sides of the K–H potential; $\mathcal{E} = 36$, $\omega = 3$.

analogues of the Mittleman's trial function[10]. Note, that this is the first evidence that such states are dynamically accessible! This is illustrated in Fig. 5, where we display the time dependence of the positions of both electrons.

RESULTS IN 3–D

The situation is somewhat similiar in 3–D, although it is more difficult to achieve stabilization here. As initial distributions we use here either Wannier orbits with a given energy, or dynamical distributions, consisting typically of 900 points. In Fig. 6 we present results of MC simulation for ionization probabilities as a function of the electric field strength for the case of ionization from the ground state of He atom ($r_{reg} = 0.75$, $Z = 2$ and $E = -3$); The ground state has been modelled as Wannier orbit. Those results were analysed using the method of detection of individual electron energies.

Fig. 6a has been obtained for $\omega = 6$, and the same pulse shape as used in the previous Section. As we see, we cannot identify the regime B here (dominance of two–electron ionization). Two–electron ionization turns on for $\mathcal{E} = 600$, i.e. $\alpha \simeq 8$. We follow here the curves to very large values of α in order to show the tendency observed in Ref. 9, that the system looses its stability for too large values of α. For smaller $\omega = 4.5$ the results are quite analogous to those obtained in 1–D., except that for large values of α practically all final states are doubly ionized. The results of Fig. 6 will be modified quantitatively if we use the detection criterium based on global energy, since most of two–photon bound states from Fig. 6 are in fact autoionizing states. However, the qualitative behavior of the curves remains the same.

The trajectories of two–electron bound states in the considered regime of parameters look as follows. Typically, the electrons start in the Mittleman state, but

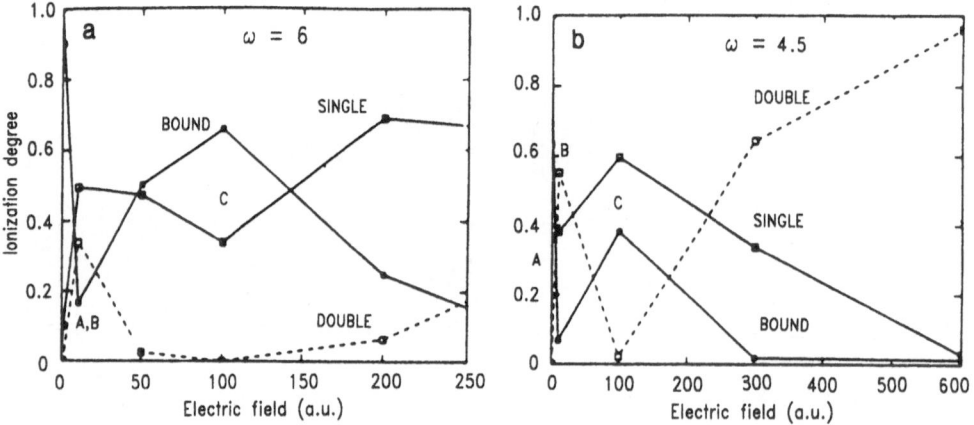

Figure 6. Probabilities of single electron ionized states and two–electron ionized states as a function of $\mathcal{E}$. The laser pulse has (a) $\omega = 6$ (b) $\omega = 4.5$; $\omega \tau_D / 2\pi = 100$, $\omega \tau_{on} / 2\pi = \omega \tau_{off} / 2\pi = 5$.

then one or two of them jump into a higher Rydberg orbit. We have, however, found a strong evidence that Mittleman states play a very important role dynamically.

One possibility to obtain a nice trapping of the electrons on both sides of K–H potential is to start with initial states of higher energy, i.e. either excited states or even autoionizing states. In Fig. 7 we illustrate this point for $r_{reg} = 1.$, $Z = 2$ and $E = -2$, Wannier orbit. The initial state is thus at the edge of 1–electron ionization threshold. Laser frequency $\omega = 4.$, and $\alpha \simeq 6$.

The same method can be applied to more realistic excited states or to the ground state of the H$^-$ ion. To this end we do not use Wannier orbits, but dynamical samples in which initially individual electron energies are distinct. Obviously, in the course of evolution, the roles of the two electrons will mix, but such states contain much more electron correlation effects (when one electron is close to the nucleus, the other one is far). In the regime of highly excited initial states and low frequencies of the laser (lying just above the first ionization threshold) we do not observe any absolutely stable two electron states with $E_{fin} < E_1$. There are, however, many final autoionizing states that are created via Mittleman and Gavrila mechanism.

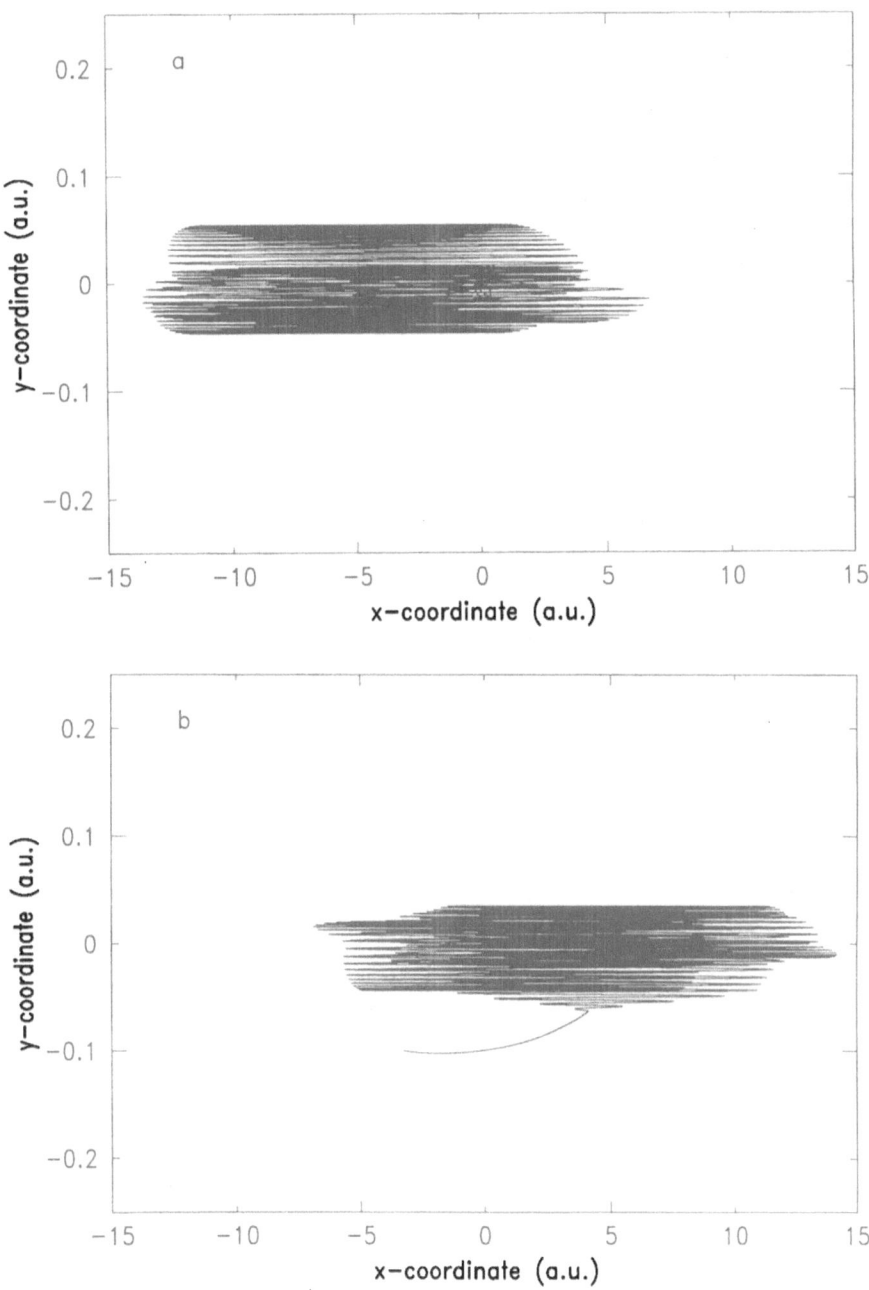

Figure 7. Projections of trajectories of (a) the first and (b) the second electron trapped on the two opposite sides of the K–H potential on the $x - y$ plane; x is the polarization direction; $\mathcal{E} = 100$, $\omega = 2$ and $E = -2$.

CONCLUSIONS

We have developed the method of classical phase space averaging for multielectron systems. We have used smoothed potentials for this aim which, on one hand, has a physical motivation, and, on the other hand, allows to solve problems of constructing the initial distributions and detecting final states. We have shown the evidence of the new type of stabilization with two–electrons on both sides of K–H potential. We have identified regimes of parameters in which stabilization is dynamically accessible.

Finally, we might speculate about further prospects: 3 electron states "crystallized" in the vertices of the circularly shaped K–H potential in a circularly polarized field, multielectron states accessible in K–H potential forming a Lissajou like curve resulting from applying multimode field with two–commensurate frequencies etc.

ML thanks Centre d'Etudes de Saclay (SPAM) the French Ministry of Science and Technology for financial support. KR and ML would like to acknowledge the support of the KBN Grant.

REFFERENCES

1. J.I. Gersten and M.H. Mittleman, Phys. Rev. $A10$, 74 (1974); *ibid.* $A11$, 1103 (1975);
2. M. Gavrila and J. Kamiński, Phys. Rev. Lett. 52, 613 (1984).
3. Q. Su, J.H. Eberly, and J. Javanainen, Phys. Rev. Lett. 64, 862 (1990); Q. Su and J.H. Eberly, J. Opt. Soc. Am. $B7$, 564 (1990); Phys. Rev. $A43$, 2474 (1991); V.C. Reed, P.L. Knight, and K. Burnett, Phys. Rev. Lett. 67, 1415 (1991).
4. K.C. Kulander, K.J. Schafer, and J.L. Krause, Phys. Rev. Lett. 66, 2601 (1991).
5. M. Dörr, R.M. Potvliege, and R. Shakeshaft, Phys. Rev. Lett. 64, 2003 (1990); M. Pont and M. Gavrila, Phys. Rev. Lett. 65, 2362 (1991); M. Pont and R. Shakeshaft, Phys. Rev. $A44$, R4110 (1991).
6. J. Mostowski and J.H. Eberly, J. Opt. Soc. Am. $B8$, 1212 (1991).
7. C.K. Law, Q. Su and J.H. Eberly, Phys. Rev. $A44$, 7844 (1991).
8. R. Grobe and J.H. Eberly, Phys. Rev. Lett. 68, 2905 (1992).
9. R. Grobe and J.H. Eberly, preprint University of Rochester (1992).
10. M.H. Mittleman, Phys. Rev. $A42$, 5645 (1990).
11. K.C. Kulander, Phys. Rev. $A36$, 2726 (1987).
12. J. Grochmalicki, M. Lewenstein, and K. Rzążewski, Phys. Rev. Lett. 66, 1038 (1991); M. Gajda, J. Grochmalicki, M. Lewenstein and K. Rzążewski, Phys. Rev. $A46$, 1638 (1992).
13. R. Grobe and C. K. Law, Phys. Rev. $A44$, R4114 (1991).
14. T. Menis, R. Taieb, V. Veniard, and A. Maquet, J. Phys. $B25$, L263 (1992).
15. P.R. Lerner, K. LaGattuta, and J.S. Cohen, Los Alamos preprint LA–UR–92–3525, (1992).
16. C.L. Kirshbaum and L. Wilets, Phys. Rev. $A21$, 834 (1980); D.W. Wasson and S.E. Koonin, Phys. Rev. $A39$, 5676 (1989); this approach has been also adopted in Ref. 15.
17. Note that the extension of the scaling invariance to quantum mechanics requires that $\hbar \rightarrow \alpha^{-1/3}\hbar$.

MULTIPHOTON DETACHMENT OF HYDROGEN NEGATIVE ION

L.Dimou and F.H.M.Faisal

Fakultät für Physik
Universität Bielefeld
Bielefeld, Germany

INTRODUCTION

Investigation of detachment of electrons from negative ions in intense laser fields has recently experienced a renaissance. The first multiphoton detachment experiment was performed by Hall, Robinson and Branscomb[1] more than a quarter century ago. Theoretical investigation of the detachment process began at about the same period which lead to the first non-perturbative theory ,the so called KFR-theory[2-4],of laser atom interaction.The recent upsurge of theoretical interest[5-13] in the problem is generated by the discovery of the above-threshold ionization (ATI) process[14] and the progress made both in experimental techniques, measurement of detachment rates[15-20] and theoretical methods for investigating laser-atom interaction in the non-perturbative domains of field strengths. The above threshold detachment (ATD) process in negative ions,which is analogous to the ATI process in neutral atoms, has been observed very recently[18-20] in several negative ions.The energy spectrum of both ATI and ATD show characteristic peaks separated by the constant energy of the incident photons.These spectra are <u>universal</u> in character in the sense that their overall structures are independent of the atom, or negative ion, involved. Much interests have arisen in this connection recently to get a better understanding of the similarities and differences between ATI and the ATD processes. Perhaps the most fundamental difference between the two systems lies in the asymptotic nature of the interaction potential seen by the electron. The atomic electron in the final state experiences a long range Coulomb potential ($-r^{-1}$) whereas the active electron in the negative ion experiences short range ('static' and $-r^{-4}$-polarization) potentials. It is some times believed that during the ATI process the so called 'excess' photons are absorbed only because the interaction between the ionized electron in the continuum and the residual ion can be mediated by the long range Coulomb potential,which extends far beyond the core region.This might suggest that 'excess' photon absorption would be unlikely in the case of negative ions due to the lack of long range Coulomb potential in the final state. Theoretically, however,from the very beginning of non-perturbative investigations, the KFR-like models[3-5,21,22](which should be more directly applicable for the negative ions than for the neutral atoms) predicted unequivocally absorption of an arbitrary number of photons above the threshold by the active electron in a short range (even zero-range) potential. The recent observations of ATD-peaks in a number of negative ions[18-20], of course, have left no room for any doubt regarding the ability of short range potentials to cause absortion of 'excess' photons. It appears, therefore, quite appropriate to assume that the process of absorption of photons ,both in ATI and ATD, occurs essentially in the vicinity of the core, which facilitates the necesssary conservation of momentum in the total absorption process. The energy conservation is readily met by the absorbed photon energies.

In the present paper we shall confine ourselves to the investigation of the response of the hydrogen negative ion to a monochromatic electromagnetic field, and consider the total detachment rates as a function of frequency as well as the intensity of the field. We shall be interested both in above-threshold and in below-threshold frequencies. The probability of detachment per unit time is directly related to the line width of the initial state and can be determined quantitatively from the theory of resonance line width. This will be formulated within the frame work of the general S-matrix theory, albeit, on appropriately extending the latter theory from its usual collisional context into the context of quantum decay process (from an initillay prepared bound state) in the external radiation field. Since the laser field-density is generally high it may be treated accurately as a time dependent classical electromagnetic field, while the atom must be treated fully quantum mechanically. In this case it is convenient and useful to apply the general Floquet transformation[23-25] to convert the time dependent Schrödinger problem into a stationary one. It can be shown that the final result obtained in this way is equivalent to the result obtained using a quantized radiation field, in the occupation number representation, for virtually all lasers. This is, in fact, an expression of the quantum-classical correspondence [25] for high density of the field.

In this work we shall use the ab initio numerical method of radiative or Floquet close coupling theory, which is based on the Floquet expansion and the appropriate form of the S-matrix theory in configuration space, as developed originally by Dimou and Faisal[26,27] and by Guisti-Suzor and Zoller[28]. This method has since been applied usefully in various problems of laser-atom interaction physics by an increasing number of authors[29-33]. Another non-perturbative method, which is very similiar to the Floquet close coupling theory, has recently been developed [34,35] within the R-matrix formalism.

THE FLOQUET CLOSE -COUPLING EQUATIONS

The Schrödinger equation of hydrogen negative ion interacting with a monochromatic laser field can be written in the space translated reference frame[36,37,3] as (in atomic units, $e=\hbar=m=1$):

$$id/dt\ \Psi(t) = [\ -1/2\Delta + V(\mathbf{r}-\alpha_0(t))\ ]\ \Psi(t) \qquad (1)$$

Notice that Eq.(1) implies that we are dealing with the evolution of the active outer electron whose motion is governed by the average potential due to the fields of the core electron and the nucleus as well as the polarization potential due to the polarization of the core by the field of the active electron, which acts back on to the active electron. These potentials can be self consistently derived from the full two-electron Hamiltonian within the well known 'static-polarization' approximation[38,39]. In view of the dominance of these potentials at short distances and in the asymptotic region, the present one-electron analysis is expected to be a good approximation for multiphoton detachment processes in H⁻, except perhaps under special choice of the frequency to match with a core transition, or at ultra-high intensities where H⁻ should be treated fully as a two-electron system. The potential $V(\mathbf{r})$ discussed above has the explicit form:

$$V(r) = - [(1+1/r) e^{-2r} + 9/4 (r_0^2 + r^2)^{-2}] \qquad (2)$$

where the Buckingham parameter r_0 is chosen to reproduce the electron affinity of H⁻ exactly($r_0 = 1.7033$ a.u. , $\varepsilon_b = 0.754$ eV). In Eq.(1) ,in the space translated frame ,the potential (2) appears with $\mathbf{r}$ shifted by

$$\alpha_0 (t) = -1/c \int^t \mathbf{A}(t')dt' ,$$

which is the so-called instantaneous 'quiver radius' of the free electron. $\mathbf{A}(t)$ is the vector potential of the monochromatic laser field, which we choose to be circularly polarized in the present investigations.

Eq.(1) is in general a (3+1)-dimensional partial differential equation and as such poses formidable difficulty in solving it directly by numerical means. This is specially so for adiabatically changing pulses which require a very large space time grid for direct integration.Since most ,if not all ,of the high intensity pulses are of much longer duration than the period of their corresponding carrier frequency, they are effectively switched on and off in an adiabatic fashion during the interaction.The Floquet close-coupling method can handle this situation particularly conveniently in view of the fact that adiabatic pulses in the long wavelength approximation lead to an effective periodicity of the Hamiltonian, as for a uniform constant amplitude periodic electromagnetic field.The radiative or Floquet close-coupling technique essentially converts the partial differential Eq.(1) into an infinite set of ordinary differential equations for the so-called channel wave functions,$F_{nlm}(r)$.This is achieved by expanding the total wave function in Floquet plus partial waves as follows[26-28].

$$\Psi (\mathbf{r},t) = \Sigma_{nlm} \exp\{i[E-n(\omega t+\delta)]\} F_{nlm}(r) Y_{lm}(\theta,\phi) \qquad (3)$$

where E is the characteristic Floquet exponent which can be identified with the total energy of the interacting quantum system (atom + field), the Fourier index n can be identified with the <u>change</u> in the occupation number of the field, (l,m) are the angular momentum quantum number and its projection on the quantization axis, ω is the frequency and δ is an arbitrary initial phase of the field.It is convenient for practical calculations to choose the quantization axis along the polarization direction for a linearly polarized field,and along the field propagation direction for a circularly (or elliptically) polarized field.Substituting (3) in (1), equating coefficients of $\exp[in(\omega t+\delta)]$ on both sides and projecting onto the spherical harmonics, we readily obtain the system of <u>Floquet close-coupling equations</u>[26-28].

$$[d^2/dr^2 +k_n^2 - l(l+1)/r^2 +2/r] F_{nlm}(r) = -2 \Sigma_{n'l'm'}V_{nlm,n'l'm'}(r) F_{n'l'm'}(r) \qquad (4)$$

where $V_{nlm,n'l'm'}(r)$ are the channel coupling potentials and $k_n = (2(E-n\omega))^{1/2}$ are the channel wave numbers; the set of quantum numbers {nlm} defines a channel completely. For the problem of decay of a bound state the system of equations can be solved by imposing[31,32] Siegert's out-going wave boundary conditions in the open channels, with

E - $n\omega$ > o, and exponentially decreasing asymptotic boundary conditions in the closed channels,with E-$n\omega$ <o. It is one of the most satisfactory aspects of the close-coupling theories in general, and hence of the present theory, that the wave function in the open channels which correspond to the continuum motion of the electron, is extended to infinite distances by the simple matching procedure to the asymptotic channel wave functions analytically. Integrating the coupled system of equations (4) from both ends, i.e. from zero outwards and from any point in the asymptotic region inwards and matching the logarithmic derivative-matrix of these inwards and outwards propagated solutions at an intermediate radius, r_m ,we get the energy determining secular equation :

$$\det |L_{in}(r_m) - L_{out}(r_m)| = 0 \qquad (5)$$

where the log-derivative matrix L(r) is given by :

$$L(r) = d/dr\, F(r)\, /\, F(r) \qquad (6)$$

It is of course obligatory to check that the converged solutions of the system are independent of change in the value of the intermediary matching point r_m. Note that the secular equation (5) has in general solutions in the complex E-plane. The appearence of complex energy eigen values of the system, where one started with a hermitian Hamiltonian, has at times led to confused discussions in the past. The appearence of complex eigenvalues ,despite the hermitian nature of the Eq.(1), is due to the imposition of the complex boundary conditions, in the form of purely out-going waves in the open channel i.e. the Siegert's boundary condition for a decaying state.The real part of the eigenvalue corresponds to the shifted energy while the imaginary part corresponds to the width of the level which is correlated to the unperturbed level in the limit of vanishing field strength. In other words, for a complex zero $E_j = E_r - i\Gamma_j/2$ the rate of detachment of the state, j , is given by the line width $\Gamma_j = -2\text{Im}(E_j)$. In actual computations we start with a finite number of channels in Eq.(4) and successively increase the number of channels for a given set of field parameters until the convergence of the final results is ensured. In the calculations reported here the maximum number of channels found necessary to ensure the convergence was around tweenty.The complex roots of Eq.(5) are found by an iterative (complex) root searching procedure where the iteration was started with the unperturbed value of the initial energy and the field frequency (and/or strength) was increased by small steps.The next set of iterations was started with the complex root obtained at the previous step and the process repeated until the highest value of the frequency (and/or field strength) investigated was reached.

RESULTS AND DISCUSSIONS

In Fig.1 we present the result of rates of detachment of H^- as a function of photon energy for several intensities of the laser, ranging from 1.0 GW/cm^2 to 4.0 GW/cm^2,and in the frequency range ω =0.3 eV to ω =1.0 eV. It is observed that with incresing intensity the rates increse dramatically, as may be expected in the domain of intensities considered here.Note the characteristic step-like structures (seen for all the three intensities) as a function of the frequency. These steps are related to the opening of increasingly lower order photodetachment channels of H^- with the increase of the photon frequency. Thus the steps seen near the lower frequencies in these curves correspond to the opening of the two-photon detachment channel -on the right of the step- from a three-photon (and higher order) detachment channel, on the left. With further increase of the frequency there arises a second step (in all the curves) associated with the opening of the one-photon channel, on the right, from the two-photon (and higher order) channel on the left. We note that the positions of the opening of a given channel (i.e the channel threshold) are seen to differ somewhat at different intensities ;this is a consequence of the change in the electron affinity of H$^-$ in the presence of the field due to the dynamic Stark-effect.Note that the rates at the lower order thresholds change more sharply compared to the higher order thresholds.This is a consequence of two interrelated

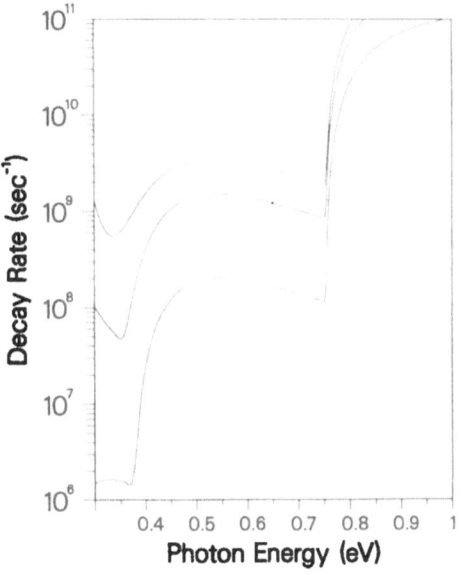

Fig.1 Rate of detachment of H$^-$ at intensities of 1.0 GW/cm^2 (lowest curve), 2.7 GW/cm^2 (middle curve) and 4.0 GW/cm^2 (upper curve), as function of the laser frequency in the range 0.3 eV to 1.0 eV.

factors: (a) the change of the threshold (typically of the order of the quiver energy) is more sensitive to the change in frequency for the higher order thresholds,and (b) the relatively stronger coupling due to the field (typically given by the quiver radius) near the higher order thresholds introduces relatively greater change in the rates , than the same relative change in the frequency does to the rates near the lower order thresholds.One should recall, in this context, that both the quiver energy and the quiver radius are inversely proportional to the square of the laser frequency and,of course,that the higher order thresholds occur at lower photon energies.

In Fig.2 we consider the dependence of the rates of detachment of H⁻ in much greater detail as a function of the frequency in the interval ω =0.34 eV to 0.42eV (in a scale of the order of meV). In this figure we present the theoretical results for the peak intensity I=2.7 GW/cm^2. We also show the experimental data[40] of Bryant and collaborators in the same interval of frequency for the experimental intensity estimated to be (2.7 ± 0.3) GW/cm^2. It will be seen that the theoretical result and the experimental data do not directly agree but run essentially parallel to each other. At the laboratory intensity of 2.7 GW/cm^2 and the laboratory wave length of λ= 9540 nm for the circularly polarized field the ponderomotive shift can be calculated to be U_p= 23 meV. Note that the ponderomotive energy is relativistically invariant and has the same value in the laboratory frame as in the frame of the negative hydrogen ions in the experiment, which move with relativistic velocity in a beam. In Fig. 3. we compare the results, taking into accout the ponderomotive shift of 23 meV, which provides a good agreement between the theory and the experiment.

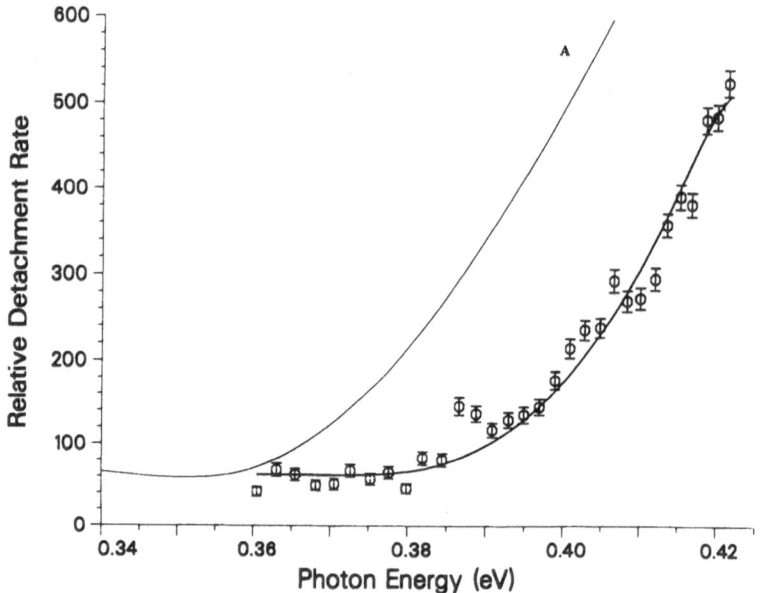

Fig.2. Detachment of H⁻ in the (c.m.) frequency domain 0.34 eV to 0.42 eV in expanded scale. Theoretical results are given by the solid curve A for I=2.7 GW/cm^2 and the experimental data are for I=(2.7 ± 0.3) GW/cm^2 . Note that the increase of the experimental curve (fitted by the experimental group to their data) is essentially parallel to that of the theoretical curve A.

In Fig.4 we investigate the role of the polarization-potential which is the dominat potential in the asymptotic region in the final state. The results shown are for the 'static+polarization'-potential (curve A) and for a Yukawa-potential (curve B). It is clearly seen that the (exponentially decreasing) Yukawa-potential alone give approximately the same order of magnitude for the rates of detachment as the 'static+polarization'-potential, but the trends of the two sets of data (in the scale of the order of meV) show a clear difference.The results obtained with simply Yukawa-potential change appreciably more slowly than the results of 'static+polarization'-potential and/or the experimetal data (c.f.Fig.3).It seems clear that the effect of core-polarization (which is accounted for by the polarization potential) for multiphoton detachment of H⁻ can not be neglected for quantitative comparisons.

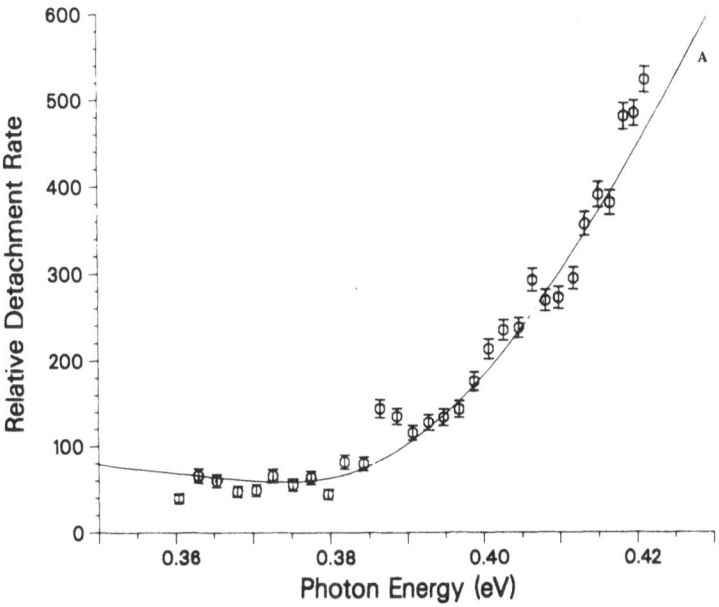

Fig.3. Comparison of the theoretical results shifed by the ponderomotive energy of 23 meV,corresponding to a laboratory intensity $I = 2.7$ GW/cm^2 and laboratory wavelength $\lambda = 9540$ nm, which shows a good agreement with the experimental data. Note that the poderomotive energy is a relativistically invariant quantity and is the same in the laboratory frame as in the c.m. frame of the (relativistically moving) ions.

Finally,in Fig.5. we present the rates of detachment of H⁻ as a function of the field strength (here shown in terms of the quiver radius , $\alpha_0 = E_0 / \omega^2$) at a fixed frequency, ω = 1.17 eV (fundamental of Nd-laser).We note that the rates of photodetachment of H⁻ increases rapidly with increasing field strength and can readily lead to saturation of the probability of detachment at high intensities. We may point out that at still higher intensities the description of the behavior of the system in terms of the rate as a parameter can not be ensured, in view of either a possible saturation of probabilities at the higher intensities or due to the break-down of the adiabaticity condition or both.The intensity dependence of the detachment of hydrogen negative ion predicted in this figure can be tested with the currently available Nd-laser in the laboratory.

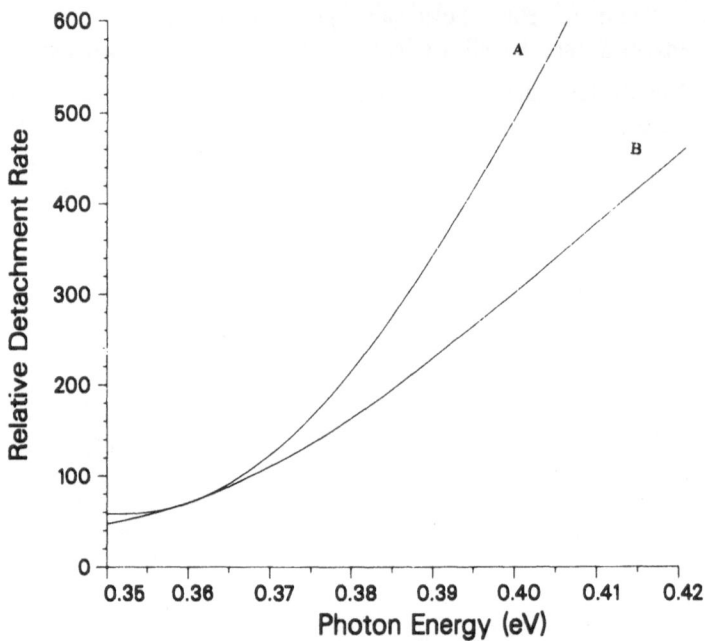

Fig.4. Comparison of detachment rates with short range 'static+polarization'-potential (curve A) and the Yukawa-potential (curve B) .Note the increase of the detachment rates in B, which is much slower compared to that of A as well as of the experimental data (c.f. figs.2,3).

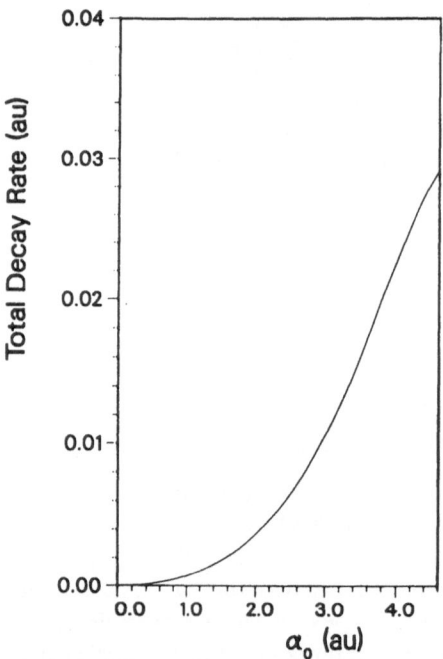

Fig.5. Dependence of the rate of detachment of H⁻ on the quiver radius (proportinal to the field strength), at a fixed frequency , $\omega = 1.17$ eV, the fundamental frequency of Nd-laser.

REFERENCES

1. J.H.Hall,E.J.Robinson,and L.M.Branscomb,Phys.Rev.Lett.14:1013(1965)
2. L.V.Keldysh,Zh.Ekps.Teor.Fiz.47:1945(1965)[Sov.Phys.JETP20:1307(1965)]
3. F.H.M.Faisal,J.Phys.B5:L89(1973)
4. H.R.Reiss,Phys.Rev.A22:1786(1980)
5. B.Chen,F.H.M.Faisal,S.Jetzke,H.O.Lutz,and P.Scanzano,Phys.Rev.A36:4091(1987)
6. Th.Mercouris and C.A.Nicolaides,J.Phys.B21,L285(1988)
7. W.Becker,J.K.McIver and M.Confer,Phys.Rev.A40:6904(1989)
8. M.Dörr,R.M.Potvliege,D.Proulx and R.Shakeshaft,Phys.Rev.A42:4138(1990)
9. F.H.M.Faisal,P.Filipowicz and K.Rzazewski,Phys.Rev.A41:6176(1990);44:2210(1991)
10. W.G.Greenwood and J.H.Eberly,Phys.Rev.A43:525(1991)
11. S.Geltman,Phys.Rev.A43:4930(1991)
12. C.-R. Liu,B.Gao and A.Starace,Phys.Rev.A46:5985(1992)
13. F.H.M.Faisal and P.Scanzano,Phys.Rev.Lett.68:2909(1992)
14. P.Agostini,F.Fabre,G.Mainfray,G.Petite and N.K.Rahman,
Phys.Rev.Lett. 42:1127 (1979)
15. C.Y.Tang,P.G.Harris,A.H.Mohagheghi,H.C.Bryant,C.R.Quick,J.B.Donahue,
R.A.Reeder,S.Cohen,W.W.Smith and J.E.Stewart,Phys.Rev.A39:6068(1989)
16. C.Y.Tang,H.C.Bryant,P.G.Harris,A.H.Mohagheghi,R.A.Reeder,H.Sharafian,
H.Tootoonchi, C.R.Quick,J.B.Donahue,S.Cohen and W.W.Smith,
Phys.Rev.Lett.66:3124(1991)

17. C.Blondel,M.Crance,C.Delsart and A.Giraud,J.de Phys.II, N°.4:4839(1992)
18. C.Blondel,M.Crance,C.Delsart and A.Giraud,J.Phys.B24:3575(1991)
19. H.Stapelfeldt,P.Balling,C.Brink and H.K.Haugen,Phys.Rev.Lett.67:1731(1991)
20. M.D.Davidson,H.G.Muller,H.B.van Linden van den Heuvel,
Phys.Rev.Lett.67:1713(1991)
21. I.J.Berson,J.Phys.B8:3078(1975)
22. F.H.M.Faisal,Il Nuovo Cimento 33B:775(1976)
23. J.H.Shirley,Phys.Rev.138:979(1965)
24. S-I.Chu and W.P.Reinhardt,Phys.Rev.Lett.39:1195(1977)
25. F.H.M.Faisal,Theory of Multiphoton Processes, Plenum Press,N.Y.(1987)
26. L.Dimou and F.H.M.Faisal,in:Photon and Continuum States of Atoms,
eds. N.K.Rahman,C.Guidotti and M.Allegrini,Springer,Berlin(1987)
27. L.Dimou and F.H.M.Faisal,Phys.Rev.Lett.59:872(1987)
28. A.Guisti-Suzor and P.Zoller,Phys.Rev.A36:5178(1987)
29. P.Marte and P.Zoller,Phys.Rev.A43:1512(1991)
30. L.A.Collins and G.Csanak,Phys.Rev.A44:R5343(1991)
31. L.Dimou and F.H.M.Faisal,Phys.Rev.A46:4442(1992)
32. L.Dimou and F.H.M.Faisal,Phys.Lett.A171:211(1992)
33. A.Guisti-Suzor and F.H.Mies, in: Multiphoton Processes,eds. G.Mainfray and
P.Agostini,Service de Phys.,Saclay,France (1990)p.169
34. P.G.Burke ,F.Franken and C.J.Joachain, Europhys.Lett.13:617(1990) ;
J.Phys.B24:761(1991)
35. M.Dörr,M.Terao-Dunseath,J.Purvis,C.S.Noble,P.G.Burke and C.J.Joachain,
J.Phys.B25:2809(1992)
36. H.A.Kramers,in:Collected Scientific Papers,North Holland,Amsterdam (1956) p.262
37. W.Henneberger,Phys.Rev.Lett.21:838(1968)
38. N.F.Mott and H.S.W.Massey,The Theory of Atomic Collisions,3rd.Ed.,Clanderon
Press,Oxford,(1965),Chap.XVII.
39. R.A.Buckingham,Proc.Roy.Soc.,A160:94(1937)
40. E.P.Mackerrow (personal communication ,at SILAP-II, Big Sky,Montana,1991)

STRONG FIELD PREDICTIONS FOR NEGATIVE IONS
IN SHORT LASER PULSES

J.H. Eberly and R. Grobe

Department of Physics and Astronomy
University of Rochester
Rochester, NY 14627

We review our recent studies of a two-electron system in an intense short-pulse laser field. This system is one-dimensional in space but it retains many of the quantum mechanical and electron correlation properties of a negative ion. We use numerical techniques to determine exactly the system's fully correlated two-electron wave functions $\Psi(x_1, x_2, t)$ and from these we can determine the system's response to laser pulses. For our model system we discuss detachment and ionization probabilities including aspects of stabilization, as well as photoelectron spectra and a novel two-electron coherence effect. These permit one to make qualitative and quantitative predictions for laboratory experiments on negative ions.

1. INTRODUCTION

Strongly perturbed two-electron systems are of great interest in quantum physics. In atomic physics the fundamental two-electron systems are neutral helium (and its isoelectronic analogs), the atomic hydrogen negative ion, and the molecular hydrogen positive ion. The main objective of this paper is to review new theoretical findings[1-4] regarding a quantum mechanical 2-electron system in a strong, pulsed laser field of optical or higher frequency.

Behavior at very high intensities is particularly interesting. It has been found in recent theoretical studies that the rate of photoionization of a one-electron atom in a very intense laser field can be a decreasing function of the laser intensity. This behaviour has been called atomic stabilization. We have addressed the question whether stabilization can occur at all in the context of a multi-electron system. For short laser pulses we find that in the low frequency regime the weakly bound valence electron can stabilize in a small window in laser

Super-Intense Laser-Atom Physics, Edited by
B. Piraux *et al.*, Plenum Press, New York, 1993

intensity for which the more deeply bound core electron is dynamically passive. We have also investigated time and energy resolved photoelectron spectra and predict a new differential growth rate. Finally we propose the experimental detection of coherence of core electrons by e-e correlation.

2. NEGATIVE ION MODEL, WAVE FUNCTIONS, AND ELECTRON REMOVAL PROBABILITIES

Our two-electron system is characterized by a nucleus of positive charge fixed at the origin at x=0 and two electrons whose spatial coordinates are x_1 and x_2. This model can be thought of as a one-dimensional analog of H⁻, the hydrogen negative ion.[1,5] Indeed, for simplicity we will refer here after to our one dimensional system as H⁻, and to its neutral atomic core as H. For this H⁻ the bare Hamiltonian is

$$H_0 = \frac{1}{2}p_1^2 + \frac{1}{2}p_2^2 + V(x_1) + V(x_2) - V(x_1-x_2) \tag{2.1}$$

where the same quasi-coulombic soft core potential

$$V(x) = -1 / \sqrt{(1 + x^2)} \tag{2.2}$$

describes the attractive electron-proton interaction as well as the mutual electron-electron repulsion. Here and elsewhere we will generally use atomic units, for which e=m=ℏ=1. We restrict our analysis to spin singlet states such that the two-electron wave function $\Psi(x_1,x_2,t)$ is symmetric under the exchange of both coordinates. In all calculations the ground state is the initial state which can be obtained by integrating the Schrödinger equation in imaginary time. The details of the numerical algorithms will be presented in a separate paper.[6]

Our one-dimensional H⁻ is realistic in a number of important respects. The bare Hamiltonian has only one bound state, as is characteristic of negative ions in general[7]. The ground state eigenenergy is -0.73 a.u. and its underlying core system (the neutral one-dimensional hydrogen atom[8]) has its ground state at -0.67 a.u. It follows that the outer electron is relatively very weakly bound, with a photodetachment potential of 0.06 a.u. The ratio of approximately 10 between ionization and detachment energies appears to be important for several of our results, and in the "real world" such a ratio is characteristic of all of the alkali negative ions from lithium to cesium. It is about 18 for hydrogen. All of these atoms appear to provide suitable systems in which the semi-quantitative predictions that come from our model can be tested.

The time-dependent interaction of the two-electron system with the laser field is described by the Schrödinger equation

$$i\partial\Psi(x_1,x_2,t)/\partial t = [H_0 + (x_1+x_2)\, \mathcal{E}(t) \sin \omega t] \, \Psi(x_1,x_2,t) \tag{2.3}$$

where $\mathcal{E}(t)$ is the laser field envelope and ω the frequency. In our studies the amplitude has been turned on and off smoothly, in some cases with the envelope $\sin^2(\pi t/T)$ and in some cases with a trapezoidal pulse shape with a two-cycle ramp. The latter has proved to be very helpful for strong field calculations because it combines the advantages of a rapid turn-on while retaining the displacement- and drift-free electron motion of a truly adiabatic pulse[9].

The latter is especially important in the super strong field stabilization regime in which unwanted ionization can be caused simply by the large drift velocity arising from a too abrupt turn-on.

The time-dependent Schrödinger equation (2.3) has been solved numerically on a 1024×1024 spatial lattice grid[10] to obtain the two-electron wave function $\Psi(x_1, x_2, t)$. With this wave function in hand we can calculate several physical observables of interest.

The probability $P1(t)$ that at least one electron is detached can be calculated from the overlap integral of the ground state $\Psi_g(x_1, x_2)$ with the time dependent wave function $\Psi(x_1, x_2, t)$.

$$P1(t) \equiv 1 - |< \Psi_g(x_1, x_2) \,|\, \Psi(x_1, x_2, t)>|^2 \qquad (2.4)$$

Another quantity of physical interest is the two-electron removal (detachment plus ionization) probability $P2(t)$ which can be defined by the probability that neither of the two electrons is in a bound state of the corresponding hydrogenic core. The latter is obviously the complement of the probability that at least one electron is in a hydrogen bound state, which we denote by $\psi_n(x)$ corresponding to the principal quantum number n.

$$P2(t) = 1 - \left\{ 2 \sum_n \int dx_2 \,|\int dx_1 \, \psi_n(x_1) \, \Psi(x_1, x_2, t)|^2 \right.$$

$$\left. - \sum_n \sum_m |\int dx_1 \int dx_2 \, \psi_n(x_1) \, \psi_m(x_2) \, \Psi(x_1, x_2, t)|^2 \right\} \qquad (2.5)$$

The terms in the first sum describe the probabilities that the electron with coordinate x_1 is in bound state ψ_n independent of the other electron. The factor 2 reflects the symmetry of the wave function under exchange of the electrons' coordinates. In order to avoid double counting we have to subtract in the double sum the probability that both electrons are bound.

3. LOW FREQUENCY DETACHMENT AND IONIZATION

By "low" laser frequencies we mean frequencies sufficiently high for one-photon detachment of the outer (weakly) bound electron but much lower than the binding potential of the core electron. For our system the value $\omega = 0.08$ a.u. satisfies the stated criterion: 0.67 >> 0.08 > 0.06 a.u. In this situation, for laser intensities not too great, one finds the motion of the inner core electron to be frozen while the outer electron detaches. For higher intensities the situation is more interesting, as the core electron becomes active.

We present in Fig. 1 our results for the one-electron ionization probability $P1(t)$ as a function of time for several laser field strengths $\mathcal{E}$ and a 20-cycle laser pulse. For the smallest of the three field strengths, $\mathcal{E} = 0.0025$ a.u., there is only about a 65% probability of detachment after the pulse. We should mention that this field strength is already so strong that the decay cannot be described by a simple exponential. This indicates that either this regime is already beyond the range of applicability of ordinary perturbation theory or the final energy continuum has some structure due to resonances.

For the next higher field strength, $\mathcal{E} = 0.005$ a.u., the detachment probability is more than 95% at the end of the pulse. For an even larger intensity, $\mathcal{E} = 0.01$ a.u., the two-electron bound state probability has already decayed to 50% during the turn-on, and after only 8 cycles the electron has fully detached. It will be important later to note that the two-electron

447

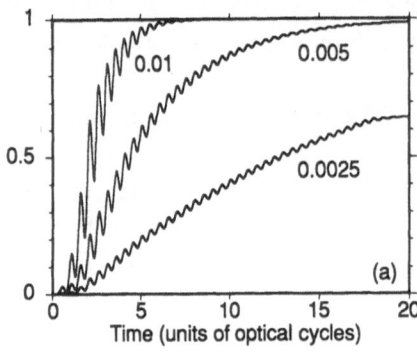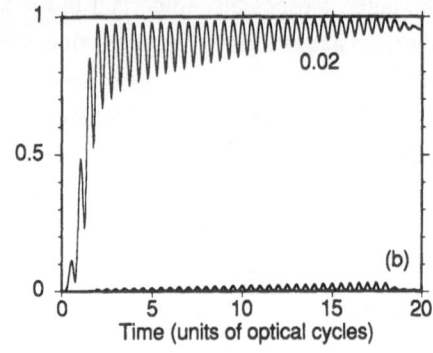

Figure 1. (a) The one-electron ionization probability (detachment probability) $P1(t)$ as a function of time for three values of the electric field strength $\mathcal{E}$. (b) The one-electron ionization probability $P1(t)$ (upper curve) and double ionization probability $P2(t)$ (lower curve) for electric field strength $\mathcal{E}=0.02$ a.u. ($\omega=0.08$ a.u., 2 optical cycle turn on and off).

ionization probability is negligible for all three laser intensities. The core electron can be regarded as basically frozen. It is clear that the detachment probability increases with increasing laser intensity throughout the regime of field strengths shown in Fig. 1a.

Fig. 1b shows the result after we have further doubled the laser field strength to $\mathcal{E}=0.02$ a.u. There is clearly evident a suppression of the detachment process. Compared with 100% for $\mathcal{E}=0.01$ a.u. after 8 cycles, we find now "only" an 80% probability of detachment and only 95% by the end of the pulse. Please note that the Kramers-Henneberger parameter $\alpha \equiv \mathcal{E}/\omega^2$ equals 3.1 for this highest laser intensity, and this agrees roughly with the atomic size and with the standard description of one-electron stabilization.[11]

The strong oscillations in the decay curves are another manifestation of stabilization. They occur with exactly twice the laser frequency and are evidence of the oscillatory motion of the electron in the field. Qualitatively similar behavior has been reported in the context of single electron system studies.[11] Finally, the second (lower) curve in Fig. 1b represents the two-electron ionization probability $P2(t)$. It is weakly oscillating with twice the laser frequency and is never greater than a few per cent. After the end of the pulse the probability that both electrons have ionized is quite negligible. We can conclude for our negative ion model that there exists an intensity window (around $\mathcal{E} \approx 0.02$ a.u.) for which the outer electron can become dynamically stabilized while the inner electron remains unaffected.

This is the first indication of dynamical stabilization in a multi-electron system for which the electron-electron correlation has been fully taken into account. To stress our finding further we show in Fig. 2a the spatial distribution of the two-electron wave packet $|\Psi(x_1,x_2,t)|^2$ after 13 optical cycles for $\mathcal{E}=0.01$ a.u. (top) and $\mathcal{E}=0.02$ a.u. (bottom). Two characteristic features of the distributions can be immediately read off the graphs. First, both distributions are concentrated along the coordinate lines ($x_1=0$ and $x_2=0$), which indicates that only one electron has been detached, that the double ionization probability is negligible, and that the core electron is rather passive. Second, for the weaker of the two intensities we find almost no probability close to the nucleus (at $x_1=x_2=0$) indicating rather complete decay of the bound state. The probability peak at the origin for the larger intensity manifests spatially the suppression of the bound state decay.

We conclude our review of the low-frequency regime of stabilization by showing in

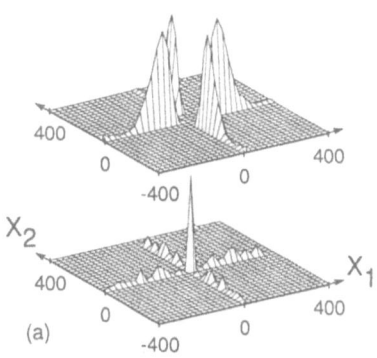
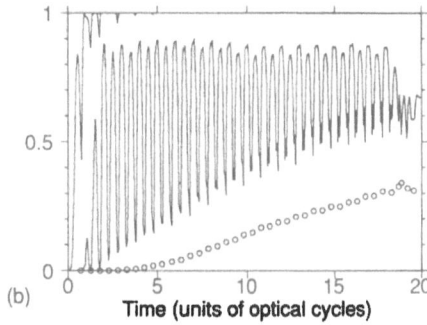

Figure 2. (a) The spatial distribution of the two-electron wave function $|\Psi(x_1,x_2,T)|^2$ after the interaction with the laser field for 13 optical cycles. $\mathcal{E}=0.01$ a.u (top), the absence of probability at the origin indicates the decay of the H⁻-bound state. $\mathcal{E}=0.02$ a.u (bottom), the peak at the origin corresponds to the H⁻-bound state. (b) The ionization probability for single ionization P1(t) (upper curve) and for double ionization P2(t) (lower curve) for electric field strength $\mathcal{E}=0.08$ a.u. The disconnected circles correspond to the ionization probability for the neutral hydrogen system. The wide oscillations evident every half-cycle also occur, but we have shown only the minima for graphical clarity ($\omega=0.08$ a.u., 2 optical cycle turn on and off).

Fig. 2b the single and double ionization probabilities for an even stronger field, $\mathcal{E}=0.08$ a.u. It is clear that stabilization of the outer electron cannot be sustained at such strong fields. In Fig. 2b the one-electron detachment probability reaches 100% after only 6 optical cycles, and the two-electron ionization probability reaches 50% by the end of the pulse. The core electron has become significantly active and it now prevents the stabilization of the outer electron. Given the existence of some kind of destabilizing e-e correlation, the question naturally arises whether the outer electron can produce an observable back reaction on the inner electron?

The answer to this question is yes, which is perhaps surprising given the weak binding of the outer electron and its speed of detachment for $\mathcal{E} = 0.08$ a.u. In order to measure the effect of the back reaction from e-e correlation on the core electron we simply compared the two-electron removal probability for H⁻ with the ionization probability for neutral H, i.e., we compared with the action of the same laser pulse on the same system except that the outer electron doesn't exist at all. The ionization probability for neutral H is represented by the open dots included in Fig. 2b. [For graphical clarity we have shown only the minima of the probability curve.] The conclusion is clear: the probability for *double* ionization of H⁻ is clearly approximately twice as high as the corresponding *single* ionization probability of neutral H. One might say that as a consequence of final state e-e collisions [correlation] the fully detached outer electron is still able to promote or assist *double* ionization. This stresses again the destabilizing effect of the e-e correlation on the negative ion.

4. SUPPRESSION OF DOUBLE IONIZATION AND HIGH-FREQUENCY STABILIZATION

In this section we will discuss a higher frequency regime ($\omega = 1$ a.u.) for which both electrons can ionize under the absorption of only 1 photon. We would expect that in this regime both electrons are equally active and any frozen core approximation would break down even for small laser intensities. In the section after this one we will present photoelectron spectra predicted for this high frequency regime.

In Figs. 3a-c we present the time dependence of the probability for one-photon single and double ionization for three different laser intensities. Single and double ionization occur now on roughly the same time scale. Fig. 3a shows that after 10 cycles P1(t) is roughly 90% and P2(t) is already 30%. This is for laser field strength $\mathcal{E}$=0.5 a.u.[α = 0.5], a case comparable to the two lower curves in Fig. 1a [$\alpha \approx 0.4$ and 0.8], where we found the strongly contrasting result that P1(t) can be greater than 50% while the inner electron is completely passive [P2(t) is negligible].

Fig. 3b shows that for the field strength of $\mathcal{E}$=1.0 a.u. [α = 1.0] one-electron detachment is already complete after the first few cycles of the turn on, comparable to the case shown in the upper curve of Fig. 1a [$\alpha \approx 1.6$]. The difference here is that at the end of the pulse the probability for double ionization is 82% instead of negligible.

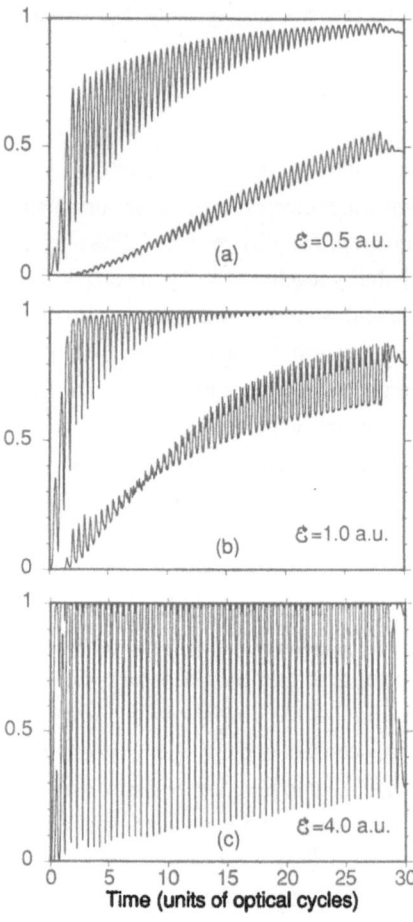

Figure 3. The ionization probabilities for single ionization (upper curve) and for double ionization (lower curve) for several electric field strengths. In all cases, ω=1.0 a.u. with 2 optical cycle turn on and off.

Evidence for the onset of core stabilization then appears for higher fields. Fig. 3c shows the results for $\mathcal{E} = 4.0$ a.u. [$\alpha = 4.0$]. A decrease in double ionization is clear. After the pulse has turned off P2(T) is only 30%, compared with the 82% shown in 3b. For the even stronger field $\mathcal{E} = 8.0$ a.u. the double ionization probability reduces to only 20% and at these superstrong fields we begin again to observe a small amount of two-electron stabilization.

The transition just described, from single to double ionization and the subsequent suppression of double ionization with increasing field strength is also apparent from the two-electron spatial probability distribution $|\Psi(x_1,x_2,t)|^2$. In Figs. 4a-c we show this distribution after the system has interacted with the field for 30 optical cycles. We have already mentioned that a high concentration of probability along the nodal lines, as shown in 4a, corresponds to predominant single ionization. In Fig. 4b the crosslike structure has almost disappeared, indicating a strong double ionization in agreement with Fig. 3b. Then for a super-strong field we find a reversal in behavior, as shown in Fig. 4c. Only one electron has irreversibly escaped and the probability along the nodal lines is much larger than for the weaker intensity of graph 4b. Note that the probability along the line of largest e-e repulsion ($x_1=x_2$) is very small in all distributions. This confirms that both electrons prefer to escape in opposite directions from the nucleus.

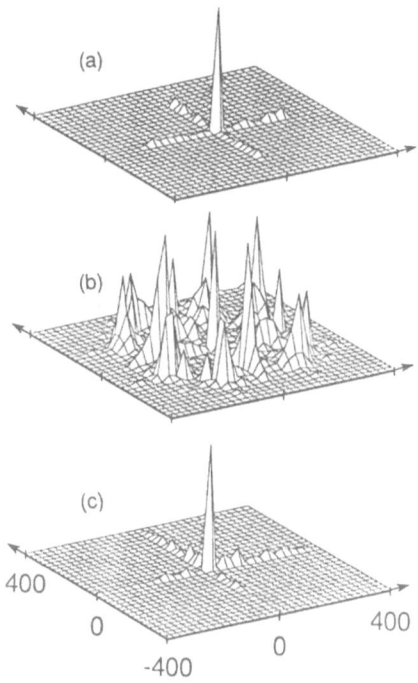

Figure 4. The spatial distribution of the two-electron wave function $|\Psi(x_1,x_2,T)|^2$ after the interaction with the laser field for 30 optical cycles. (a) $\mathcal{E}=0.5$ a.u., (b) $\mathcal{E}=1.0$ a.u., (c) $\mathcal{E}=4.0$ a.u. ($\omega=1.0$ a.u., 2 optical cycle turn on and off).

5. PHOTOELECTRON SPECTRA

We now review our results[1] for the photoelectron energy spectra. They show spectral structure, different ionization rates for different channels, and several independent ATI series.

In order to calculate the electron energies following the laser pulse the ground state contribution was removed from the final wave function and the remainder $\psi(x_1,x_2,T)$ was Fourier-decomposed into its momentum amplitude $\phi(k_1,k_2)$

$$\phi(k_1,k_2) = \iint dx_1\, dx_2 \exp[\, i\, (k_1x_1+k_2x_2)\,]\ \psi(x_1,x_2,T) \tag{5.1}$$

where k_1 and k_2 denote the momenta. An effective one-electron energy density $P(E)$ can be defined as the integral of $|\phi(k_1,k_2)|^2$ over one of the momenta, say k_2, and a subsequent transformation to energy via $E \equiv k_1^2/2$

$$P(E) = (2E)^{-1/2} \int dk_2\ |\phi(k_1,k_2)|^2. \tag{5.2}$$

To aid in interpretation we adopt an extreme independent-electron view of the system's energy levels, as in Fig. 5a. The weakly bound outer electron is assigned the binding energy given by the photodetachment threshold energy, $E = -0.06$, while the inner electron is assigned the binding energy of the core ground state, $E = -0.67$. This over-simplified picture predicts that photopeaks can arise either from the ionization of the outer electron or of the inner electron (following "channels" A and B in the figure).

In Fig. 5b the low-energy photoelectron spectrum $P(E)$ is presented for $\mathcal{E} = 0.5$ and $\omega = 1$ [here ω was deliberately chosen large enough to exceed the two-electron threshold]. According to the energy levels of Fig. 5a, and depending on whether the outer or inner electron is removed, there should be a peak at $1-0.06 = 0.94$ or $1-0.67 = 0.33$ in the photospectrum. As Fig. 5b shows both peaks occur (labeled A1 and B1) and they are the strongest peaks. In addition, the sum of their kinetic energies is $0.94 + 0.33 = 1.27$, and this equals $E_g + 2\omega = -0.73 + 2$, as it should.

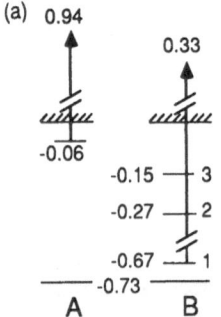
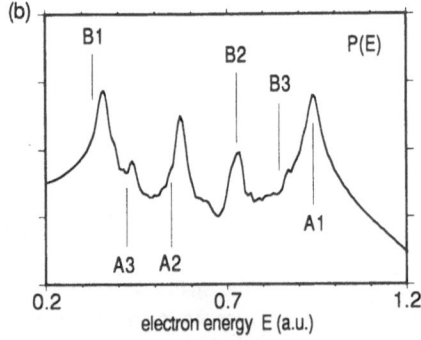

Figure 5. (a) An energy level diagram showing approximate independent-electron ionization channels for the weakly bound "outer" and strongly bound "inner" electron. (b) The one-electron photo-spectrum $P(E)$ for the laser pulse described in the text. The transition leading to the energies of the peaks A1 and B1 are indicated by the arrows in (a), and the smaller peaks have energies corresponding to rearrangements within the core.

In other words, peaks A1 and B1, taken together, could be interpreted as a simultaneous two-photon double-ionization process. Whether this is appropriate or whether the 2-electron process is step-wise cannot be answered by inspecting the long-time photo-spectrum.

However, one of the advantages of our method is that it allows the wave functions we have computed to be time-resolved as well as energy-resolved. We now show that time resolution allows one to see unambiguously that only one interpretation of peaks A1 and B1 can be supported. Let us write two possible stepwise routes to the same final doubly-ionized state $|E_{A1}; E_{B1}>$ as follows (the notation follows Fig.5):

$$|g> + \omega \rightarrow |E_{A1}; n=1> \quad \text{and} \quad |E_{A1}; n=1> + \omega \rightarrow |E_{A1}; E_{B1}> \tag{5.3}$$

$$|g> + \omega \rightarrow |n^*; E_{B1}> \quad \text{and} \quad |n^*; E_{B1}> + \omega \rightarrow |E_{A1}; E_{B1}> \tag{5.4}$$

In both of routes (5.3) and (5.4) there is a discrete intermediate state. In route (5.3) the outer electron is immediately detached while the inner electron "waits" in the core ground state n=1. In route (5.4) the outer electron "waits" in a high-lying Rydberg state n* while the inner electron is ejected immediately. The primary evidence is shown in Fig. 6a where snapshots are displayed of the photo-electron spectrum for 20, 40 and 60 cycle pulses. They indicate that the A peaks grow on a different time scale (more rapidly) than the B peaks.

The behavior in Fig. 6 can be compared with an analysis similar to that of Crance and Aymar.[12] From rate equations describing first route (5.3) and then route (5.4) one can compute the ratio of the population in the B1 peak to the population in the A1 peak. The corresponding predictions are shown against the computed data in Fig. 6b, clearly favoring a definite time ordering that is equivalent to step-wise ejection of the inner electron, and route (5.4) can be ruled out. On this basis it is clear that the satellite peaks A2, A3, etc., are one-

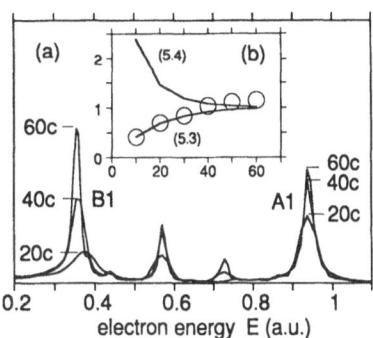

electron energy E (a.u.)

Figure 6. The same one-electron spectrum P(E) shown in 5b, but for three different laser pulse durations: T=20c, 40c, and 60c. Channel A peaks are prompt, growing rapidly initially, whereas channel B peaks start more slowly and are still growing rapidly between 40c and 60c. In the inset we show the predictions of rate equations appropriate to route (5.3) and (5.4). In both cases the transition rate coefficients were taken equal to the observed exponential ground state decay rate. The time dependence of the ratio of the areas of the peaks B1 and A1 in the numerical photoelectron spectra is indicated by circles. Route (5.3) is clearly preferred.

photon 2-electron shake-up peaks. That is, they arise from the detachment of the outer electron by one photon accompanied by shake-up of the inner electron into an excited core state. The satellite B peaks arise basically from subsequent ionization from these core excited states.

So far we have restricted our discussion to photopeak energies comparable to or smaller than the laser photon energy $\omega=1$ a.u. It is natural to ask whether above-threshold peaks are also predicted, and the answer is positive, as we have shown elsewhere.[1]

6. COHERENCE TRANSFER BY TWO-ELECTRON CORRELATION

In the present section we describe a new two-electron coherence-imprinting process that is related to the one-electron Autler-Townes effect.[13] This process exhibits unexpected features that are not found in the one-electron effect, or are the reverse of known one-electron features. In the photoelectron spectrum we find a new doublet in addition to the one predicted by Knight.[13,14] There is a novel dressed-state reversal, i.e., the *higher*-energy peak in the new doublet comes from the *lower*-energy dressed bound state. In addition, the ground (initial) state probability decays *without* Rabi oscillations.

For later reference, in Fig. 7 we have sketched again the bare energy levels for our negative ion, showing schematically what we believe to be a useful indication of the

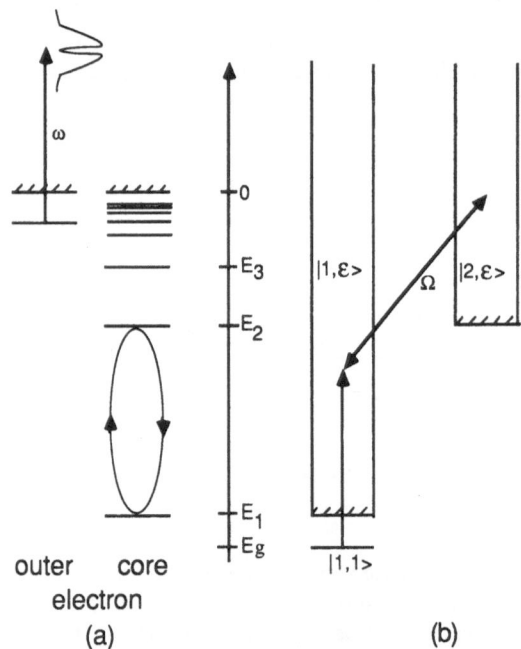

Figure 7. (a) An energy level diagram for the negative ion model based on an independent-electron picture for the weakly bound "outer" and strongly bound "inner" electron. (b) Total energy diagram, but only the ground state at $E_g = -0.73$ a.u. and the first two continua beginning at $E_1 = -0.67$ a.u. and $E_2 = -0.275$ a.u. have been displayed.

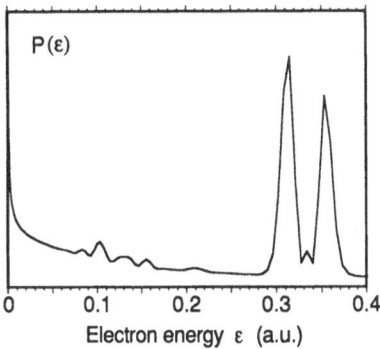

Figure 8. Photoelectron spectral structure created by a sine-squared incident laser pulse with $\mathcal{E}$ = 0.05 a.u., ω = 0.395 a.u., and T = 40 cycles. Details near to the higher split peak are shown again in Fig. 9b.

dynamics leading to our experimental data. On the left side of part (a) we show the energy levels that we associate with the weakly bound outer electron (photodetachment potential of 0.06 a.u.) and on the right side of part (a) we show the lowest several discrete levels of the core electron. In part (b) of Fig. 7 we give again the alternative level scheme of Fig. 5a. The continua above the first and second thresholds are associated with the core-electron in its ground state at E_1 = - 0.670 a.u. and its first excited state at E_2 = - 0.275 a.u. Between parts (a) and (b) we have drawn a vertical energy scale applicable to both parts. In the experiments described here we limited the laser field strength to $\mathcal{E}$ = 0.05 a.u. [laser intensity slightly less than 10^{14} W/cm^2], and scanned the frequency of the incident laser. The most interesting frequencies are in the neighborhood of the core electron's n=1 to n=2 resonance.

The resonant Rabi cycling indicated in Fig. 7a for the core electron suggests[14] that a doublet should be found in the core electron's photospectrum with energy ω above the n=2 state, i.e., with kinetic energy given by $\varepsilon = \omega + E_2 \approx (E_2 - E_1) + E_2 = 0.395 + (-0.275) \approx$ 0.120 a.u., and this is indeed the case. The splitting of the peaks should be equal to the resonant Rabi frequency and this is also the case.[15] Thus the familiar one-electron Autler-Townes effect is present. However, a much stronger doublet is found centered at a different higher energy, $\varepsilon \approx 0.335$ a.u. These two photoelectron doublets are shown in Fig. 8. The stronger and higher doublet has not previously been predicted.

In Fig. 9 we show three photoelectron spectra. The typical case finds a single photopeak and its energy can be seen to be $\varepsilon \approx \omega$ - 0.06. Since 0.06 a.u. is exactly the binding energy of the *outer* electron it is compelling to conclude that this is simply the normal photopeak associated with detachment of the outer electron. The exceptional case is the double peak in Fig. 9b. The question is how to understand both its location and its splitting since the outer electron has no bound-bound resonances at all [and no auto-ionizing resonances in the spectral region involved here].

We propose to interpret the peak splitting common to Figs. 8 and 9b as arising from *two-electron coherence transfer* from the core electron to the outer electron when the core electron is resonantly excited. We can picture the transfer mechanism as arising from the periodic change of the outer electron's binding potential due to the Rabi oscillations of the core electron. Or we can say that the core-electron's resonant transitions periodically change

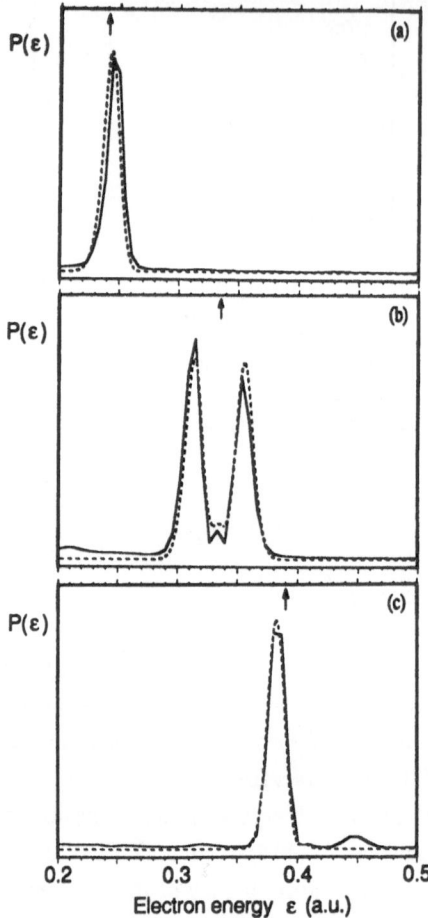

Figure 9. Three photoelectron spectra. The arrow indicates the kinetic energy predicted for outer electron photodetachment: $\varepsilon = -0.06$ a.u.$+\omega$. The dashed curves are the analytical predictions. The incident laser frequencies are: (a) $\omega = 0.30$ a.u.; (b) $\omega = 0.395$ a.u. $\approx E_{21}$; and (c) $\omega = 0.45$ a.u. Note that Fig. 9b is identical with the upper part of Fig. 8.

the type of continuum which characterizes the photodetachment state. In any event it is remarkable that the coherence transfer is as strong as it is.

We have found that a relatively simple theoretical model can be developed to support our interpretation of two-electron coherence transfer. Our model can be solved analytically and could be applied to any two-electron system. It accounts for the spectral phenomena of Fig. 9 as well as associated temporal phenomena.[4] The resulting predictions for photoelectron
spectra are shown as the dotted lines in Fig. 9. The evident agreement between dotted and solid curves is striking.[16] In addition, we note that *all atomic parameters were determined once and for all by our hydrogen negative ion model potential V(x)*. That is, the numerical values of the atomic energies E_g, E_1, E_2, the dipole matrix element $<1|x|2>$ and the ground state decay rate are not adjustable in any way to improve the fit of the curves in Fig. 9.

We have also studied the time-dependent effects associated with the peak splitting shown in Figs. 8 and 9. Again in agreement with our analytical predictions[4], our exact wave

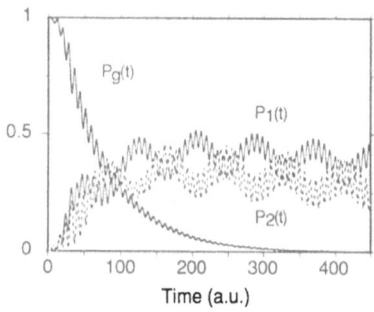

Figure 10. The time dependence of the ground state probability $P_g(t) = |<g|\Psi(x_1, x_2, t)>|^2$, and the probabilities $P_1(t)$ and $P_2(t)$ to find at least one electron in the lowest two hydrogenic states $\psi_1(x)$ and $\psi_2(x)$ (const. intensity laser pulse with $\mathcal{E}=0.075$ a.u., two cycle linear turn on, $\omega=0.395$ a.u., the Rabi period $2\pi/\Omega \approx 80$ a.u. corresponds to ≈ 6 optical cycles).

functions show strong Rabi oscillations of the n=1 and n=2 core state probabilities and it is easy to check that the period of these Rabi oscillations is consistent with the numerical Rabi frequency mandated by the negative ion potential: $2\pi/\Omega \approx 80$ a.u. for $\omega = 0.395$ a.u. [see Fig. 10] However, in contrast to the one-electron case[14], we find no Rabi oscillations in the decay of the negative ion ground state under the condition of exact core resonance. The only oscillations present in the ground state decay curve occur at twice the laser frequency and can be associated with counterrotating terms.

7. SUMMARY AND CONCLUSIONS

We have reviewed several results of wave function calculations[1-4] for a two-electron system interacting with strong short laser pulses. The wave functions are exact numerical solutions of the two-electron Schrödinger equation, fully symmetrized and correlated.

We have found that for low-frequency excitation, with laser photon energies corresponding to one-photon detachment of the outer electron but many-photon ionization of the core electron, one-electron removal can be suppressed and the outer electron can be stabilized in a small range of laser intensities.[2,3] At higher intensities the core electron becomes active and can promote detachment of the outer electron, even though 100% detached. The double ionization probability of the negative ion in this situation can be twice as great as the ionization probability of the neutral atom. This gives, in principle, the possibility of investigating stabilization by detecting an enhanced double ionization for intensities at which one would usually expect only single ionization. In the regime of higher frequencies, where one photon is sufficient for double ionization, we have shown that a strong suppression of double ionization can be induced, i.e., one electron can be stabilized as is typical for single-electron systems.

Time- and energy-resolved photoelectron spectra have indicated[1] that different ionization channels can be associated with different ionization rates, which allows one to distinguish between direct and stepwise ionization routes for the inner electron.

A further exploration of the time-dependent two-electron wave functions predicts the existence of a previously unreported two-electron coherence effect.[4] This effect has a number of unexpected features. There is peak splitting in the photodetachment spectrum of the outer electron, which has no corresponding bound-bound resonance. The ground state decays without resonant Rabi oscillations. The outer electron's novel properties arise from a core electron resonance. In particular, the Rabi frequency of the core resonance agrees with the outer electron's photopeak splitting.

These observations are the first, to the best of our knowledge, of induced core electron coherence in the presence of an outer electron, i.e., core coherence that imprints itself on outer electron dynamics. On the basis of elementary similarities between the structure of our artificial atom and that of alkali negative ions generally we suggest that a direct laboratory observation of this core electron coherence might be carried out with Na^-. The Na^- core resonance (sodium D_1 or D_2 line) has an oscillator strength greater than unity, so dye laser intensities in the 10^{13} W/cm^2 range should be more than sufficient to split the predicted strong doublet by 10 meV, which is comparable to resolution currently reported for ATI photoelectron studies

Acknowledgements

This research was supported by the National Science Foundation through PHY 92-00542 and PHY 89-20108. RG acknowledges partial support through a Feodor Lynen Fellowship from the Alexander von Humboldt Foundation of Germany and a Research Associateship in Computer Science and Engineering from the National Science Foundation. We also acknowledge assistance with computing resources from the Pittsburgh Supercomputing Center and the Allied-Signal Foundation.

REFERENCES

1. R. Grobe and J.H. Eberly, *Phys. Rev. Lett.* 68, 2905 (1992).

2. R. Grobe and J.H. Eberly, *Phys. Rev. A* 47, RXXXX (1993)

3. R. Grobe and J.H. Eberly, *Las. Phys.* (in press, 1993).

4. R. Grobe and J.H. Eberly, *Phys. Rev. A* (submitted,1992).

5. For an investigation of the one-dimensional helium system see M.S. Pindzola, D.C. Griffin and C. Bottcher, *Phys. Rev. Lett.* 66, 2305 (1991).

6. R. Grobe and J.H. Eberly, *Phys. Rev. A* (in preparation).

7. For recent overviews of the status of negative ions and multi-electron atoms in laser fields, see P.A. Golovinskii and I.Yu. Kiyan, *Sov. Phys. Uspekhi* 33, 453 (1990), and C.A. Nicolaides, Th. Mercouris and G. Aspromallis, *J. Opt. Soc. Am. B* 7, 494 (1990).

8. For structural properties of the one-dimensional hydrogen atom see J. Javanainen, J.H. Eberly and Q. Su, *Phys. Rev. A* 38, 3430 (1988); J.H. Eberly, *Phys. Rev. A* 42, 5750 (1990); and Q. Su and J.H. Eberly, *Phys. Rev. A* 44, 5997 (1991).

9. R. Grobe and M.V. Fedorov, *Phys. Rev. Lett.* 68, 2592 (1992) and R. Grobe and M.V. Fedorov, *J. Phys. B* (in press, 1993).

10.   The time-dependent Schrödinger equation has been solved with a split-operator FFT method which has been discussed in detail in M.D. Feit, J.A. Fleck, Jr. and A. Steiger, *J. Compt. Phys.* 47, 412 (1982).

11.   Q. Su, J.H. Eberly and J. Javanainen, *Phys. Rev. Lett.* 64, 862 (1990); Q. Su and J.H. Eberly, *J. Opt. Soc. Am. B* 7, 564 (1990); J.H. Eberly and Q. Su, in "*Multiphoton Processes*" edited by G. Mainfray and P. Agostini (CEA Press, Paris, 1991), p. 71; Q. Su and J.H. Eberly, Phys. Rev. A 43, 2474 (1991); and J.H. Eberly, R. Grobe, C.K. Law and Q. Su, in "*Atoms in Intense Radiation Fields*", edited by M. Gavrila (Academic, Orlando, 1992).

12.   See, for example, the analysis of helium with the aid of rate equations by M. Crance and M. Aymar, *J. Physique* 46, 1887 (1985), and the perturbative analyses in X. Tang and P. Lambropoulos, *Phys. Rev. Lett.* 58, 108 (1987), and A. L'Huillier and G. Wendin, *Phys. Rev. A* 36, 5632 (1987).

13.   S.L. Autler and C.H. Townes, *Phys. Rev.* 100, 703 (1955). See also P.L. Knight and P.W. Milonni, *Phys. Rep.* 66, 21 (1980), and P.L. Knight, M.A. Lauder and B.J. Dalton, *Phys. Rep.* 190, 1 (1991).

14.   P.L. Knight, *Opt. Comm.* 22, 172 (1977) and *J. Phys. B* 11, L511 (1978).

15.   The multi-peak structure in the lower-energy "normal" doublet arises because the interaction is pulsed rather than cw. This effect was predicted for autoionization by K. Rzazewski, *Phys. Rev. A* 28, 2565 (1983) and for resonant multiphoton ionization by D. Rogus and M. Lewenstein, *J. Phys. B* 19, 3051 (1986).

16.   On the right of the main peak in Fig. 9c there is a small secondary peak in the "experimental" data. This is not related to the peak splitting of Fig. 9b but has a second interesting origin. It arises when two-photon absorption is accompanied by core rearrangement, which shifts the normal photodetachment peak from $\varepsilon = \omega - 0.06 \approx 0.39$ to $\varepsilon = 2\omega - 0.06 - E_{21} \approx 0.90 - 0.06 - 0.395 \approx 0.445$, in agreement with the figure. Rearrangement into core level 3 is also possible, in which case E_{21} must be replaced by $E_{31} = -0.52$, giving a photoelectron energy $\varepsilon \approx 0.32$, and this peak is even smaller but still evident in Fig. 9c. The effects of core rearrangement on photodetachment spectra have been discussed before in ref. [1].

H⁻ IN INTENSE LASER FIELDS: LASER - INDUCED

EXCITED STATES AND DICHOTOMY

M. Gavrila

F.O.M. Institute for Atomic and Molecular Physics
Amsterdam, The Netherlands
and
Institute for Theoretical Atomic and Molecular Physics
Harvard-Smithsonian Center for Astrophysics
Cambridge, MA 02138, USA

I. Introduction

Substantial progress has been achieved in recent years in the theoretical understanding of the nonperturbative interaction of one-electron atoms (e.g., hydrogen) with intense laser fields. (For an overall survey, see the volume quoted in Ref.1.) In contrast, the behavior of two- and several-electron atoms under these circumstances is at an incipient stage of study. The general case, when nonperturbative field effects are comparable in magnitude with electron correlation, is a formidable problem indeed, for which best suited theoretical approaches and computational methods still need to be worked out.

In this article we extend to the two-electron case the *high-frequency Floquet theory (HFFT)* of laser-atom interactions, we have developed earlier for one-electron atoms. (A review of the one-electron case is contained in Ref.2.) This represents a stationary approach to the interaction problem, describing the steady atomic ionization occuring in constant intensity fields. With certain limitations it is applicable to laser pulses as well. The theory is valid at sufficiently high frequencies. However, as at high intensities all frequencies become sufficiently high to insure the validity of the theory (see Ref.2), the HFFT is particularly well adapted for the high intensity regime.

The two-electron HFFT is then applied to the case of H$^-$. This choice is motivated by the fact that H$^-$ is a system well suited to be treated by HFFT, it contains electron correlation to highest degree, and it is of experimental interest.

Field-free H$^-$ is a loosely bound system, with the small electron affinity (binding energy of the second electron to the hydrogen atom) of $J = 0.751$ eV $= 0.0227$ a.u. The quantum numbers of the ground state are: total angular momentum $L = 0$, total spin $S = 0$, even parity. A highly correlated eigenfunction is needed to get a good result for the energy, and much effort has been made to achieve this by Pekeris and others[3]. H$^-$ has the interesting feature that only one bound state (namely the ground state) has been observed[4].

The one- and multiphoton ionization of H$^-$ of in weak fields has been studied extensively by lowest order perturbation theory. A recent survey of these calculations was given by Geltman[5]. Nonperturbative, Floquet-type calculations at low frequencies and rather low intensities (10^{10} - 10^{11} W / cm^2) have been done by Mercouris and Nicolaides[6], and by Dörr, Potvliege, Proulx and Shakeshaft[7]. At higher intensities Mittleman[8], using concepts of the HFFT, has explored the dichotomy regime and has shown that under these circumstances the electron affinity of the H atom equals its binding energy. Experiments have been carried out since 1989 by the Los Alamos group (Bryant, Smith, Tang and others; e.g., see Ref.9), with relativistic H$^-$ beams obliquely crossed by laser beams to study multiphoton ionization at intensities around 10^{10} W / cm^2.

In Section II we develop the extension of the HFFT to two electrons. (The case of one electron was discussed in detail in Ref.2.) The basic equations for solving the structure problem (energy eigenfunctions and eigenvalues) are given, as well as formulas for the ionization rates. However, we shall be interested here only in the structure problem of H$^-$. In the high-frequency regime this depends on the field properties (frequency and intensity) via one characteristic parameter α_0. We consider first, in Section III, the limiting case when α_0 is large (large intensities). In this limit the two-electron atom undergoes dichotomy (splits into two), just as its one-electron counterpart. We also show that it has an infinite set of "light-induced excited states", some of them doubly excited. These results have been obtained in collaboration with J. Shertzer[10]. In Section IV we report on a calculation with H.G. Muller[11] on the structure of H⁻ at intermediate values of α_0. This is a fully correlated computation of the configuration-interaction type. We follow the α_0 - dependence of the ground state, from perturbative α_0 to the large-α_0 limit, and study the appearance and evolution with α_0 of the light-induced excited states. (For a photon of energy of 5.0 eV, the first state appears at intensity $4.9 \ 10^{14}$ W / cm².) We illustrate the importance of electron correlation, and stress the inadequacy of one-electron models in this regime.

II. High-frequency Floquet theory for two electrons

We start from the familiar two-electron equation in the laboratory reference frame and the momentum gauge

$$[\frac{1}{2m}(\mathbf{P}_1 - \frac{e}{c}\mathbf{A}_1)^2 + \frac{1}{2m}(\mathbf{P}_2 - \frac{e}{c}\mathbf{A}_2)^2 - \frac{Z e^2}{r_1} - \frac{Z e^2}{r_2} + \frac{e^2}{r_{12}}]\Phi_L = i\,\hbar\,\frac{\partial \Phi_L}{\partial t},$$
(1)

where we do not specify the nuclear charge Z. By assuming the dipole approximation, $\mathbf{A}_1$ and $\mathbf{A}_2$ become equal to a function of t only: $\mathbf{A}_1 = \mathbf{A}_2 = \mathbf{A}(t)$. The $\mathbf{A}^2$ terms can now be removed as in the one electron case. Consequently, the continuum of the atomic problem will start at W = 0 independently of the field intensity (i.e., will not shift ponderomotively). In the following we shall be dealing with *linear polarization*. Hence, we can take: $\mathbf{A}(t) = a\,\mathbf{e}\,\sin \omega\,t$, where **e** is the polarization vector[12]. By performing the space translation of vector

$$\alpha(t) = -\frac{e}{mc}\int^t \mathbf{A}(t')\,dt' = \alpha_0\,\mathbf{e}\,\cos \omega t \quad , \quad \alpha_0 = -(ea/mc\omega) ,$$
(2)

on both electrons (i.e., by moving over to an oscillating reference frame), we obtain the space-translated Schrödinger equation[13]:

$$[\frac{1}{2m}\mathbf{P}_1^2 + \frac{1}{2m}\mathbf{P}_2^2 - \frac{Ze^2}{|r_1+\alpha(t)|} - \frac{Ze^2}{|r_2+\alpha(t)|} + \frac{e^2}{|r_1-r_2|}]\Psi = i\,\hbar\,\frac{\partial\Psi}{\partial t}.$$
(3)

Note that the inter-electron repulsion term is not affected by the translation.

Although we have neglected spin dynamically, we need take it into account kinematically to satisfy the Pauli principle. This implies that the position wave functions of interest $\Psi(\mathbf{r}_1, \mathbf{r}_2, t)$ should be either symmetric at the interchange of $\mathbf{r}_1$ and $\mathbf{r}_2$ for the spin singlet case (S = 0), or antisymmetric for the spin triplet (S = 1) case.

We adopt the *stationary approach* for solving Eq.(3), and seek the resonance states of the system (see Ref.2, Sec.III). These are characterized by the Floquet expansion

$$\Psi(\mathbf{r_1},\mathbf{r_2},t) = e^{-i(Et/\hbar)} \sum_n \Phi_n(\mathbf{r_1}, \mathbf{r_2}) e^{-in\omega t} \quad , \tag{4}$$

where E is the complex initial energy of the ionizing atom, and by appropriate boundary conditions to be discussed shortly. Inserting Eq.(4) and the Fourier expansion of the oscillating potential

$$-\frac{Ze^2}{|\mathbf{r_j} + \alpha_0 \, e \cos\omega t|} = \sum_n V_n(\alpha_0,\mathbf{r_j}) e^{-in\omega t} \, , \quad (j = 1, 2) \tag{5}$$

into Eq.(3), we obtain the system of coupled equations for the Floquet components Φ_n :

$$[H(1,2) - (E + n\hbar\omega)]\Phi_n = -\sum_m{}' [V_{n-m}(\mathbf{r_1}) + V_{n-m}(\mathbf{r_2})]\Phi_m \, , \tag{6}$$

containing the two-electron Hamiltonian

$$H(1,2) = \frac{1}{2m}[\mathbf{P}_1^2 + \mathbf{P}_2^2] + V_0(\alpha_0 ; \mathbf{r_1}) + V_0(\alpha_0 ; \mathbf{r_2}) + \frac{e^2}{r_{12}} \, . \tag{7}$$

In Eq.(6), and in the following, a primed sum $\sum'$ indicates that the term containing V_0 should be omitted from the sum. Eq.(5) yields

$$V_n(\alpha_0 ; \mathbf{r_j}) = -\frac{1}{2\pi} \int_0^{2\pi} \frac{Ze^2}{|\mathbf{r_j} + \alpha_0 \, e \cos\chi|} e^{in\chi} d\chi \, . \tag{8}$$

We now develop the *high-frequency theory* for solving the Floquet system of coupled equations, Eq.(6). (We adopt here a somewhat different approach from that in Ref.2.) The presence of the large $n\hbar\omega$ term in all equations, except the one for n = 0, indicates that Φ_0 is the dominant component at large ω and fixed α_0. Thus, to *lowest order in* $1/\omega$, Eq.(6) reduces to

$$H(1,2) \Phi_0 = W \Phi_0 \, . \tag{9}$$

The solution of Eq.(9) will be denoted by $\Phi_0^{(0)}$ to emphasize that it is the lowest order in $1/\omega$ approximation to Φ_0 . Eq.(9) determines the structure of the atomic system in the high-frequency limit. Its eigenvalues W are real and depend only on α_0 . In place of the original Coulomb electron-nucleus attraction, H(1,2) contains the "dressed potential" $V_0(\alpha_0 ; \mathbf{r})$ given by Eq.(8), which is the time average of the oscillating potential $(-Z e^2 / |\mathbf{r} + \alpha(t)|)$, see

Eq.(5). V_0 can also be viewed as the potential generated by an electrostatic "line of charge", (non-uniformly) distributed from $-\alpha_0$ to $+\alpha_0$ on the axis directed along $\mathbf{e}$ and passing through the origin. For linear polarization, V_0 can be expressed as

$$V_0(\alpha_0, \mathbf{r}) = - (2Ze^2 / \pi) (r_+ r_-)^{-1/2} K\left[2^{-1/2} (1 - \hat{\mathbf{r}}_+ \cdot \hat{\mathbf{r}}_-)^{1/2} \right], \tag{10}$$

where $\mathbf{r}_\pm = \mathbf{r} \pm \alpha_0 \mathbf{e}$, and K is the complete elliptic integral of the first kind. A discussion of its properties was given in Ref.1, Sec.V. A.

We now pass to the *next approximation* within the theory. This is obtained by treating the Φ_n terms on the right-hand-side (rhs) of Eq.(6) as corrections to the $\Phi_0^{(0)}$ term obtained from the previous approximation [all $V_n(\alpha_0; \mathbf{r})$ are of order 1 with respect to $1/\omega$]. This yields

$$[H(1,2) - E]\, \Phi_0 = - \sum_m{}' [V_{-m}(\mathbf{r}_1) + V_{-m}(\mathbf{r}_2)]\, \Phi_m(\mathbf{r}_1, \mathbf{r}_2), \tag{11}$$

$$[H(1,2) - (E + n\hbar\omega)]\, \Phi_n = - [V_n(\mathbf{r}_1) + V_n(\mathbf{r}_2)]\, \Phi_0^{(0)}(\mathbf{r}_1, \mathbf{r}_2). \quad (n \neq 0) \tag{12}$$

Eqs.(11) and (12) are to be solved with boundary conditions appropriate for ionization. Those for single-electron ionization should reflect the fact that the ejected electron ($r_1 \to \infty$) has absorbed n photons ($n \geq N$, where N is the minimum needed to go to the continuum), whereas the electron left behind (r_2 finite) can remain attached in an arbitrary state α of the residual atom (ion). (We are not considering here the case of multiphoton double ionization, characterized by $r_1 \to \infty$ and $r_2 \to \infty$.) We shall assume that also the residual atom (ion) can be described by HFFT, i.e. is represented by the Hamiltonian

$$H(2) = \frac{1}{2m}\, P_2^2 + V_0(\alpha_0 ; \mathbf{r}_2). \tag{13}$$

This has stationary eigenstates given by the equation

$$H(2)\, u_\alpha(\mathbf{r}_2) = W_\alpha\, u_\alpha(\mathbf{r}_2). \tag{14}$$

Under these circumstances the boundary condition is

$$\Phi_n(\mathbf{r}_1, \mathbf{r}_2) \xrightarrow[r_1 \to \infty]{} S_\alpha\, u_\alpha(\mathbf{r}_2)\, [e^{i (k_{n\alpha} r_1 - \gamma_{n\alpha} \ln 2k_{n\alpha} r_1)} / r_1]\, f_{n\alpha}(\hat{\mathbf{r}}_1); \tag{15}$$

here $\gamma_{n\alpha} = - [(Z - 1) e^2 m / \hbar^2 k_{n\alpha}]$. The expression for $\gamma_{n\alpha}$ reflects the fact that when $Z \geq 2$ the electron feels asymptotically a Coulomb field of charge $Z - 1$, whereas for $Z = 1$ (H$^-$) the asymptotic field is short range. Note that, because of the asymmetry of the situation, the rhs of Eq.(15) is asymmetric, although Φ_n is symmetric. Conservation of energy requires that

$$E + n \hbar \omega = W_\alpha + (\hbar^2 k_{n\alpha}^2 / 2m) . \qquad (E < W_\alpha < 0) \qquad (16)$$

(The imaginary part of E is ignored, since, according to the fundamentals of resonance theory, see Ref.2, Sec.III B, this should be infinitesimal.)

$f_{n\alpha}(\hat{r})$ in Eq.(15) is the ionization amplitude for absorption of n - photons with the one-electron atom being left in state α. The corresponding differential rate is given by (see Ref.2, Sec.III.B, and p.466):

$$\frac{d\Gamma_{n\alpha}}{d\Omega} = \frac{\hbar^2 k_{n\alpha}}{m} \mid f_{n\alpha} (\hat{r}) \mid^2 , \qquad (17)$$

if we assume that $u_\alpha(r_2)$ in Eq.(15) is normalized to 1.

Eq.(12) is an inhomogeneous equation for Φ_n, with a known rhs, and we solve it by the Green's function method (see Ref.2, Sec.IV for the one-electron case). The Green's function associated with the two-electron Hamiltonian H(1,2) is defined by

$$[E - H(r_1, r_2)] \, G(E; r_1,r_2; r'_1,r'_2) = \delta(r_1 - r'_1) \, \delta(r_2 - r'_2) , \qquad (18)$$

and the requirement that it vanish sufficiently rapidly when $r_1 \to 0$ or $r_2 \to 0$, and when $r_1 \to \infty$ or $r_2 \to \infty$. It can be shown that for $r_1 \to \infty$ and r_2 finite, the Green's function for $E + i\varepsilon$ ($E > 0$; $\varepsilon > 0$, infinitesimal) has the asymptotic behavior

$$G^{(+)}(E; r_1,r_2; r'_1,r'_2) \to$$

$$- \frac{1}{2\pi} \, S_\alpha \, u_\alpha(r_2) \, [e^{\, i \, (k_\alpha r_1 - \gamma_\alpha \ln 2k_\alpha r_1)} / r_1] \, [\phi^{(-)}_{\alpha k_\alpha}(r_1', r_2')]^* , \qquad (19)$$

where $E = (\hbar^2 k_\alpha^2 / 2m) + W_\alpha$, $\gamma_\alpha = - [(Z - 1) e^2 m / \hbar^2 k_\alpha]$, and $\phi^{(-)}_{\alpha k_\alpha}$ is the continuum eigenfunction of H(1,2), Eq.(7), for energy E, defined by the asymptotic behavior

$$\phi^{(-)}_{\alpha k_\alpha}(r_1, r_2) \to$$

$$\qquad (20)$$

$$u_\alpha(r_2) \{ e^{\, i \, [\, k_\alpha r_1 - \gamma_\alpha \ln (k \, r_1 + k_\alpha r_1)]} + f(\hat{r}_1) \, (e^{\, -i \, [k_\alpha r_1 - \gamma_\alpha \ln (k_\alpha r_1 + k_\alpha r_1)]} / r_1) \}.$$

Eq.(20) contains the residual atom eigenfunctions Eq.(14), and the asymptotic form of the Coulomb scattering wave function $u_k^{(-)}$ for charge (Z - 1).

The Green's function defines an integral operator $G(E)$ in position space via

$$< r_1', r_2' \mid G(E) \mid r_1, r_2 > \ = \ G(E ; r_1, r_2; r_1', r_2') \ . \tag{21}$$

Using Eq.(18), a special solution of Eq.(12) is

$$\Phi_n(r_1,r_2) \ =$$
$$\iint G(E + n \hbar \omega; r_1,r_2; r_1',r_2') \ [V_n(r_1') + V_n(r_2')] \ \Phi_0^{(0)}(r_1',r_2') \ dr_1' dr_2' \ . \tag{22}$$

With Eq.(19), the asymptotic behavior of this solution is

$$\Phi_n(r_1, r_2) \ \underset{r_1 \to \infty}{\to} \ \mathbf{S}_\alpha \ u_\alpha(r_2) \ [e^{\, i \, (k_{n\alpha} \, r_1 - \gamma_{n\alpha} \, \ln 2 k_{n\alpha} \, r_1)} / r_1] \ f_{n\alpha}(\hat{r}_1) \ , \tag{23}$$

where

$$f_{n\alpha}(\hat{r}) \ =$$
$$- \frac{1}{2\pi} \ \iint [\phi_{\alpha, \, k_{n\alpha}}^{(-)}(r_1,r_2)]^{*} \ [V_n(r_1) + V_n(r_2)] \ \Phi_0^{(0)}(r_1,r_2) \ dr_1 dr_2 \ , \tag{24}$$

is the n - photon ionization amplitude. Here $k_{n\alpha} = k_{n\alpha} \, \hat{r}$, and $k_{n\alpha}$ is given by Eq.(16). As apparent from Eq.(23), Φ_n has the desired asymptotic form Eq.(15), and is therefore is the solution we are seeking.

With the operator notation introduced in Eq.(21), Eqs.(22) and (24) can be written in more compact form as

$$\mid \Phi_n > \ = \ G(E + n \hbar \omega) \ [V_n(1) + V_n(2)] \mid \Phi_0^{(0)} > \ . \tag{25}$$

and

$$f_{n\alpha}(\hat{r}) = \ - \frac{1}{\pi} \ < \phi_{\alpha, k_{n\alpha}}^{(-)} \mid V_n \mid \Phi_0^{(0)} > \ ; \tag{26}$$

in the latter equation we have also taken advantage of symmetry.

Entering Eq.(25) into Eq.(11) we find

$$[H(1,2) - E] \mid \Phi_0 >$$
$$= - \sum_n{}' \ [V_{-n}(1) + V_{-n}(2)] \ G(E + n \hbar \omega) \ [V_n(1)+V_n(2)] \mid \Phi_0^{(0)} > \ . \tag{27}$$

Following an argument similar to that for the one-electron case (Ref.2, p.465), the operator $G(E + n \hbar \omega)$ containing the complex E can be replaced by $G^{(+)}(W + n \hbar \omega)$ containing the real W resulting from the lowest order approximation, Eq.(9). Moreover, treating the rhs of Eq.(27) as a small perturbation on the lhs, yields the correction ΔE to W:

$$\Delta E = \sum_n{}' < \Phi_0^{(0)} \mid [V_{-n}(1)+V_{-n}(2)] \, G^{(+)}(W + n \hbar \omega) \, [V_n(1)+V_n(2)] \mid \Phi_0^{(0)} > . \tag{28}$$

Because of symmetry this can be written further as

$$\Delta E = 4 \sum_n{}' < \Phi_0^{(0)} \mid V_{-n} \, G^{(+)}(W + n \hbar \omega) V_n \mid \Phi_0^{(0)} > . \tag{29}$$

ΔE is complex. By defining as usual $\Delta E = \Delta W - i \, (\, \Gamma / 2 \,)$, and proceeding as for the one-electron case (see Ref.2, Sec.IV C), we can obtain from Eq.(29) the expression for the level shift ΔW, and the total ionization rate Γ.

To conclude the presentation of the HFFT for two electrons, we give the expression in the lab frame of our approximate solution for the oscillating frame. As Eq.(4) reduces to,

$$\Psi(r_1, r_2, t) \cong \Phi_0^{(0)}(r_1, r_2) \, e^{-i \, (E \, t / \hbar)} , \tag{30}$$

by performing the translation back to the lab frame we find

$$\Phi_L(r_1, r_2, t) \cong \Phi_0^{(0)}(r_1 - \alpha(t), r_2 - \alpha(t)) \, e^{-i \, (E \, t / \hbar)} . \tag{31}$$

Thus, in this approximation, $\Phi_L(r_1, r_2, t)$ oscillates harmonically with amplitude α_0, and so do the probability densities derived from it.

We now briefly address the question of the validity of the HFFT, i.e. the conditions under which the scheme of successive approximations, presented here to the first two lowest orders, can be expected to converge in some pragmatic sense. This issue was discussed for the one-electron case in some detail in Ref.2, Sec.IV.D. For linear polarization the conclusion was that the photon energy ω should be large with respect to the average excitation energy of the lowest lying state of the manifold to which the initial state belongs; this energy is in general of the order of magnitude of the binding energy of that lowest state. The previous consideration applies to the atom *in the field*, characterized by parameter α_0. The same arguments can be extended to the two-electron case, to obtain as sufficient condition for the validity of the HFFT that ω be large with respect to the total binding energy $|E(\alpha_0)|$ of the system at the value of α_0 considered. Note that intense fields correspond to large α_0, and $|E(\alpha_0)|$ is then substantially reduced from its original, unperturbed value (see

Sections III and IV). The condition can thus be met at relatively low frequencies, even in the infrared.[14]

Further in this article we are interested only in the structure of H^- in intense fields, given by Eq.(9), leaving the calculation of the ionization rates for a later time. First we shall make some comments on the *symmetry manifolds* of the solutions of Eq.(9).

Since $V_0(\alpha_0; r)$ is axially symmetric around the axis defined by e, which we take along the z axis, the projection of the total angular momentum on it, $L_z = L_{1z} + L_{2z}$, is conserved. The associated magnetic quantum number is denoted by m. Note that, because of the time-reversal invariance of $H(1,2)$, the eigenvalues of Eq.(9) depend in fact only on $|m|$, and, hence, there is a twofold degeneracy for $m \neq 0$. The fact that $V_0(\alpha_0; r)$ is an even function implies that parity $P = 0, 1$ is conserved; $P = 0$ corresponds to even (gerade), and $P = 1$ to odd (ungerade) states.

The operator $H(1,2)$ being symmetric in the variables of the two electrons, admits a complete set of symmetrized (symmetric or antisymmetric) wave functions in r_1 and r_2. The symmetric wave functions can be combined with the spin $S = 0$ eigenstate, and the antisymmetric wave functions with the spin $S = 1$ eigenstates, to give fully antisymmetric position-spin set. Hence, the possibility of classifying the states by the total spin $S = 0, 1$.

We can summarize the classification of the manifolds by the notation $^{2S+1}\Lambda^P$, where $\Lambda = |m|$, and of the eigenvalues by $^{2S+1}W^P_{\Lambda j}$, where j is a labeling index. This classification is akin to that for homonuclear diatomic molecules.

In the next two Sections we shall consider the solution of the eigenvalue problem Eq.(9) for H^- (Z=1) in different ranges of values of α_0: first in the large-α_0 limit, where electron correlation does not play a dynamical role, and second, at small and intermediate α_0 values, where electron correlation is essential.

III. Structure of H^- at large - α_0. Dichotomy

At high frequencies, large α_0 implies high intensities. Under the circumstances, the two-electron case is closely related to that of one-electron. The dominant large - α_0 behavior of the latter was discussed in an earlier publication[15] (see also Ref.2, Sec.V.A.3), and will be briefly summarized here.

In the following we shall be using atomic units; the a.u. of intensity is $3.51 \ 10^{16} \ W / cm^2$. The expression of α_0, Eq.(2), then becomes

$$\alpha_0 = I^{1/2} \ \omega^{-2} \quad a.u. \tag{32}$$

The *one-electron case* is described to lowest order in the HFFT by the eigenvalue equation Eqs.(13),(14). Here too, parity P, and magnetic quantum number m are conserved. It was shown in Ref.15 that in the large - α_0 limit the solution of the problem reduces to one in which the potential $V_0(\alpha_0; r)$ is replaced by its simplest approximation close to one of the end points of the "line of charge", for example $+\alpha_0 \ e$; we call this the "end-point potential", $\tilde{V}_0(\alpha_0, r_-)$. Changing variables according to $\xi = \alpha_0^{-1/3} \ r_-$ leads to the eigenvalue equation

$$\left[-\tfrac{1}{2} \Delta_\xi - \frac{2^{1/2}}{\pi \xi^{1/2}} \ K(\sin \tfrac{\chi}{2}) \right] v = \mathcal{W} \ v \ , \tag{33}$$

where Δ_ξ refers to the ξ variables, and χ is the angle between ξ and e. Because the potential in Eq.(33) is axially symmetric but has no parity, only the magnetic quantum number m is conserved now. We denote the eigenvalues and eigenfunctions of Eq.(33) by $\mathcal{W}_{|m|j}$ and v_{mj}; j is a labeling subscript within the m manifold. $v_{mj}(\xi)$ has a finite extension for bound states, and decreases exponentially at large distances; we assume it is normalized to 1. The connection with the original problem, Eqs.(13),(14), is given by

$$W_{|m|j} \cong \alpha_0^{-2/3} \ \mathcal{W}_{|m|j} \ , \qquad u_{mj}(\mathbf{r}) \cong N_{mj}(\alpha_0) \ v_{mj}(\mp \alpha_0^{-1/3} \mathbf{r}_\pm) \ . \tag{34}$$

$N_{mj}(\alpha_z)$ is a normalization constant. Both eigenfunctions Eq.(34) correspond to the same eigenvalue, so that we are dealing with a twofold degeneracy. One of the eigenfunctions is defined in the vicinity of $+\alpha_0 e$ (upper signs), the other in the vicinity of $-\alpha_0 e$ (lower signs). To establish the connection with the atomic problem at finite α_0, we need solutions of well defined parity. As easily seen, a normalized u_{mj}^P of parity P can be found in the form:

$$u_{mj}^P(\alpha_0, \mathbf{r}) = (2\alpha_0)^{-1/2} \left[v_{mj}(\alpha_0^{-1/3} \ \mathbf{r}_-) + (-1)^P \ v_{mj}(-\alpha_0^{-1/3} \ \mathbf{r}_+) \right] \ . \tag{35}$$

The degeneracy now appears in the form of the two possible functions for P = 0, 1 , both associated with the same $W_{|m|j}$; this has been termed the "gerade-ungerade degeneracy" of the large-α_0 limit. At large (but finite) α_0 the energies $W_{|m|j}^P(\alpha_0)$ of the two states Eq.(35) will become different, which yields the gerade-ungerade level splitting. This splitting decreases exponentially with α_0.[15]

Eqs. (34) and (35) contain the important results that the binding energies of the levels vanish at large α_0 as $\alpha_0^{-2/3}$, and that this is accompanied by a dilatation of the eigenfunctions around the end points $\pm \alpha_0 \mathbf{e}$, as $\alpha_0^{1/3}$. The $\alpha_0^{1/3}$ dilatation of the wave functions cannot compensate recession of the end points, whose distance is $2\alpha_0$.

The splitting of the atomic wave function at large values of α_0 (in the oscillating frame of reference) into two non-overlapping parts located around the end points $\pm\alpha_0 \mathbf{e}$, can be described as atomic *dichotomy* . The change in the numerical wave functions leading to dichotomy, at increasing values of α_0, was illustrated in Fig.4 of Ref.2.

Here we are also interested in the *next order correction in $1/\alpha_0$* to the eigenvalues Eq.(34).This originates in the deviation of the exact dressed potential $V_0(\alpha_0; \mathbf{r})$ in the vicinity of $\alpha_0 \mathbf{e}$ from the "end-point potential" $\tilde{V}_0(\alpha_0; \mathbf{r}_-)$ approximation, and can be evaluated by perturbation theory. Combining the result with Eq.(34) we get

$$W_{|m|j} \cong \alpha_0^{-2/3} \, \mathcal{W}_{|m|j} \; + \alpha_0^{-4/3} \, \mathcal{U}_{|m|j} \; + \; O(\alpha_0^{-2}), \qquad (36)$$

where $\mathcal{U}_{|m|j}$ can be expressed as an integral over $v_{mj}^{(\pm)}(\mathbf{r})$. The gerade-ungerade degeneracy persists in Eq.(36); this is due to the fact that the correction we have calculated is a power of $1/\alpha_0$, whereas the gerade-ungerade splitting, which appears only when departures from dichotomy are taken into account, vanishes exponentially with α_0.

To treat the *two-electron case* let us first consider Eqs.(7) and (9) with the r_{12}^{-1} term omitted. This corresponds to two independent electrons which, at large α_0, are in dichotomous situations. Let us assume that they are in states m_1, j_1 and m_2, j_2 . The combined non-interacting system is characterized by the configuration $(m_1, j_1; m_2, j_2)$, having energy $W_{|m_1| j_1} + W_{|m_2| j_2}$. Taking advantage of the existing degeneracy, we can assume that the corresponding wave functions have symmetry and parity; they form a complete set. Note that, due to completeness, there exists a set of exact eigenfunctions of Eq.(9), having given m at finite α_0, that can be connected continuously to the states with given configuration $(m_1, j_1; m_2, j_2)$ of the large-α_0 limit $(m = m_1 + m_2)$.

As the dominant term in the expression Eq.(36) of $W_{|m_1| j_1} + W_{|m_2| j_2}$ is of order $O(\alpha_0^{-2/3})$, for any configuration leading to a bound state of the interacting system we can treat the inter-electron repulsion term by perturbation theory; indeed, its dominant contribution is only $1/2\alpha_0$. It is not difficult to calculate higher order corrections: it turns out that the first one is $\mathcal{O}(\alpha_0^{-5/3})$ and the following one $\mathcal{O}(\alpha_0^{-7/3})$. By combining the results, we find for the eigenvalues of Eq.(9) the large-α_0 expansion:

$$W_{|m_1|j_1, |m_2|j_2} = \alpha_0^{-2/3}(\mathcal{W}_{|m_1|j_1} + \mathcal{W}_{|m_2|j_2}) + \frac{1}{2\alpha_0} + \alpha_0^{-4/3}(\mathcal{U}_{|m_1|j_1} + \mathcal{U}_{|m_2|j_2})$$
$$+ \alpha_0^{-5/3}(\mathcal{S}_{|m_1|j_1} + \mathcal{S}_{|m_2|j_2}) + \mathcal{O}(\alpha_0^{-2}), \qquad (37)$$

where all coefficients can be calculated from v_{mj}. This approximate energy eigenvalue of the two-electron atom corresponds to the case when it is in a state of *dichotomy*: one electron is localized around $+\alpha_0\,\mathbf{e}$, the other around $-\alpha_0\,\mathbf{e}$. To $\mathcal{O}(\alpha_0^{-2/3})$, Eq.(37) contains the result by Mittleman for the ground state ($m_1 = m_2 = 0$, $j_1 = j_2 = 1$), that the binding energy of H^- is the double of that of H.

We now turn to the *numerical solution* of the eigenvalue equation Eq.(33), yielding the eigenvalues $\mathcal{W}_{|m|j}$. Note that the potential it contains is dominated by the power $\xi^{-1/2}$. Besides its singularity at the origin, the potential also has a logarithmic singularity along the negative z axis ($\chi = \pi$), due to the elliptic function $K(\sin(\chi/2))$. After eliminating the azimuthal angle dependence $e^{im\phi}$, we are left with a 2D nonseparable equation. This was solved by a very accurate finite-element diagonalization program, developed by Schertzer et al.[16]

The results for the lowest lying $\mathcal{W}_{|m|j}$ eigenvalues of the m = 0 to 5 manifolds are presented in Table 1. Several striking features are noted. Thus, the eigenvalues tend to accumulate to the continuum threshold at $\mathcal{W} = 0$. This is related to the overall $\xi^{-1/2}$ behavior of the potential, for it is known that a purely radial potential of this type can support an infinite number of bound states[17]. The angular factor $K(\sin(\chi/2))$ does not apparently modify this behavior. Moreover, in contrast to the Coulomb potential, the spacing of the lower-lying eigenvalues $\mathcal{W}_{mj}$ (as well as of the physical $W_{|m|j}$), is much closer; this, as we shall see shortly, is of far reaching consequence.

Let us consider the values of $(\mathcal{W}_{|m_1|j_1} + \mathcal{W}_{|m_2|j_2})$. In view of Eq.(37), at sufficiently large α_0 this quantity will give information on which configurations $(m_1, j_1; m_2, j_2)$ will lead to bound states when the inter-electron repulsion is switched on. The requirement is that this quantity be *less than* $\mathcal{W}_{0\,1} =$

−0.5596, which marks the incipience of the first continuum of the scaled problem, Eq.(33). Otherwise, when switching on the inter-electron repulsion, the configuration would become a resonance in the continuum. We want to group the values of $(\mathcal{W}_{|m_1| j_1} + \mathcal{W}_{|m_2| j_2})$ according to the good quantum number m characterizing the exact states. For this we have to take care that $m = m_1 + m_2$; for m = 0, this means: $m_2 = -m_1$, and for m = 1 : $m_2 = 1 - m_1$, where m_1 is arbitrary. Table 2, based on the results of Table 1, gives for m = 0, 1, the lowest-lying values of $(\mathcal{W}_{|m_1| j_1} + \mathcal{W}_{|m_2| j_2})$ leading to bound states. The results are arranged in groups, separated by solid horizontal lines, according to $|m_1|$, $|m_2|$; all possible combinations leading to bound states are given. For m = 0, the combinations of $|m_1|$, $|m_2|$ are: 0 , 0 , and 0 , 1; for m = 1 they are 0 , 1 , and 1 , 2 . Within each group $|m_1|$, $|m_2|$, there are subgroups characterized by values of j_1 ; at all places where not all existing j_2 eigenvalues of a subgroup have been given, a line of asterisks ∗ has been drawn.

Several important conclusions can be inferred from Table 2. Firstly, for both m = 0, 1 there is a *large (probably infinite) set of singly excited bound states* (having $j_1 = 1$ and increasing j_2) which tend to accumulate towards the first continuum limit at −0.5596. Secondly, there is also a *finite (albeit large) set of doubly excited bound states* (having $2 \leq j_1 \leq j_2$). Their number is finite because so is the number of relevant j_2 values for each $j_1 \geq 2$, and, obviously, the maximum relevant value of j_1 is bounded. These conclusions are valid for other m values as well. However, whereas the first one holds for all m, the second one is valid for a limited number of m, i.e. only for a limited number of m values are there doubly excited states.

These are arresting results because, as well known, field-free H⁻ has only one bound state in ground state manifold. Hence, we are confronted here with the existence of *light-induced excited states*, some of which are doubly excited. None of these states are subject to autodetachment, as they lie below the first continuum threshold. However, all of them undergo multiphoton detachment but, according to HFFT, this will be strongly quenched at large frequencies.

The previous analysis pertains to the large-α_0 limit. Natural questions arise. How does this situation develop from the one at small α_0 (i.e., small intensity)? When do the light-induced states appear? What is the role electron correlation? Such questions will be answered in the following Section.

TABLE 1

Eigenvalues $\mathcal{W}_{|m|j}$ of the large - α_0 limit. Asterisks indicate that higher j states for the |m| considered have been omitted. (From J. Shertzer and M. Gavrila, Ref.10)

|m|	j	$\mathcal{W}_{|m|j}$	|m|	j	$\mathcal{W}_{|m|j}$
0	1	-0.5596	3	1	-0.2114
0	2	-0.4517	3	2	-0.1968
0	3	-0.3932	3	3	-0.1853
0	4	-0.3547	3	4	-0.1761
0	5	-0.3267	*	*	*
0	6	-0.3055	4	1	-0.1817
0	7	-0.2943	4	2	-0.1712
0	8	-0.2864	4	3	-0.1628
*	*	*	4	4	-0.1558
1	1	-0.3410	*	*	*
1	2	-0.2999	5	1	-0.1605
1	3	-0.2726	5	2	-0.1527
1	4	-0.2527	5	3	-0.1462
*	*	*	5	4	-0.1407
2	1	-0.2575	*	*	*
2	2	-0.2348			
2	3	-0.2182			
2	4	-0.2053			
*	*	*			

IV. Structure of H$^-$ at intermediate - α_0. Appearance of light-induced excited states

We shall now describe a fully correlated computation of the eigenvalue problem, Eq.(9), carried out for H$^-$ by H.G. Muller and the author[12]. This was done using prolate spheroidal coordinates ξ, η, ϕ, centered on the end points $\pm \alpha_0$ **e.** In terms of the coordinates $r_{\pm}$ introduced before, they are defined by $\xi = (r_+ + r_-)/2\alpha_0$, $\eta = (r_+ - r_-)/2\alpha_0$, and ϕ is the azimuthal angle around the axis of **e** . (For $r \to \infty$, these coordinates tend to become polar: $\xi \to r/\alpha_0$ and $\eta \to \cos\theta$).

A basis set of the form $(\xi - 1)^p \eta^q [(1 - \eta^2)(\xi^2 - 1)]^{m/2} e^{im\phi} e^{-\gamma\xi}$, where p, q, m are positive integers, was introduced for each of the electrons. The two-electron basis was constructed from appropriately symmetrized products of one-electron basis functions. Our calculation can be considered, therefore, as being of the *configuration-interaction* type. The resulting energy

(From J. Shertzer and M. Gavrila, Ref.10)

TABLE 2

Two-electron eigenvalues $\mathcal{W}_{|m_1|j_1} + \mathcal{W}_{|m_2|j_2}$ for various $|m_1|, |m_2|, j_1, j_2$. Solid lines separate eigenvalues with different $|m_1|, |m_2|$. Asterisks indicate that higher j_2 states belonging to such a group have been omitted.

m = 0

$|m_1|$	$|m_2|$	j_1	j_2	$\mathcal{W}_{|m_1|j_1} + \mathcal{W}_{|m_2|j_2}$	$|m_1|$	$|m_2|$	j_1	j_2	$\mathcal{W}_{|m_1|j_1} + \mathcal{W}_{|m_2|j_2}$
0	0	1	1	-1.1193	0	0	7	7	-0.5886
0	0	1	2	-1.0114	0	0	7	8	-0.5807
0	0	1	3	-0.9528	*	*	*	*	*
0	0	1	4	-0.9143	0	0	8	8	-0.5728
0	0	1	5	-0.8863	*	*	*	*	*
0	0	1	6	-0.8652	1	1	1	1	-0.6820
0	0	1	7	-0.8539	1	1	1	2	-0.6409
0	0	1	8	-0.8460	1	1	1	3	-0.6136
*	*	*	*	*	1	1	1	4	-0.5937
0	0	2	2	-0.9035	*	*	*	*	*
0	0	2	3	-0.8450	1	1	2	2	-0.5998
0	0	2	4	-0.8064	1	1	2	3	-0.5725
0	0	2	5	-0.7785					
0	0	2	6	-0.7573					
0	0	2	7	-0.7461					
0	0	2	8	-0.7382					
*	*	*	*	*					
0	0	3	3	-0.7864					
0	0	3	4	-0.7479					
0	0	3	5	-0.7199					
0	0	3	6	-0.6987					
0	0	3	7	-0.6875					
0	0	3	8	-0.6796					
*	*	*	*	*					
0	0	4	4	-0.7093					
0	0	4	5	-0.6814					
0	0	4	6	-0.6602					
0	0	4	7	-0.6490					
0	0	4	8	-0.6411					
*	*	*	*	*					
0	0	5	5	-0.6534					
0	0	5	6	-0.6322					
0	0	5	7	-0.6210					
0	0	5	8	-0.6131					
*	*	*	*	*					
0	0	6	6	-0.6110					
0	0	6	7	-0.5998					
0	0	6	8	-0.5919					
*	*	*	*	*					

m = 1

$|m_1|$	$|m_2|$	j_1	j_2	$\mathcal{W}_{|m_1|j_1} + \mathcal{W}_{|m_2|j_2}$
0	1	1	1	-0.9006
0	1	1	2	-0.8595
0	1	1	3	-0.8322
0	1	1	4	-0.8123
*	*	*	*	*
0	1	2	2	-0.7516
0	1	2	3	-0.7243
0	1	2	4	-0.7044
*	*	*	*	*
0	1	3	3	-0.6658
0	1	3	4	-0.6459
*	*	*	*	*
0	1	4	4	-0.6074
*	*	*	*	*
1	2	1	1	-0.5985
1	2	1	2	-0.5758
*	*	*	*	*

matrix was diagonalized, and convergence obtained by including up to 100 basis functions from a total of 200, selected by an optimization procedure devised by H.G. Muller. We estimate that the eigenvalues $W(\alpha o)$ of Eq.(9) were calculated with a relative accuracy of better than $5 \ 10^{-4}$, and the electron affinity $J(\alpha_0)$ to better than 1% .

We present our numerical results for the detachment energies of H^- in Fig.1. We define the detachment energy of a state by $^{2S + 1}D^P_{\Lambda j}(\alpha_0) = E(\alpha_0) - ^{2S + 1}W^P_{\Lambda j}(\alpha_0)$, where $E(\alpha_0)$ is the ground state energy of H in the field. (For the ground state of H^- the detachment energy coincides with the electron affinity.) $E(\alpha_0)$ was recomputed, for consistency, in the spheroidal one-electron basis. We recall that $|E(\alpha_0)|$ is a rapidly decreasing function of α_0 (see Fig.1 of Ref.2). With the exception of one state, represented by curve b, we show in Fig.1 only states A-E that belong to the $^1\Sigma_g$ manifold.

The behavior of the ground state $A^1\Sigma_g$, the only one to exist for small α_0, is rather peculiar. As α_0 increases from 0, the electron affinity decreases from its field-free value of 0.75 eV to a minimum value of 0.6 eV, around $\alpha_0 = 2.5$. As α_0 increases further, the affinity increases to an absolute maximum of 1.1 eV, around $\alpha_0 = 17$, and thereafter decreases slowly. At $\alpha_0 = 45$, the largest value we have considered, $J(\alpha_0) = 0.817$ eV is still larger than the field-free value, whereas $|E(\alpha_0)|$ has reduced to 0.044 a.u. = 1.2 eV. This clearly illustrates how drastically the nonperturbative field effects occurring at $\alpha_0 > 1$ can modify the electron correlation.

Fig.1 also shows the appearance of the *light-induced excited states* B - E. After becoming bound at some value of α_0, their detachment energy increases to a maximum, and then decreases afterwards. Note that, once formed, these states cannot disappear any more, because of the asymptotic α_0-dependence given by Eq.(37). The first light-induced excited state to appear, for α_0 just above 3.50, belongs to the $^3\Sigma_u$ manifold, see curve b . (For $\alpha_0 = 3.55$ its detachment energy is at least 10 meV.) Curve b merges with curve A just above $\alpha_0 = 10$ because of singlet-triplet coalescence.

At $\alpha_0 = 45$ the ground state $A^1\Sigma_g$, as well as $b^3\Sigma_u$, is already in the fully dichotomous (large-α_0 limit) regime described in Sec.III. This is not yet the case for the excited states at $\alpha_0 = 45$. In order to analyze their character (singly or doubly excited) for $\alpha_0 < 45$, we have carried out an approximate one-electron calculation to define orbitals, which could serve as basis for a configuration description for the two-electron problem at intermediate α_0 . One of the electrons, located in the vicinity of one end point, say $+\alpha_0$ e, feels approximatively a potential of the form:

$$V_0(\alpha_0, \mathbf{r}) + \frac{1}{r_+} \quad , \tag{38}$$

where the first term represents the attractive potential of the proton, while the second term simulates the effect of the repulsion by the second electron located in the vicinity of the other end point, $-\alpha_0\ \mathbf{e}$. Solving the one-electron Schrödinger equation for this potential yields orbitals which form an adequate basis for a configuration description of the two-electron system at intermediate α_0.[18] Properly symmetrized products of such orbitals were compared in terms of shape and nodal structure to the eigenfunctions of the exact computation, Eq.(9). An unambiguous identification of the character of the latter could be made at $\alpha_0 \approx 30$. Thus $(B, C)^1\Sigma_g$ and $b^3\Sigma_u$ could be identified as singly excited states for all α_0 considered. For $\alpha_0 < 39$, $D^1\Sigma_g$ is a singly excited state too, whereas $E^1\Sigma_g$ is a *doubly excited state*. However, the curves D and E collide in the α_0 interval 39.0 ± 0.5. Although symmetry considerations (non-crossing rule) require that the crossing be avoided, the two curves come so close that we could not tell them apart at $\alpha_0 = 39$, within the accuracy of our calculation. We have checked that the shape of the eigenfunctions is transferred diabatically from one to the other at the crossing as α_0 grows, i.e., the doubly excited configuration is transferred to the state $D^1\Sigma_g$, and the singly excited one to the state $E^1\Sigma_g$.

A large (probably infinite) set of excited states is expected to appear, connecting to the large-α_0 spectrum discussed in the previous Section. On the other hand, the corresponding energy curves will undergo a large number of avoided crossings before this limit is reached.

At this point we would like to mention the chronology of the concept of light-induced excited states. These have been found before, in one-dimensional short range potential models, first by Bhatt, Piraux and Burnett[19], and then by Bardsley, Comella, and Szöke[20] (see also the more recent work by G. Yao and Shih-I Chu[21]); their theoretical interpretation was discussed by Dörr and Potvliege[22]. For a real atom (hydrogen) the existence of such states was discovered by Dörr, Potvliege, Proulx and Shakeshaft[23]. We are signaling their existence in a two-electron problem; note that they appear here due to a different mechanism, namely the interplay of nonperturbative field effects and electron correlation.

In Fig.2 we illustrate the molecular character of the eigenstates $\Phi(\mathbf{r}_1, \mathbf{r}_2)$ and the electron correlation arising from it. We consider specifically the $C^1\Sigma_g$

state at $\alpha_0 = 30$, and represent the (unnormalized) two-electron probability density $|\Phi(\mathbf{r}_1, \mathbf{r}_2)|^2$ as function of $\mathbf{r}_2$, when $\mathbf{r}_1$ is fixed at various points $\mathbf{r}_1 = -\lambda\,\alpha_0\,\mathbf{e}$ ($0 < \lambda < 1$), situated on the line of charges. Since, under the circumstances, $|\Phi|^2$ is axially symmetric with respect to $\mathbf{r}_2$ around the line of

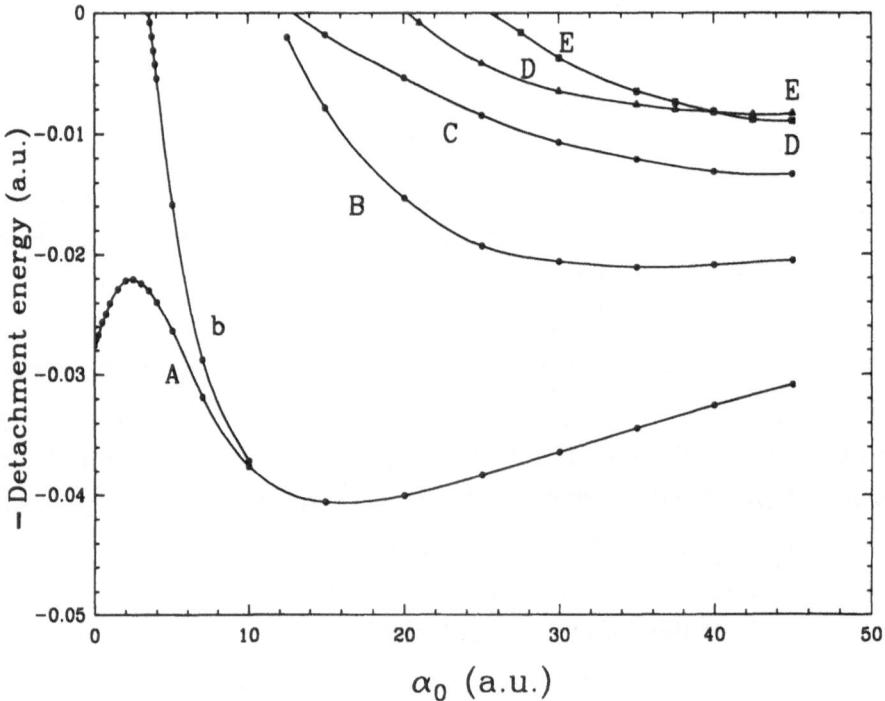

Fig. 1. Negative of the detachment energies of the ground state (A), and the lowest lying light-induced excited states (B - E) of the $^1\Sigma_g$ manifold of H⁻ as function of $\alpha_0 = I^{1/2}\,\omega^{-2}$. (All quantities are in a.u.). The lowest state of the $^3\Sigma_u$ manifold, $b\,^3\Sigma_u$, is also represented. E is a doubly excited state, for $\alpha_0 < 39$. Note the avoided crossing of the D and E curves around $\alpha_0 = 39$. (From H.G. Muller and M. Gavrila, Ref.11.)

charge taken as z axis, it is sufficient to represent it in an arbitrary plane passing through this axis. As apparent the probability density has sizable values over a linear extension along the z axis of about $2\alpha_0 = 60$, i.e. the H⁻ atom can become extremely spread out. When $\mathbf{r}_1$ moves in from the end-point

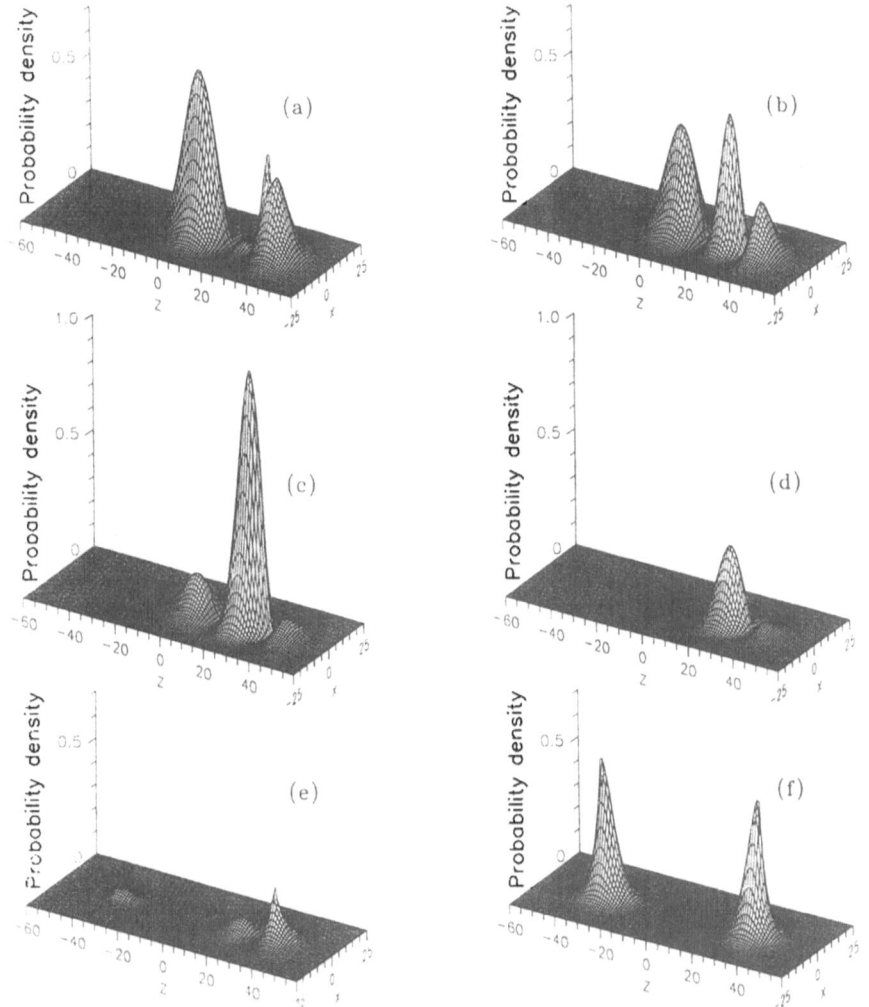

Fig. 2. Unnormalized probability density $|\Phi(\mathbf{r}_1, \mathbf{r}_2)|^2$ at $\alpha_0 = 30$ for the state $C^1\Sigma_g$ as function of $\mathbf{r}_1$, when $\mathbf{r}_1$ is fixed at various points $\mathbf{r}_1 (0, 0, -\lambda \alpha_0)$ on the z axis taken along the "line of charge". Due to the axial symmetry, it is sufficient to represent $|\Phi(\mathbf{r}_1, \mathbf{r}_2)|^2$ in a plane passing through this axis. The plots correspond to: (a) $\lambda = 1$, (b) $\lambda = 0.875$, (c) $\lambda = 0.75$, (d) $\lambda = 0.50$, (e) $\lambda = 0.375$, (f) $\lambda = 0$. The normalized probability density is obtained by multiplication with $3.475 \, 10^{-11}$. All quantities are in a.u. (From H.G. Muller and M. Gavrila, Ref.11.)

value ($\lambda = 1$) towards the origin ($\lambda = 0$), there is a drastic change in the shape of $|\Phi|^2$, as it tends to be pushed away from the location of $\mathbf{r}_1$ by the electron - electron repulsion. At the same time this illustrates the inadequacy of one-electron models with short range potentials in the regime of intensities considered here.

Acknowledgments

The author has benefitted from stimulating discussions with A. Dalgarno, H.G. Muller, J. Shertzer, and W. van der Kaaij. This work was partially supported by the European Community Scientific Stimulation Program under Contract SCI - 0103 - C.

References

1. *Atoms in Intense Laser Fields*, Ed. M. Gavrila (Academic Press, 1992).

2. M. Gavrila, *Atomic structure and decay in high-frequency fields*, in Ref.1, p.435.

3. For information on structure calculations for H$^-$, see H. Hotop and W. C. Lineberger, J. Phys.Chem.Ref.Data **4**, 3 (1975); **14**, 731 (1985), and C. F. Bunge and A. V. Bunge, Int.Journ.Quant.Chem.: Quant.Chem.Symp. **12**, 345 (1978).

4. It has been proven mathematically that there exists only one bound state in the manifold of the ground state. However, another bound state has been computed in the manifold ^3P_g (L = 1, S = 1, even parity); this is the doubly excited state $(2p)^2$ ^3P_g, having the very small affinity of 0.0095 eV (with respect to the H(2p) threshold).

5. S. Geltman, Phys.Rev.A **43**, 4930 (1991).

6. Th. Mercouris and C.A. Nicolaides, J.Phys.B **21**, L285 (1988); **24**, L57 (1991); **24**, L165 (1991); Phys.Rev.A **45**, 2116 (1992).

7. M. Dörr, R. M. Potvliege, D. Proulx, and R. Shakeshaft, Phys.Rev.A **42**, 4138 (1990).

8. M.H. Mittleman, Phys.Rev.A **42**, 5645 (1990).

9. W.W. Smith et al, J.Opt.Soc.Am.B **8**, 17 (1991)

10. J. Shertzer and M. Gavrila, submitted for publication.

11. H.G. Muller and M. Gavrila, to appear in Phys.Rev.Lett.

12. We take a as constant. For the transition to a laser pulse (variable $a(t)$) see the discussion in Ref.2, p. 444.

13. W. Pauli and M. Fierz, Nuovo Cimento **15**, 167 (1938); H.A. Kramers, *Collected Scientific Papers* (North Holland,Amsterdam, 1956), p.866; W.C. Henneberger, Phys.Rev.Lett. **21**, 838 (1968); see also F.H. Faisal, J.Phys B **6**, L89 (1973).

14. This condition refers to a constant intensity field. A related issue is that of the survival of the atom during the rise of the intensity in a realistic laser pulse, so that it can actually be exposed to the highest intensities in the pulse. This issue will be discussed elsewhere.

15. M. Pont, N. Walet and M. Gavrila, Phys.Rev.A **41**, 477 (1990).

16. F.S. Levine and J. Shertzer, Phys.Rev.A **32**, 3285 (1985); J. Shertzer, Phys.Rev.A **39**, 3833 (1989).

17. C. Quigg and J.L. Rosner, Physics Reports **56**, 167 (1979).

18. With these at hand one could carry out a Hartree-Fock type calculation, that would represent an alternative way to solve the eigenvalue problem at intermediate α_0, where electron correlation is not too important.

19. R. Bhatt, B. Piraux, and K. Burnett, Phys.Rev.A **37**, 98 (1988).

20. J.N. Bardsley, A. Szöke and M. J. Comella, J.Phys.B **21**, 3899 (1988); J. N. Bardsley and M. J. Comella, Phys.Rev.A **39**, 2252 (1989).

21. G. Yao and Shih-I Chu, Phys.Rev.A **45**, 6735 (1992).

22. M. Dörr and R. M. Potvliege, Phys.Rev.A **41**, 1472 (1990).

23. M. Dörr, R. M. Potvliege, D. Proulx, and R. Shakeshaft, Phys.Rev.A **43**, 3729 (1991).

ELECTRON-ENERGY RESOLVED EXPERIMENTS ON MULTIPHOTON DETACHMENT FROM NEGATIVE IONS

Marc D. Davidson,[1] Hugo W. van der Hart,[1] Doug Schumacher,[2]
Phil H. Bucksbaum,[2] Harm Geert Muller,[3]
and H. Ben van Linden van den Heuvell[1,3]

[1]Van der Waals-Zeeman Laboratory, University of Amsterdam,
 Valckenierstraat 65, 1018 XE Amsterdam, The Netherlands
[2]Department of Physics, University of Michigan,
 Ann Arbor, Michigan 48109, USA
[3]FOM-Institute for Atomic and Molecular Physics,
 Kruislaan 407, 1098 SJ Amsterdam, The Netherlands

INTRODUCTION

In many experiments on atoms in strong laser fields, it is the nature of the interaction of the atom with the radiation field which is the subject of investigation, more than the specific structure of the atom which is used. An example of the phenomena that have been discovered in this kind of experiments is the effect of excess-photon ionization (EPI), also called above-threshold ionization (ATI). It is found that electrons can absorb more photons than required to overcome the threshold for ionization and this manifests itself for example in electron-energy spectra consisting of several peaks separated by the photon energy. Since the first observation by Agostini et al[1] in 1979, the effect of EPI has been thoroughly investigated[2]. It was probably initially assumed that in experiments performed at high light intensities the structure of the particular atom was of no importance, however, in later experiments it was found that the structure of the atom leaves its mark on the measurement by means of resonance enhancement of the EPI[3].

Experiments on negative ions can lead to more insight in the interpretation of experiments on atoms and can also lead to the discovery of new phenomena. The short-range potential in which the additional electron is bound in a negative ion gives, in general, rise to only one bound state. The short-range potential is particularly important for the interpretation of the effect of EPI or EPD (excess-photon detachment). Although theoretical calculations show the possibility of EPI, it is not clear from experiments on neutral atoms alone whether the additional photons have not been absorbed *after* ionization. The latter possibility is unlikely in the case of EPD since there, the electron no longer has any interaction, after detachment, with the neutral atom. Since the combined conservation of energy and momentum forbid the absorption of photons by a free non-relativistic electron, EPD from negative ions

Super-Intense Laser-Atom Physics, Edited by
B. Piraux *et al.*, Plenum Press, New York, 1993

shows that the absorption of the additional photons has occured as an integral part of the detachment process.

Only recently EPD from negative ions has been observed[4-9]. A question can then be asked as to why there has not been the same activity in the field of multiphoton detachment from negative ions as there has been in the field of multiphoton ionization of atoms. An answer to this question lies partly in the difficulty in producing useful densities of negative ions. It has also been shown that it is difficult to combine multiphoton detachment from negative ions with electron-energy diagnostics. In this contribution we will discuss experiments in which the experimental problems have been solved by building a Penning ion trap in a magnetic-bottle electron spectrometer. In the Penning trap negative chlorine ions have been confined, which have been irradiated by laser light with various wavelengths (532 nm, 1064 nm, 1908 nm) and polarization (linear, circular).

EXPERIMENTAL METHOD

One of the goals in the experiments discussed in this paper, is an accurate determination of the energy of the detached electrons. For this reason we have chosen to use a magnetic-bottle electron spectrometer[10], which combines a high efficiency with a high energy resolution (in the order of 15 meV). Since this spectrometer already contains a magnetic field, the obvious choice to obtain a useful density of negative ions is a Penning trap[11]. The magnetic field of 1 T, which is present in the interaction region of the spectrometer, confines charged particles along the direction perpendicular to the axis of the magnetic field. In order to confine the charged particles along the axis of the magnetic field, an electric quadrupole field is added by means of four flat ring electrodes mounted on top of the tips of the pole pieces (figure 1). The straightforward shape of the ring electrodes makes the interaction region easy accessible for laser beams, but gives rise to higher multipole moments. Since the time needed to trap the ions is small (of the order of a few milliseconds), the deviation from a perfect quadrupole field does not give rise to any problems. The procedure for filling the ion trap is given in figure 2. The Cl^- ions are produced by dissociative attachment of slow electrons to CCl_4 molecules[12]. The slow electrons are produced by means of ionization of CCl_4 by the frequency-doubled light of a Nd:YAG laser. The ionization potential of CCl_4 is 11.47 eV[13] and therefore, ionization by five 2.33 eV photons produces 0.2 eV electrons. A continuous flow of CCl_4 gas provides a pressure of 1×10^{-6} mbars, which is about one order of magnitude larger than the background pressure. The ion trap confines all negatively charged particles and consequently, the slow electrons are produced in the center of the trap, and stay there until they have reacted with the CCl_4 gas. At the CCl_4 pressure used, no electrons are left 500 μs after the 532-nm ionization pulse. In the range of pressures between 10^{-6} and 10^{-5} mbars we find a trapping time of the Cl^- ions of $(2 \times 10^{-7}$ s$)$/(pressure (in mbars)), which is in agreement with trapping times found in experiments performed in conventionally shaped ion traps at much lower pressures[14]. We estimate the number of trapped Cl^- ions to be about a few thousand in a region of about 1 mm^3.

A laser pulse from a second laser, which is used for the detachment from the negative ions, is focussed on the cloud of negative ions every 2 ms after the laser pulse that is used for filling of the trap. About 20 ns before this second laser pulse arrives the ion trap is electrically switched off. After detachment the photoelectrons are energy analyzed on the basis of their time-of-flight over a 50-cm flight path.

MULTIPHOTON DETACHMENT ELECTRON-ENERGY SPECTRA

In the first experiments[7,9] a mode-locked Nd:YAG laser has been used for the

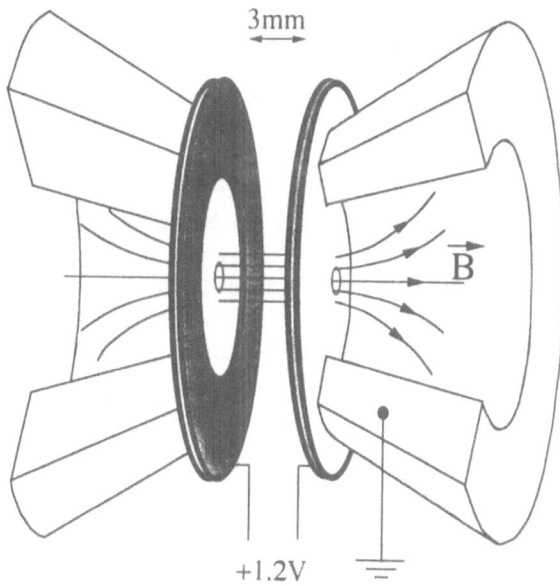

Figure 1. A detail of the experimental set up showing the Penning ion trap. The magnetic field between the pole pieces is 1 T. On top of the tip of the pole pieces gold ring electrodes are mounted. The grounded ring electrodes, connected directly to the pole pieces, have an outer diameter of 15 mm and and a hole of 1.5 mm. The ring electrodes mounted on top of these plates have an outer diameter of 15 mm and an inner diameter of 7 mm.

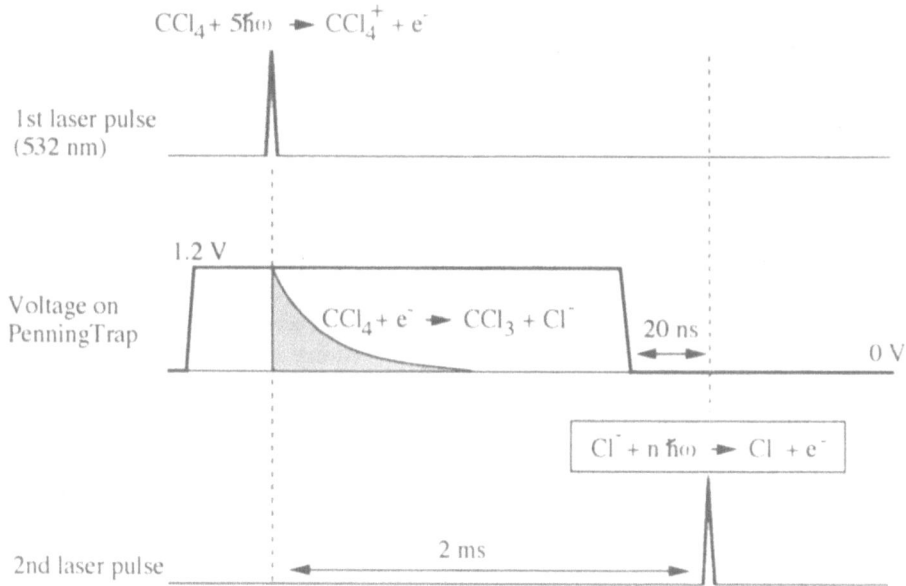

Figure 2. A schematic representation of the procedure for filling the ion trap. Ionization by means of the first laser pulse fills the trap with low-energy electrons, which react in a fraction of a millisecond with the background gas under formation of Cl$^-$. After 2 ms the Cl$^-$ ions are set free and a second laser pulse detaches the electrons. The whole cycle is repeated every 33-100 ms.

detachment from the ions. This laser produces pulses with a wavelength of 1064 nm and a pulse duration of 35 ps. In addition to the fundamental wavelength, we have used the second harmonic with a wavelength of 532 nm, and the first Stokes line generated in a Raman cell, with a wavelength of 1908 nm. The Raman cell consists out of a 2 meter long cell filled with hydrogen gas under a pressure of 8 bars and has a conversion efficiency of about 10 % of the fundamental light in the first Stokes line. The choice of 1908 nm light is determined by the combination of two considerations. Firstly, one needs longer wavelengths to increase the contribution of EPD in the total detachment yield. Secondly, at too long wavelengths the multiphoton detachment (MPD) spectra are blurred due to light-induced shifts of the threshold. Figure 3 shows the electron-energy spectra obtained at the wavelengths 532 nm, 1064, nm, and 1908 nm for both linear and circular polarization.

As expected, the spectra show an increasing contribution of excess-photon absorption with increasing wavelength. While at the shortest wavelength the two fine-structure components are still resolvable, at the longer wavelengths this is not possible due to the light-induced shifts. At a wavelength of 1908 nm the different EPD channels start to overlap.

In addition, the influence of circularly-polarized light on the detachment process can be seen in the spectra. When circularly-polarized light is used, only transitions are allowed with a change in the magnetic quantum number of one (or minus one depending on right- or left-handed polarization). Consequently, with the absorption of many photons the electron is forced to make a transition to a state with a high angular momentum. This increase in angular momentum results in a poor overlap between the initial and final state and therefore, in a reduction of the probability for detachment. Nevertheless, an increment of the number of absorbed photons also means an increase of the kinetic energy of the electron, which makes it easier to overcome the centrifugal barrier[15]. The combination of these two effects gives rise to an optimum, which does not necessarily lies at the channel for detachment by means of the lowest number of photons. In the case of two-photon detachment by 532-nm light no difference can be seen between linear and circular polarization, since in both cases the same transitions can be used: P (m=-1) -> S (m=0) -> P (m=1). Only the background is suppressed by the use of circular polarization. In the case of detachment by 1064-nm light, an increase can be seen of the EPD channels when circularly-polarized light is used and, in the case of 1908-nm light, the non-perturbative regime is reached where the first peak is significantly weaker than the EPD peaks.

THRESHOLD BEHAVIOUR

In atoms an enhancement of the probability for excess-photon ionization occurs when a step in the multiphoton absorption process is resonant with an atomic level[3]. Although for a given photon energy the ionization process can be non-resonant at low light intensities, an atomic level is shifted into resonance by the light field at a certain higher intensity. This means that if electron-energy spectra are obtained with very strong ultra-short laser pulses, sharp structures can be seen instead of broad peaks. In negative ions these kind of resonances do not occur, simply since there are no bound states other than the ground state. Structures in EPD spectra are nevertheless expected. Faisal and Scanzano[16] have shown that singularities in the detachment rate can occur when a step in the multiphoton absorption process coincides with the threshold for detachment. They calculated spectra for detachment from a model H- ion in an intense laser field (figure 4). In the electron-energy spectra the different EPD peaks are lifetime broadened which causes them to overlap. Since electrons appear in the spectrum with every energy, one can directly see what happens at the threshold. In particular, an enhancement of the probability for EPD is found at electron energies which are a multiple of the photon energy.

As a result of the low detachment energy of negative ions, it is difficult to reach the

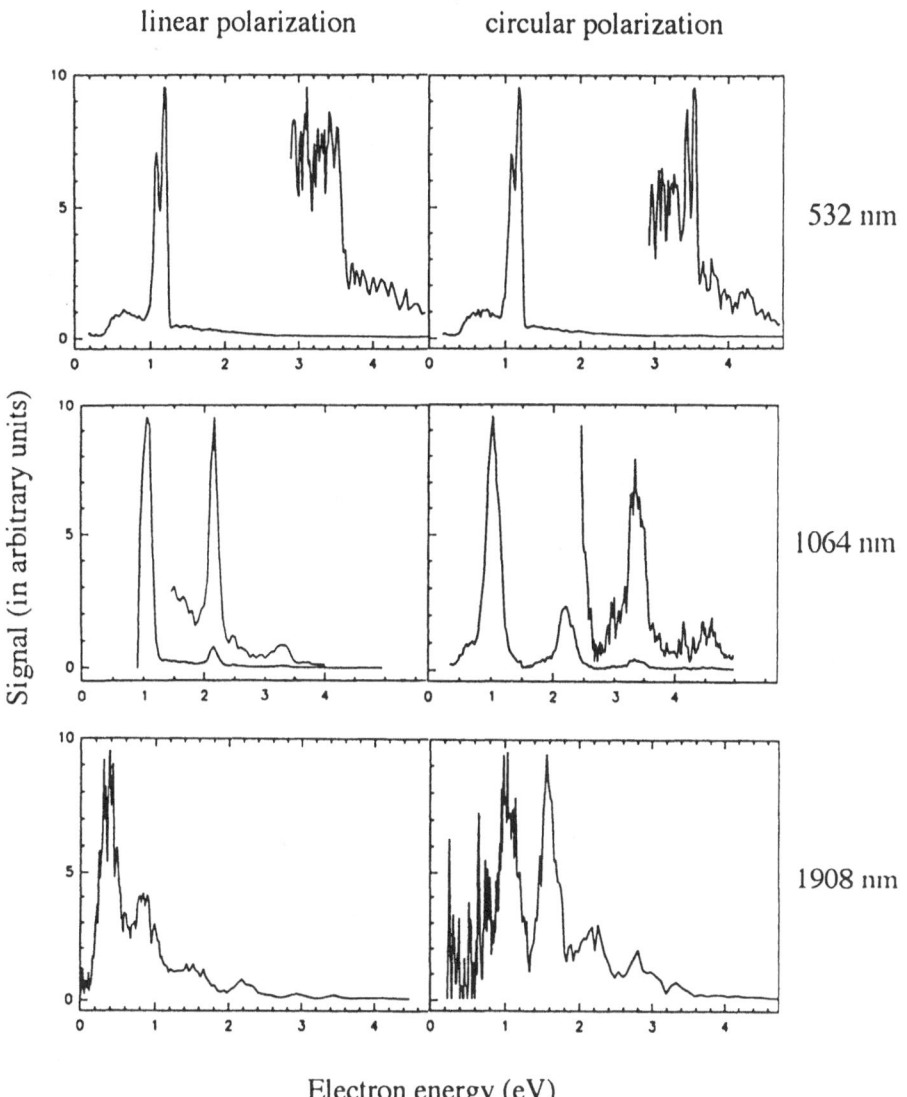

Figure 3. Electron energy spectra measured after multiphoton detachment from Cl⁻ by means of linearly- and circularly-polarized light at three different wavelengths: 532 nm, 1064 nm, and 1908 nm. The pulse duration is in the order of 35 ps.

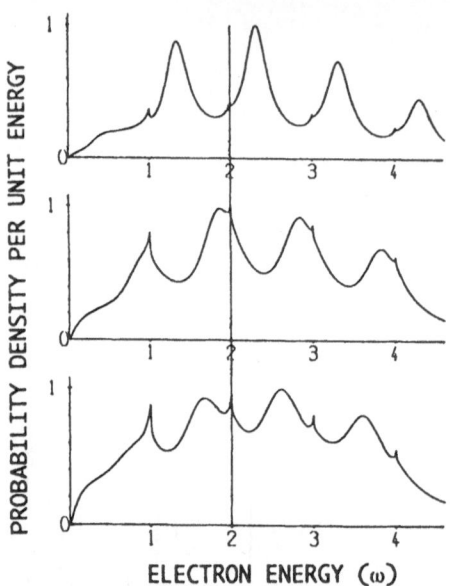

Figure 4. Wigner comb structure in calculated spectra for EPD from H⁻ by 0.35 eV photons and light-pulse duration in the femtosecond regime. The light intensity is from the top: 8.5 x 10^{15} W/m², 1.1 x 10^{16} W/m², 1.2 x 10^{16} W/m². Note the light-induced shifts in the spectra and the lifetime broadening by the high intensity. Nevertheless, the Wigner comb remains attached to the various thresholds (ref. 16).

intensities in experiments used in the calculations by Faisal and Scanzano. Depletion of the ground state occurs before the high intensities are reached. However, instead of using one fixed laser frequency, one can also scan the photon energy over the threshold. The same threshold structures should be visible in the cross sections for detachment as a function of the photon energy. We therefore performed calculations on a simple model consisting of an electron in a one-dimensional zero-range potential. Results from these calculations will show most resemblance with transitions in negative ions with an l=0 bound state like H⁻ to the s-wave continuum. Figure 5 shows the cross section for absorption of two and four photons as a function of the photon energy. In the model calculation the threshold energy of the H⁻ ion has been used. To make a qualitative comparison calculations have been performed[17] for 'real' H⁻. In these calculations a lowest-order perturbation approach has been used, based on the use of B-splines, with electron correlations fully included (figure 6). Obviously, the calculations for H⁻ generally lead to more realistic values of the detachment cross sections, but tend to be inaccurate at the thresholds for detachment. Nevertheless, when the energy of one photon is near the threshold energy, the strong decrease of the cross section for detachment by means of two photons to the s-wave continuum can be seen in both type of calculations. At present we are not able to perform experiments on H⁻ and have therefore performed an experiment on Cl⁻ to look for threshold structures. In figure 7 the electron signal is shown when the additional electron is detached from Cl⁻ by means of two photons. The strong decrease of the electron signal which occurs at the moment that the photon energy is larger than the threshold energy, is due to the onset of one-photon detachment and subsequent depletion of the

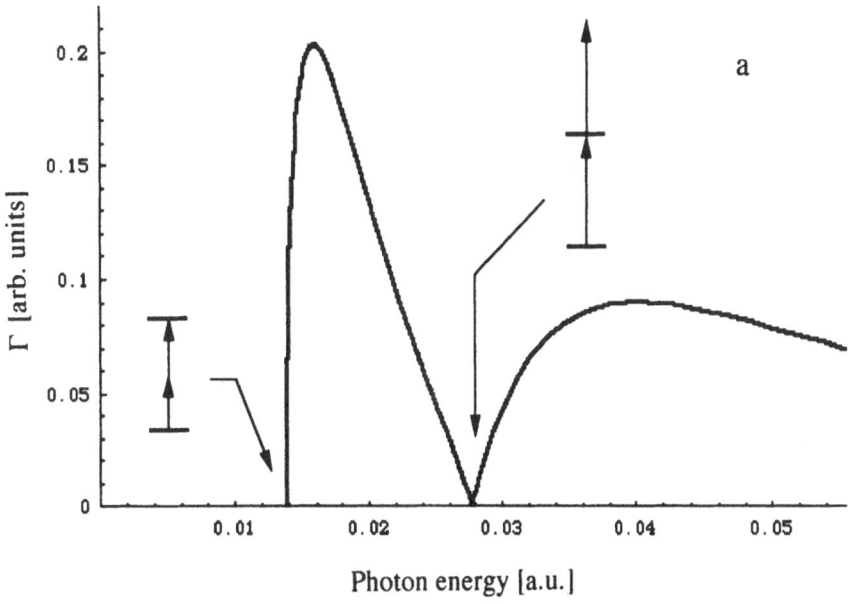

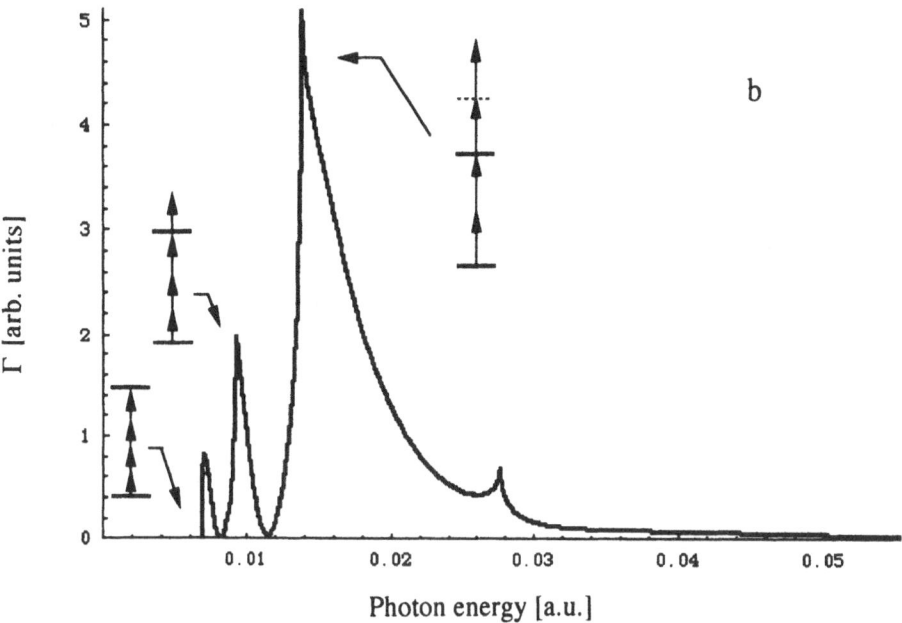

Figure 5. The square of the transition matrix elements for detachment of an electron bound in a one-dimensional zero-range potential by means of two photons (a) and four photons (b) as a function of the photon energy.

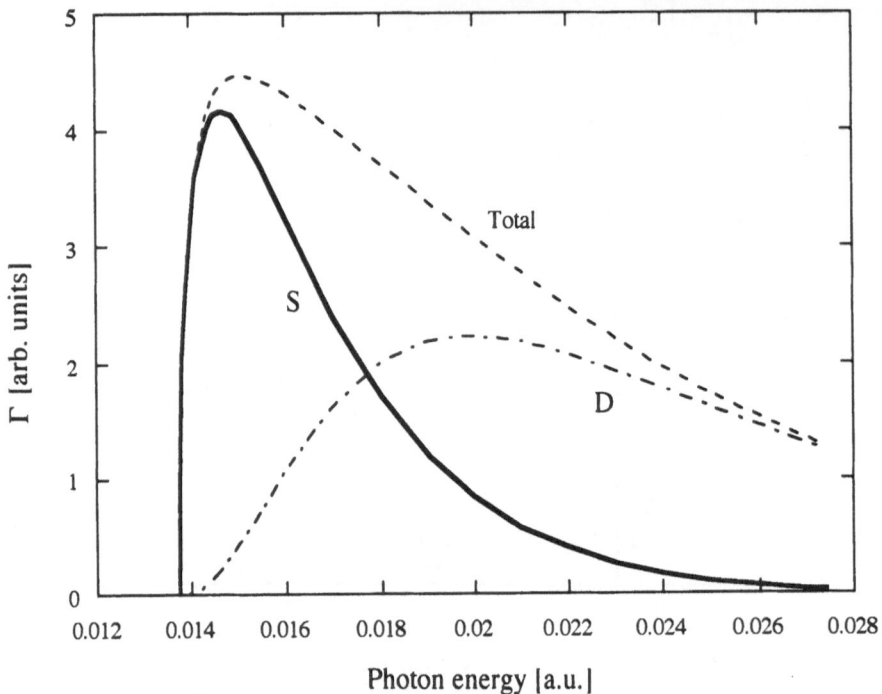

Figure 6. The square of the transition matrix elements for two-photon electron-detachment from H⁻ to the s- and d-continuum as a function of the photon energy.

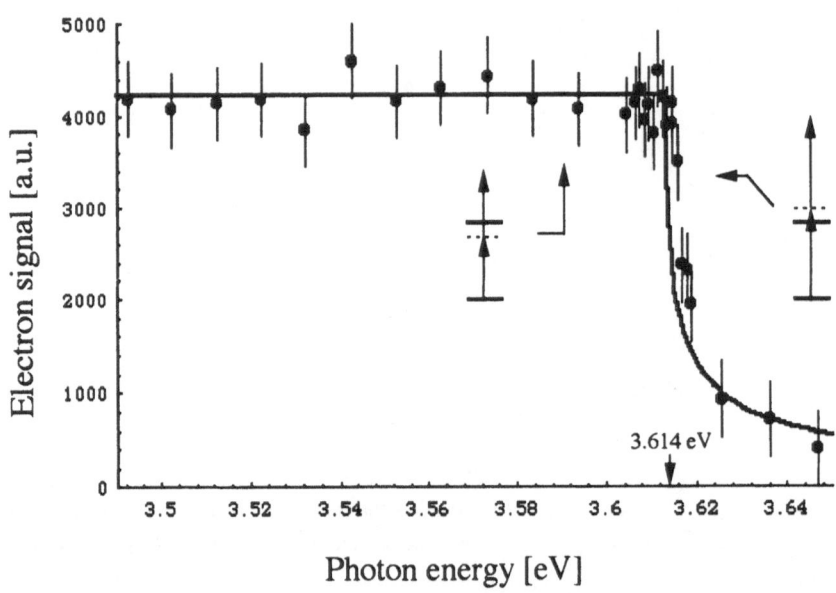

Figure 7. Two-photon electron-detachment signal of Cl⁻ as a function of the photon energy. The experimental points are fitted with a calculation which assumes a cross section for the two-photon process, constant over the full photon-energy range. The decrease of the signal above the threshold is due to saturation of one-photon electron-detachment.

groundstate. The experimental data is fitted with a calculation which assumes a cross section for two-photon detachment, which is independent of the photon energy in the investigated region (under the threshold as well as above the threshold). These preliminary measurements do not show evidence for threshold structures.

REFERENCES

1. P. Agostini, F. Fabre, G. Mainfray, G. Petite, and N.K. Rahman, *Phys. Rev. Lett.* 42:1127 (1979).

2. See, e.g., *Atoms in Intense Laser Fields* , in "Advances in Atomic, Molecular, and Optical Physics", M. Gavrila, ed., Academic Press, 1992.

3. R.R. Freeman, P.H. Bucksbaum, H. Milchberg, S. Darack, D. Schumacher, and M.E. Geusic, *Phys. Rev. Lett.* 59:1092 (1987).

4. C. Blondel, M. Crance, C. Delsart, and A. Giraud, *J. Phys. B* 24:3575 (1991).

5. H. Stapelfeldt, P. Balling, C. Brink, and H.K. Haugen, *Phys. Rev. Lett.* 67:1731 (1991).

6. H. Stapelfeldt and H.K. Haugen, *Phys. Rev. Lett.* 69:2638 (1992).

7. M.D. Davidson, H.G. Muller, and H.B. van Linden van den Heuvell, *Phys. Rev. Lett.* 67:1712 (1991).

8. M.D. Davidson, B. Broers, H.G. Muller, and H.B. van Linden van den Heuvell, *J. Phys. B* 25:3093 (1992).

9. M.D. Davidson, D. Schumacher, P.H. Bucksbaum, H.G. Muller, and H.B. van Linden van den Heuvell, *Phys. Rev. Lett.* 69:3459 (1992).

10. P. Kruit, and F.H. Read, *J. Phys. E* 16:313 (1983).

11. H. Dehmelt, *in* "Advances in Atomic and Molecular Physics", D.R. Bates and I. Estermann, ed., Academic, New York, 1967, Vol. 3; *ibid.* 1969, Vol. 5.

12. A. Chutjian and S.H. Alajajian, *Phys. Rev. A* 31:2885 (1985).

13. A.A. Radzig, and B.M. Smirnov, 1985, "Reference Data on Atoms, Molecules, and Ions", J.P. Toennies, ed., Springer Series in Chemical Physics Vol. 31, Springer-Verlag, Berlin.

14. W.A.M. Blumberg, R.M. Jopson, and D.J. Larson, *Phys. Rev. Lett.* 40:1320 (1978).

15. P.H. Bucksbaum, M. Bashkansky, R.R. Freeman, T.J. McIlrath, and L. Di Mauro, *Phys. Rev. Lett.* 56:2590 (1986).

16. F.H.M. Faisal and P. Scanzano, *Phys. Rev. Lett.* 68:2909 (1992).

17. H.W. van der Hart, to be published (1993).

OBSERVATION OF RESONANT EXCESS PHOTON DETACHMENT VIA A WINDOW RESONANCE IN THE NEGATIVE CESIUM ION

Henrick Stepelfeldt[1], Peter Balling[1]", and Harold K. Haugen[2]

[1]Institute of Physics and Astronomy, University of Aarhus
 DK-800 Aarhus C, Denmark
[2]IMR and McMaster Accelerator Laboratory, McMaster University
 1280 Main Street West, Ontario, Canada L8S 4 M1

INTRODUCTION

Excess photon absorption (EPA) has been a subject of intense investigation over the past decade in strong field atomic physics (see overviews by Ref. [1-2]). In EPA the atom absorbs more photons than the minimum number required to reach the ionization limit. The process manifests itself in the energy spectrum of the ejected photoelectrons as a series of peaks separated in energy by the photon energy. In the case of neutral atoms the process is often denoted Above Threshold Ionization (ATI) and was observed for the first time in 1979 by Agostini and coworkers [3]. The similar process in negative ions, known as excess photon detachment (EPD), represents a qualitatively new situation in strong-field atomic physics since the additional electron in a negative ion is bound in a short-range potential in contrast to the long-range Coulomb potential. The requirement of conservation of both energy and momentum forbids the absorption of photons by a free electron. Hence, the absorption of any excess photons has to occur as an integral part of the detachment process, where the electron is still able to exchange momentum with the remaining atomic core. In the case of photodetachment of negative ions this is a much more stringent requirement than in the case of photoionization in which the long-range Coulomb potential makes photon-electron interactions possible to rather large distances from the atomic core. Also, the short-range potential implies that negative ions can exist only in a finite number of bound states, in general only in the ground state. This lack of excited bound states removes complications

Super-Intense Laser-Atom Physics, Edited by
B. Piraux *et al.*, Plenum Press, New York, 1993

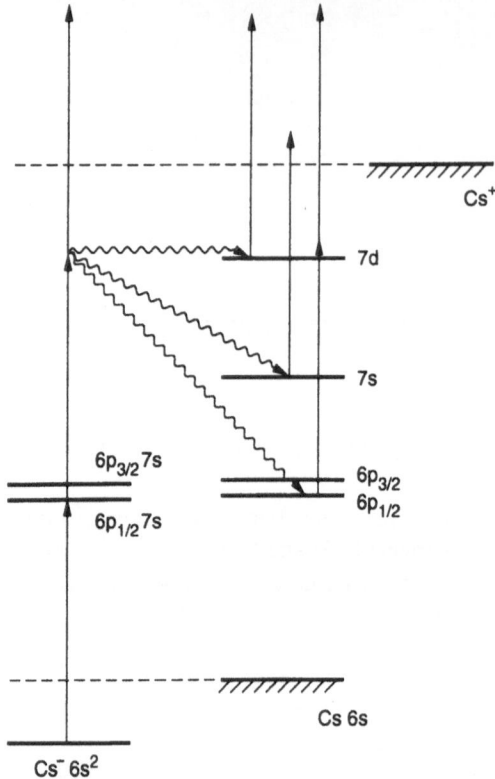

Figure 1 Simplified energy level diagram for Cs⁻ and Cs illustrating EPD via the Cs⁻(6p$_{1/2}$7s) Feshbach state.

in the multiphoton-detachment process due to transient resonances [1]. Furthermore, the final state of the detachment process is to a good approximation described as a free electron state [4].

The first experimental investigations of excess photon detachment of negative ions were reported autumn 1991 by Blondel et al. [5], Davidson et al. [6], and Stapelfeldt et al. [7]. The three groups observed nonresonant EPD from three different negative ions at the fundamental wavelength of the Nd:YAG laser (1064 nm). The experiments employed peak powers in the TW/cm^2 regime and pulse durations between ~40 ps and ~500 ps. Under these conditions the absorption of up to two excess photons could be observed by means of electron-spectroscopy. More recently this first generation of EPD experiments has been extended to include studies with 100 fs pulses [8] and studies with longer wavelengths [9].

The purpose of the present work is to show that despite the fact that negative ions lack excited bound states multiphoton detachment may still be significantly influenced by resonances due to the presence of doubly excited states above the detachment limit. In this work we have studied irradiation of the negative cesium ion in the limit where the photon energy greatly exceeds the binding energy (Fig. 1). It is shown how EPD from the ground state of Cs⁻ is strongly enhanced when the photon energy coincides with the Cs⁻(6p$_{1/2}$7s) Feshbach state. At this energy, the cross section for single photon detachment drops two orders of magnitude over an energy range of 20 cm^{-1} [10-12]. The Feshbach resonance

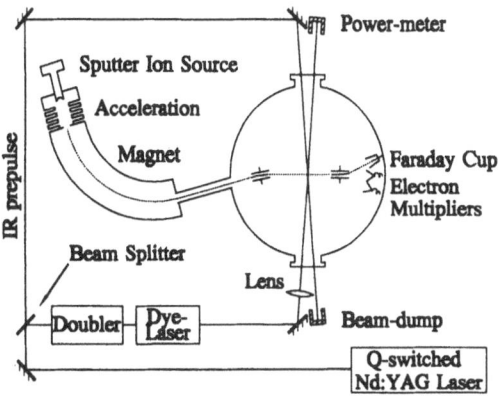

Figure 2 Schematic of the experimental setup showing the negative ion production and transport, the UHV interaction chamber and the laser system.

suppresses depletion due to single photon detachment and is at the same time expected to enhance the probability for absorption of one or more excess photons.

EXPERIMENTAL SETUP

A schematic of the experimental setup is shown in Fig. 2. The negative ions are produced in the ion source and accelerated to 25 keV. The accelerated ion beam is magnetically analyzed and collimated before entering the ultrahigh vacuum (UHV) chamber. The ion beam is crossed at 90^0 with the focused laser beam in the middle of the chamber. Following the interaction region, a set of vertical deflection plates provides electrostatic analysis of the charge states in the fast beam; the positive and neutral particles are detcted by electron multipliers, and a Faraday cup measures the negative ion current. Typically, the current is between 5 and 10 nA. A Nd:YAG pumped dye laser is used in the visible range between 14800 cm^{-1} and 15600 cm^{-1}. It is focused into the chamber by a 32-cm-focal-length lens giving a maximum intensity of ~6×10^9 W/cm^2. The interaction time between the negative ions and the laser pulse is given by the transit time of the ions through the laser focus rather than the duration of the laser pulse. In the present experimental configuration the transit time is ~0.7 ns. In some parts of the experiment a weak unfocused infrared (1064 nm) pulse is sent into the chamber from the opposite side of the chamber and overlapped temporally and spatially with the visible pulse. Most data are taken in an analog regime utilizing a gated integrator and a boxcar averager.

EXPERIMENTAL RESULTS

The experimental results are shown in Fig. 3. The upper and lower curves show the Cs and the Cs^+ signals respectively as the wavelength of the laser is scanned. The expected window resonance is seen in the neutral signal. Because of saturation the depth of the window is reduced as compared to similar scans at lower intensities [10-11]. In the Cs^+

spectrum two strong resonances are observed at ~14949 cm^{-1} and at the window position. The enhancement at 14949 cm^{-1} is expained as a sequential removal of the two valence electrons of Cs$^-$: First, Cs$^-$ is single photon detached to the ground state of neutral cesium, Cs(6s), and secondly, Cs(6s) is (2+1)-photon ionized via Cs(11d). The fine structure observed in this peak at lower intensities (Fig. 3(a)) is due to the fine structure of the intermediate Cs(11d) state. At the highest intensity the (2+1)-photon transition is strong enough that depletion of the Cs ground state occurs. This effect is seen as a hole in the Cs spectrum (Fig. 3(b)).

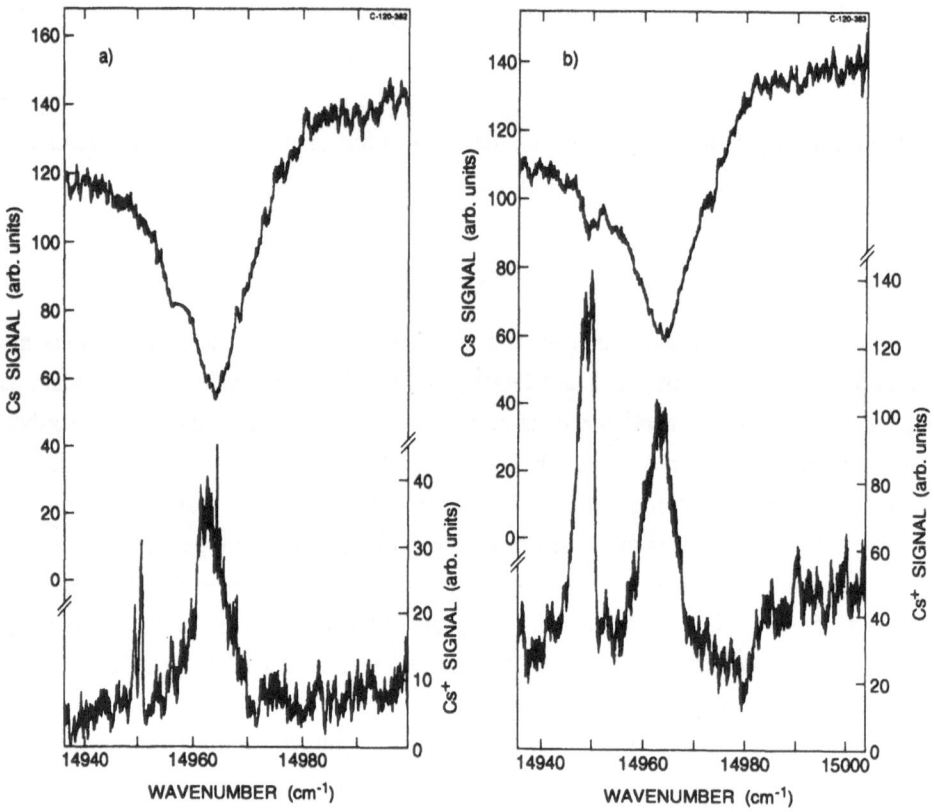

Figure 3 The Cs$^+$ signal (lower curves) and the Cs signal (upper curves) versus the laser wavelength. (a) I=6x10^9 W/cm^2, (b) I=1x10^9 W/cm^2.

A similar sequential scheme cannot explain the Cs$^+$ resonance at 14963 cm^{-1} since single photon detachment is at the minimum and no resonant enhancement in 3-photon ionization of Cs(6s) occurs at this wavelength. Instead, Cs$^-$ absorbs at least two photons via the Cs$^-$(6p$_{1/2}$7s) state. Cs$^+$ is produced either by a sequential removal of the two electrons via some excited Cs state or they are ejected simultanously under absorption of three photons from the Cs$^-$ ground state. In this first work no conclusions could be drawn on which mechanism is the dominating one. To further resolve this issue we are now extending the studies to include electronspectroscopy.

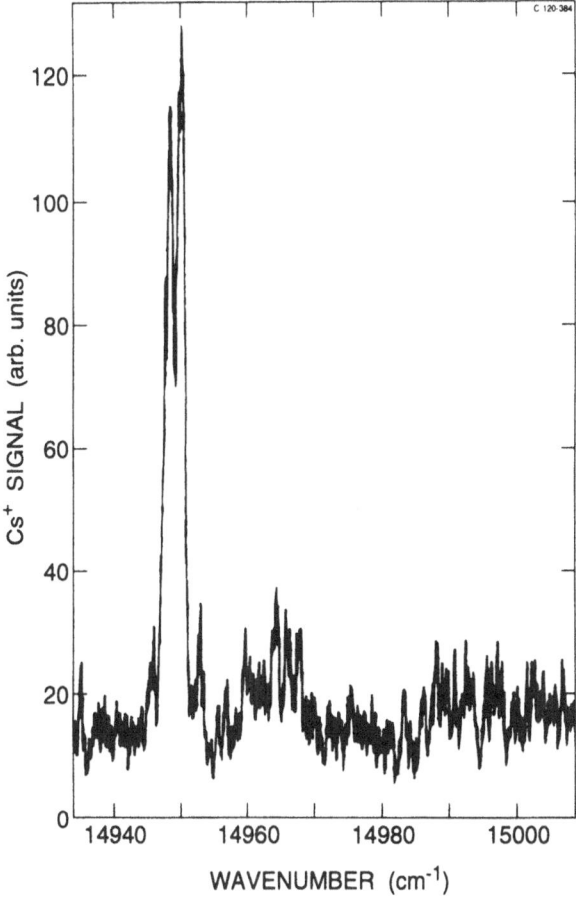

Figure 4 The Cs$^+$ signal versus the laser wavelength when an infrared prepulse is included.

To verify our interpretation of resonant EPD we recorded the Cs$^+$ signal when the negative cesium ions are exposed to a weak infrared pulse (~9398 cm^{-1}) prior to the intense visible pulse (Fig. 4). The infrared prepulse transforms all Cs$^-$ ions to Cs(6s) so that Cs$^+$ is solely produced from subsequent multiphoton ionization of Cs(6s). The enhancement at 14949 cm^{-1} does not disappear, establishing that it originates from resonant multiphoton ionization of the ground state of Cs. In contrast, the broad feature at 14963 cm^{-1} completely vanishes, demonstrating that this resonance indeed results from structure in the Cs$^-$ ion.

THE EXCITATION MECHANISM OF Cs$^-$

The experimental results show that at least two photons are absorbed from the ground state of Cs$^-$ when the photon energy coincides with the position of the $6p_{1/2}7s$ Feshbach resonance. In order to develop an understanding of the excitation mechanism of Cs$^-$ at this energy we consider the four possible excitation paths for the two-photon absorption process. For simplicity we assume that two-photon absorption leads to direct excitation of the continuum. Hence, we neglect the possible influence of doubly excited states after absorption of the second photon. The four path are illustrated in Fig. 5. Path (1) consists

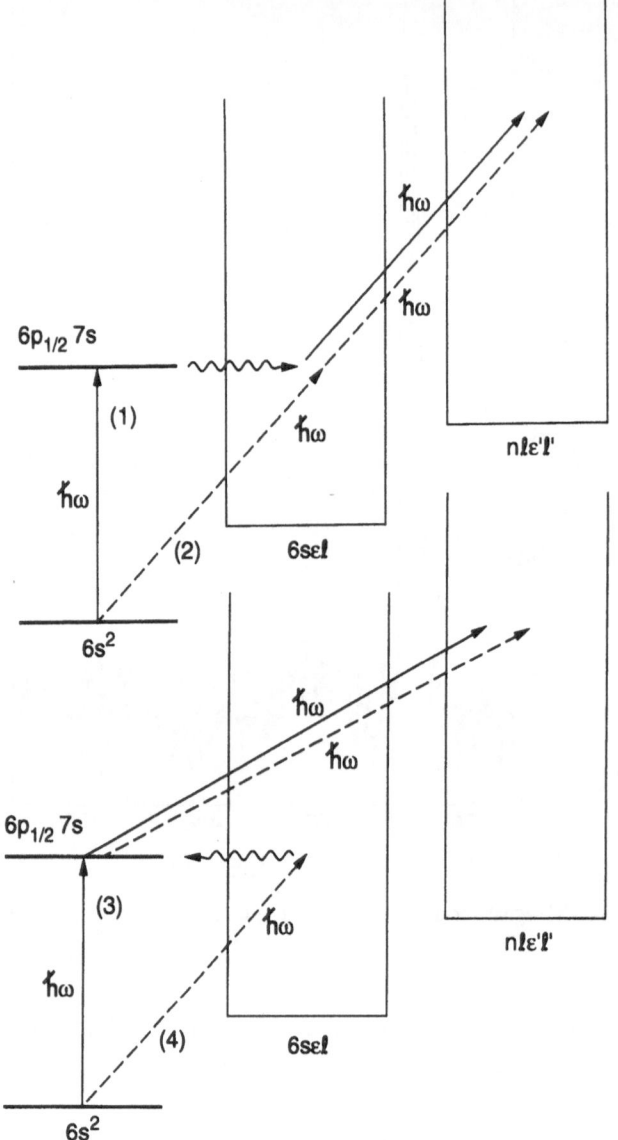

Figure 5 Schematic representation of the four possible two-photon excitation paths (see text).

of excitation of $6p_{1/2}7s$, subsequent autodetachment to $6s\varepsilon l$, and finally a continuum-continuum transition. In path (2) $6s\varepsilon l$ is excited directly followed by the continuum-continuum transition. Path (3) denotes excitation of $6p_{1/2}7s$, and subsequently a transition to the $n l\varepsilon'l'$continuum. Path (4) comprises excitation of $6s\varepsilon l$, followed by inverse autodetachment to $6p_{1/2}7s$, and finally excitation of the $n l\varepsilon'l'$. We expect that path (1) and (2) have negligible amplitude because a continuum electron is not likely to absorb a photon [13]. Hence, the structure of the two-photon cross section is a result of interference between path (3) and path (4). Furthermore, it seems likely that it is primarily the $6p_{1/2}$ electron that is excited during absorption of the second photon, since the wavelength is close to the $Cs(6p_{1/2}) \rightarrow Cs(7d)$ transition. The 7s electron is then probably "shaken off" into the continuum or else the ion may be left in an autodetaching state. To gain more insight into the details of the excitation mechanism an experiment, which include electron spectroscopy, is necessary.

Another point of interest is the somewhat counterintuitive observation of an enhancement in the two-photon absorption at the position of the Feshbach resonance despite the fact that there is a window in the one-photon cross section at that position. It can, however, be shown that a peak will almost always appear in the two-photon absorption via an autodetaching state independent of the shape of the one-photon autoionizing resonance [14]. Window resonances in the one-photon cross section are not likely to enhance the two-photon cross section as much as resonant peaks in the one-photon cross section, but on the other hand, depletion of the target will be more pronounced in the latter case. An extension of the present experimental work to other negative ions should provide excellent tests of the dependence of EPD on the Fano profile of the single photon detachment cross section.

DOUBLE IONIZATION : SEQUENTIAL OR DIRECT PROCESS

The present work also touches on the long-standing question as to whether a double ionization regime, where a simultaneous ejection process dominates, can be achieved. This question has been studied extensively for neutral atoms where the overall conclusion is that the dominant mechanism is a sequential removal of the electrons via a low-lying state of the singly charged ion. Negative ions may, however, provide good candidates for the observation of direct double ionization by utilizing detachment minima similar to the one described in this work. A negative ion where absorption of one additional photon via such a minimum brings the system above the second ionization limit might possibly give the first clear evidence of simultaneous ejection of the two electrons in a multiphoton absorption process. A possible experimental scheme is based on the negative silicon ion, where a recent publication predicts the existence of a relatively deep minimum in the single photon detachment cross section at a photon energy of 4.84 eV [15]. The minimum is due to the interference between the two excitation channels :

$$Si^-(3s^23p^3\ {}^4S) + \hbar\omega \rightarrow Si(3s^23p^2\ {}^3P) + \varepsilon\ell$$
$$Si^-(3s^23p^3\ {}^4S) + \hbar\omega \rightarrow Si^-(3s3p^4)$$

In contrast to the situation in Cs^-, absorption of a second photon through the Si^- minimum exceeds the double ionization threshold by 0.14 eV. Hence, observation of an enhancement in the the Si^- production at this energy indicates that the two electrons are removed directly from the ground state of Si^-. However, a sequential production via some excited state in neutral cesium is not precluded, since absorption of the second photon might possibly just detach one electron rather than two electrons. To distinguish between the two mechanisms for double ionization the detection of positive ions has to be supplemented with electron spectroscopy. If the two electrons are ejected simultanously a continous electron energy distribution with maximum energy 0.14 eV is expected. In contrast, the sequential production gives rise to discrete peaks at energies of typically a few eV.

CONCLUSION

A study of excess photon detachment of the negative cesium ion has been presented. EPD occured efficiently under relatively long and weak pulses in contrast to the situation encountered in previous EPD experiments. This was due to the presence of a window resonance in Cs^- which suppressed depletion from single photon detachmnet and at the same time enhanced the probability for absorption of excess photons. Future electron spectroscopy and extension of the studies of resonant EPD to other negative ions may provide clear evidence for a significant contribution of direct double ionization.

Acknowledgements

We would like to acknowledge very helpful discussions with F. Robicheaux and C. H. Greene.

REFERENCES

[1]   R. R. Freeman, and P. H. Bucksbaum, J. Phys. B **24**, 325 (1991).

[2]   N. B. Delone and M. V. Fedorov, Prog. Quantum Electron. **13**, 267 (1989).

[3]   P. Agostini et al., Phys. Rev. Lett. **42**, 1127 (1979).

[4]   M. Crance, J. Phys. B **21**, 3559 (1988).

[5]   C. Blondel, M. Crance, C. Delsart, and A. Giraud, J. Phys. B **24**, 3575 (1991).

[6]   M. D. Davidson, H. G. Muller, and H. B. van Linden van den Heuvell, Phys. Rev. Lett. **67**, 1712 (1991).

[7]   H. Stapelfeldt, P. Balling, C. Brink, and H. K. Haugen, Phys. Rev. Lett. **67**, 1731 (1991).

[8]   M. D. Davidson, B. Broers, H. G. Muller, and H. B. van Linden van den Heuvell, J. Phys. B **25**, 3093 (1992).

[9]   M. D. Davidson et al., Phys. Rev. Lett. **69**, 3459 (1992).

[10]   J. Slater, F. H. Read, S. E. Novick, and W. C. Lineberger, Phys. Rev. A **17**, 201 (1978).

[11]   T. A. Patterson et al., Phys. Rev. Lett. **32**, 189 (1974).

[12]   C. H. Greene, Phys. Rev. A **42**, 1405 (1990).

[13]   A. Guisti-Suzor and P. Zoller, Phys. Rev. A **36**, 5178 (1987).

[14]   F. Robicheaux and B. Gao, submitted to Phys. Rev. A.

[15]   G. F. Gribakin, A. A. Gribakina, B. V. Gul'tsev, and V. K. Ivanov, J. Phys. B **25**, 1757 (1992).

AUTHOR INDEX

SUBJECT INDEX

Connection machine, 195, 204, 395
Correlated eigenfunction, 462
    two-electon wavefunction, 409, 445
Correlation function, 185-193
Coulomb
    explosion, 330, 338
    scattering wavefunction, 466
    singularity, 226
    Sturmian function, 234, 235
Crank-Nicholson algorithm, 111, 113,
    187

Dc-Stark effect, 296, 367;
    see also Lo-Surdo
Death valley, 247
Decay rate, 195, 197, 247, 276
Delta potential, 10
Detachment rate for H$^-$, 378, 439-442
Diatomic molecule, 251-259
    H$_2^+$, 329-339, 365-366
Dichotomy, 196, 461-463, 471
Difference frequency generation, 43-48
Dipole
    acceleration, 120
    approximation, 137, 187, 193, 199-203,
      234, 272, 330, 393, 463
    moment, 12, 55, 63-70, 98, 120,
      174-175, 246, 253
    operator, 173
Dissociation
    decay, 254
    dynamics, 341-349
    over (under) the barrier, 366
    suppression, 254, 365
Double ionization, 165-170, 385-388,
    396-400, 449-451, 499-500;
    see also double photoionization
Double photoionization
    angular distribution, 385-387,
      397-400
    knockout mechanism, 376, 388
    shakeoff mechanism, 375, 386-387
    suppression, 449-451
    photon sharing mechanism, 376, 388
Double quantum well, 70
Doubly excited state, 403-410, 473, 477
Dressed
    atom picture, 155
    potential, 464
    state model, 332-333, 342

Drift
    motion, 158
    velocity, 158

Ehrenfest's theorem, 120
Electron
    affinity, 437, 439, 478
    correlation, 203, 375, 392, 400,
      403-410, 413, 423, 425,
      428, 454-457, 461, 469, 476-477
    energy spectrum,
      see above-threshold ionization
    relativistic, 125
Electronic state, 252, 336
Epsilon algorithm, 190
Essential state model, 65
Excess photon
    absorption, 493
    detachment, 483-486, 493,
      494
    ionization, 483
Exponential decay, 272-273, 334-338,
    335-339

Fermi
    golden rule, 271-273
    molecular dynamics, 413-423
Feshbach resonance, 365, 375,
    380, 494, 497
Finite difference
    equation, 394
    integration, 98, 330
Finite element diagonalization, 472
Floquet calculation, 63-70, 130, 134,
    155, 162, 173, 185-191,
    364-367, 436-438
    high frequency theory, 193,
      203, 461-474
Fragmentation of molecules, 351-362
Franck-Condon principle, 252, 342
Free-free scattering, 194
Frozen core approximation, 407,
    409, 449

Gauge
    length, 55, 215, 303-305,
      318, 329
    momentum (velocity), 173-174,
      303, 305-308, 329, 463
Generalized parity operator, 64